£38.99
JP
6/11

Aircraft
Electricity & Electronics

D1494642

Aviation Technology Series

Aircraft Powerplants, Seventh Edition, Kroes/Wild
Aircraft Maintenance and Repair, Sixth Edition, Kroes/Watkins/Delp
Aircraft Basic Science, Seventh Edition, Kroes/Rardon
Aircraft Electricity and Electronics, Fifth Edition, Eismin
Aircraft Gas Turbine Engine Technology, Second Edition, Treager

GLENCOE Aviation Technology Series

Aircraft
Electricity & Electronics

Fifth Edition

Thomas K. Eismin

GLENCOE

Macmillan/McGraw-Hill

New York, New York   Columbus, Ohio   Mission Hills, California   Peoria, Illinois

AIRCRAFT ELECTRICITY AND ELECTRONICS, Fifth Edition
International Editions 1994

8 9 10 SLP FC 20 9 8 7 6 5 4 3 2

Library of Congress Cataloging-in-Publication Data
Eismin, Thomas K.
    Aircraft electricity and electronics / Thomas K. Eismin. - 5th ed.
        p.   cm.-(Avaiation technology series)
    Includes index.
    ISBN 0-02-801859-1
    1. Airplanes-Electric equipment.   2. Airplanes-Electronic equipment.   3. Avionics.
I. Title.   II. Series.
TL690.E37   1994
629.135'4-dc20   93-43328
                                                        CIP

When ordering this title, use ISBN 0-07-113286-4

Printed in Singapore

Contents

v

Preface

In the past decade the aviation industry has truly gone through an electronics revolution. The miniaturization of digital electronics has allowed manufacturers to do more in less space and weight than with conventional systems. Electronic circuits are found on virtually every system of a modern aircraft. Large transport category aircraft utilize a variety of computers for navigation and flight management. Today, an aircraft technician must possess a thorough understanding of both the basic electrical theory and advanced electronic systems. *Aircraft Electricity and Electronics* provides the reader with practical knowledge that can be used by students and technicians alike.

In this edition of *Aircraft Electricity and Electronics* several new technologies are introduced to the reader. The chapter on digital systems has been expanded. Modern digital data bus systems, such as the ARINC 629, are presented. The chapter on aircraft communication and navigation has been expanded to include state-of-the-art systems such as satellite communications, global positioning systems, and data link equipment. Modern central maintenance computer systems are discussed, and much more.

The fifth edition has also improved some of the basic information necessary to build a proper foundation for understanding aircraft electrical systems. The current Federal Aviation Regulations concerning the certification of the Airframe and Powerplant (A & P) mechanics are still a vital component of this text. The text also presents information well beyond these basic A & P requirements, thus providing the student with a thorough understanding of the theory, design, and maintenance of current aircraft electrical and electronic systems.

The text is written with the assumption that the reader possesses no prior knowledge of electricity and electronics; and yet, the text may also be used by experienced technicians to gain a better understanding of advanced systems. In chapters 1 through 5 basic electrical theory and concepts are discussed. These chapters include the fundamentals necessary for a strong understanding of the FAA's regulations as they pertain to aircraft electrical systems. Chapters 6 through 12 contain vital information on the design and maintenance of specific systems. In these chapters, digital concepts are presented, computerized power distribution systems are discussed, and electrical test equipment is introduced. Chapters 13 through 17 introduce the reader to the advanced electronic systems such as communication and navigation equipment, autoflight systems, fly-by-wire components, and built-in-test equipment.

Aircraft Electricity and Electronics is one text of the Aviation Technology Series published by the Glencoe Division of Macmillan/McGraw-Hill School Publishing Company. The other books in this series are *Aircraft Basic Science, Aircraft Maintenance and Repair, Aircraft Powerplants,* and *Aircraft Gas Turbine Engine Technology.* Used together, these texts provide information concerning all areas of aircraft maintenance technology.

Thomas K. Eismin

Acknowledgments

The author wishes to express appreciation to the following organizations for their generous assistance in providing illustrations and technical information for this text: Airbus Industrie, U.S. Division, New York, New York; AiResearch Manufacturing Company, Division of the Garrett Corporation, Torrance, California; AMP Products Corporation, Southeastern, Pennsylvania; B & K-Precision/Dynascan Corporation, Chicago, Illinois; Amprobe Instrument Division, Lynbrook, New York; Atlantic Instruments, Inc., Melbourne, Florida; Beech Aircraft Company, Wichita, Kansas; The Bendix Corporation, Aerospace Electronics Sector, Arlington, Virginia, Avionics Division, Fort Lauderdale, Florida, Communications Division, Baltimore, Maryland, Electric Power Division, Eatontown, New Jersey, Flight Systems Division, Teterboro, New Jersey; Boeing Commercial Airplane Company, a Division of the Boeing Company, Seattle, Washington; Cessna Aircraft Company, Wichita, Kansas; Christie Corporation, Gardena, California; Clarostat Manufacturing Co., Plano, Texas; Collins Divisions, Rockwell International, Cedar Rapids, Iowa; Concord Battery Corp., W. Covina, California; Daniels Manufacturing Corporation, Orlando, Florida; Dayton-Granger, Inc., Fort Lauderdale, Florida; Delco Remy Division, AC Spark Plug Division, General Motors Corporation, Anderson, Indiana; The Deutsch Co., Banning, California; Federal Aviation Administration, Washington, D.C.; GC Electronics, Rockford, Illinois; Global Specialties, an Interplex Electronics Company, New Haven, Connecticut; Government Electronics Division, Motorola, Inc., Scottsdale, Arizona; Honeywell, Inc., Minneapolis, Minnesota; International Rectifier Company, El Segundo, California; John Fluke Mfg. Co., Inc., Everett, Washington; Keith and Associates, Lafayette, Indiana; King Radio Corporation, Olathe, Kansas; Lear Sieglar, Inc., International Division, Standford, Connecticut; Lockheed California Company, Burbank, California; Marathon Power Technologies, a Subsidiary of Marathon Manufacturing Companies, Inc., Waco, Texas; McDonnell Douglas Corporation, Douglas Aircraft Division, Long Beach, California; Narco Avionics, Inc., Fort Washington, Pennsylvania; Piper Aircraft Corporation, Vero Beach, Florida; Prestolite Corporation, an Allied Corporation, Toledo, Ohio; Prestolite Wire Division, an Allied Corporation, Port Huron, Michigan; Purdue University, Aviation Technology Department, West Lafayette, Indiana; Ralston Purina Co., St. Louis, Missouri; Saft America, Incorporated, Valdosta, Georgia; Simpson Electric Company, Elgin, Illinois; Sperry Aerospace and Marine Group Corporation, Phoenix, Arizona; Stacoswitch, Inc., Costa Mesa, California; Sundstrand Service Corporation, Sundstrand Corporation, Rockford, Illinois; Sundstrand Data Control, Inc., Redmond, Washington; Sun Electric Corporation, Crystal Lake, Illinois; Teledyne Battery Products, Redlands, California; Texas Instruments, Inc., Dallas, Texas; Thomas and Betts Corporation, Raritan, New Jersey; Westinghouse Electric Corporation, Electrical Systems Division, Lima, Ohio; Weston Instruments Division, Sangamo Weston, Incorporated, Newark, New Jersey; 3M Center, Aviation Safety and Security Systems Division, St. Paul, Minnesota.

Fundamentals of Electricity 1

This present period in history may well be called the "age of electronics" or the "electronics revolution" because electricity and electronics have become vital in every facet of modern life. The food you eat, your clothes, even the air you breathe, virtually everything you take for granted during a typical day has been affected by the modern age of electronics. This is particularly true in the aviation and aerospace fields because all modern aircraft and space vehicles are largely dependent upon electronics and electricity for communications, navigation, and control. **Electronics** is merely a special application of electricity wherein precise manipulation of electrons is employed. However, since electricity is considered to be the movement of electrons, with relatively low precision, the term *electronics* is usually thought to include the field of electricity.

Since electricity and/or electronics is so often used in conjunction with the mechanical systems of modern aircraft, today's technician must possess a thorough understanding of all facets of electronics. Typically this knowledge would be used during inspection, installation, and repair of systems on board the aircraft. Once electronic equipment is removed, the repair, overhaul, and testing of such equipment is usually performed by avionic specialists.

Previous to the last century, little was known concerning the nature of electricity. However, modern theoretical concepts, mathematics, and basic physical laws have explained how electricity acts. We can now predict with extreme accuracy virtually all aspects of electricity, either through mathematics or by observation and documentation of electrical actions. The precise reasons why electricity acts as it does may be debated until the next century; meanwhile, we will continue to make electricity a useful tool by predicting its actions.

On modern aircraft, electricity performs many functions, including the ignition of fuels in turbine engines, the operation of communication and navigation systems, the movement of flight controls, and analysis of system performance. There are literally thousands of electrical connections controlling hundreds of electrical devices, each of which was installed and will be maintained by an aircraft technician. In the nose section alone of a common DC-10 there are over 55 miles of wire. The enormous increase in the use of electronic systems has made it essential for the aircraft technician to obtain a thorough understanding of electricity and electronics.

THE ELECTRON THEORY

The atomic structure of matter dictates the means for the production and transmission of electrical power. All matter contains microscopic particles made of electrons and protons. The forces that bind these particles together to create matter are the same forces that create electrical current flow and produce electrical power. Every aircraft generator, alternator, and battery, virtually all electrical components, react according to the **electron theory.** The electron theory describes specifically the internal molecular forces of matter as they pertain to electrical power. The electron theory is therefore a vital foundation upon which to build an understanding of electricity and electronics.

Molecules and Atoms

Matter is defined as anything that occupies space; hence, everything that we can see and feel constitutes matter. It is now universally accepted that matter is composed of molecules, which, in turn, are composed of atoms. If a quantity of a common substance, such as water, is divided in half, and the half is then divided, and the resulting quarter divided, and so on, a point will be reached where any further division will change the nature of the water and turn it into something else. The smallest particle into which any compound can be divided and still retain its identity is called a **molecule.**

If a molecule of a substance is divided, it will be found to consist of particles called **atoms.** An atom is the smallest possible particle of an element. An **element** is a single substance that cannot be separated into different substances except by nuclear disintegration.

There are more than 100 recognized elements, several of which have been artificially created from various radioactive elements. Common elements are iron, oxygen, aluminum, hydrogen, copper, lead, gold, silver, and so on. The smallest division of any of these elements will still have the properties of that element.

A **compound** is a chemical combination of two or more different elements, and the smallest possible particle of a compound is a molecule. For example, a molecule of water (H_2O) consists of two atoms of hydrogen and one atom of oxygen. A diagram representing a water molecule is shown in Figure 1–1.

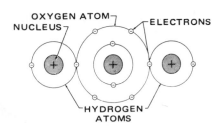

FIGURE 1–1 A water molecule.

Electrons, Protons, and Neutrons

Many discoveries have been made that greatly facilitate the study of electricity and provide new concepts concerning the nature of matter. One of the most important of these discoveries has dealt with the structure of the atom. It has been found that an atom consists of infinitesimal particles of energy known as electrons, protons, and neutrons. All matter consists of two or more of these basic components. The simplest atom is that of hydrogen, which has one electron and one proton, as represented in the diagram of Figure 1–2a. The structure of an oxygen atom is indicated in Figure 1–2b. This atom has eight protons, eight neutrons, and eight electrons. The protons and neutrons form the **nucleus** of the atom; electrons revolve around the nucleus in orbits varying in shape from elliptical to circular and may be compared to the planets as they move around the sun. A **positive** charge is carried by each proton, no charge is carried by the neutrons, and **negative** charge is carried by each electron. The charges carried by the electron and the proton are equal in magnitude but opposite in nature. An atom that has an equal number of protons and electrons is electrically neutral; that is, the charge carried by the electrons is balanced by the charge carried by the protons.

It has been explained that an atom carries two opposite charges: protons in the nucleus have a positive charge, and electrons have a negative charge. When the charge of the nucleus is equal to the combined charges of the electrons, the atom is neutral; but if the atom has a shortage of electrons, it will be **positively charged.** Conversely, if the atom has an excess of electrons, it will be **negatively charged.** A positively charged atom is called a **positive ion,** and a negatively charged atom is called a **negative ion.** Charged molecules are also called ions. It should be noted that protons remain within the nucleus; only electrons are added or removed from an atom, thus creating a positive or negative ion.

Atomic Structure and Free Electrons

The path of an electron around the nucleus of an atom describes an imaginary sphere or shell. Hydrogen and helium atoms have only one shell, but the more complex atoms have numerous shells. Figure 1–2 illustrates this concept. When an atom has more than two electrons, it must have more than one shell, since the first shell will accommodate only two electrons. This is shown in Figure 1–2b. The number of shells in an atom depends on the total number of electrons surrounding the nucleus.

The atomic structure of a substance is of interest to the electrician because it determines how well the substance can conduct an electric current. Certain elements, chiefly metals, are known as **conductors** because an electric current will flow through them easily. The atoms of these elements give up electrons or receive electrons in the outer orbits with little difficulty. The electrons that move from one atom to another are called **free electrons.** The movement of free electrons from one atom to another is indicated by the diagram in Figure 1–3, and it will be noted that they pass from the outer shell of one atom to the outer shell of the next. The only electrons shown in the diagram are those in the outer orbits.

As shown in Figure 1–3, the movement of free electrons does not always constitute electric current flow. There are often several free electrons randomly drifting through the atoms of any conductor. It is only when these free electrons move in the same direction that electric current exists. A power supply, such as a battery, typically creates a potential difference from one end of a conductor to another. A strong negative charge on one end of a conductor and a positive charge on the other is the means to create a useful electron flow.

An element is a conductor, nonconductor (insulator), or semiconductor depending on the number of electrons in the valence orbit of the material's atoms. The **valence orbit** of any atom is the outermost orbit (shell) of that atom. The electrons in this valence orbit are known as **valence electrons.** All atoms desire to have their valence orbit completely full of electrons, and the fewer valence electrons in an atom, the easier it will accept extra electrons. Therefore, atoms with fewer than half of their valence electrons tend to easily accept (carry) the moving electrons of an electric current flow. Such materials are called **conductors.** Materials that have more than half of their valence electrons are called **insulators.** Insulators will not easily accept extra electrons. Materials

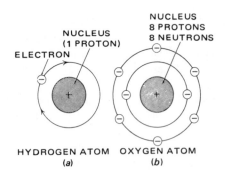

FIGURE 1–2 Structure of atoms.

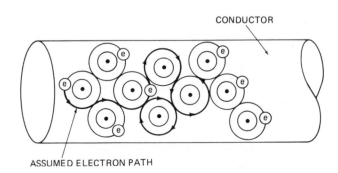

FIGURE 1–3 Random movement of free electrons.

with exactly half of their valence electrons are **semiconductors.** Semiconductors have very high resistance to current flow in their pure state; however, when exact numbers of electrons are added or removed, the material offers very low resistance to electric current flow.

Two of the best conductors are gold and silver; their valence orbits are nearly empty, containing only one electron each. Two of the best insulators are neon and helium; their atoms contain full valence orbits. We commonly substitute other "less perfect" materials for conductors and insulators to reduce costs and increase workability. Common conductors are copper and aluminum; common insulators are air, plastic, fiberglass, and rubber. The two most common semiconductors are germanium and silicon; both of these materials have exactly four electrons in their valence orbits. In general, atoms with four valence electrons are semiconductors; atoms with fewer than four valence electrons are conductors; those with more than four valence electrons are insulators.

Simply being a conductor does not create electron movement. There must be an external force in addition to the molecular forces present inside the conductor's atoms. On the aircraft the external forces are usually supplied by the battery or generator. The atoms' internal forces are caused by the repulsion of two similar charged bodies, such as two electrons or two protons, and the attraction of two dissimilar charged bodies, such as one electron and one proton.

When two electrons are near each other and are not acted upon by a positive charge, they repel each other with a relatively tremendous force. It is said that if two electrons could be magnified to the size of peas and were placed 100 ft apart, they would repel each other with tons of force. It is this force that causes electrons to move through a conductor. Remember, the attraction force of the protons in their nucleus to the electrons in their orbits creates stability in an atom whenever a neutral charge is present. If an extra electron enters the atom's outer orbit, the atom becomes very unstable. It is this unstable repelling force between the orbiting electrons that causes the movement of any extra electron through the conductor. When an extra electron enters the outer orbit of an atom, the repelling force immediately causes another electron to move out of the orbit of that atom and into the orbit of another. If the material is a conductor, the electrons move easily from one atom to another.

Direction of Current Flow

It has been shown that an electric current is the result of the movement of electrons through a conductor. Since a negatively charged body has an excess of electrons and a positively charged body a deficiency of electrons, it is obvious that the electron flow will be **from** the negatively charged body **to** the positively charged body when the two are connected by a conductor. It can therefore be said that electricity flows from negative to positive.

In many cases it is assumed that electric current flows from positive to negative. Since the polarities of electric charges were arbitrarily assigned names (*positive* and *negative*), the actual direction of current flow is difficult to distinguish without the true nature of electric current being considered. When studying the molecular nature of electricity, it is necessary to consider the true direction of electron flow, but for all ordinary electrical applications, the direction of flow can be considered to be in either direction as long as the theory is used consistently. Many texts adhere to the conventional theory that current flows from positive to negative; however, it is the purpose of this text to consider all current flow as moving from negative to positive. Electrical rules and diagrams are arranged to conform to this principle in order to prevent confusion and to give the student a true concept of electrical phenomena. The Federal Aviation Administration (FAA) adheres to the concept that current flows from negative to positive; therefore, the majority of the aviation industry also follows this convention.

It is important not to let this concept of current flow direction confuse your understanding of electricity. The actual direction of current flow is not important when troubleshooting aircraft electrical systems. It is often important to know if current is flowing or not; however, the direction of flow is irrelevant. Simply be consistent in your approach to direction of current flow, and remember while reading this text or any FAA material, *current flows from negative to positive.*

One of the latest theories that defines the direction of current flow states that electrons flow in one direction and holes flow in the opposite direction. A **hole** is the space created by the absence of an electron. As electrons would move from negative to positive, holes would move from positive to negative. This concept is often used when studying the internal current flow of semiconductors; however, for general applications of current flow, holes need not be considered.

STATIC ELECTRICITY

Electrostatics

The study of the behavior of static electricity is called **electrostatics.** The word **static** means stationary or at rest, and electric charges that are at rest are called **static electricity.**

A material with atoms containing equal numbers of electrons and protons is electrically neutral. If the number of electrons in that material should increase or decrease, the material is left with a static charge. An excess of electrons creates a negatively charged body; a deficiency of electrons creates a positively charged body. This excess or deficiency of electrons can be caused by the friction between two dissimilar substances or by contact between a neutral body and a charged body. If friction produces the static charge, the nature of that charge is determined by the types of substances. The following list of substances is called the **electric series,** and the list is so arranged that each substance is positive in relation to any one that follows it, when the two are in contact.

1. Fur	6. Cotton	11. Metals
2. Flannel	7. Silk	12. Sealing wax
3. Ivory	8. Leather	13. Resins
4. Crystals	9. The body	14. Gutta percha
5. Glass	10. Wood	15. Guncotton

If, for example, a glass rod is rubbed with fur, the rod becomes negatively charged, but if it is rubbed with silk, it becomes positively charged.

When a nonconductor is charged by rubbing it with a dissimilar material, the charge remains at the points where the friction occurs because the electrons cannot move through the material; however, when a conductor is charged, it must be insulated from other conductors or the charge will be lost.

An electric charge may be produced in a conductor by induction if the conductor is properly insulated. Imagine that the insulated metal sphere shown in Figure 1–4 is charged negatively and brought near one end of a metal rod that is also insulated from other conductors. The electrons constituting the negative charge in the sphere repel the electrons in the rod and drive them to the opposite site end of the rod. The rod then has a positive charge in the end nearest the charged sphere and a negative charge in the opposite end. This may be shown by suspending pith balls in pairs from the middle and ends of the rod by means of conducting threads. At the ends of the rod, the pith balls separate as the charged sphere is brought near one end; but the balls near the center do not separate because the center is neutral. As the charged sphere is moved away from the rod, the balls fall to their original positions, thus indicating that the charges in the rod have become neutralized.

The force that is created between two charged bodies is called the **electrostatic force.** This force can be either attractive or repulsive, depending on the object's charge. Like charges repel each other. Unlike charges attract each other. The electrostatic force is similar to those forces that exist inside of an atom between electrons and protons. However, the electrostatic force is considered to be on a much larger scale, dealing with entire objects, not minute atomic particles. The amount of static charge contained within a body will determine the strength of the electrostatic field. Weak charges produce weak electrostatic fields and vice versa. Precisely, the strength of an electrostatic field between two bodies is directly proportional to the strength of the charge on those two bodies. Figure 1–5a demonstrates this concept. The strength of the electrostatic force is also affected by the distance between the two charged bodies. If the distance between the two charged substances increases, the electrostatic force decreases; conversely, if the distance decreases, the force increases. Precisely, the electrostatic force between two charged bodies is inversely proportional to the square of the distance between those two bodies. That is, as the distance becomes twice as large between the bodies, the electrostatic force is one-fourth as great. This concept is demonstrated in Figure 1–5b.

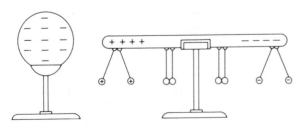

FIGURE 1–4 Charging by induction.

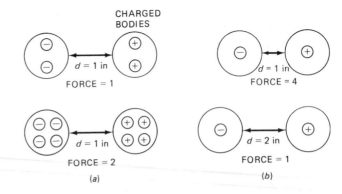

FIGURE 1–5 The strength of an electrostatic force. (a) Twice the static charge equals twice the static force. (b) Twice the distance equals one-fourth the static force.

Static electrical discharge will eventually occur to all charged bodies. Any unbalance of charge strives for equilibrium. Usually contact is made with another object to neutralize the static charge. If a charged body contacts a neutral body, both objects will then share the original charge. If the neutral body is large enough, such as the earth, virtually all the charge will become neutralized, or absorbed, by the large body.

UNITS OF ELECTRICITY

Current

An electric **current** is defined as a flow of electrons through a conductor. Earlier in this chapter it was shown that the free electrons of a conducting material move from atom to atom as the result of the attraction of unlike charges and the repulsion of like charges. If the terminals of a battery are connected to the ends of a wire conductor, the negative terminal forces electrons into the wire and the positive terminal takes electrons from the wire; hence as long as the battery is connected, there is a continuous flow of current through the wire until the battery becomes discharged.

Because each electron has mass and inertia, electron flow is capable of doing work such as turning motors, lighting lamps, and warming heaters. Just as moving water can turn a primitive paddle wheel to grind wheat, moving electrons can do the same. Even at the speed of light, a single electron could not turn a paddle wheel owing to its minute size; however, if enough electrons could be hurled at the paddle wheel, indeed it would turn. This is only a hypothetical situation, for in practical terms we know electrons must travel through a conductor. Whenever it becomes hard to accept that moving electrons can perform useful functions, consider that any mass with inertia can perform work.

It is said that an electric current travels at the speed of light, more than 186,000 miles per second (mps) [299,000 km/s]. Actually, it would be more correct to say that the effect, or force, of electricity travels at this speed. Individual electrons move at a comparatively slow rate from atom to atom in a conductor, but the influence of a charge is "felt" through the entire length of a conductor instantaneously. A simple illustration will explain this phenomenon. If we completely fill a tube with tennis balls, as shown in Figure 1–6, and then push an additional ball into one end of the tube, one ball will fall out the other end. This is similar to the effect of electrons as they are forced into a conductor. When electrical pressure is applied to one end of the conductor, it is immediately effective at the other end. It must be remembered, however, that under most conditions, electrons must have a complete conducting path before they will enter or leave the conductor.

When it is necessary to measure the flow of a liquid through a pipe, the rate of flow is often measured in **gallons per minute.** The gallon is a definite quantity of liquid and may be called a unit of quantity. The unit of quantity for electricity is the **coulomb** (C), named for Charles A. Coulomb (1736–1806), a French physicist who conducted many experiments with electric charges. One coulomb is the amount of electricity that, when passed through a standard silver nitrate solution, will cause 0.001118 gram (g) of silver to be deposited upon one electrode. (An electrode is a terminal, or pole, of an electric circuit.) A coulomb is also defined as 6.28×10^{18} electrons, that is, 6.28 billion billion electrons.

The rate of flow for an electric current is measured by the number of coulombs per second passing a given point in a circuit. Instead of designating the rate of flow in coulombs per second, a more convenient unit called the **ampere** (A) is used. **One ampere is the rate of flow of 1 coulomb per second.** The ampere was named in honor of the French scientist André M. Ampère (1775–1836). The term **current** is symbolized by the letter **I.** Current is measured in amperes, which is often abbreviated **amps.**

Voltage Potential Difference and Electromotive Force

Just as water flows in a pipe when there is a difference of pressure at the ends of the pipe, an electric current flows in a conductor because of a difference in electrical pressure at the ends of the conductor. If two tanks containing water at different levels are connected by a pipe with a valve, as shown in Figure 1–7, water flows from the tank with the higher level to the other tank when the valve is open. The difference in water pressure is due to the higher water level in one tank.

It may be stated that in an electric circuit, a large number of electrons at one point will cause a current to flow to another point where there is a small number of electrons if the two points are connected by a conductor. In other words,

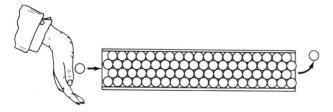

FIGURE 1–6 Demonstration of current flow. One electron into the conductor instantaneously means one electron out of the conductor.

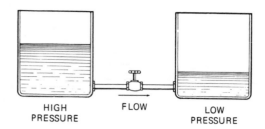

FIGURE 1–7 Difference of pressure.

when the electron level is higher at one point than at another point, there is a difference of potential between the points. When the points are connected by a conductor, electrons flow from the point of high potential to the point of low potential. There are numerous simple analogies that may be used to illustrate potential difference. For example, when an automobile tire is inflated, a difference of potential (pressure) exists between the inside of the tire and the outside. When the valve is opened, the air rushes out. In this case the air inside the tire represents an excess of electrons, a high potential, or a negative charge. The air outside the tire represents a deficiency of electrons, a low potential, or a positive charge.

The force that causes electrons to flow through a conductor is called **electromotive force,** abbreviated emf, or electron-moving force. The practical unit for the measurement of emf or potential differences is the **volt** (V). The word *volt* is derived from the name of the famous electrical experimenter, Alessandro Volta (1745–1827), of Italy, who made many contributions to the knowledge of electricity.

One volt is the emf required to cause current to flow at the rate of 1 ampere through a resistance of 1 ohm. The term *ohm* is defined later in this chapter. Electromotive force and potential difference may be considered the same for all practical purposes. When there is a potential difference, or difference of electrical pressure, between two points, it simply means that a field of force exists that tends to move electrons from one point to the other. If the points are connected by a conductor, electrons will flow as long as the potential difference exists.

With reference to Figure 1–8 it can be seen that voltage, the potential difference in a battery, creates an electron flow, just as pressure inside a balloon, a potential air pressure difference, creates an air flow. Voltage (electrical pressure) causes electrons to flow through a conductor. This is no mystery. Any object that is either extremely large or as small as an electron will tend to move when pressure is applied in a certain direction.

Electromotive force, which is the force that causes electrons to move, could also be considered electrical potential or pressure. The term **voltage,** which is measured in volts, is typically substituted for emf. Voltage is symbolized by the letter **E,** and volts is symbolized by the letter **V.**

Resistance

Resistance is that property of a conductor which tends to hold, or restrict, the flow of an electric current; it is encountered in every circuit. Resistance may be termed *electrical*

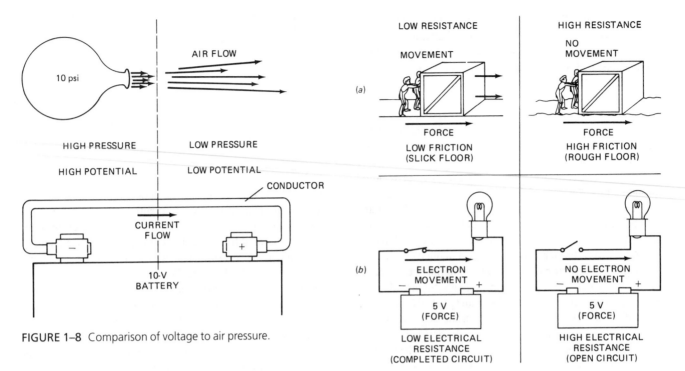

FIGURE 1–8 Comparison of voltage to air pressure.

FIGURE 1–9 Comparison of resistance to friction.

friction because it affects the movement of electricity in a manner similar to the effect of friction on mechanical objects. For example, if the interior of a water pipe is very rough because of rust or some other material, a smaller stream of water will flow through the pipe at a given pressure than would flow if the interior of the pipe were clean and smooth. The rough pipe offers greater resistance, or friction, than the smooth pipe.

The unit used in electricity to measure resistance is the **ohm.** The ohm is named for the German physicist Georg S. Ohm (1789–1854), who discovered the relationship between electrical quantities known as **Ohm's law.** Resistance is opposition to current flow and is symbolized by the letter **R.** Resistance is measured in ohms, which is symbolized by the Greek letter omega, Ω.

Earlier it was explained that materials with a small number of valence electrons, fewer than four, are conductors. Conductors have a relatively low resistance because they accept extra electrons (current flow) easily. If a voltage is applied to a conductor, an electric current will flow, assuming a complete circuit is present. As seen in Figure 1–9*a,* if a heavy wooden crate is pushed on a highly polished floor, the crate will slide easily because the floor offers low resistance, or low opposition, to movement. If the same crate is placed on a rough concrete floor and pushed again with the same force, little or no movement will take place owing to the high resistance offered by the rough floor. Now compare the crate in Figure 1–9*b* with the circuit in Figure 1–9*b.* A circuit of low resistance with an applied 5 V will easily move electrons. The same 5 V applied to a circuit of high resistance—an open switch, for example—is capable of moving no electrons.

Insulators are materials that have more than four valence electrons. Insulators will not accept the extra electrons of current flow easily and therefore are considered to have rela-

tively high resistance. If a moderate voltage is applied to an insulator, no electric current will flow. There are no perfect insulators, but many substances have such high resistance that for practical purposes they may be said to prevent the flow of current. Substances having good insulating qualities are dry air, glass, mica, porcelain, rubber, plastic, asbestos, and fiber compositions. The resistance of these substances varies to some extent, but they may all be said to block the flow of current effectively.

According to the electron theory, the atoms of an insulator do not give up electrons easily. When a voltage is applied to such a substance, the outer electron orbits are distorted; but as soon as the voltage is removed, the electrons return to their normal positions. If, however, the voltage applied is so strong that it strains the atomic structure beyond its elastic limit, the atoms lose electrons and the material becomes a conductor. When this occurs, the material is said to be ruptured.

THEORY OF MAGNETISM

The Magnet

Almost everyone has witnessed the effects of magnetism, and many have owned simple permanent magnets such as that illustrated in Figure 1–10. However, few people realize the importance of magnetism and its relationship to electricity. It would be hard to refute the fact that electricity would not exist without magnetism. A **magnet** may be defined as an object that attracts such magnetic substances as iron or steel. It produces a magnetic field external to itself that reacts with magnetic substances.

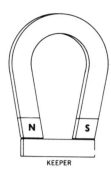

FIGURE 1–10 A permanent magnet.

A **magnetic field** is assumed to consist of invisible lines of force that leave the **north** pole of a magnet and enter the **south** pole. The direction of this force is assumed only in order to establish rules and references for operation. Whether there is any actual movement of force from the north pole to the south pole of a magnet is not known, but it is known that the force acts in a definite direction. This is indicated by the fact that a north pole will repel another north pole but will be attracted by a south pole. Like poles repel; unlike poles attract. **A permanent magnet** is one that maintains an almost constant magnetic field without the application of any magnetizing force. Some magnetized substances show practically no loss of magnetic strength over a period of several years.

A **natural magnet** is one found in nature; it is called a **lodestone,** or *leading stone.* The natural magnet received this name because it was used by early navigators to determine direction. The lodestone is composed of an oxide of iron called magnetite.

When first discovered, the lodestone was found to have peculiar properties. When it was freely suspended, one end always pointed in a northerly direction. For this reason, one end of the lodestone was called the *north-seeking* and the other the *south-seeking* end. These terms have been shortened to *north* and *south,* respectively. The reason that a freely suspended magnet assumes a north-south position is that the earth is a large magnet and the earth's magnetic field exists over the entire surface. The suspended magnet's lines of force interact with the earth's magnetic field and align the magnet accordingly. According to definition, the magnetic pole near the earth's north geographic pole is actually the earth's south magnetic pole. This can be demonstrated by suspending a magnet on a string and noting the direction in which the north pole points. The magnet's north pole points to the earth's geographic north, but by definition, north should repel north; therefore, the earth's south magnetic pole is actually nearest the earth's geographic north. This concept is demonstrated in Figure 1–11. To eliminate confusion, the direction in which a magnet's north pole points is called the earth's north pole. In reality it is magnetic south.

The magnetic poles of the earth are not located at the geographic poles. The magnetic pole in the northern hemisphere is located east of geographic north. The magnetic south pole is located west of geographic south, as illustrated in Figure 1–11. The difference between the geographic and magnetic

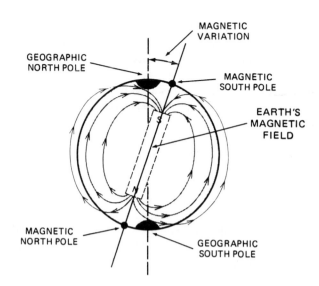

FIGURE 1–11 The earth's magnetic field.

poles is called **magnetic variation.** Magnetic variation is sometimes referred to as *magnetic declination.* In general, this principle of magnetic variation does not affect electrical phenomena; however, it becomes very important when navigating aircraft using a magnetic compass.

The true nature of magnetism is not clearly understood, although its effects are well known. One theory that seems to provide a logical explanation of magnetism assumes that atoms or molecules of magnetic substances are in reality small magnets. It is reasoned that electrons moving around the nucleus of an atom create minute magnetic fields. In magnetic substances such as iron it is assumed that most of the electrons are moving in one general direction around the nuclei; hence these electrons produce a noticeable magnetic field in each atom, and each atom or molecule becomes a tiny magnet. When the substance is not magnetized, the molecules lie in all positions in the material, as shown in Figure 1–12a, and their fields tend to cancel one another. When the substance is placed in a magnetic field, the molecules align themselves with the field, and the fields of the molecules add to the strength of the magnetizing field. A diagram of a magnetized substance is shown in Figure 1–12b.

When a piece of soft iron is placed in a magnetic field, almost all the molecules in the iron align themselves with the

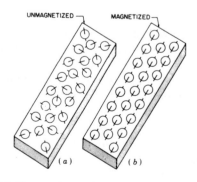

FIGURE 1–12 Theory of magnetism.

field, but as soon as the magnetizing field is removed, most of the molecules return to their random positions, and the substance is no longer magnetized. Because some of the molecules tend to remain in the aligned position, every magnetic substance retains a slight amount of magnetism after having been magnetized. This retained magnetism is called **residual magnetism.**

Certain substances, such as hard steel, are more difficult to magnetize than soft iron because of the internal friction among the molecules. If such a substance is placed in a strong magnetic field and is struck several blows with a hammer, the molecules become aligned with the field. When the substance is removed from the magnetic field, it will retain its magnetism; hence it is called a **permanent magnet.** Hard steel and certain metallic alloys—such as Alnico, an alloy containing nickel, aluminum, and cobalt—that have the ability to retain magnetism are able to do so for the same reason that they are difficult to magnetize; that is, the molecules do not shift their positions easily. When the molecules are aligned, all the north poles of the molecules point in the same direction and produce the north pole of the magnet. In like manner, the south poles of the molecules produce the south pole of the magnet.

Many substances have no appreciable magnetic properties. The atoms of these substances apparently have their electron orbits in positions such that their fields cancel one another. Among these substances are copper, silver, gold, and lead.

The ability of a material to become magnetized is called **permeability.** A material with high permeability is easy to magnetize or demagnetize. A material with low permeability is hard to magnetize or demagnetize. Materials with high permeability, such as soft iron, are most useful as temporary magnets. Materials with low permeability, such as Alnico, are best suited for permanent magnets.

Properties of Magnetism

The field of force existing between the poles of a magnet is called a **magnetic field.** The pattern of this field may be seen by placing a stiff paper over a magnet and sprinkling iron filings on the paper. As shown in Figure 1–13, the iron filings will line up with the lines of magnetic force. It will be noted that the lines directly between the poles are straight, but the lines farther from the direct path are curved. This curving is due to the repulsion of lines traveling in the same direction. If iron filings are sprinkled on a paper placed over two north poles, the field will have the pattern shown in Figure 1–14. Here the lines of force from the two poles come out and curve away from one another.

Magnetic force, which is also called **magnetic flux,** is said to travel from north to south in invisible lines. We cannot say that this is literally true, but by assuming a direction, we provide a reference by which calculations can be made and effects determined. Since iron filings in a magnetic field arrange themselves in lines, it is logical to say that magnetic force exists in lines.

The space or substance traversed by magnetic lines of

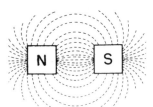

FIGURE 1–13 A magnetic field.

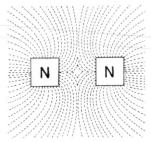

FIGURE 1–14 Magnetic field between two like magnetic poles.

force is called the **magnetic circuit.** If a soft-iron bar is placed across the poles of a magnet, almost all the magnetic lines of force (flux) go through the bar, and the external field will be very weak.

The external field of a magnet is distorted when any magnetic substance is placed in that field because it is easier for the lines of force to travel through the magnetic substance than through the air (see Figure 1–15). The opposition of a material to magnetic flux is called **reluctance** and compares to resistance in an electric circuit. The symbol for reluctance is R and the unit is **rel.** As with electric current, the material that will completely resist magnetic flux lines is unknown. However, some materials will accept flux lines more easily than others.

In review, the properties of magnets are as follows: (1) The pole that tends to point toward the earth's geographic north is called the magnet's north pole. The opposite end is called the south pole. (2) Like magnetic poles repel each other, and unlike poles attract each other. (3) A magnetic field surrounds each magnet and contains magnetic flux lines. These flux lines are directly responsible for the magnetic properties of the material. (4) The strength of any magnet is directly proportional to the density of the flux field. That is, a stronger magnet will have a relatively larger number of flux

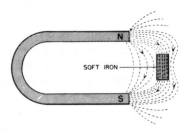

FIGURE 1–15 Field distorted by a magnetic substance.

lines concentrated in a given area. (5) Magnetic fields are strongest near the poles of the magnet. This is due to the concentration of flux lines at each pole. (6) By definition, magnetic flux lines flow from the north to the south pole of any magnet. This property becomes important when studying certain relationships of magnetism. (7) Flux lines never intersect. This is because flux lines repel each other with relatively tremendous force. (8) Magnetic flux lines always take the path of least resistance, such as when they distort in order to travel through a piece of soft iron as opposed to traveling through air.

MAGNETIC DEVICES

Electromagnets

Electromagnets, in various forms, are very useful items and have become commonplace on modern aircraft. **Electromagnets,** as the name implies, are produced by using an electric current to create a magnetic field. Around every conductor carrying current a magnetic field exists. This magnetic field is created owing to the movement of electrons through the conductor. Typically this magnetic field is so small it is unnoticed. However, if the current is very strong or the conductor is formed into a coil, the magnetic field strength increases. Most electromagnet conductors are wound into coils to create the desired magnetic field strength.

In Figure 1–16, the shaded circle represents a cross section of a conductor with current flowing in toward the paper. The current is flowing from negative to positive. When the current flows as indicated, the magnetic field is in a counterclockwise direction. This is easily determined by the use of the left-hand rule, which is based upon the true direction of current flow. When a wire is grasped in the left hand with the thumb pointing from negative to positive, the magnetic field around the conductor is in the direction that the fingers are pointing.

If a current-carrying wire is bent into a loop, the loop assumes the properties of a magnet; that is, one side of the loop will be a north pole, and the other side will be a south pole. If a soft-iron core is placed in the loop, the magnetic lines of force will traverse the iron core, and it becomes a magnet. When a wire is made into a coil and connected to a source of power, the fields of the separate turns join and thread through

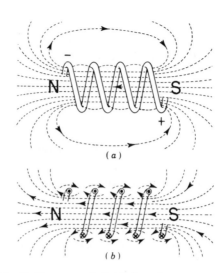

FIGURE 1–17 The magnetic field of a coil.

the entire coil, as shown in Figure 1–17a. Figure 1–17b shows a cross section of the same coil. Note that the lines of force produced by one turn of the coil combine with the lines of force from the other turns and thread through the coil, thus giving the coil a magnetic polarity. The polarity of the coil is easily determined by the use of the **left-hand rule for coils:** *When a coil is grasped in the left hand with the fingers pointing in the direction of current flow, that is, from negative to positive, the thumb will point toward the north pole of the coil.*

When a soft-iron core is placed in a coil, an electromagnet is produced. Of course, the wire in the coil must be insulated so that there can be no short circuit between the turns of the coil. A typical electromagnet is made by winding many turns of insulated wire on a soft-iron core that has been wrapped with an insulating material. The turns of wire are placed as close together as possible to help prevent magnetic lines of force from passing between the turns. Figure 1–18 is a cross-sectional drawing of an electromagnet.

The strength of an electromagnet is directly proportional to the current carried by the wire coil and to the number of turns in that coil. That is, as either the current through the coil or the number of wire wraps around the coil increases, the electromagnet's strength also increases. Also, use of a core

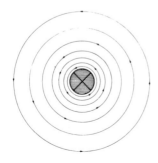

FIGURE 1–16 Magnetic field around a conductor.

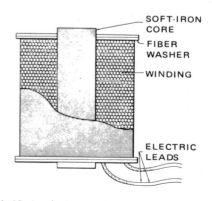

SOFT-IRON CORE
FIBER WASHER
WINDING
ELECTRIC LEADS

FIGURE 1–18 An electromagnet.

material of high permeability will increase an electromagnet's strength. The same electromagnet using a core of low permeability would have a decreased magnetic strength. Other factors also affect an electromagnet's strength, although they are negligible for most general-purpose applications.

The force exerted upon a magnetic material by an electromagnet is inversely proportional to the square of the distance between the pole of the magnet and the material. For example, if a magnet exerts a pull of 1 lb [0.4536 kg] upon an iron bar when the bar is $\frac{1}{2}$ in. [1.27 cm] from the magnet, then the pull will only be $\frac{1}{4}$ lb [0.1134 kg] when the bar is 1 in. [2.54 cm] from the magnet. For this reason, the design of electric equipment using electromagnetic actuation requires careful consideration of the distance through which the magnetic force must act.

Solenoids

It has been explained that a coil of wire, when carrying a current, will have the properties of a magnet. Such coils are frequently used to actuate various types of mechanisms. If a soft-iron bar is placed in the field of a current-carrying coil, the bar will be magnetized and will be drawn toward the center of the coil, thus becoming the core of an electromagnet. By means of suitable attaching linkage, the movable core may be used to perform many mechanical functions. An electromagnet with a movable core is called a **solenoid.**

A solenoid typically uses a split core; one part of the core is a nonmagnetic outer sleeve fixed permanently inside the coils. The other portion of the core is allowed to slide inside this fixed outer sleeve, as demonstrated in Figure 1–19. The spring typically holds the movable core partially extended from one end of the electromagnetic coil. When the coil is energized, the electromagnet's force pulls the movable core into the hollow sleeve opposing the spring force. This imparts motion through a connecting rod to the mechanical linkage.

Solenoids are commonly used to operate electrical contacts, valves, circuit breakers, and several types of mechanical devices. The chief advantage of solenoids is that they can be placed almost anywhere in an airplane and can be controlled remotely by small switches or electronic control units. Although the use of solenoids is limited to operations where only a small amount of movement is required, they have a

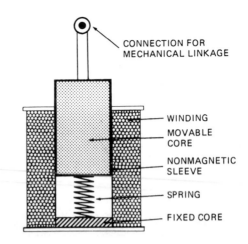

FIGURE 1–19 A solenoid.

much greater range of movement, quicker response, and greater strength than fixed-core electromagnets.

Relays

Electromagnets that contain a fixed core and a pivoting mechanical linkage are called **relays.** Relays are usually used for low-current switching applications. Figure 1–20a illustrates a typical switching relay.

The part of the relay attracted by the electromagnet to close the contact points is called the **armature.** There are several types of armatures in electrical work, but in every case it will be found that an armature consists, in part, of a bar or core of material that may be acted upon by a magnetic field. In a relay, the armature is attracted to the electromagnet, and the movement of the armature either closes or opens the contact points. In some cases, the electromagnet operates several sets of contact points simultaneously.

There is much confusion surrounding the terminology of relays and solenoids because of their similarities. Relays are often called solenoids and vice versa. For the purpose of this text, and as generally accepted in the aircraft industry, a *solenoid* is an electromagnet with a movable core material, and a *relay* is an electromagnet with a fixed core. These definitions hold true whether the electromagnet is used for electrical switching or other mechanical functions. Figures 1–20a and 1–20b illustrate the difference between switching relays and

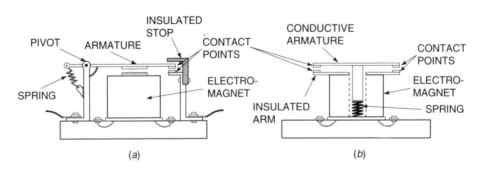

(a)   (b)

FIGURE 1–20 Electromagnetic switches. (a) Relay; (b) solenoid.

solenoids. Many aircraft manufacturers have substituted the term *contactor* or *breaker* for electrical switching solenoids or relays.

As discussed earlier, *voltage* is the force, or pressure, that creates electron movement through conductors and hence creates useful electric power. Voltage must be present in all circuits in order to produce current flow, but what creates voltage? Voltage is created by limited means, and only two methods produce nearly 100 percent of all electric power consumed by typical aircraft.

Friction is a method of producing voltage by simply rubbing two dissimilar materials together. This usually produces **static electricity,** which is not typically a useful form of power. In fact, most static electricity found on the aircraft becomes a nuisance to both communication and navigation systems as well as advanced electronic devices.

Pressure is another means of producing voltage. **Piezoelectricity** means electricity created by applying pressure to certain types of crystals. Since only small amounts of power are produced using piezoelectricity, applications are limited. Some microphones used for radio communications employ the piezoelectric effect to convert sound waves into electric power. Most piezoelectric devices use crystalline materials like quartz to produce voltage. When a force is applied to certain crystals, their molecular structure distorts and electrons may be emitted into a conductor.

Light is a source of energy that also can be converted into electricity. The **photoelectric effect** produces a voltage when light is emitted onto certain substances. Zinc is a typical photosensitive material. If exposed to ultraviolet rays, under the correct conditions, zinc will produce a voltage. Although photoelectric devices are limited in modern aircraft, spacecraft and satellites rely heavily on photo cells and the sun for a source of electric power.

Heat can also be used to produce voltage. Electricity produced by subjecting two dissimilar metals to above normal temperatures is called the **thermoelectric effect.** For example, copper and zinc held firmly together will produce voltage when subjected to heat. This combination of two dissimilar metals is called a **thermocouple.** Thermocouples are used in virtually any electronic temperature sensor found on aircraft. These include exhaust gas and cylinder head temperature sensors, electronic equipment temperature monitors, and some fire detectors.

Chemical action is often used to produce electricity for aircraft systems. A battery is found on virtually all aircraft, producing voltage because of the reaction of two or more different chemicals. When two or more of the correct chemicals come in contact, their structures are altered and voltage is produced. Most aircraft contain a battery used for engine starting and emergency procedures.

Magnetism is used to produce the majority of all electric power. **Electromagnetic induction** is the process where voltage is produced by moving a conductor through a magnetic field.

Basic Principles

The transfer of electric energy from one circuit to another without the aid of electric connections is called **induction.** When electric energy is transferred by means of a magnetic field, it is called **electromagnetic induction.** This type of induction is universally employed in the generation of electric power. Electromagnetic induction is also the principle that makes possible the operation of electric transformers and the transmission of radio signals.

Electromagnetic induction occurs whenever there is a relative movement between a conductor and a magnetic field, provided that the conductor is cutting across (linking with) magnetic lines of force and is not moving parallel to them. The relative movement may be caused by a stationary conductor and a moving field or by a moving conductor with a stationary field. A moving field may be provided by a moving magnet or by changing the value of the current in an electromagnet.

The two general classifications of electromagnetic induction are **generator action** and **transformer action.** Both actions are the same electrically, but the methods of operation are different. Transformer action will be discussed in a later chapter of this text.

Generator Action

The basic principle of generator action is shown in Figure 1–21. As the conductor is moved through the field, a voltage is induced in it. The same action takes place if the conductor is stationary and the magnetic field is moved. The direction of the induced voltage depends on the direction of the field and may be determined by using the **left-hand rule for generators:** *Extend the thumb, forefinger, and middle finger of the*

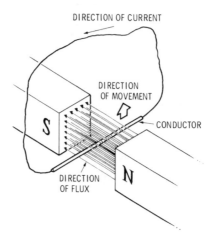

FIGURE 1–21 Generator action.

FIGURE 1–22 Left-hand rule for generators.

left hand so that they are at right angles to one another, as shown in Figure 1–22. Turn the hand so that the index finger points in the direction of the magnetic field and the thumb points in the direction of conductor movement. Then the middle finger will be pointing in the direction of the induced voltage.

Figure 1–23 illustrates another kind of generator action. Here a bar magnet is pushed into a coil of wire. A sensitive meter connected to the leads from the coil shows that a current flows in a certain direction as the magnet moves into the coil. As soon as the magnet stops moving, the current flow stops. When the magnet is withdrawn, the meter shows that the current is flowing in the opposite direction. The current induced in the coil is caused by the field of the magnet as it cuts across (links with) the turns of wire in the coil. The induced current is always in such a direction that its magnetic field opposes any change in the existing magnetic field.

In Figure 1–23a it will be seen that the north pole of the coil is adjacent to the north pole of the bar magnet; hence it opposes the insertion of the magnet into the coil. At the instant that the magnet begins to move out of the coil, current induced in the coil changes to the opposite direction; hence the field of the coil is reversed. The south pole of the coil field is now adjacent to the north pole of the bar magnet and opposes the withdrawal of the magnet (see Figure 1–23b).

Generally speaking, to produce a voltage through electromagnetic induction, there must be a magnetic field, a conductor, and relative motion between the two. The magnetic field can be produced by a permanent magnet or an electromagnet. Typically, electromagnets are used because of their advantages of increased magnetic strength. The conductor used is usually wrapped in the form of a coil, which produces a greater induced voltage. The motion can be created by moving either the magnet or the conductor. Typically, this is done by rotating a coil inside a magnetic field or by rotating a magnetic field inside a wire coil.

REVIEW QUESTIONS

1. Describe the properties of a permanent magnet.
2. What is the difference between substances required for permanent magnets and those used for temporary magnets?
3. Define *permeability; reluctance.*
4. When the direction of current flow through a coil is known, how do you determine the polarity of the coil?
5. How does the pull of a magnet on a piece of steel at 1-in. distance compare with the pull at 2-in. distance?
6. Compare a solenoid with an electromagnet.
7. Describe a relay.
8. What conditions are necessary to produce electromagnetic induction?
9. How do you determine the direction of current flow?
10. If a positively charged terminal is connected to a negatively charged terminal, in which direction will the electrons flow?
11. What undesirable effects are caused by static electricity during the operation of an airplane?
12. Define *molecule* and *atom.*
13. What particles are found in an atom?
14. What is an element in matter?
15. What is another name for a charged atom?
16. What makes some substances conductors, nonconductors, or semiconductors?
17. What force is required to cause electrons to move through a conductor?
18. Explain the nature of static charges.
19. What is an electric current?
20. What name is given to the unit of electromotive force?
21. To what physical force may electromotive force be compared?
22. What is the unit of electric current flow?
23. What is the unit of electrical quantity?
24. Define *resistance* and give the unit of resistance.
25. What factors determine the resistance of a conductor?

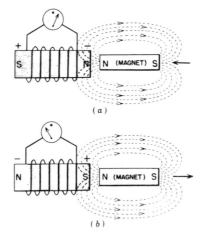

FIGURE 1–23 Current induced by a changing magnetic field.

Applications of Ohm's Law 2

We have already studied the three fundamental elements of electricity: voltage, amperage, and resistance. **Ohm's law,** first presented by the German physicist Georg Simon Ohm (1787–1854), describes the relationships between these elements. These relationships are the foundation upon which all electrical concepts are based. The mathematical relationships presented in Ohm's law explain the otherwise mysterious link between voltage, amperage, and resistance for virtually all direct-current (dc) electrical circuits. As you progress through this text, you will understand just how important Ohm's law is in the design and repair of aircraft electrical systems. For example, understanding Ohm's law is necessary to determine the correct size and length of wire to be used in a circuit, the proper sizes of fuses and circuit breakers, and many other details of a circuit and its components. It is the purpose of this chapter to introduce the concepts of Ohm's law and present their mathematical relationships.

OHM'S LAW

Definitions

In mathematical problems, emf is expressed in volts, and the symbol E is used to indicate the emf until the actual number of volts is determined. R is the symbol for resistance in ohms, and I is the symbol for current, or amperage. The letter I may be said to represent the **intensity** of current. The letter symbols $E, R,$ and I have an exact relationship in electricity given by Ohm's law. This law may be stated as follows: **The current in an electric circuit is directly proportional to the emf (voltage) and inversely proportional to the resistance.** Ohm's law is further expressed by the statement: **1 volt causes 1 ampere to flow through a resistance of 1 ohm.** The equation for Ohm's law is

$$I = \frac{E}{R}$$

which indicates that the current in a given circuit is equal to the voltage divided by the resistance.

An equation is defined as a proposition expressing equality between two values. It may take as many forms as those shown for Ohm's law in Figure 2–1. The different forms for the Ohm's law equation are derived by either multiplication or division. For example,

$$R(I) = R\left(\frac{E}{R}\right) \qquad \text{becomes} \qquad RI = \frac{RE}{R}$$

Then

$$RI = E \qquad \text{or} \qquad E = IR$$

In a similar manner, if both sides of the equation $E = IR$ are divided by I, we arrive at the form

$$R = \frac{E}{I}$$

Thus we find it simple to determine any one of the three values if the other two are known. Ohm's law may be used to solve any common dc circuit problem because any such circuit, when operating, has voltage, amperage, and resistance. To solve alternating-current (ac) circuit problems, other values must be taken into consideration. These will be discussed in the chapter on alternating current.

From the study of Ohm's law, it has been seen that the current flowing in a circuit is directly proportional to the voltage and inversely proportional to the resistance. If the voltage applied to a given circuit is doubled, the current will double. If the resistance is doubled and the voltage remains the same, the current will be reduced by one-half (see Figure 2–2). The circuit symbol for a battery that is the power source for these circuits and the circuit symbol for a resistor or resistance are indicated in the illustration.

OHM'S LAW

CURRENT = $\dfrac{\text{ELECTROMOTIVE FORCE}}{\text{RESISTANCE}}$

$I = \dfrac{E}{R}$ AMPERES = $\dfrac{\text{VOLTS}}{\text{OHMS}}$

RESISTANCE = $\dfrac{\text{ELECTROMOTIVE FORCE}}{\text{CURRENT}}$

$R = \dfrac{E}{I}$ OHMS = $\dfrac{\text{VOLTS}}{\text{AMPERES}}$

ELECTROMOTIVE FORCE = CURRENT × RESISTANCE

$E = IR$ VOLTS = AMPERES × OHMS

FIGURE 2–1 Equations for Ohm's law.

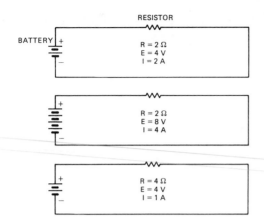

FIGURE 2–2 Effects of current and voltage.

The equations of Ohm's law are easily remembered by using the simple diagram shown in Figure 2–3. By covering the symbol of the unknown quantity in the diagram with the hand or a piece of paper, the known quantities are found to be in their correct mathematical arrangement. For example, if it is desired to find the total resistance of a circuit in which the voltage is 10 and the amperage is 5, cover the letter R in the diagram. This leaves the letter E over the letter I; then

$$R = \frac{E}{I} = \frac{10}{5} \quad \text{or} \quad R = 2\ \Omega$$

If it is desired to find the voltage in a circuit when the resistance and the amperage are known, cover the E in the diagram. This leaves I and R adjacent to each other; they are therefore to be multiplied according to the equation $E = IR$.

One of the simplest descriptions of the Ohm's law relationships is the water analogy. Water pressure and flow, along with the restrictions of a water valve, respond in a manner similar to the relationships of voltage, amperage, and resistance in an electric circuit. As illustrated in Figure 2–4, an increase in voltage (electrical pressure) creates a proportional increase in current (electrical flow), just as an increase in water pressure creates an increase in water flow. Figure 2–5 shows the relationship between resistance and current. As the resistance of a circuit increases, the current decreases, assuming that the voltage remains constant. Water responds similarly. As the water valve is closed (increasing resistance), the water flow decreases.

The water analogy of Ohm's law is a simple comparison.

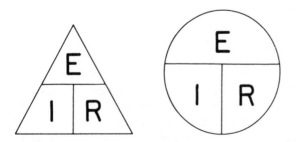

FIGURE 2–3 Diagram for Ohm's law.

Use the analogy to gain a better understanding of the relationships between voltage, amperage, and resistance.

Electric Power and Work

Power means the rate of doing work. One horsepower (hp) [746 watts (W)] is required to raise 550 pounds (lb) [249.5 kilograms (kg)] a distance of 1 ft [30.48 cm] in 1 s. When 1 lb [0.4536 kg] is moved through a distance of 1 ft, 1 foot-pound (ft·lb) [13.82 cm·kg] of work has been performed; hence 1 hp is the power required to do 550 ft·lb [7601 cm·kg] of work per second. The unit of power in electricity and in the SI metric system is the **watt (W)**, which is equal to 0.00134 hp. Conversely, 1 hp is equal to 746 W. In electrical terms, **1 watt is the power expended when 1 volt moves 1 coulomb per second through a conductor; that is, 1 volt at 1 ampere produces 1 watt of power.** The formula for electric power is

$$P = EI \quad \text{or} \quad \text{Power} = \text{voltage} \times \text{amperage}$$

The power equation can be combined with the Ohm's law equations to allow more flexibility when determining power in a circuit. The following are the three most common varieties of the power equations:

$$P = EI \qquad P = I^2R \qquad P = \frac{E^2}{R}$$

The derivatives of the basic power equations are found as follows:
If

$$P = EI \qquad \text{and} \qquad E = IR$$

then, substituting for E,

$$P = (IR)I \quad \text{or} \quad P = I^2R$$

Of course, these equations can be arranged to solve for E, I, or R:

$$E^2 = PR \qquad I^2 = \frac{P}{R} \qquad \text{and} \qquad R = \frac{P}{I^2}$$

When power is lost in an electric circuit in the form of heat, it is called the *IR loss* because the heat produced is a function of a circuit's current and resistance. The equation $P = I^2R$ best represents the heat energy loss of any dc circuit, where P equals the lost power, measured in **watts.**

Power in an electric circuit is always additive. That is, total power equals the sum of the powers consumed by each individual unit. The power consumed by any individual load can be found using the equation

$$P = I^2R \quad \text{or} \quad P = IE$$

While determining power of any portion of a circuit, be sure to apply the I, E, or R (current, voltage, or resistance) that applies to the load being calculated.

Since we know the relationship between power and electrical units, it is simple to calculate the approximate amperage to operate a given motor when the efficiency and operating voltage of the motor are known. For example, if it is desired to install a 3-hp [2.238 kilowatt (kW)] motor in a

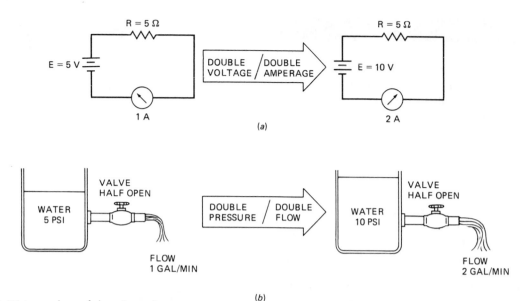

FIGURE 2–4 Water analogy of changing voltage.

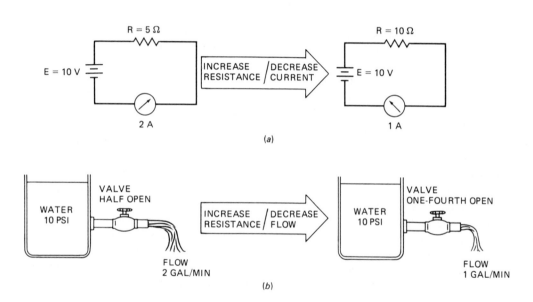

FIGURE 2–5 Water analogy of changing resistance.

24-V system and the efficiency of the motor is 75 percent, we proceed as follows:

$$1 \text{ hp} = 746 \text{ W}$$
$$P = 3 \times 746 = 2238 \text{ W}$$
$$I = \frac{2238}{24} = 93.25 \text{ A}$$

Since the motor is only 75 percent efficient, we must divide 93.25 by 0.75 to find that approximately 124.33 A is required to operate the motor at rated load. Thus, in a motor that is 75 percent efficient, 2984 W of power is required to produce 2238 W (3 hp) of power at the output.

Another unit used in connection with electrical work is the joule (**J**), named for James Prescott Joule (1818–1889), an English physicist. **The joule is a unit of work, or energy, and represents the work done by 1 watt in 1 second.** This is equal to approximately 0.7376 ft·lb. To apply this principle, let us assume that we wish to determine how much work in joules is done when a weight of 1 ton is raised 50 ft. First we multiply 2000 by 50 and find that 100,000 ft·lb of work is done. Then, when we divide 100,000 by 0.7376, we determine that approximately 135,575 J of work, or energy, was used to raise the weight.

It is wise for the technician to understand and have a good concept of the joule because this is the unit designated by the metric system for the measurement of work or energy. Other units convertible to joules are the British thermal unit (Btu), calorie (cal), foot-pound, and watthour (Wh). All these units represent a specific amount of work performed.

To cause a current to flow in a conductor, a difference of potential must be maintained between the ends of the conductor. In an electric circuit this difference of potential is normally produced by a battery or a generator; so it is obvious that both ends of the conductor must be connected to the terminals of the source of emf.

Figure 2–6 shows the components of a simple circuit with a battery as the source of power. One end of the circuit is connected to the positive terminal of the battery and the other to the negative terminal. A switch is incorporated in the circuit to connect the electric power to the load unit, which may be an electric lamp, bell, or relay or any other electric device that could be operated in such a circuit. When the switch in the circuit is closed, current from the battery flows through the switch and load and then back to the battery. Remember that the direction of current flow is from the negative terminal to the positive terminal of the battery. The circuit will operate only when there is a continuous path through which the current may flow from one terminal to the other. When the switch is opened (turned off), the path for the current is broken, and the operation of the circuit ceases.

Since airplanes are usually constructed of metal, the airplane structure may be used as an electric conductor. In the circuit in Figure 2–6, if one terminal of the battery and one terminal of the load are connected to the metal structure of the airplane, the circuit will operate just as well as with two wire conductors. A diagram of such a circuit is shown in Figure 2–7. When a system of this type is used in an airplane, it is called a **grounded** or **single-wire** system. The ground circuit is that part of the complete circuit in which current passes through the airplane structure. Any unit connected electrically to the metal structure of the airplane is said to be grounded. When an airplane employs a single-wire electric system, it is important that all parts of the airplane be well bonded to provide a free and unrestricted flow of current

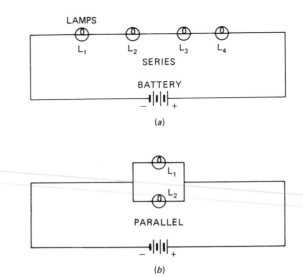

FIGURE 2–8 Two basic methods of connecting units in an electric circuit: (a) Series—if one lamp opens, all lamps stop illuminating, (b) parallel—if one lamp opens, the other is unaffected.

throughout the structure. This is particularly important for aircraft in which sections are joined by adhesive bonding.

There are two general methods for connecting units in an electric system. These are illustrated in Figure 2–8. The first diagram shows four lamps connected in series. A **series circuit** contains only one electron path. In a series circuit or series portion of a circuit, all the current must pass through each unit of that circuit. Therefore, if one unit of a series circuit should burn out, or open, the entire circuit will no longer receive current. For example, in Figure 2–8a, if lamp 1 should open, the other lamps of that circuit will also stop illuminating.

In a **parallel circuit** there are two or more paths for the current, and if the path through one of the units is broken, the other units will continue to function. The units of an aircraft electric system are usually connected in parallel; hence the failure of one unit will not impair the operation of the remainder of the units in the system. A simple parallel circuit is illustrated in the diagram of Figure 2–8b.

A circuit that contains electrical units in both parallel and series is called a **series-parallel circuit** (see Figure 2–9). Most complex electrical systems, such as communication radios, flight computers, and navigational equipment, consist of several series-parallel circuits. Ohm's law can be used to determine the electrical values in any common circuit, even though it may contain a number of different load units. In

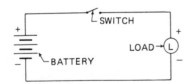

FIGURE 2–6 A simple dc circuit.

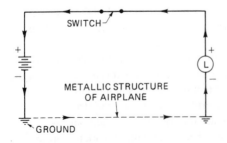

FIGURE 2–7 A single-wire electrical system.

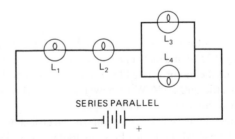

FIGURE 2–9 A series-parallel circuit diagram.

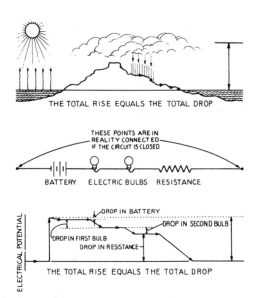

FIGURE 2–10 Water analogy of voltage drops.

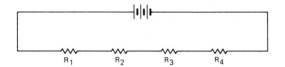

FIGURE 2–11 A series circuit with four separate loads.

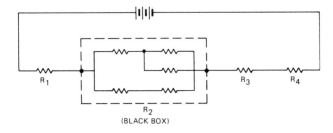

FIGURE 2–12 A series circuit containing a series-parallel load.

order to solve such a circuit, it is necessary to know whether the units are connected in series, in parallel, or in a combination of the two methods. When the type of circuit is determined, the proper formula may be applied.

Voltage Drop

When a current flows through a resistance, a voltage or pressure drop is created. This loss of voltage, known as a **voltage drop** (V_x), is equal to the product of current and resistance. An individual voltage drop is expressed as $V_x = IR$, where V_x is measured in volts, I in amps, and R in ohms. *Note:* The subscript (x) is used here to represent a number that applies to a specific voltage drop, such as voltage drop #1 (V_1) or voltage drop #2 (V_2). In a series circuit, the sum of the individual voltage drops is equal to the applied voltage. This may be expressed as

$$E_t = V_1 + V_2 + V_3$$

for a circuit containing three resistors.

Figure 2–10 shows this concept using the water analogy. Notice that with either the water or electrical circuit, the total pressure rise is equal to the total pressure drop; that is, the electrical pressure increase created by the battery is equal to the total pressure drop across both lamps and the resistor. This can be expressed mathematically as

$$E_t = V_{L1} + V_{L2} + V_R$$

SOLVING SERIES CIRCUITS

As explained previously, a series circuit consists of only one current path. When two or more units are connected in series, the entire quantity of moving electrons (current) must pass through each unit to complete the circuit. Therefore, each unit of a series circuit receives the same current flow, even though their individual voltage drops may vary.

Two or more units do not have to be adjacent to each other in a circuit to be in series. In the circuit of Figure 2–11, it can be seen that the current flow through each unit in the circuit must be the same, regardless of the direction of current flow. If we replace the load resistor R_2 with an electronic system or device contained in a *black box* as shown in Figure 2–12, the current flow in each resistor will still be the same, provided that the total resistance of the black-box load is the same as it was for R_2. In this case, we regard the black box as a single unit rather than concern ourselves with the separate components within the black box. Thus we see that there is only one path for current flow in a series circuit; however, an individual load unit may consist of more than one component within itself. Note that the black box in Figure 2–12 is shown with several resistances connected in a network within the box. In the series circuit under consideration, we are only concerned with the total resistance of the black-box unit.

The load units adjacent to each other in a circuit are connected in series if there are no electrical junctions between the two units. This is illustrated in Figure 2–13. In circuit *a*, R_1 and R_2 are connected in series because there is no electrical junction between them to take a part of the current, and all the current flowing through R_1 must also pass through R_2. In

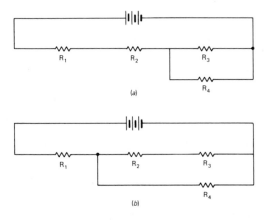

FIGURE 2–13 A circuit diagram showing load units connected in both series and parallel.

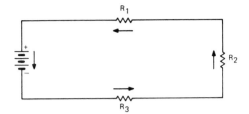

FIGURE 2–14 Current flow in a series circuit. Each load receives equal current.

circuit *b*, R_1 and R_2 are not connected in series because the current that flows through R_1 is divided between R_2 and R_4. Note, however, that R_2 and R_3 are in series because the same current must pass through both of them.

Examine the circuit of Figure 2–14 in which R_1, R_2, and R_3 are connected in series, not only to each other but also to the power source. The electrons flow from negative to positive in the circuit and from positive to negative in the power source. The same flow, however, exists in every part of the circuit, because there is only one path for current flow. Since the current is the same in all parts of the circuit,

$$I_t = I_1 = I_2 = I_3$$

That is, the total current is equal to the current through R_1, R_2, or R_3.

Resistance and Voltage in a Series Circuit

In a series circuit, the total resistance is equal to the sum of all the resistances in the circuit; hence,

$$R_t = R_1 + R_2 + R_3 + \cdots$$

The voltage (potential difference) measured between any two points in a series circuit depends on the resistance between the points and the current flowing in the circuit. Figure 2–15 shows a circuit with three resistances connected in series. The difference in potential maintained by the battery between the ends of the circuit is 24 V.

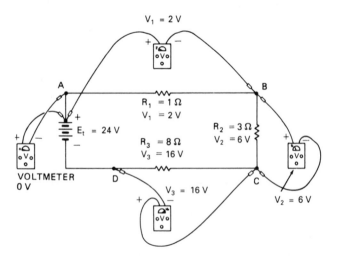

FIGURE 2–15 The summation of voltage drops.

As previously explained in the discussion of Ohm's law, the voltage between any two points in a circuit can be determined by the equation

$$E = IR$$

That is, the voltage is equal to the current multiplied by the resistance. In the circuit of Figure 2–15, we have given a value of 1 Ω to R_1, 3 Ω to R_2, and 8 Ω to R_3. According to our previous discussion, the total resistance of the circuit is expressed by

$$R_t = R_1 + R_2 + R_3$$

or

$$R_t = 1 + 3 + 8$$
$$= 12 \ \Omega$$

Since the total voltage E_t for the circuit is given as 24, we can determine the current in the circuit by Ohm's law, using the form

$$I = \frac{E}{R}$$

Then

$$I_t = \frac{24 \text{ V}}{12 \ \Omega}$$
$$= 2 \text{ A}$$

Since we know that the current in the circuit is 2 A, it is easy to determine the voltage across each load resistor. Since $R_1 = 1 \ \Omega$, we can substitute this value in Ohm's law to find the voltage difference across R_1.

$$V_1 = I_1 R_1$$
$$= 2 \times 1$$
$$= 2 \text{ V}$$

In like manner,

$$V_2 = I_2 R_2$$
$$= 2 \times 3$$
$$= 6 \text{ V}$$

and

$$V_3 = I_3 R_3$$
$$= 2 \times 8$$
$$= 16 \text{ V}$$

When we add the voltages in the circuit, we find

$$E_t = V_1 + V_2 + V_3$$
$$V_t = 2 + 6 + 16$$
$$= 24 \text{ V}$$

We have determined by Ohm's law that the total of the voltages (voltage drops) across units in a series circuit is equal to the voltage applied by the power source, in this case the 24-V battery.

In a practical experiment, we can connect a **voltmeter** (voltage-measuring instrument) from the positive terminal of the battery in a circuit such as that shown in Figure 2–15 to point A, and the reading will be zero. This is because there is no appreciable resistance between these points. When we connect the voltmeter between the positive terminal of the battery and point B, the instrument will give a reading of 2 V. By similar use of the voltmeter, we measure between points B and C and obtain a reading of 6 V, and between points C and D for a reading of 16 V. In a circuit such as that shown, we can assume that the resistance of the wires connecting the resistors is negligible. If the wires were quite long, it would be necessary to consider their resistances in analyzing the circuit.

As we have shown, in a series circuit, the voltage drop across each resistor (load unit) is directly proportional to the value of the resistor. Since the current through each unit of the circuit is the same, it is obvious that it will take a higher electrical pressure (voltage) to push the current through a higher resistance, and it will require a lower pressure to push the same current through a lower resistance.

The voltage across a load resistor is a measure of the work required to move a unit charge (given quantity of electricity) through the resistor. Electric energy is consumed as current flows through a resistor, and the electric energy is converted to heat energy. As long as the power source produces electric energy as rapidly as it is consumed, the voltage across a given resistor will remain constant.

Students who have mastered Ohms's law and the three fundamental formulas for series circuits can apply their knowledge to the solution of any series circuit where sufficient information is given. The following examples are shown to illustrate the techniques for solution:

Example A: Figure 2–16.

$$E_t = 12 \text{ V}$$
$$I_1 = 3 \text{ A}$$
$$R_2 = 2 \text{ }\Omega$$
$$R_3 = 1 \text{ }\Omega$$

Since I_1 is given as 3 A, it follows that I_t, I_2, and I_3 are also equal to 3 A, because current is constant in a series circuit. Then

$$R_t = \frac{E_t}{I_t}$$
$$= \frac{12}{3}$$
$$= 4 \text{ }\Omega$$
$$V_2 = 2 \times 3$$
$$= 6 \text{ V}$$
$$V_3 = 1 \times 3$$
$$= 3 \text{ V}$$

Since $R_1 + R_2 + R_3 = R_t$, we can easily determine that $R_1 = 1 \text{ }\Omega$. By using the formula $E = IR$, we find that $V_1 = 3$ V.

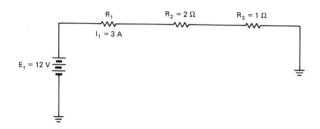

FIGURE 2–16 Series circuit for Example A.

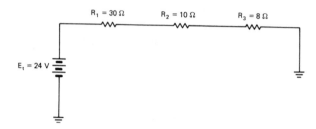

FIGURE 2–17 Series circuit for Example B.

The solved problem may then be expressed as follows:

$$E_t = 12 \text{ V} \qquad I_t = 3 \text{ A} \qquad R_t = 4 \text{ }\Omega$$
$$V_1 = 3 \text{ V} \qquad I_1 = 3 \text{ A} \qquad R_1 = 1 \text{ }\Omega$$
$$V_2 = 6 \text{ V} \qquad I_2 = 3 \text{ A} \qquad R_2 = 2 \text{ }\Omega$$
$$V_3 = 3 \text{ V} \qquad I_3 = 3 \text{ A} \qquad R_3 = 1 \text{ }\Omega$$

Example B: Figure 2–17.

$$V_t = 24 \text{ V}$$
$$R_1 = 30 \text{ }\Omega$$
$$R_2 = 10 \text{ }\Omega$$
$$R_3 = 8 \text{ }\Omega$$

Then

$$R_t = 30 + 10 + 8$$
$$= 48 \text{ }\Omega$$

$$I_t = \frac{E_t}{R_t}$$
$$= \frac{24}{48}$$
$$= 0.5 \text{ A}$$

V_1, V_2, and V_3 are determined by multiplying each resistance value by 0.5 A, the current value of the circuit. The solved circuit is shown in Figure 2–18.

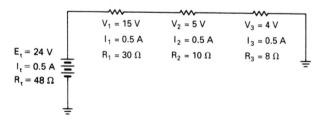

FIGURE 2–18 Simplified circuit for Example B.

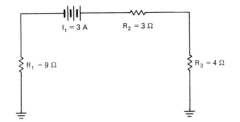

FIGURE 2–19 Series circuit for Example C.

Example C: Figure 2–19. This circuit presents the case where current and resistance are known, and it is required to find the individual and total voltages. The known circuit values are as follows:

$$I_t = 3 \text{ A}$$
$$R_1 = 9 \text{ } \Omega$$
$$R_2 = 3 \text{ } \Omega$$
$$R_3 = 4 \text{ } \Omega$$

From the values given, we can easily determine that the total resistance is 16 Ω. The voltages can then be determined by Ohm's law:

$$E = IR$$
$$E_t = I_t \times R_t$$
$$= 3 \times 16$$
$$= 48 \text{ V}$$

The values of the solved circuit are then as shown below:

$E_t = 48$ V	$I_t = 3$ A	$R_t = 16 \text{ } \Omega$
$V_1 = 27$ V	$I_1 = 3$ A	$R_1 = 9 \text{ } \Omega$
$V_2 = 9$ V	$I_2 = 3$ A	$R_2 = 3 \text{ } \Omega$
$V_3 = 12$ V	$I_3 = 3$ A	$R_3 = 4 \text{ } \Omega$

It will be noted in all the circuits presented thus far that the values are always in accordance with Ohm's law formulas. It is recommended that the student check the problems given to verify the results.

Example D: Figure 2–20. The values for the circuit shown are indicated in the illustration. It is left up to the student to work out the solution. Remember that the total resistance for a series circuit is equal to the sum of the individual resistances.

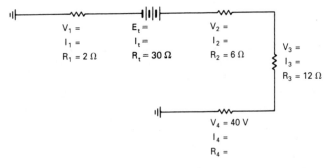

FIGURE 2–20 Series circuit for Example D.

SOLVING PARALLEL CIRCUITS

A parallel circuit always contains two or more electric current paths. When two or more units are connected in parallel, each unit will receive a portion of the circuit's total current flow. That is, the circuit's total current divides at one or more points, and a portion travels through each resistance of the circuit (see Figure 2–21).

Typically, when we analyze a circuit of this type, we assume that the resistance of a wire is negligible and the power source has no internal resistance. A parallel circuit always contains more than one path for current to flow; therefore, the current can "choose" which load unit to travel through. Current always tries to take the path of least resistance and will divide proportionately through a parallel circuit containing load units of different resistances. In a parallel circuit, each load unit will receive a portion of the total current flow. The unit with the highest resistance will receive the least current flow. The unit with the lowest resistance will receive the highest current flow. Equal resistors receive equal current flows.

Typically, load units of an aircraft are arranged in parallel with respect to the power source and to each other. This is done to allow a different current path through each unit; therefore, the resistance of each unit will determine the current flow through that unit. An example is a flap motor using 30 A, a navigation light using 2 A, and the landing light, with the switch turned off, using 0 A. This type of current flexibility is a necessity for almost every electrical system.

The resistors (load units) do not need to be arranged as in Figure 2–21 to be connected in parallel. The three circuits of Figure 2–22 show loads connected in parallel. Circuits *a* and *b* are identical to the circuit of Figure 2–21, and circuit *c* has an additional load unit connected in parallel. A careful examination of the circuits will reveal that the connections are in common for each side of the power source. There is a direct connection (current path) without resistance from any one negative terminal of a load unit to the negative terminal of any other load unit and to the negative terminal of the power source. The same condition is true with respect to all positive terminals.

There may be some junctions between two or more resistors connected in parallel, but these junctions do not change the fact that the resistances are still connected in parallel. It

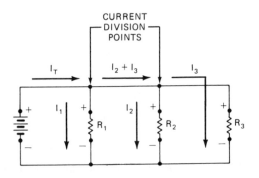

FIGURE 2–21 Current flow through a parallel circuit.

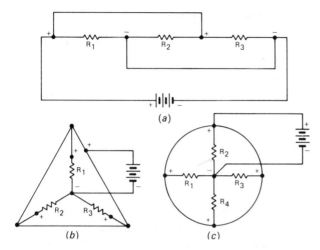

FIGURE 2–22 Different arrangements of parallel circuits.

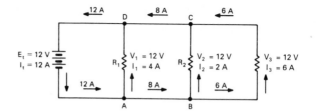

FIGURE 2–25 Current flow in a parallel circuit.

will be noted in Figure 2–23 that three of the resistances, R_1, R_2, and R_3, have common terminals with one another, even though there are other resistances connected between their common terminals and the power source. It will further be noted that R_4 and R_5 are connected in parallel because they have positive terminals connected together and negative terminals connected together. The resistance R_6 is in series, not with any other single resistance, but with the parallel groups.

The voltage across any resistance in a parallel group is equal to the voltage across any other resistance in the group. Note in Figure 2–24 that the voltage of the source is 12 V. Since the terminals of the source are connected directly to the terminals of the resistances, the difference in potential across each resistance is the same as that of the battery or source. By testing with a voltmeter, it would be found that the potential difference across each resistance in the circuit would be 12 V. The formula for voltage in a parallel circuit is

$$E_t = V_1 = V_2 = V_3 = V_4 \cdots$$

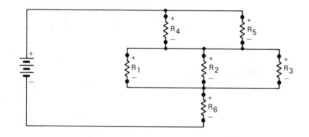

FIGURE 2–23 Parallel grouping of resistors.

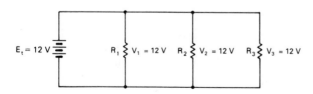

FIGURE 2–24 Voltages in a parallel circuit.

This formula states that a consistent voltage will be applied to each unit of a parallel circuit. The ability to apply an equal voltage to all power users is another important reason that the entire aircraft electrical system (not necessarily individual electrical components) is wired in parallel. As described earlier, the current in a parallel circuit divides proportionately among each resistance (load unit).

In the circuit of Figure 2–25, the current through R_1 is given as 4 A, the current through R_2 is 2 A, and the current through R_3 is 6 A. To supply this current flow through the three resistances, the power source must supply $4 + 2 + 6$, or a total of 12 A to the circuit. It must be remembered that the power source does not actually manufacture electrons, but it does apply the pressure to move them. All the electrons that leave the battery to flow through the circuit must return to the battery. The power source for a circuit can be compared to a pump that moves liquid through a pipe.

An examination of the circuit in Figure 2–25 reveals that a flow of 12 A comes from the negative terminal of the battery, and at point A the flow divides to supply 4 A for R_1 and 8 A for the other two resistors. At point B the 8 A divides to provide 2 A for R_2 and 6 A for R_3. On the positive side of the circuit, 6 A joins 2 A at point C, and the resulting 8 A joins 4 A at point D before returning to the battery. The formula for current in a parallel circuit is then seen to be

$$I_t = I_1 + I_2 + I_3 + \cdots$$

Since the current flow and voltage are given for each resistor in Figure 2–25, it is easy to determine the value of each resistance by means of Ohm's law; that is,

$$R = \frac{E}{I}$$

Then

$$R_1 = \frac{12}{4} = 3 \ \Omega$$

$$R_2 = \frac{12}{2} = 6 \ \Omega$$

$$R_3 = \frac{12}{6} = 2 \ \Omega$$

$$R_t = \frac{12}{12} = 1 \ \Omega$$

Remember when solving for R_1 to be sure to use the voltage drop for resistor 1 and current through resistor 1 (V_1 and I_1). However, E_t can be substituted for V_1 because voltage is constant in a parallel circuit.

The formula for the total resistance in a parallel circuit can be derived by use of Ohm's law and the formulas for total voltage and total current. Since

$$I_t = I_1 + I_2 + I_3$$

and

$$I = \frac{E}{R}$$

we can replace all the values in the preceding formula for total current with their equivalent values in terms of voltage and resistance. Thus we arrive at the equation

$$\frac{E_t}{R_t} = \frac{V_1}{R_1} + \frac{V_2}{R_2} + \frac{V_3}{R_3}$$

In a parallel circuit, $E_t = V_1 = V_2 = V_3$. Therefore, we can divide all the terms in the previous equation by E_t and arrive at the formula

$$\frac{1}{R_t} = \frac{1}{R_1} + \frac{1}{R_2} + \frac{1}{R_3}$$

Solving for R_t, the equation becomes

$$R_t = \frac{1}{1/R_1 + 1/R_2 + 1/R_3}$$

This equation can be expressed verbally as follows: *The total resistance in a parallel circuit is equal to the reciprocal of the sum of the reciprocals of the resistances.*

The **reciprocal** of a number is the quantity 1 divided by that number. For example, the reciprocal of 3 is $\frac{1}{3}$. When the reciprocal of a number is multiplied by that number, the product is always 1.

If the formula for total resistance in a parallel circuit is applied to the circuit problem of Figure 2–25, we find

$$R_t = \frac{1}{1/3 + 1/6 + 1/2}$$
$$= \frac{1}{0.33 + 0.167 + 0.5}$$
$$= \frac{1}{1}$$
$$= 1\ \Omega$$

If some or all of the resistances in a parallel circuit are of the same value, the resistance value of one can be divided by the number of equal-value resistances to obtain the total resistance value. For example, if a circuit has four 12-Ω resistors connected in parallel, the value 12 can be divided by the number 4 to obtain the total resistance value of 3 Ω for the four resistances.

When two resistances are connected in parallel, we can use a formula derived from the general formula for R_t to determine the total resistance. The formula is as follows:

$$R_t = \frac{1}{1/R_1 + 1/R_2}$$

Inverting,

$$\frac{1}{R_t} = \frac{1}{R_1} + \frac{1}{R_2}$$

Using a common denominator,

$$\frac{1}{R_t} = \frac{R_2}{R_1 \times R_2} + \frac{R_1}{R_1 \times R_2}$$

Combining,

$$\frac{1}{R_t} = \frac{R_1 + R_2}{R_1 \times R_2}$$

Inverting,

$$R_t = \frac{R_1 \times R_2}{R_1 + R_2}$$

From the foregoing formula, we find that when two resistors are connected in parallel, the total resistance is equal to the product of the two resistance values divided by their sum. If a 5-Ω resistance is connected in parallel with a 6-Ω resistance, we apply the formula thus:

$$R_t = \frac{5 \times 6}{5 + 6}$$
$$= \frac{30}{11}$$
$$= 2.73\ \Omega$$

Another fact of parallel resistor groups is that the total resistance of the group is always less than the smallest resistance of that group. For example, if $R_1 = 3\ \Omega$, $R_2 = 6\ \Omega$, and $R_3 = 2\ \Omega$, then R_t will be less than 2 Ω. As previously stated,

$$R_t = \frac{1}{1/R_1 + 1/R_2 + 1/R_3}$$

or

$$R_t = \frac{1}{1/3 + 1/6 + 1/2}$$
$$= \frac{1}{0.33 + 0.167 + 0.5}$$
$$= \frac{1}{1}$$
$$= 1\ \Omega$$

The R_t of 1 Ω is indeed less than 2 Ω, the smallest resistance of the group.

The rules to determine voltage, current, and resistance for parallel circuits have numerous applications. For example, a parallel circuit having some resistance values unknown, but at least one current value given with a known resistance value, can be solved through the use of Ohm's law and the formula for total resistance. See Figure 2–26.

An examination of this circuit reveals that $I_2 = 8$ A and

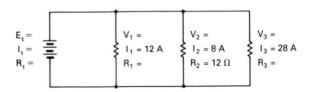

FIGURE 2–26 Diagram of a parallel circuit.

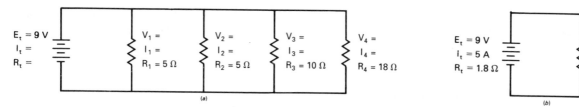

FIGURE 2–27 A parallel circuit and its simplified equivalent.

(a) Complete circuit showing all resistors, 1–4. (b) The simplified circuit showing effective resistor 1–4.

$R_2 = 12 \ \Omega$. With these values it is apparent that the voltage across R_2 is equal to 96 V. That is,

$$V_2 = I_2 \times R_2$$
$$= 8 \times 12$$
$$= 96 \ V$$

Since the same voltage exists across all the load resistors in a parallel circuit, we know that E_t, V_1, and V_3 are all equal to 96 V. We can then proceed to find that $R_1 = \frac{96}{12}$ or 8 Ω and $R_3 = \frac{96}{28}$ or 3.43 Ω. Since total current is equal to the sum of the current values, $I_t = 12 + 8 + 28$ or 48 A. The total resistance is then $\frac{96}{48} = 2 \ \Omega$, since $R_t = E_t/I_t$.

In any circuit where a number of load units are connected in parallel or in series, it is usually possible to simplify the circuit in steps and derive an equivalent circuit. A sample parallel circuit and its simplified equivalent are illustrated in Figure 2–27.

The first step used to solve this parallel problem is to combine all individual resistors using the formula

$$R_t = \frac{1}{1/R_1 + 1/R_2 + 1/R_3 + 1/R_4}$$

or

$$R_t = \frac{1}{1/5 + 1/5 + 1/10 + 1/18}$$

or

$$R_t = 1.8 \ \Omega$$

The second step is to solve for I_t.

$$I_t = \frac{E_t}{R_t}$$
$$= \frac{9 \ V}{1.8 \ \Omega}$$
$$= 5 \ A$$

The third step is to find the individual current flows through each resistor. Since voltage is constant in a parallel circuit, E_t can be substituted for each individual voltage drop.

$$I_1 = \frac{E_1}{R_1} \qquad I_1 = \frac{9 \ V}{5 \ \Omega} \qquad I_1 = 1.8 \ A$$

$$I_2 = \frac{E_2}{R_2} \qquad I_2 = \frac{9 \ V}{5 \ \Omega} \qquad I_2 = 1.8 \ A$$

$$I_3 = \frac{E_3}{R_3} \qquad I_3 = \frac{9 \ V}{10 \ \Omega} \qquad I_3 = 0.9 \ A$$

$$I_4 = \frac{E_4}{R_4} \qquad I_4 = \frac{9 \ V}{18 \ \Omega} \qquad I_4 = 0.5 \ A$$

The fourth step should be to check the calculations. In a parallel circuit, current is additive to find total current. Therefore, if the sum of the individual current flows equals the total current, the calculations were done correctly. The check would be as follows:

$$I_t = I_1 + I_2 + I_3 + I_4$$
$$= 1.8 + 1.8 + 0.9 + 0.5$$
$$= 5.0 \ A$$

Since 5 A is the calculated total current flow, one can assume that the calculations are correct.

Another quick check can be done by comparing the calculated total resistance with the smallest resistance value of the parallel group. As stated earlier, the total resistance of a parallel group must always be less than the lowest-value resistor. If this is not true for your calculations, it must be assumed that a mistake was made.

SERIES-PARALLEL CIRCUITS

As the name implies, a series-parallel circuit is one in which some load units are connected in series and some are connected in parallel. Such a circuit is shown in Figure 2–28. In this circuit it is quickly apparent that the resistances R_1 and R_2 are connected in series and the resistances R_3 and R_4 are connected in parallel. When the two parallel resistances are combined according to the parallel formula, one resistance, $R_{3,4}$, is found, and this value is in series with R_1 and R_2 as shown in Figure 2–29. The total resistance R_t is then equal to the sum of R_1, R_2, and $R_{3,4}$.

If certain values are assigned to some of the load units in the circuit of Figure 2–28, we can solve for the unknown val-

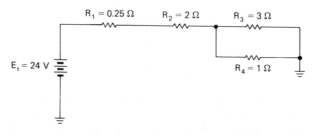

FIGURE 2–28 A simple series-parallel circuit.

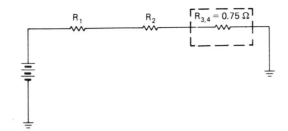

FIGURE 2–29 A series equivalent of the series-parallel circuit of Figure 2–28.

ues and arrive at a complete solution for the circuit. For the purposes of this problem, the following are known:

$$E_t = 24 \text{ V}$$
$$R_1 = 0.25 \ \Omega$$
$$R_2 = 2 \ \Omega$$
$$R_3 = 3 \ \Omega$$
$$R_4 = 1 \ \Omega$$

To solve for the unknown values, the following steps must be taken.

The first step is to combine all parallel resistors, such as in Figure 2–29. To combine the parallel resistors R_3 and R_4, use the formula

$$R_{3,4} = \frac{1}{1/R_3 + 1/R_4}$$
$$= \frac{1}{1/3 + 1/1}$$
$$= 0.75 \ \Omega$$

Second, combine all series resistors using the formula

$$R_t = R_1 + R_2 + R_{3,4}$$
$$= 0.25 + 2 + 0.75$$
$$= 3 \ \Omega$$

In this case, the resistance total was found by using only two steps. More complex circuits may require that these steps be performed in opposite order and/or several times to determine the value of R_t.

Third, compute total current using the formula

$$I_t = \frac{E_t}{R_t}$$
$$= \frac{24}{3}$$
$$= 8 \text{ A}$$

Fourth, compute the voltage drop across the series resistors. The formula $V_x = IR$ will be used twice in this case, once for R_1 and once for R_2. *Note:* Because I_1 and I_2 have not yet

been calculated, I_t must be substituted for their values. This is possible because both R_1 and R_2 are in series.

$$V_1 = I_1 R_1$$
$$= 8 \times 0.25$$
$$= 2 \text{ V}$$

$$V_2 = I_2 R_2$$
$$= 8 \times 2$$
$$= 16 \text{ V}$$

Fifth, calculate the voltage drop across the parallel resistors using the formula $V_x = IR$. This can only be done for the entire group of parallel resistors ($R_{3,4}$) because the current flow through the individual resistors is yet unknown. Since voltage is constant in parallel, the voltage drop across $R_{3,4}$ is equal to the voltage drop across R_3 and R_4 individually.

$$V_{3,4} = I_{3,4} R_{3,4}$$
$$= 8 \times 0.75$$
$$= 6 \text{ V}$$

Therefore,

$$V_3 = 6 \text{ V} \quad \text{and} \quad V_4 = 6 \text{ V}$$

Note: I_t was substituted for the unknown value $I_{3,4}$ because the effective resistor $R_{3,4}$ is in series (see Figure 2–29).

Sixth, calculate current flow through the parallel resistors using $I = V/R$.

$$I_3 = \frac{V_3}{R_3}$$
$$= \frac{6}{3}$$
$$= 2 \text{ A}$$

$$I_4 = \frac{V_4}{R_4}$$
$$= \frac{6}{1}$$
$$= 6 \text{ A}$$

The entire circuit has now been analyzed using the basic elements of Ohm's law. The completed solution is listed below.

$E_t = 24$ V	$I_t = 8$ A	$R_t = 3 \ \Omega$
$V_1 = 2$ V	$I_1 = 8$ A	$R_1 = 0.25 \ \Omega$
$V_2 = 16$ V	$I_2 = 8$ A	$R_2 = 2 \ \Omega$
$V_3 = 6$ V	$I_3 = 2$ A	$R_3 = 3 \ \Omega$
$V_4 = 6$ V	$I_4 = 6$ A	$R_4 = 1 \ \Omega$

It should be considered that the previous series-parallel circuit was relatively simple and therefore easy to solve. In many cases where several groups of series and parallel resistances are combined, the calculations above must be repeated and/or performed in different order.

The solution of a series-parallel circuit such as that shown in Figure 2–30 is not difficult provided that the load-unit

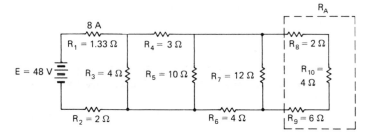

FIGURE 2–30 Series-parallel circuit.

(resistance) values are kept in their correct relationships. To determine all the values for the circuit shown, we must start with R_8, R_9, and R_{10}. Since these resistances are connected in series with each other, their total value is $2 + 4 + 6 = 12\ \Omega$. We shall call this total R_A; that is, $R_A = 12\ \Omega$. The circuit can then be drawn as in Figure 2–31, which is the equivalent of the original circuit.

In the circuit of Figure 2–31 it can be seen that R_7 and R_A are connected in parallel. The formula for two parallel resistances can be used to determine the resistance of the combination. We shall call this combination R_B. Then

$$R_B = \frac{R_7 \times R_A}{R_7 + R_A}$$

$$= \frac{12 \times 12}{12 + 12}$$

$$= \frac{144}{24}$$

$$= 6\ \Omega$$

Now an equivalent circuit can be drawn as in Figure 2–32 to further simplify the solution. In this circuit we combine the

two series resistances, R_B and R_6, to obtain a value of $10\ \Omega$ for R_C. The equivalent circuit is then drawn as in Figure 2–33.

Since the new equivalent circuit shows that R_5 and R_C are connected in parallel and that each has a value of $10\ \Omega$, we know that the combined value is $5\ \Omega$. We designate this new value as R_D and draw the circuit as in Figure 2–34. R_D is connected in series with R_4; hence the total of the two resistances is $8\ \Omega$. This is designated as R_E for the equivalent circuit of Figure 2–35. In this circuit we solve the parallel combination of R_3 and R_E to obtain the value of $2.67\ \Omega$ for R_F. The final

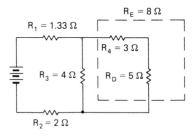

FIGURE 2–33 Third simplification step.

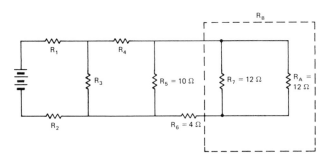

FIGURE 2–31 First simplification step.

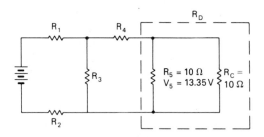

FIGURE 2–34 Fourth simplification step.

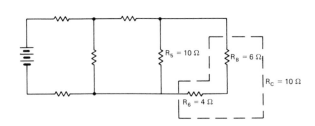

FIGURE 2–32 Second simplification step.

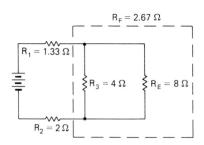

FIGURE 2–35 Fifth simplification step.

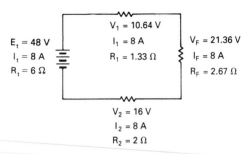

FIGURE 2–36 Final simplified version of Figure 2–30.

equivalent circuit is shown in Figure 2–36 with R_1, R_F, and R_2 connected in series. These resistance values are added to find the total resistance for the circuit.

$$R_t = 1.33 + 2.67 + 2 = 6 \ \Omega$$

With the total resistance known and E_t given as 48 V, it is apparent that $I_t = 8$ A ($I_t = E_t/R_t$). The values for the entire circuit can be computed using Ohm's law and proceeding in a reverse sequence from that used in determining total resistance.

First, since $I_t = 8$ A, I_1, I_F, and I_2 must each be 8 A because the resistances are shown to be connected in series in Figure 2–36. By Ohm's law ($E = IR$) we find that $V_1 = 10.64$ V, $V_F = 21.36$ V, and $V_2 = 16$ V. Referring to Figure 2–35, it can be seen that 21.36 V exists across R_3 and R_E. This makes it possible to determine that $I_3 = 5.33$ A and $I_E = 2.67$ A. In Figure 2–34 we note that I_4 and I_D must both be 2.67 A because the two resistances are connected in series. Then $V_4 = 8$ V and $V_D = 13.35$ V. Since V_D is the voltage across R_5 and R_C in the circuit of Figure 2–33, it is easily found that $I_5 = 1.33$ A and $I_C = 1.33$ A. In the circuit of Figure 2–32 it is apparent that 1.33 A must flow through both R_B and R_6 because they are connected in series and we have already noted that $I_C = 1.33$ A. Then $V_B = 8$ V and $V_5 = 13.35$ V.

Since $V_B = 8$ V, we can apply this voltage to the circuits as shown in Figures 2–30 and 2–31 and note that both V_7 and V_A are 8 V. Then $I_7 = 0.67$ A and $I_A = 0.67$ A. Since R_8, R_9, and R_{10} are connected in series and the same current, 0.67 A, flows through each, $V_8 = 1.33$ V, $V_9 = 4$ V, and $V_{10} = 2.67$ V.

The completely solved circuit is shown in Figure 2–37. A check of all the values given will reveal that they comply with the requirements of Ohm's law. *Note:* Some minor error may exist due to rounding of the numbers during calculation.

KIRCHHOFF'S LAWS

The circuits in this chapter are all solvable by means of Ohm's law as demonstrated. There are, however, many circuits that are more complex, which cannot be solved by Ohm's law alone. For these circuits, Kirchhoff's laws may provide the necessary techniques and procedures.

Kirchhoff's laws were discovered by Gustav Robert Kirchhoff, a German physicist of the nineteenth century. The two laws may be stated as follows:

Law No. 1. *In a series circuit, the algebraic sum of the voltage drops in that circuit must be equal to the source voltage.* Kirchhoff's law of voltage drops may also be applied to any portion of a circuit that is connected in series.

Law No. 2. *In a parallel circuit, the algebraic sum of the currents entering a point is equal to the algebraic sum of the currents leaving that point.* Kirchhoff's parallel law of current flows may also be applied to any portion of a circuit that is connected in parallel.

Kirchhoff's law for series voltage drops can be expressed algebraically as follows:

$$E_t - V_1 - V_2 - V_3 = 0$$

or

$$E_t = V_1 + V_2 + V_3$$

Figure 2–38 shows a circuit to illustrate the principle of Kirchhoff's second law. In this circuit it can be noted that I_t, the current flowing to point A, is equal to $I_1 + I_2 + I_3$, the current flowing away from point A. Kirchhoff's law of parallel current flows can be expressed by the following equations:

$$I_t - I_1 - I_2 - I_3 = 0$$

or

$$I_t = I_1 + I_2 + I_3$$

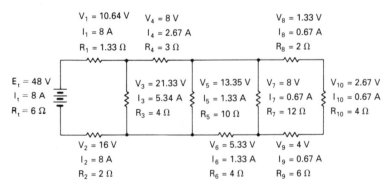

FIGURE 2–37 The completely solved version of Figure 2-30.

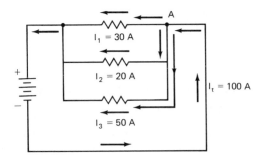

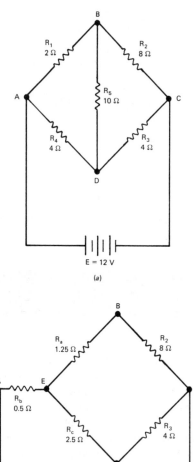

FIGURE 2–38 Diagram to illustrate Kirchhoff's second law. The current to a point is equal to the current from that point.

Both of Kirchhoff's laws become very useful tools in finding solutions to complex electric circuits. In general, when you are solving series-parallel circuits and you are forced to solve an equation with more than one unknown, remember the following: (1) In series circuits or series portions of a circuit, the sum of the voltage drops is equal to the voltage applied across the entire group of series resistors. (2) The current flow through a series circuit is constant and equal to the total current flow through the entire series portion of the circuit.

In parallel circuits or parallel portions of a circuit: (1) The voltage applied to each resistance is constant and equal to the voltage applied to the entire parallel portion of the circuit. (2) The sum of the current flows through each parallel resistance is equal to the total current entering that parallel portion of the circuit. With these four basic principles and the correct substitution procedures there should be no circuit too difficult to solve.

SOLUTION OF A RESISTANCE BRIDGE CIRCUIT

When resistances are connected in a bridge circuit as shown in Figure 2–39a, it will be noted that two Δ (delta) circuits are formed. These circuits share the resistance R_5. Because of this, it is not possible to solve the circuit by the methods we have explained previously. A mathematical method has been devised whereby the circuit can be solved by converting one of the Δ circuits to an equivalent Y circuit.

Figure 2–39b represents an equivalent circuit where the Δ circuit ABD of Figure 2–39a has been converted to the equivalent Y circuit ABD in Figure 2-39b. This conversion is accomplished with formulas as follows:

$$R_a = \frac{R_1 \times R_5}{R_1 + R_4 + R_5}$$

$$R_b = \frac{R_1 \times R_4}{R_1 + R_4 + R_5}$$

$$R_c = \frac{R_4 \times R_5}{R_1 + R_4 + R_5}$$

FIGURE 2–39 Circuit to illustrate the conversion of a delta circuit to an equivalent Y circuit and the solution of a resistance bridge circuit.

The circuit of Figure 2–39b is a simple series-parallel type and can be solved as we have explained previously.

For an example of how the circuit of Figure 2–39 can be solved, we shall first assign resistance values to the resistors in Figure 2–39a. $R_1 = 2\ \Omega, R_2 = 8\ \Omega, R_3 = 4\ \Omega, R_4 = 4\ \Omega$, and $R_5 = 10\ \Omega$. Then

$$R_a = \frac{2 \times 10}{2 + 4 + 10} = \frac{20}{16} = 1.25\ \Omega$$

$$R_b = \frac{2 \times 4}{2 + 4 + 10} = \frac{8}{16} = 0.5\ \Omega$$

$$R_c = \frac{4 \times 10}{2 + 4 + 10} = \frac{40}{16} = 2.5\ \Omega$$

In the circuit of Figure 2–39b, R_a and R_2 are in series, and R_c is connected in series with R_3. Since series circuit values are added to determine the value of the total, we add the series resistances in this case. Then $R_a + R_2 = 1.25 + 8 = 9.25$ Ω, and $R_c + R_3 = 2.5 + 4 = 6.5$ Ω. The combination of $R_a + R_2$ is in parallel with the combination of $R_c + R_3$; hence we use the parallel formula for two resistances to determine the equivalent value.

$$R_t = \frac{6.5 \times 9.25}{6.5 + 9.25} = \frac{60.125}{15.75} = 3.82 \ \Omega$$

Since the parallel circuit is in series with R_b, we add the total of the parallel resistances (3.82 Ω) to R_B (0.5 Ω) to obtain the combined equivalent resistance for the circuit; that is,

$$0.5 + 3.82 = 4.32 \ \Omega$$

Since 12 V is applied to the bridge circuit, the current through the circuit is $12/4.32 = 2.78$ A.

PRACTICAL APPLICATIONS OF OHM'S LAW

For an aircraft technician, there are countless uses for the material contained in this chapter. Ohm's law can be used during the installation, repair, and inspection of various electrical units; in the acquisition of electrical components; in determining wire sizes for a given application; and in basic electric circuit design. Some examples of these applications are stated in the following problems. It should be noted that owing to the brevity of these examples, they may not fully illustrate the complexity of a given situation that might be encountered during actual aircraft maintenance.

Problem No. 1. During an annual inspection it was noticed that the bus bar (the main electrical distribution connection) had been replaced by the previous aircraft owner. One way for the technician to verify the airworthiness of this bus bar is to determine its actual load-carrying capability and compare it with the aircraft's actual total load. It was determined from the part number of the bus bar that the maximum amperage allowable to enter this part was 60 amps. Is the bus bar within its amperage limit?

Solution. By applying Kirchhoff's law for parallel circuits, it was determined that the current flowing from the bus bar was also the current flowing through the bus bar. The maximum allowable current through the bus is 60 amps; therefore, the total aircraft load could not exceed this value. Since all aircraft circuits are connected in parallel to the bus, the total current was determined using

$$I_t = I_1 + I_2 + I_3 + I_4 + I_5 + I_6 + I_7$$

If the loads on the aircraft are as follows, it is a simple process to determine if the bus bar is electrically overloaded.

Navigation lights	10 A
Navigation radio	4 A
Communication radio	3 A
Pitot heat	12 A
Flap motor	8 A
Hydraulic pump motor	16 A
Fuel pump motor	6 A

Simply sum the individual current flows to find the total current flow.

$$I_t = 10\,A + 4\,A + 3\,A + 12\,A + 8\,A + 16\,A + 6\,A$$
$$= 59\,A$$

Since the aircraft's total load is only 59 amps and the bus bar can handle 60 amps, the bus installation can be considered within its current limit.

Problem No. 2. What size generator must be placed on the aircraft used in Problem 1? The approved generators for that particular airplane are rated at 30, 60, and 90 A.

Solution. Once again, since we know that the current to a point is equal to the current from that point, we can determine that the 59 A "pulled" from the aircraft's bus bar must be "pushed" into the bus bar by the generator. Therefore, the 60-A generator would be required as a minimum. However, the 59 A calculated earlier does not include the current needed to charge the battery after starting the aircraft engine. (*Note:* On this aircraft, the battery current does not feed through the bus; it is received directly from the generator.) Since battery charging current can often exceed 20 A for short periods, the 90-A generator should be installed.

Problem No. 3. While a new electric fuel pump is installed on an aircraft, the fuel flow adjustment must be made by changing the voltage to the pump motor. This change in voltage changes the rpm of the pump motor, hence changing the fuel flow through the pump. To accomplish this voltage change, the aircraft system contains an adjustable resistor in series with the fuel pump motor. If the aircraft manual calls for 8 V to be applied to the pump motor and the aircraft system voltage is 14 V, at what resistance must the variable resistor be set?

Solution. Since voltage drops are additive in a series circuit, the voltage drop of the resistor plus the voltage drop of the fuel pump must equal 14 V (system voltage); or, 14 V − 8 V = resistor voltage drop. The voltage drop of the resistor is therefore 6 V. The equation $R = E/I$ can be used to determine the resistor's value. According to the data plate of the fuel pump, the motor draws 2 A at 8 V. Since the motor and resistor are in series, 2 A must also flow through the variable resistor. Using

$$R_r = \frac{V_r}{I_r}$$

where R_r = resistance of the resistor in ohms
V_r = the voltage drop over the resistor (6 V)
I_r = the current flow through the resistor (2 A)

$$R_r = \frac{6\,V}{2\,A}$$
$$= 3\ \Omega$$

The variable resistor should be set for 3 Ω in order to produce the correct fuel flow.

REVIEW QUESTIONS

1. Define Ohm's law.
2. What letter is used to represent electric current?
3. What name is given to the unit of electromotive force?
4. How is emf expressed during calculations of Ohm's law?
5. What is the relationship between E, R, and I?
6. What is the basic equation for Ohm's law?
7. What simple analogy can be used to help understand the concepts of Ohm's law?
8. Water pressure can be compared to what element of Ohm's law?
9. Give the formula for the general rule of resistance.
10. What is meant by a single-wire power system?
11. Explain the difference between series circuits and parallel circuits.
12. Give the three forms for the formula of Ohm's law.
13. Define *watt*.
14. Compare watts with horsepower.
15. What horsepower is expended in a circuit in which the voltage is 110 V and the current is 204 A?
16. Show that the power expended in a given circuit is proportional to the square of the voltage.
17. What amperage is required to drive a 5-hp motor in a 110-V circuit when the motor has an efficiency of 60 percent?
18. Define current flow in a series circuit.
19. Define *voltage drop*.
20. Explain the relationship of voltage drops in a series circuit.
21. Explain the current flows in a parallel circuit.
22. Explain how voltage is applied to various components in a parallel circuit.
23. What is the total resistance when resistances of 3, 4, 6, and 8 Ω are connected in parallel? in series?
24. What resistance would have to be connected in series with a 3-V lamp in a 28-V circuit when the operating current of the lamp is 0.5 A? What is the operating resistance of the lamp?
25. Explain Kirchhoff's law of voltage drops.
26. Explain Kirchhoff's law of parallel current flows.
27. Give the equation to find total resistance in a parallel group of resistors.
28. Give the equation to find total resistance in a series group of resistors.
29. In a parallel circuit, total resistance is always less than what value?

3 Aircraft Storage Batteries

There are literally hundreds of types and sizes of batteries and cells currently in use. An increase in the various forms of electronic devices has created a demand for a variety of batteries. The aircraft technician may find several types of cells used to power monitoring or test equipment; however, there are currently two types of batteries used on nearly all aircraft, the nickel-cadmium and lead-acid battery.

All battery cells produce dc voltage. The actual voltage level is a function of the chemicals used to form the cell. The direct current supplied by a battery is a function of the chemicals used to produce the cells and the size and number of cells forming the battery. These concepts must be considered when designing a circuit and choosing the power source for that circuit. This chapter will examine the theory and construction of several types of batteries and their cells.

DRY CELLS AND BATTERIES

Voltaic Cells

In an earlier portion of this text, it was explained that various dissimilar substances have opposite polarities with respect to one another and that when two such substances are rubbed together, one will have a positive charge and the other a negative charge. Dissimilar metals also have this property, and when two such metals are placed in contact with each other, there will be a momentary flow of electrons from the one having a negative characteristic to the one having a positive characteristic. If two plates of dissimilar metals are placed in a chemical solution called an *electrolyte,* opposite electric charges will be established on the two plates.

An **electrolyte** is technically defined as a compound that, when molten or in solution, conducts electric current and is decomposed by it. In simple terms, an electrolyte is a solution of water and a chemical compound that will conduct an electric current. The electrolyte in a typical aircraft storage battery consists of sulfuric acid and water. Various salts dissolved in water will also form electrolytes.

An electrolyte will conduct an electric current because it contains positive and negative ions. When a chemical compound is dissolved in water, it separates into its component parts. Some of these parts carry a positive charge, and others carry a negative charge.

The action of an electrolyte will be clear if a specific case is considered. When a rod of carbon and a plate of zinc are placed in a solution of ammonium chloride, the result is an elementary voltaic cell (see Figure 3–1). The carbon and zinc elements are called **electrodes.** The carbon, which is the positively charged electrode, is called the **anode,** and the zinc plate is called the **cathode.** The combination of two electrodes surrounded by an electrolyte will form a **cell.**

As soon as the zinc (Zn) plate is placed in the electrolyte, zinc atoms begin to go into solution as ions, each leaving two electrons at the plate. An **ion** is an atom or molecule that is either positively or negatively charged. A positively charged ion has a deficiency of electrons, and a negatively charged ion has an excess of electrons. The zinc atoms going into solution as positive ions cause the zinc plate to become negatively charged. The zinc ions in the solution are positive because each one lacks the two electrons left at the plate. This positive charge causes the zinc ions to remain near the zinc plate because the plate has become negative. The effect of the zinc ions gathered near the plate is to stop the decomposition of the zinc plate for as long as the negative charge of the plate is balanced by the positive charge of the zinc ions in solution.

The ammonium chloride in solution in the electrolyte apparently separates into positive hydrogen ions and a combination of ammonium and chlorine that is negatively charged. When the two electrodes are connected by an external conductor, the free electrons from the zinc plate flow to the carbon rod; and the hydrogen ions move to the carbon rod, where each ion picks up one electron and becomes a neutral

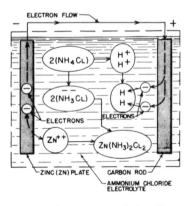

FIGURE 3–1 Chemical action in a voltaic cell.

hydrogen atom. The positive zinc ions combine with the negative ammonium chloride to take the place of the hydrogen ions released into solution. The effect of these chemical actions is to remove electrons from the carbon rod and to liberate free electrons at the zinc plate. This results in a continuous supply of electrons available at the negative (zinc) electrode. When the two electrodes are connected, the electrons will flow to the carbon rod, where the hydrogen ions become hydrogen atoms as the result of their neutralization by the electrons. Eventually, hydrogen gas bubbles form on the carbon rod and insulate it from the solution. This is called **polarization** and will cause the current flow to stop until the hydrogen is removed. For practical voltaic cells, it is necessary to employ a method of depolarization.

The standard **dry cell** used in flashlights and for other purposes for which a low-voltage dc supply is desired employs a compound called manganese dioxide (MnO_2) to prevent the accumulation of hydrogen at the positive electrode in the cell. Figure 3–2 is a drawing of this type of cell. A dry cell is so called because the electrolyte is in the form of a paste; the cell may therefore be handled without the danger of spillage. The zinc can is the negative electrode, and the paste electrolyte is held in close contact with the zinc by means of a porous liner. The space between the carbon rod and the zinc can is filled with manganese dioxide saturated with electrolyte. Graphite is mixed with the manganese dioxide to reduce the internal resistance of the cell. The top of the cell is sealed with a wax compound to prevent leaking and drying of the electrolyte. Many cells are encased in a tin-plated steel can to make them more durable; a layer of insulating material is then placed between the inner zinc can and the outer can to prevent short circuiting.

The voltage developed by a zinc-carbon cell is approximately 1.5 V. The voltage of any cell depends on the materials used as the electrodes. A lead-acid secondary cell, such as those employed in storage batteries, develops a voltage of 2.1 V. The electrodes (plates) are composed of lead for the negative and lead peroxide for the positive. As previously stated, dissimilar metals always have a definite polarity with respect to one another. For example, if nickel and aluminum are placed in an electrolyte, the nickel will be positive and the aluminum negative. However, if nickel and silver are acted upon by the same electrolyte, the nickel will be negative and the silver positive. The more active a metal is chemically, the greater its negative characteristic.

In a **secondary cell,** the chemical action that produces the electric current can be reversed; in other words, secondary cells can be recharged. This is accomplished by applying a voltage higher than that of the cell to the cell terminals; this causes a current to flow through the cell in a direction opposite to that in which the current normally flows. The positive terminal of the charging source is connected to the positive terminal of the cell, and the negative terminal of the charging source is connected to the negative terminal of the cell. Since the voltage of the charger is higher than that of the cell, electrons flow into the negative plate and out of the positive plate. This causes a chemical action to take place that is the reverse of the one that occurs during operation of the cell; the elements of the cell return to their original composition. At this time, the cell is said to be *charged*. Secondary cells can be charged and discharged many times before they deteriorate to the point at which they must be discarded.

A cell that cannot be recharged satisfactorily is called a **primary cell.** The elementary voltaic cell described previously in this section is a primary cell. Some of the elements deteriorate as the cell produces current; hence the cell cannot be restored to its original condition by charging. The common flashlight cell is a familiar example of a primary cell. The negative plate of a primary cell deteriorates because the material goes into solution with the electrolyte. In the secondary cell, the material of the plates does not go into solution but remains in the plates, where it undergoes a chemical change during operation.

Alkaline and Mercury Cells

Voltaic cells utilizing an alkaline electrolyte are usually termed **alkaline cells.** The electrolyte consists primarily of a potassium hydroxide solution. A variety of alkaline cells are currently available, as seen in Figure 3–3. Potassium hydroxide (KOH) is a powerful caustic similar to household lye and can cause severe burns if it comes into contact with the skin. The electrodes of such cells can be of several different types of materials, such as manganese dioxide and zinc, silver oxide and zinc, silver oxide and cadmium, mercuric oxide and zinc, or nickel and cadmium. These various electrode materials will determine if the alkaline cell is a rechargeable secondary cell or a nonrechargeable primary cell. The different electrodes will also determine the cell's voltage output. Most common alkaline cells produce approximately 1.5 V without a load applied to the cell.

Mercury cells are another common type of dry cell used for a variety of applications. A mercury cell consists of a positive electrode of mercuric oxide mixed with a conductive material and a negative electrode of finely divided zinc. The electrodes and the caustic electrolyte are assembled in sealed steel cans. Some electrodes are pressed into flat circular shapes, and others are formed into hollow cylindrical shapes, depending on the type of cell for which they are made. The electrolyte is immobilized in an absorbent material between

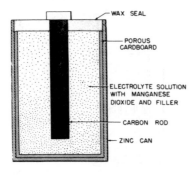

FIGURE 3–2 Construction of a simple dry cell.

WAX SEAL

POROUS CARDBOARD

ELECTROLYTE SOLUTION WITH MANGANESE DIOXIDE AND FILLER

CARBON ROD

ZINC CAN

FIGURE 3–3 A variety of dry cell batteries. *(Eveready Battery Co.)*

the electrodes. Mercury cells are often used for small *button* batteries found in miniature equipment, such as watches and calculators.

Nickel-Cadmium Cells

Nickel-cadmium electric cells and batteries have been developed to a high degree of efficiency and dependability. They are used in small devices that formerly used carbon-zinc dry cells and in other devices where carbon-zinc cells cannot meet the load requirements. They are also being manufactured in large sizes for use in aircraft where large load requirements are present. The service and maintenance of nickel-cadmium aircraft batteries are discussed later in this chapter.

Nickel-cadmium cells are made with various electrode designs, but the active elements remain the same. In a charged state, the negative electrode consists of metallic cadmium (Cd), and the positive electrode is nickel oxyhydroxide (NiOOH). During discharge the electrodes alter chemical composition. The negative electrode becomes cadmium hydroxide (CdOH), and the positive electrode becomes nickel hydroxide [$Ni(OH)_2$].

The most common electrode designs for nickel-cadmium cells consist of perforated steel pockets to hold the active materials or perforated nickel plates or woven nickel screens into which the active materials are impregnated by sintering. **Sintering** is a process of heating finely divided metal particles in a mold to approximately melting temperature. The metal particles weld together where they are in contact with other particles, and this results in a porous material. In the case of nickel-cadmium electrodes, the sintered material is nickel or nickel carbonyl for the positive plates and cadmium

for the negative plates. A nickel-cadmium cell that has been cut away to show construction is illustrated in Figure 3–4.

As mentioned previously, a secondary cell is one that can be charged and discharged repeatedly without appreciable deterioration of the active elements. An advantage of the nickel-cadmium secondary cell is that it can stand in a discharged condition indefinitely at normal temperatures without deterioration. If a lead-acid battery is left in the discharged condition for a substantial period of time, sulfation of the plates occurs, and the cells lose much of their capacity.

During the discharge of a nickel-cadmium cell, electrons are released in the negative material as chemical change takes place. These electrons flow through the outer electric circuit and return to the positive electrode. Positive ions in the electrolyte remove the electrons from the positive electrode. During charge, the reverse action takes place, and the negative electrode is restored to a metallic cadmium state.

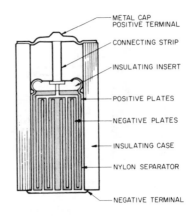

FIGURE 3–4 Ni-cad cutaway to show construction.

Nickel-cadmium cells generate gas during the latter part of a charge cycle and during overcharge. Hydrogen is formed at the negative electrode, and oxygen is formed at the positive electrode. In vented-type batteries, the hydrogen and oxygen generated during overcharge are released to the atmosphere together with some electrolyte fumes. In a sealed dry cell, it is necessary to provide a means for absorbing the gases. This is accomplished by designing the cadmium electrode with excess capacity. This makes it possible for the positive electrode to become fully charged before the negative electrode. When this occurs, oxygen is released at the positive electrode, while hydrogen cannot yet be generated because the negative electrode is not fully charged. The cell is so designed that the oxygen can travel to the negative electrode, where it reacts to form chemical equivalents of cadmium oxide. Thus, when a cell is subject to overcharge, the cadmium electrode is oxidized at a rate just sufficient to offset input energy, and the cell is kept at equilibrium at full charge.

If a cell is charged at the recommended rate, overcharging can occur for as long as 200 or 300 charge cycles without damage to the cell. If the charge rate is too high, the oxygen pressure in the cell can become so great that it will rupture the seal. For this reason, charge rates must be carefully controlled.

Open- and Closed-Circuit Voltages

There are two different ways to measure the voltage of a battery or cell. Voltage measured when there is no load applied to the battery is called the **open-circuit voltage** (OCV). The voltage measured while a load is applied to the battery is called the **closed-circuit voltage** (CCV). The OCV is always higher than the CCV because a battery can maintain a higher pressure (voltage) when there is no current flow leaving the battery. The OCV of a fully charged aircraft battery may reach 13.2 V; however, when even a small load is applied, the CCV will measure near 12 V. This battery would typically be referred to as a 12-V battery. The CCV of a battery is usually a function of the load applied and the state of charge of that battery. If a battery is connected to a heavy load, the CCV will be lower than if that battery was connected to a light load. If a battery is near total discharge, the CCV will be lower than if that same battery was fully charged. The OCV of a battery is typically affected very little by its state of charge until the battery reaches near complete discharge. Figure 3–5 illustrates the relationships between OCV and CCV for various loads and battery states of charge.

Internal Resistance

The resistance present inside of a battery while connected to a load is called **internal resistance** (IR). IR restricts the movement of current inside of any power source, including batteries. In the case of a battery, the IR is determined by the load applied and the battery's state of charge.

A battery's IR is equal to the difference between the OCV and the CCV, divided by the applied load. That is, IR = (OCV − CCV)/load amperage. This equation is derived from Ohm's law, $R = E/I$. A battery's internal resistance can be determined as follows: If the OCV = 14 V and the CCV = 12 V with a 100-A load applied, the IR is 0.02 Ω. The calculation is

$$IR = \frac{OCV - CCV}{I \text{ load}}$$
$$= \frac{14 \text{ V} - 12 \text{ V}}{100 \text{ A}}$$
$$= \frac{2 \text{ V}}{100 \text{ A}}$$
$$= 0.02 \text{ } \Omega$$

A battery's internal resistance always becomes greater as the battery becomes discharged. This is due to the lowering of a battery's CCV as the battery becomes weaker. The OCV remains nearly constant while the CCV drops; therefore, the difference between these two voltages increases. Hence IR increases.

The IR of a battery becomes very significant when a power source is chosen or a delicate circuit is designed. However, for general-purpose applications, a battery's internal resistance will not adversely affect an aircraft electrical system until that battery becomes over 75 percent discharged. When the battery reaches this low state of charge, its internal resistance becomes too high and the CCV lowers. This low CCV obviously affects circuit performance.

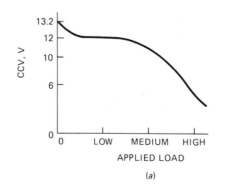

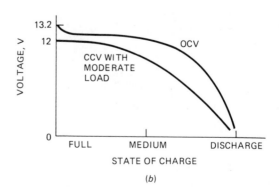

FIGURE 3–5 (a) Closed-circuit voltage versus applied load; (b) open- and closed-circuit voltage versus a battery's state of charge.

LEAD-ACID STORAGE BATTERIES

The term **storage battery** has been used for many years as the name for a battery of secondary cells and particularly for lead-acid and nickel-cadmium batteries.

Two types of **lead-acid** batteries currently being used in aviation are (1) the **vented cell** and (2) the **sealed (recombinant gas)** battery (Figure 3–6). The modern sealed-cell lead-acid batteries are more powerful and require less maintenance than the older vented lead-acid aircraft batteries. For this reason, lead-acid batteries are being used to replace the more expensive nickel-cadmium battery in some turbine-powered aircraft. On turbine-powered aircraft, however, the installation of lead-acid batteries typically requires that external power be readily available for engine starting, and the lead-acid batteries require more frequent replacement. Despite the great strides made to improve lead-acid batteries, they are still unable to deliver the current generated by nickel-cadmium batteries; therefore, nickel-cadmium batteries will remain a practical power source for aircraft. Nickel-cadmium batteries are discussed later in this chapter.

Lead-acid secondary cells consist of lead-compound plates immersed in a solution of sulfuric acid and water, which is the **electrolyte.** Each cell has an OCV of approximately 2.1 V when fully charged. When connected to a substantial load, the voltage is approximately 2 V. Aircraft storage batteries of the lead-acid type are generally rated at

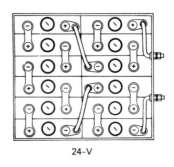

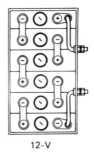

12-V   24-V

FIGURE 3–7 Arrangement of cells in a lead-acid storage battery.

12 or 24 V; that is, they have either 6 or 12 cells connected in series (see Figure 3–7).

Actually, the voltage of a 12-V battery is near 12.6 V (6 cells × 2.1 V/cell) when the battery is fully charged. A 24-V battery actually provides 25.2 V (12 cells × 2.1 V/cell). Figure 3–7 illustrates how the individual cells of a battery can be connected by external connector plates. On modern lead-acid batteries, cell connectors are actually housed inside the battery, therefore limiting the possibility of accidental cell shorting.

Schematic diagrams of cells connected in series and parallel are shown in Figure 3–8. In the series diagram, four 2-V cells are connected in series to produce 8 V. If the same four cells are connected in parallel, as shown in Figure 3–8b, the total voltage is the same as that of one cell; however, the capacity of a group, in amperes, is four times the capacity of a single cell. To increase both the voltage and the amperage by combining single cells, the cells are connected in a series-parallel circuit like that shown in Figure 3–9. When 16 cells are connected in this manner, the voltage is four times as great as that of a single cell, and the current capacity of the combined cells is four times as great as that of a single cell.

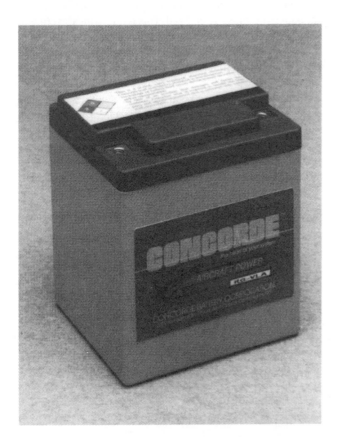

FIGURE 3–6 A sealed lead-acid aircraft battery. *(Concord Battery Corporation)*

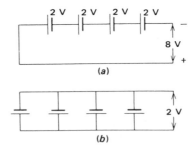

FIGURE 3–8 Series and parallel cell connections.

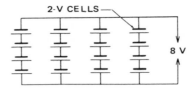

FIGURE 3–9 Series-parallel cell connections.

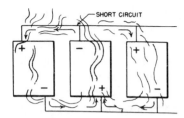

FIGURE 3–10 Incorrect battery connections.

When batteries or cells are connected incorrectly, they may be damaged. For example, if a technician intends to connect three batteries in parallel and connects one of them incorrectly, as shown in Figure 3–10, there will be a short circuit from the center battery to the two end batteries. This will either burn out the wiring or discharge the batteries and possibly damage them beyond repair. It is essential that the technician understand well the characteristics of battery circuits and the proper methods for connecting batteries and cells.

Storage batteries are convenient for aircraft use because their weight is not excessive for the power developed, and they can be kept in a nearly fully charged state by means of an engine-driven dc alternator. It must be remembered that the aircraft storage battery is used only when other sources of electric power are not available.

On light aircraft, the battery is used during initial engine starting, for intermittent loads that exceed alternator output, and in emergency situations (alternator failure). Large turbine-powered aircraft typically use the storage battery only for emergency power; any current required for starting the engines is supplied by a separate ground power unit. The storage battery on most commercial jets would supply approximately 30 minutes of emergency power in the case of a complete alternator system failure.

Theory of the Lead-Acid Cell

The lead-acid secondary cell used in a storage battery consists of positive plates filled with lead peroxide (PbO_2); negative plates filled with pure spongy lead (Pb); and an electrolyte consisting of a mixture of 30 percent sulfuric acid and 70 percent water, by volume (H_2SO_4).

A chemical action takes place when a battery is delivering current as shown in Figure 3–11. The sulfuric acid in the electrolyte breaks up into hydrogen ions (H_2) carrying a positive charge and sulfate ions (SO_4) carrying a negative charge. The SO_4 ions combine with the lead plate and form lead sulfate ($PbSO_4$). At the same time, they give up their negative charge, thus creating an excess of electrons on the negative plate.

The H_2 ions go to the positive plate and combine with the oxygen of the lead peroxide (PbO_2), forming water (H_2O), and during the process they take electrons from the positive plate. The lead of the lead peroxide combines with some of the SO_4 ions to form lead sulfate on the positive plate. The result of this action is that the positive plate has a deficiency of electrons and the negative plate has an excess of electrons.

When the plates are connected together externally by a conductor, the electrons from the negative plate flow to the positive plate. This process will continue until both plates are coated with lead sulfate and no further chemical action is possible; the battery is then said to be discharged. The lead sulfate is highly resistant to the flow of current, and it is chiefly this formation of lead sulfate that gradually lowers the capacity of the battery until it is discharged.

During the charging process, current is passed through the storage battery in a reverse direction. A dc supply is applied to the battery with the positive pole connected to the positive plate of the battery and the negative pole connected to the negative plate. The voltage of the source is greater than the voltage of the battery. This causes the current to flow in a direction to charge the battery. The SO_4 ions are driven back into solution in the electrolyte, where they combine with the H_2 ions of the water, thus forming sulfuric acid. The plates then return to their original composition of lead peroxide and spongy lead. When this process is complete, the battery is charged.

Figure 3–12 shows the chemical changes that occur to a lead-acid battery during charge and discharge.

Lead-Acid Battery Construction

A storage battery consists of a group of lead-acid cells connected in series and arranged somewhat as shown in Figure 3–7. Under moderate load the closed circuit voltage (CCV) of the 6-cell battery is approximately 12 V, and that of a 12-cell battery is about 24 V. As stated earlier, CCV is the voltage of the battery when connected to a load.

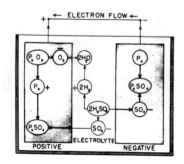

FIGURE 3–11 Chemical action in a lead-acid secondary cell.

	CHARGED STATE	CHEMICAL CHARGE	DISCHARGE
POSITIVE PLATE	PbO_2	LOOSES O_2 GAINS SO_4	$PbSO_4$
NEGATIVE PLATE	Pb	GAINS SO_4	$PbSO_4$
ELECTROLYTE	H_2SO_4	LOOSES SO_4 GAINS O_2	H_2O

FIGURE 3–12 Chemical changes of a lead-acid battery during charge and discharge.

Each cell of a storage battery has positive and negative plates arranged alternately and insulated from each other by separators. Each plate consists of a framework, called the **grid,** and the **active material** held in the grid. A standard formula for the grid material is 90 percent lead and 10 percent antimony. The purpose of the antimony is to harden the lead and make it less susceptible to chemical action. Other metals, such as silver, are also used in some grids to increase their durability.

A typical grid is illustrated in Figure 3–13. The heavy border adds strength to the plate, and the small horizontal and vertical bars form cavities to hold the active material. The structural bars also act as conductors for the current, which is distributed evenly throughout the plate. Each plate is provided with extensions, or feet, which rest upon ribs on the bottom of the cell container. These feet are arranged so the positive plates rest upon two of the ribs and the negative plates upon the two alternative ribs. The purpose of this arrangement is to avoid the short-circuiting that could occur as active material is shed from the plates and collects at the bottom of the cell.

The plates are made by applying a lead compound to the grid. The paste is mixed to the proper consistency with diluted sulfuric acid, magnesium sulfate, or ammonium sulfate and is applied to the grid in much the same manner as plaster is applied to a lath wall. The paste for the positive plates is usually made of red lead (Pb_3O_4) and a small amount of litharge (PbO). In the case of the negative plates, the mixture is essentially litharge with a small percentage of red lead. The consistency of the various materials and the manner of combining them have considerable bearing on the capacity and life of the finished battery.

In compounding the negative-plate paste, a material called an **expander** is added. This material is relatively inert chemically and makes up less than 1 percent of the mixture. Its purpose is to prevent the loss of porosity of the negative material during the life of the battery. Without the use of an expander, the negative material contracts until it becomes quite dense, thus limiting the chemical action to the immediate surface. To obtain the maximum use of the plate material, the chemical action must take place throughout the plate from the surface to the center. Typical expanding materials are lampblack, barium sulfate, graphite, fine sawdust, and ground carbon. Other materials, known as hardness and porosity agents, are sometimes used to give the positive plates desired characteristics for certain applications. One or more manufacturers reinforce the active material of the battery plates with plastic fibers 0.118 to 0.236 in. [3 to 6 mm] long. This adds substantially to the active life of the battery.

After the active-material paste is applied to the grids, the plates are dried by a carefully controlled process until the paste is hardened. They are then given a forming treatment in which a large number of positive plates are connected to the positive terminal of a charging apparatus and a like number of negative plates, plus one, are connected to the negative terminal. They are placed in a solution of sulfuric acid and water (electrolyte) and charged slowly over a long period of time. A few cycles of charging and discharging convert the lead compounds in the plates into active material. The positive plates thus formed are chocolate brown in color and of a hard texture. The negative-plate material has been converted into spongy lead of a pearl-gray color. After forming, the plates are washed and dried. They are then ready to be assembled into **plate groups.**

Plate groups are made by joining a number of similar plates to a common terminal post (see Figure 3–14). The number of plates in a group is determined by the capacity desired, inasmuch as capacity is determined by the amount (area) of active material exposed to the electrolyte.

Since increasing plate area will increase a battery's capacity, many manufacturers strive for the maximum in internal battery dimensions. That is, if the inside of the battery can be kept as large as possible, the plate area can be increased. To do this, ultrathin plastic cases have been employed to "squeeze" the maximum plate area inside the battery of a given size. It also stands to reason that increasing the battery's outer (and inner) dimensions could be a means to increase capacity. However, for aircraft use, we typically strive for the smallest, lightest battery with a relatively high capacity.

Each plate is made with a lug at the top to which the **plate strap** is fused. A positive-plate group consists of a number of positive plates connected to a plate strap, and a negative group is a number of negative plates connected in the same manner. The two groups meshed together with separators between the positive and negative plates constitute a **cell element** (see Figure 3–15). It will be noted in the illustrations that there is one less positive plate than negative plates. This

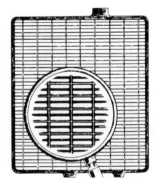

FIGURE 3–13 Grid for a lead-acid cell plate.

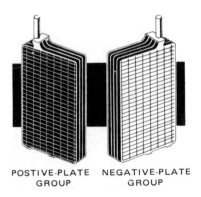

POSITIVE-PLATE GROUP   NEGATIVE-PLATE GROUP

FIGURE 3–14 Plate groups.

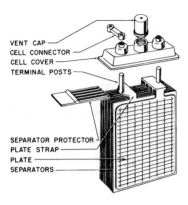

FIGURE 3–15 Cell element for a lead-acid cell.

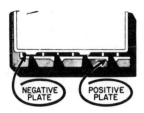

FIGURE 3–16 Sediment space in a cell container.

arrangement provides protection for the positive plates, inasmuch as they are more subject to warping and deterioration than the negative plates. By placing negative plates on each side of every positive plate, the chemical action is distributed evenly on both sides of the positive plate, and there is less tendency for the plate to warp.

The **separators** used in lead-acid storage batteries are made of fiberglass, rubber, or other insulating materials. Their purpose is to keep the plates separated and thus prevent an internal short circuit. Without separators, even if the containers were slotted to keep the plates from touching, material might flake off the positive plates and fall against the negative plates. Negative material might expand sufficiently to come in contact with the positive plates, or the positive plates might buckle enough to touch the negative plates.

The material of the separators must be very porous so that it will offer a minimum of resistance to the current passing through. The separators are saturated with electrolyte during operation, and it is this electrolyte that conducts the electric current. It is obvious also that the separators must resist the chemical action of the electrolyte.

Glass-wool separators are used by some manufacturers. Fine glass fibers are laid together at different angles and cemented on the surface with a soluble cement. The glass wool is placed in the cell adjacent to the positive plate. Because of the compressibility of glass wool, it comes into very close contact with the positive plate and prevents the loosened active material from shedding. It is claimed that batteries with this type of separator have a longer life than those without it.

Another very effective method for providing plate separation is to enclose the positive plates in microporous polyethylene pouches. This increases the efficiency of the battery because the plates are much closer together, approximately 0.05 in. [1.25 mm], than they are with other types of separators. The pouches also prevent the shedding of active material from the positive plates.

When the cell elements are assembled, they are placed in the **cell container,** which is made of hard rubber or a plastic composition. Cell containers are usually made in a unit with as many compartments as there are cells in the battery. In the bottom of the container are four ribs. Two of these ribs support the positive plates, and the other two support the negative plates. This arrangement leaves a space underneath the plates for the accumulation of sediment, thus preventing the

sediment from coming in contact with the plates and causing a short circuit. The construction of the cell bottom and of the plate-supporting ribs is shown in Figure 3–16.

The sediment space provided in storage batteries is of such capacity that it is not necessary to open the cells to clean out the sediment. When the sediment space is full to the point at which the spent material may come in contact with the plates, the cell is worn out.

The assembled cell of a storage battery has a cover made of material similar to that of the cell container. The cell cover is provided with two holes through which the terminal posts extend and a threaded hole into which is screwed the vented cell cap. When the cover is placed on the cell, it is sealed in with a special sealing compound. This is to prevent spillage and loss of electrolyte.

When a storage battery is on charge and approaching the full-charge point or is at the full-charge point, there is a liberal release of hydrogen and oxygen gases. It is necessary to provide a means whereby these gases can escape, and this is accomplished by placing a vent in the cell cap. This vent contains a lead valve that is so arranged that it will close the vent when the battery is inverted or in any other position at which there is danger of spillage. A vent cap of this type is illustrated within the circles of Figure 3–17. Another type of vent system, also illustrated (within the large arrows) in Figure 3–17, incorporates a tube that extends almost to the top of the plates. With this type of construction, the battery plates fill only slightly more than one-half the cell container. The space in the top of the container is provided to hold the electrolyte when the battery is on its side or in an inverted position. A baffle plate is placed slightly above the plates to

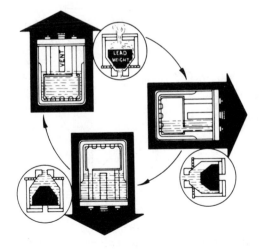

FIGURE 3–17 Battery vent caps.

Lead-Acid Storage Batteries   **37**

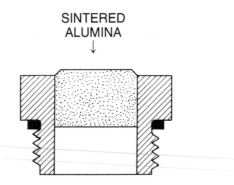

SINTERED
ALUMINA
↓

FIGURE 3–18 Vented plug with a sintered alumina barrier.

prevent splashing of the electrolyte. A hole in the baffle plate for the escape of gas and for access to the electrolyte is located to one side of the bottom of the gas-escape tube. If the electrolyte level is flush with the baffle plate, the end of the tube will always be above the electrolyte level, regardless of the position of the battery.

Another type of battery vent contains a sintered alumina (aluminum oxide) plug instead of the heavy lead one used in many other types of batteries. This plug permits the diffusion of gases through it without letting fluids pass. It is much smaller and lighter than the lead-valve plug; hence it saves both weight and space. The construction of this plug is illustrated in Figure 3–18.

A battery vent cap that is particularly well adapted to acrobatic and military aircraft is shown in Figure 3–19. As shown in the drawing, there is a valve in the bottom of the unit, and this valve is opened and closed by the action of the conical weight in the upper part of the cap. When the battery is tilted approximately 45°, the weight drops against the side of the cap, pulling up on the valve stem and closing the valve.

When the battery is brought back to a position approximately 32° from vertical, the weight centers itself again, allowing the valve stem to lower and open the vent valve.

Battery Design Features

Although the majority of lead-acid storage batteries are constructed with similar features, there are many differences in size and detail design, depending on the use to which the battery is to be put. A completely assembled metal-encased battery for aircraft is shown in Figure 3–20. The cell containers are integral with a metal shielding box coated with acid-resistant paint.

This box provides mechanical protection as well as electrical shielding and is fitted with a metal cover secured in place with hold-down rods. The design also provides an air-tight space above the cells so that the gases being emitted will not escape into the aircraft in which the battery is installed. The vent space is provided with a connection for a tube installed to carry the battery gases overboard. This is a requirement for any battery that emits gases during operation.

The main negative and positive terminals of the battery are connected to external terminals in the side of the metal case. These terminals are adequately insulated from the case by washers and bushings.

A storage battery for light aircraft is shown in Figure 3–21. This battery is made with a lightweight polystyrene case and is designed for use in an aircraft with an enclosed and ventilated battery compartment. The plates in this battery are reinforced with plastic fibers, and the positive plates are enclosed in microporous pouches to provide plate separation and protection. The intercell connectors are internal and permanently sealed with an epoxy resin. *Note:* There are 12 cell

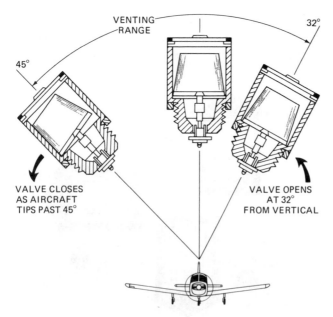

FIGURE 3–19 Vented cell cap for acrobatic aircraft. *(Teledyne Battery Products)*

FIGURE 3–20 Aircraft battery with metal case. *(Teledyne Battery Products)*

FIGURE 3–21 Battery for use in an enclosed and ventilated battery compartment. *(Teledyne Battery Products)*

caps in this battery; therefore, this battery contains 12 cells and is considered a 24-V battery (12 cells × 2 V/cell = 24 V).

A battery designed for use where there is no enclosed battery compartment is shown in Figure 3–22. A gas-collecting manifold is provided by the sealed compartment at the top of the battery. This compartment is provided with an air inlet and an air and gas outlet that are connected to an inlet air source and a discharge tube, respectively. During operation, airflow through the manifold carries gases and acid fumes overboard.

Sealed lead-acid batteries do not have the same problems with corrosive fumes and electrolyte leakage. The sealed design does not require vents to release gases generated from charging, and the microfibrous glass-mat separators absorb most of the electrolyte, allowing only a small amount of elec-

FIGURE 3–22 Battery with a sealed case for installations where there is no sealed battery compartment. *(Teledyne Battery Products)*

TABLE 3–1 Freezing points for different states of charge in a lead-acid storage battery.

| | FREEZING POINT | |
SPECIFIC GRAVITY	°F	°C
1.300	−95	−70.6
1.285	−85	−65
1.275	−80	−62.2
1.250	−62	−52.2
1.225	−35	−37.2
1.200	−16	−26.7
1.175	−4	−20.0
1.150	+5	−15.0
1.125	+13	−10.6
1.100	+19	−7.2

trolyte to exist freely outside the plate elements. The battery is designed to allow for this "free" electrolyte, and no outside leakage of electrolyte occurs.

Cold-Weather Operation

Temperature is a vital factor in the operation and life of a storage battery. Chemical action takes place more rapidly as temperature increases. For this reason, a battery will give much better performance in temperate or tropical climates than in cold climates. On the other hand, a battery will deteriorate faster in a warm climate.

In cold climates, the state of charge in a storage battery should be kept at a maximum. A fully charged battery will not freeze even under the most severe weather conditions, but a discharged battery will freeze very easily.

When water is added to a battery in extremely cold weather, the battery must be charged at once. If this is not done, the water will not mix with the acid and will freeze. Table 3–1 gives the freezing point for various states of charge.

LEAD-ACID BATTERY MAINTENANCE PROCEDURES

Precautions

Follow these precautions when servicing aircraft batteries:

1. Always wear safety glasses.
2. Remove the negative lead first and install it last.
3. Do not cause a short circuit between the battery terminals. Be cautious of jewelry and watches. Some are good conductors and may short-circuit the battery, causing severe injury to the technician.
4. Never service the batteries near an open flame or sparks.
5. Never jump-start an aircraft from another power source if the airplane's battery is discharged. The battery within the aircraft is not an airworthy battery because of its discharged state. A battery requires several hours to recharge completely when fully discharged and will be unable to sup-

port the aircraft's electrical system in the event of an emergency. During jump-starting of an airplane, a strong current flow into the airplane's battery may damage the cell plates, which will lead to premature battery failure.

Lead-Acid Battery Inspection and Service

Most aircraft are scheduled for 50-h, 100-h, annual, or periodic inspections. During these inspection periods, the battery should be inspected and serviced as required. A service schedule of 50-h of flight time or once a month (whichever comes first) will ensure that the battery will continue to perform properly.

As a general rule, one should always follow the manufacturer's maintenance instructions whenever possible. The following is offered as a general guide for inspection and service of batteries.

1. Inspect the mounting of the battery. Make sure that no part of the supporting structure is cracked or weakened in any way.

2. Remove the cover from the battery case, if it is the covered type, and inspect the interior. Look for evidence of leakage and corrosion. The top of the battery should be clean and dry. A small amount of corrosion around the terminals can be removed with a stiff brush and a mild soda solution. A wire brush should not be used because of the danger of short-circuiting the battery.

It is important to note that a battery whose top is damp with electrolyte and dirt will discharge itself quite rapidly because of the conductance of the electrolyte; a steady current flows from the negative terminal of the battery to the positive terminal. Hence it is essential that the top of any storage battery be kept clean and dry. When using a soda solution to neutralize lead-acid battery spills, take care to see that none of the solution enters the battery cells. If it does, the solution will neutralize the electrolyte, and the battery is likely to go "dead." After cleaning the battery with soda solution, rinse it with clean water and dry the top of the battery.

If a large amount of corrosion is found in the battery case, the battery should be removed and the case cleaned thoroughly. If appreciable damage has been done, either to the battery case or to the battery mounting structure, the damaged parts should be repaired or replaced.

3. Check the electrolyte level in the battery. If the liquid is below the plates of the battery cell, add clean distilled water until it is approximately $\frac{3}{8}$ in. (0.95 cm) above the plates. Some batteries have an electrolyte level indicator just above the plates. If a battery is so equipped, the electrolyte should be filled to this level. Figure 3–23 illustrates the correct electrolyte level for a lead-acid battery. Remember, always add distilled water only, never electrolyte. The proper level should be above the plates and about 1 in. below the top of the battery.

4. If the battery is suspected of being defective, perform a battery load test or a hydrometer test (lead-acid batteries only). If the battery indicates that it is weak during either test, recharge the battery and retest after the battery has stabilized (about 1 h). Remember, a hydrometer test should never be

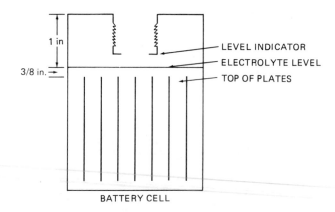

FIGURE 3–23 Cell electrolyte level.

performed on lead-acid batteries after water is added to the cells. The readings will be erroneous until the water and the electrolyte are thoroughly mixed.

5. Inspect the terminal connections. See that they are tight and free from corrosion. If a quick-disconnect plug is used on the battery, remove it and inspect the contacts. If they are dirty or corroded, clean them thoroughly and apply a small amount of terminal lubricant. Replace the plug, making sure that the handwheel is tight.

6. Inspect the battery cables for condition of insulation, evidence of chafing, and security of connections.

7. Replace the cover on the battery case, making sure that the hold-down nuts are tightened sufficiently and safetied.

8. Inspect the ventilation system of the aircraft and battery box. Be sure the vent tubes are clear and without damage. If a sump jar is used in the aircraft battery box ventilation system, see that the felt pad is covered with a soda solution. Check that all fittings are tight and free from leaks. Inspect the airplane near the area of the discharge tube exit. This area often corrodes and must be cleaned and neutralized periodically.

Hydrometer Test

For aircraft lead-acid batteries, it is typical to use a hydrometer test to determine the batteries' state of charge. A **hydrometer** is a tool used to measure the specific gravity, or density, of a liquid. The **specific gravity** of a substance is defined as **the ratio of the weight of a given volume of that substance to the weight of an equal volume of pure water at +4°C.**

The specific gravity of the electrolyte in a lead-acid cell decreases as the charge in the cell decreases. This is because the acid in the electrolyte becomes chemically combined with the active material in the plates as the battery produces current; hence less acid remains in the electrolyte. Since the specific gravity of the acid is considerably greater than that of water, the loss of acid causes the specific gravity of the electrolyte to drop.

A hydrometer is used to determine the specific gravity of the electrolyte in a lead-acid cell. A typical hydrometer used for battery testing is shown in Figure 3–24. It consists of a small sealed glass tube weighted at the end to make it float in an upright position. The amount of weight in the bottom of

FIGURE 3–24 Drawing of a typical hydrometer.

the tube is determined by the specific gravity range of the fluid to be tested. In the case of a battery hydrometer, the specific gravity range is 1.100 to 1.300. This small tube is placed inside a larger glass-tube syringe. With this arrangement, the electrolyte can be drawn from a cell into the glass tube and the reading noted. The electrolyte is then returned to the cell from which it was taken.

The specific gravity reading is taken at the fluid level on the stem of the hydrometer when it is floating freely in the electrolyte.

When a battery is tested with a hydrometer, the temperature of the electrolyte must be taken into consideration because the specific gravity readings on the hydrometer will vary from the true specific gravity as the temperature goes above or below 80°F [26.7°C]. No correction is necessary when the temperature of the electrolyte is between 70 and 90°F [21.1 and 32.2°C] because the variation is not great enough to be considered. At higher or lower temperatures it is necessary to apply a correction according to Table 3–2.

The corrections in Table 3–2 should be added to or subtracted from the reading on the hydrometer. For example, if the temperature of the electrolyte is 10°F [− 12.2°C] and the hydrometer reading is 1.250, the corrected reading will be 1.250 − 0.028, or 1.222. Notice that the correction points represent thousandths.

TABLE 3–2 Hydrometer corrections for temperature.

Electrolyte temperature		Correction, points
°F	°C	
120	48.9	Add 16
110	43.3	Add 12
100	37.8	Add 8
90	32.2	No correction
80	26.7	No correction
70	21.1	No correction
60	15.6	Subtract 8
50	10.0	Subtract 12
40	4.4	Subtract 16
30	−1.1	Subtract 20
20	−6.7	Subtract 24
10	−12.2	Subtract 28
0	−17.8	Subtract 32
−10	−23.3	Subtract 36
−20	−28.9	Subtract 40
−30	−34.4	Subtract 44

Some hydrometers are equipped with a correction scale inside the tube; the temperature correction then can be applied as the hydrometer reading is taken.

Battery Load Testers

There are various automatic **battery load testers** available; one is illustrated in Figure 3–25. This machine will test not only batteries but also the aircraft's charging system if so desired. While an automatic battery load test is being performed, the load is applied for 15 s, and the *open-circuit voltage (OCV)* and *closed-circuit voltage (CCV)* are automatically compared. The OCV is the voltage of the battery with no load applied; the CCV is the measure of battery voltage while the battery is under load. If the CCV falls below 9.6 V, the indication "bad" appears on the test unit. If the CCV is maintained above 9.6 V during the entire load test, the unit will indicate "good." Once again, discharged batteries must be charged prior to testing.

The high-rate discharge battery capacity test is probably the most common and practical test used. This test is designed to simulate the load typically placed on the battery during an engine start. This load can reach several hundred amps for a few minutes, definitely the most strenuous time for any battery.

To simulate a starting load, the test equipment, as illustrated in Figure 3–26, is connected to the battery in parallel. Then the operator applies a load approximately two or three times the battery's ampere-hour rating. (Ampere-hour rating will be discussed later in this chapter.) While the load is applied, the CCV of the battery is measured. The load should be applied for less than 2 min and the CCV monitored during this time. If the CCV drops below 11 V but remains above 10 V, the battery is probably only slightly discharged. If the battery's CCV stays between 10 and 9 V, the battery is only

FIGURE 3–25 Automatic battery load tester. *(Sun Electric Corp.)*

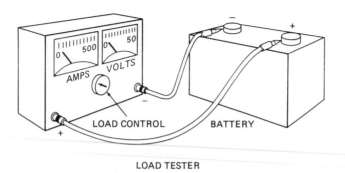

FIGURE 3–26 Battery load tester.

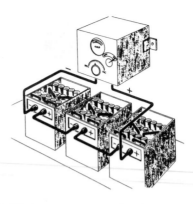

FIGURE 3–27 Constant-current charging.

slightly charged, and any CCV below 9 V indicates a very weak or dead battery. Remember, a low CCV reading does not necessarily mean a defective battery. A low reading only indicates a weak battery charge. This can be caused by a defective battery or a good battery that has been partially discharged earlier. To determine whether the battery is defective, recharge the battery and retest. If the CCV is still low, the battery is defective. If the CCV remains high after the recharge, the battery is probably airworthy.

Battery Charging

Secondary cells are charged by passing a direct current through the battery in a direction opposite to that of the discharge current. This means that the supply current's positive connection must be connected to the battery's positive connection and the negative connected to negative. Various methods of supplying the charge current are available. Onboard the aircraft, the generator or alternator will supply the charging current. Other ground-based charging equipment will convert common 115-V ac current into the dc voltage needed for battery charging. The two general types of charging equipment are **constant-current chargers** and **constant-voltage chargers.**

Constant-Current Chargers

As the name implies, a constant-current battery charger supplies a consistent current to a battery for the entire charge cycle. The charging equipment monitors current flow and varies the applied voltage in order to charge the battery. As the battery begins to charge, its voltage is lower than when the battery becomes fully charged. The constant-current charger will increase its voltage supplied to the battery during charge in order to maintain the current flow set by the operator.

Figure 3–27 illustrates the proper connection of more than one battery to a constant-current charger. The batteries are connected in series with respect to each other and the charger, thus allowing for a constant current flow through each battery. Constant-current chargers require careful supervision while in use. Because of the risk of overcharging, most constant-current chargers will automatically turn off after a predetermined time. The exact current flow and time of charge must be known and programmed into the charging equipment to prevent over- or undercharging of batteries. The specifications are normally available from the battery manufacturer; however, unless the original state of charge remaining in the battery is known, an improper charge is still likely. For this reason, constant-current chargers are often used on new batteries where the initial charge state is known, but they are seldom used on lead-acid batteries that have already been placed in service. Nickel-cadmium batteries often use constant-current charging equipment, as will be discussed later.

Constant-Voltage Chargers

As the name implies, this charging equipment supplies a constant voltage to the battery and allows current to change as the battery becomes charged. The constant-voltage charger supplies approximately 14 V for charging 12-V batteries and 28 V for charging 24-V batteries. A higher potential at the charger is necessary to ensure current flows from the charger to the battery. If the battery is nearly discharged, it will offer very little opposition to the electrons flowing into the battery. As the battery becomes charged, it will offer more resistance to the current supplied by the charger. Since that charger supplies a constant voltage, a relatively high current will flow into a discharged battery, and that current will slowly diminish as the battery becomes charged.

When the battery is fully charged, its voltage will be almost equal to the charger voltage; hence the charging current will drop to less than 1 A. When the charging current is low, the battery may remain on charge without any appreciable effect; however, the electrolyte level should be watched closely to see that it does not fall below desired levels.

Because the current supplied to the battery drops to a very low value as the battery becomes charged, constant-voltage charging is usually considered the safest method of battery charging. A constant-voltage charger is, by far, the most common type of ground-based battery charger. A constant-voltage charge is also the type supplied by the aircraft generator or alternator system.

If more than one battery is connected to a constant-voltage charger, all the batteries and the charger must be connected in parallel. This will ensure a constant voltage to each battery.

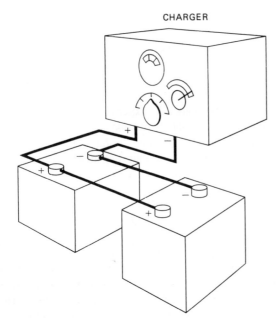

CHARGER

FIGURE 3-28 Constant-voltage charging of multiple batteries.

Charging multiple batteries is illustrated in Figure 3-28. Various types of constant-voltage chargers are available. Typically, they range from 5- to 50-A capacity, indicating the maximum current flow into a discharged battery. Each charger will lower current to about 1 A when the battery becomes charged. Some chargers come with timers or voltage monitors to shut the system off when the battery reaches the fully charged state. Typically, a low current (about 1 A) can be supplied to a fully charged battery for 24 h or less without damage to the battery. After 24 h, the liquid electrolyte level is at risk of becoming too low.

Charging Precautions

There are several precautions that should be observed when handling lead-acid batteries, especially during charging. The most dangerous problem occurs when a battery is charged and hydrogen and oxygen gases are emitted by the cells. Since this is an explosive mixture, it is essential to take precautions against igniting the gas by a spark or an open flame. Some precautions to prevent explosion are as follows:

1. Always charge batteries in a well-ventilated area. Forced-air fans to help remove any dangerous fumes are recommended; simply assuming that a large room, such as a hangar, is well ventilated is incorrect.

2. Always turn off the battery charger before disconnecting any connections between the battery and the charger. This will help eliminate the possibility of sparks at the battery terminals.

3. When removing the battery from the aircraft, always disconnect the negative lead first. When installing the battery, always connect the negative lead last. This will help prevent accidental shorts between the airframe and the battery's positive terminal.

4. Make sure that the caps of each cell of the battery are vented and that the vents are clean. If the caps appear old and dirty, soak them in plain hot water in order to clean the vents. If the vents remain clogged, replace the caps prior to charging.

5. Remove the battery from the aircraft prior to charging whenever possible. The corrosive electrolyte tends to vaporize during charging and escape through the vented battery caps. This electrolyte will corrode the aircraft if the battery is charged while in the airplane. If the battery should be charged while it remains in the aircraft, never operate any radios or other aircraft electronic equipment. A battery charger does not regulate voltage accurately enough to ensure trouble-free operation of electronic equipment.

6. Always take precautions not to spill electrolyte on skin or clothes; the liquid is very corrosive and will burn. **Always wear safety glasses** or another form of eye protection when servicing lead-acid batteries. This will protect your eyes from accidental acid contact. If electrolyte should spill from the battery, the affected area should be washed with water and neutralized with a solution of bicarbonate of soda and water, then thoroughly rinsed again with plain water. A solution of common baking soda and water is typically used to neutralize lead-acid electrolyte spills.

Placing New Lead-Acid Batteries in Service

The principal rule to observe when placing new lead-acid batteries in service is to follow the manufacturer's instructions. Because these instructions may vary considerably, care must be taken to follow them accurately.

Often, new lead-acid batteries that are stored in warehouses for long periods of time or placed in storage pending sale are not filled with electrolyte. The plates are dry-charged before assembly, and no electrolyte is placed in the cells until the battery is put in service.

When new batteries are received in the dry state, they should be filled with electrolyte having the specific gravity recommended by the manufacturer. After one or more hours, the electrolyte level should be checked, and if it has fallen, more electrolyte should be added to bring it up to the recommended level. The battery can be placed in service after the electrolyte has been in the cells for at least 1 h; if time is available, however, it is better that the battery be charged slowly for approximately 18 h. The rate of charge will depend on the type and capacity of the battery; this information is usually included in the instructions supplied with the battery.

When a lead-acid storage battery containing electrolyte is placed in storage, it should first be fully charged. All electrolyte spilled on the top of the battery should be removed, and the battery should then be washed with clean water and thoroughly dried. The terminals of the battery should be coated with terminal grease or petroleum jelly. While the battery is in storage, it should be recharged every 30 days to compensate for the self-discharge that takes place when it is not in use.

BATTERY RATINGS

A **battery** consists of a number of primary or secondary cells connected in series. The cells are arranged in this manner to increase the battery's voltage above the voltage available from only one cell. When cells are connected in series, the total battery voltage is equal to the sum of all the voltages of each cell. That is, total battery voltage = voltage of cell 1 + voltage of cell 2 + voltage of cell 3, etc. Figure 3–29 shows the cells of a lead-acid battery connected in series to obtain 12 V. Each lead-acid cell produces approximately 2 V; therefore, six cells are needed. **Capacity** is the measure of a battery's total available current. The capacity for all batteries is rated in a unit of current for a length of time. Small batteries are usually rated in **milliampere-hours** (mAh), because their load drain is usually less than 1 A for several hours. Larger batteries, typical of those found on aircraft, are usually rated in **ampere-hours.** These batteries can supply several amps for a much longer time period than smaller batteries; i.e., they have a higher capacity than smaller batteries. A battery's capacity is equal to the time required to fully discharge that battery multiplied by the current draw applied to the battery. In other words, if a battery can supply 2 A for 2 h, it has a capacity of 4 Ah (2 A × 2 h = 4 Ah), or 1 A for 4 h, or 8 A for 0.5 h. A battery with any combination of current flow and time that will give the product of 4 is considered a 4-Ah battery. The ampere-hour or milliampere-hour rating is usually determined by the manufacturer for any particular battery. This capacity rating is important when determining which battery to choose for a given load situation or when determining the charge rate for certain batteries. The recharging of a small nickel-cadmium cell should be at a rate in milliamperes that is equal to approximately 10 percent of the nominal milliampere-hour capacity. For example, a 900-mAh cell or battery should be charged at 90 mA. Cells should be charged at this rate for 14 to 16 h. This will supply 140 to 160 percent of the cell's total capacity, or 90 mA × 14 h = 1260 mAh; 1260 mAh is equal to 140 percent of 900 mAh. Aircraft nickel-cadmium batteries are typically recharged at a rate equal to 140 percent of the capacity rating. A battery with a 30-Ah capacity would require 42 Ah of charging. These charge rates will be discussed later in further detail.

The capacity of any battery will vary as a function of the time the battery is allowed to discharge. Owing to the chemical nature of all batteries, if a cell is discharged quickly, it will supply less total power than a cell that is discharged slowly. This phenomenon exists because the chemicals in a battery require time to react and produce electric power. If a quick discharge occurs, not all the chemical material will have time to react, and some of the battery's capacity will remain in the chemical state. You may have experienced this when using a flashlight for a long period of time; the batteries become weak and the light goes out. If the flashlight is turned off and the chemicals have several minutes or hours to interact, the flashlight will once again emit light when turned on. This "extra" power produced by the flashlight batteries will only last a short while; however, the quick discharge of large batteries will leave significant amounts of extra power suspended in the chemical state.

Since it becomes important, at times, to determine the exact capacity of a battery, all aircraft batteries must be discharged over a consistent time when determining capacity. The standard of a 5-h discharge rate is typically used. This means that all batteries will be discharged for 5 h to determine their capacity rating.

A battery that supplies 6 A for a period of 5 h is a 30-Ah battery at a 5-h rate (6 A × 5 h = 30 Ah). This same battery may only supply 25 Ah if it was completely discharged in only 1 h (25 A × 1 h = 25 Ah). Also, if it was discharged in only 10 min, the total capacity may drop as low as 10 Ah (60 A × $\frac{1}{6}$ h = 10 Ah).

The exact differences among capacities at different discharge rates are a function of the chemicals used in the battery, their purity, and the internal structure of the battery. It is apparent, however, that every battery will supply less total power when discharged quickly. Table 3–3 shows examples of the different discharge rates for different batteries.

Voltage Ratings

Storage batteries of all types are rated according to voltage and ampere-hour capacity. It has been pointed out that the voltage of a fully charged lead-acid cell is approximately 2.1 V when the cell is not connected to a load. A nickel-cadmium cell is rated at about 1.22 CCV.

Under a moderate load, the lead-acid cell will provide about 2 V. With an extremely heavy load, such as the operation of an engine starter, the voltage may drop to 1.6. A lead-acid cell that is partially discharged has a higher internal

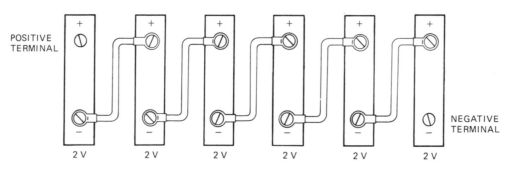

FIGURE 3–29 The connection of six 2-V cells in series to form a 12-V battery.

TABLE 3–3 The relationships of ampere-hour capacity to the length of discharge. A slower discharge rate produces a higher total capacity (ampere-hour).

VOLTAGE, V	AMPERES SUPPLIED FOR 5-H	AMPERE-HOURS AT 5-H RATE	AMPERES SUPPLIED FOR 20 MIN	AMPERE-HOURS AT 20-MIN RATE	AMPERES SUPPLIED FOR 5-MIN	AMPERE-HOURS AT 5-MIN RATE
12	5	25	48	76	140	11.7
12	7	35	66	22	180	15
12	17.6	88	145	48	370	31
24	5	25	48	16	140	12
24	7.2	36	70	23.3	180	15

resistance than a fully charged cell; hence it will have a higher voltage drop under the same load. This internal resistance is partially due to the accumulation of lead sulfate in the plates. The lead sulfate reduces the amount of active material exposed to the electrolyte; hence it deters the chemical action and interferes with the current flow.

Figure 3–30 shows the discharge characteristics of a typical aircraft battery of the lead-acid type. The OCV remains almost at 2.1 V until the battery is discharged. It then drops rapidly toward zero. The CCV gradually decreases from 2 to approximately 1.8 V as the cells discharge. Again, the voltage drops rapidly when the cell nears discharge.

Even though battery cells vary considerably in voltage under various conditions, batteries are nominally rated as 6 V (3 cells), 12 V (6 cells), and 24 V (12 cells). In replacing a battery, the technician must ensure that the replacement battery is of the correct voltage rating.

Power Ratings

As stated above under "Battery Ratings," most storage batteries are rated in ampere-hours at a 5-h discharge rate. This means that the battery was discharged to 0 V in 5 h to determine its capacity. Most 12-V batteries used for single-engine

aircraft have a capacity rating between 25 and 35 Ah; however, larger capacities are available. A direct comparison of ampere-hours alone does not indicate a battery's total power output. To determine total power, the battery's voltage must be considered because power (wattage) is the product of voltage and amperage. Two 12-V batteries can be compared on the basis of ampere-hours alone, as can two 24-V batteries. But one should remember that a 30-Ah, 12-V battery (360 Wh) contains half the power of a 30-Ah, 24-V battery (720 Wh).

If the power desired for a specific job is not available in a single battery, often two or more batteries are connected in parallel. Connecting batteries in parallel will increase the available amperage capacity and maintain a constant voltage.

Another rating applied to storage batteries is known as the 5-min discharge rate. This rating is based on the maximum current a battery will deliver for a period of 5 minutes at a starting temperature of 80°F [26.7°C] and a final average voltage of 1.2 V per cell. This applies only to lead-acid batteries. The 5-min rating gives a good indication of the battery's performance for the normal starting of engines.

When a fully charged battery is connected to a very heavy load, it apparently becomes discharged in a short time. A good example of this is the starting of an engine on a very cold morning. After turning the engine for a short time, the starter may refuse to operate. This failure occurs largely because the heavy flow of current has caused a rapid sulfation of the active material on the surface of the plates, while the material inside the plates is still in a charged condition. The lead sulfate on the surface of the plates offers a high resistance to the flow of current; hence the voltage drop within the battery becomes so large that there is not sufficient voltage to continue driving the heavy load. If the battery is allowed to remain idle for a time, it will again be able to deliver a substantial load current.

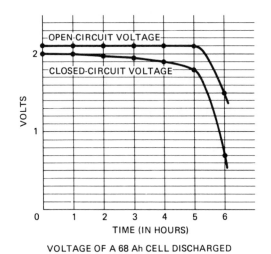

VOLTAGE OF A 68 Ah CELL DISCHARGED
AT THE RATE OF 13.6 A FOR 6 h

FIGURE 3–30 Discharge characteristics of a lead-acid cell.

Capacity Loss Due to Low Temperatures

Operating a storage battery in cold weather is equivalent to using a battery of lower capacity. For example, a fully charged battery at 80°F [26.6°C] may be capable of starting an engine twenty times. At 0°F [17.8°C], the same battery may start the engine only three times.

Low temperatures greatly increase the time necessary for charging a battery. A battery that could be recharged in 1 h at 80°F may require approximately 5 h of charging when the temperature is 0°F. These effects on a battery's capacity are caused by the slow chemical reactions created by the cold temperatures.

NICKEL-CADMIUM STORAGE BATTERIES

Aircraft nickel-cadmium storage batteries are constructed of wet cells. One advantage of the nickel-cadmium cell is that it contains a greater power-to-weight ratio than a lead-acid battery. Also, the CCV of a nickel-cadmium battery remains nearly constant during the entire discharge cycle. Nickel-cadmium batteries are much more costly than a typical lead-acid battery and therefore are usually found on turbine-powered aircraft. The extra capacity available from a nickel-cadmium battery will help prevent a hot start of the turbine engine, and thus the extra cost is justified. Nickel-cadmium batteries are often referred to as "ni-cad" (short for nickel-cadmium) batteries. Much of the material in this chapter, both general and specific, concerning nickel-cadmium batteries has been taken from the *Marathon Battery Instruction Manual* courtesy of the Marathon Power Technologies Company.

Cell and Battery Construction

The nickel-cadmium cell is a vented cell similar to that of a lead-acid battery. The cells are placed in an insulated metal or plastic case in proper order and then connected in series by the cell connectors. The end cells may be connected to external posts or to a quick-disconnect unit. A complete battery is illustrated in Figure 3–31.

Most nickel-cadmium aircraft batteries contain vented cell caps as illustrated in Figure 3–32. They are designed so that the rubber seal will allow any expanding gases inside the cell to escape. However, if the gas inside the cell is not under pressure, the rubber seal will close against the cap's outlet and seal the electrolyte from accidental spills during flight. The vents of each cell are required in case the battery becomes overcharged. Only at this time does a ni-cad cell emit gas. So little vapor is emitted during normal operation that the battery box is typically sealed. No venting is required; any and all gases remain inside the battery case, which is designed to accept and hold the gas that may be emitted from the cells.

Each cell of the battery consists of negative and positive plates, separators, electrolyte, cell container, cell cover, and vent cap. The plates are made from sintered metal plaques impregnated with the active materials for the negative and positive plates. The plaques are made of nickel carbonyl powder sintered at a high temperature to a perforated nickel-

FIGURE 3–31 Nickel-cadmium aircraft battery. *(Marathon Battery Company)*

plated steel base or a woven nickel wire base. This results in a porous material that is 80 to 85 percent open volume and 15 to 20 percent solid material. The porous plaque is impregnated with nickel salts to make the positive plates and cadmium salts to make the negative plates. After the plaques have absorbed sufficient active material to provide the desired capacity, they are placed in an electrolyte and subjected to an electric current, which converts the nickel and cadmium

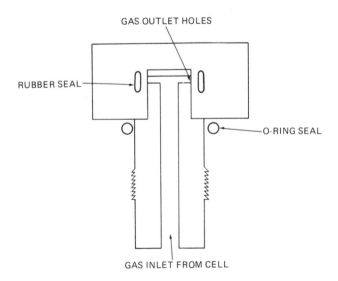

FIGURE 3–32 Nickel-cadmium vented cell cap.

salts to the final form. The plaques are then washed and dried and cut into plates. A nickel tab is welded to a corner of each plate and is the means by which the plates are joined into plate groups.

The **separator** in a nickel-cadmium cell is a thin, porous multilaminate of woven nylon with a layer of cellophane. The separator serves to prevent contact between the negative and positive plates. The separator is continuous and is interposed between the plates as each successive plate is added to the plate pack or stackup. The cellophane portion of the separator acts as a barrier membrane to keep the oxygen that is formed at the positive plates during overcharge from reaching the negative plates. Oxygen at the negative plates would recombine with cadmium and create heat that might lead to **thermal runaway;** thus the cellophane serves to inhibit thermal runaway. Thermal runaway is a condition where the battery chemicals overheat to such a degree that the battery can be destroyed or even explode.

Thermal runaway or **vicious cycling** does not accurately describe the overheating of a nickel-cadmium battery. The overheating of a nickel-cadmium battery is not self-sustaining and *can* be controlled. An aircraft charging system perpetuates this condition, and it can be stopped by isolating the battery from the charging system. A nickel-cadmium battery cannot overheat unless something internal is causing its temperature to rise. Nickel-cadmium batteries typically overheat from improper maintenance or improper use. Improper use of the battery usually occurs by drawing too much current from the battery during multiple engine starts in a short period of time. The heat retained within the battery weakens and destroys the material that separates the positive and negative plates in the cells. When this occurs, the cell's electrical resistance lowers and decreases the cell's voltage. The battery receives excessive amounts of charge current from the aircraft's constant potential charging system, which generates a great deal of heat within the battery. Temperature and/or overcurrent sensors are required by the FAA on all nickel-cadmium batteries used for engine starting. These sensors are connected to warning indicators, which allow the pilot to take corrective measures in case of extreme battery stress. Typically, if the battery reaches a thermal runaway condition, the pilot must disconnect the battery from the electrical system and land the aircraft as soon as practical.

The **electrolyte** for a nickel-cadmium battery is a solution of 70 percent distilled water and 30 percent potassium hydroxide, which gives a specific gravity of 1.3. **Specific gravities** for nickel-cadmium batteries may range between 1.24 and 1.32 without appreciably affecting battery operation. The electrolyte in a nickel-cadmium battery does not enter into the charge-discharge reaction; this is the reason you cannot determine a nickel-cadmium battery's state of charge by testing the electrolyte's specific gravity with a hydrometer. There is no change in the specific gravity of a nickel-cadmium battery.

The cell container consists of a plastic cell jar and a matching cover, which are permanently joined at assembly. It is designed to provide a sealed enclosure for the cell, preventing electrolyte leakage or contamination. The vent cap is

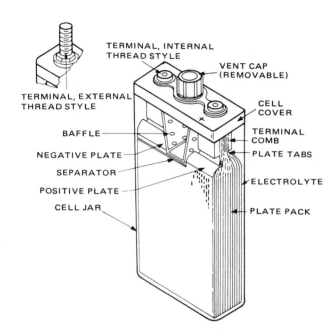

FIGURE 3–33 Nickel-cadmium cell components. *(Battery Division, General Electric Company)*

mounted in the cover of the cell and is constructed of plastic. It is fitted with an elastomer (flexible rubber or plastic) sleeve valve to permit release of gases as necessary, especially when the battery is on overcharge. The cap can be removed whenever necessary to adjust the electrolyte level. The vent valve automatically seals the cap to prevent leakage of electrolyte.

A cell core and the assembly of a complete cell for a nickel-cadmium battery are shown in the drawings of Figure 3–33.

Another type of complete nickel-cadmium cell is illustrated in Figure 3–34. The cell is assembled by welding the tabs of the plates to their respective terminal posts. The terminal and plate-pack assembly is then inserted into the cell container, and the baffle, cover, and terminal seal are

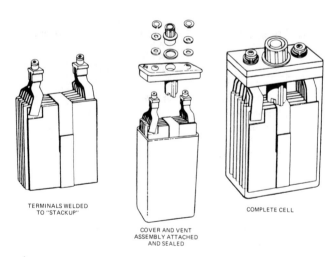

FIGURE 3–34 Nickel-cadmium cell construction. *(Marathon Battery Company)*

installed. The cover is permanently joined to the jar to produce a sealed assembly.

The cells are assembled into a battery container and connected together with stainless steel conductor links. Usually, 19 or 20 cells (depending on the total voltage required) are assembled into a battery with the correct polarity so that each cell is in series. The battery container is typically made of stainless steel, carbon steel, or a fiberglass material. All metal cases require an internal insulator. Stainless steel cases use a plastic liner, while most carbon steel cases are coated with an alkali-resistant epoxy that contains high dielectric properties. A typical battery assembly is illustrated in Figure 3–35.

Principles of Operation

The advantage of a nickel-cadmium battery is that the active materials of the cell plates change in oxidation state only, not physical state. This means that the active material is not dissolved by the electrolyte of potassium hydroxide. As a result,

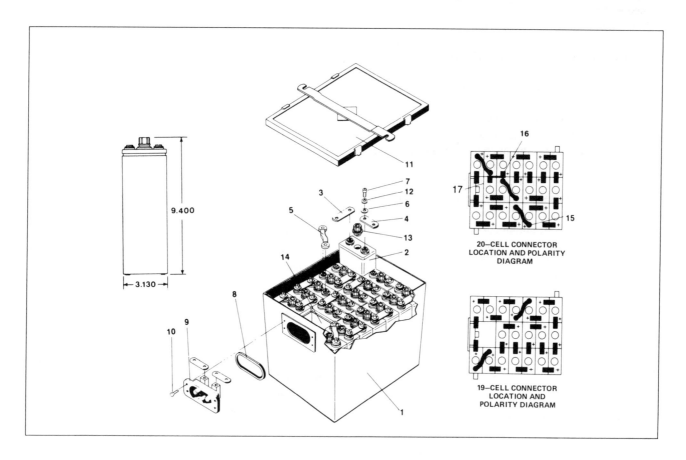

Item	Description	Part Number CA-5, CA-5-1 CA-5-20	Part Number CA-5H CA-5H-20, CA-16	Quantity
1	Can Assembly	26604	26604	1
2	Cell Assembly	36M220	36H120	19 (20)
3	Connector	16102-6	16102-6	12
4	Connector	16102-7	16102-7	6
5	Connector	16167-1	16167-3	2
6	Belleville Spring	16128-1	16128-1	40
7	Socket Head Cap Screw	10488-20	10488-20	40
8	Rectangular Ring	24583	24583	1
9	Receptacle Assembly	16163-7	16163-7	1

Item	Description	Part Number CA-5, CA-5-1 CA-5-20	Part Number CA-5H CA-5H-20, CA-16	Quantity
10	Phillips Head Screw	23084-1	23084-1	4
11	Cover Assembly	23147-3	23147-3	1
12	Double "D" Washer	23591-1	23591-1	40
13	Filler Cap & Vent Assembly	18318-1	16934-1	1
14	Spacer	27292	27292	1
15	Connector	16167-2*	16167-3*	3
16	Connector	25091*	25091*	1
17	Spacer	27291*	27291*	1
19-cell to 20-cell Conversion Kit – part number 29005				

NOTE: The batteries are 19— and 20— cell versions of Marathon's 36–40 Ah units. This figure does not necessarily represent the design of other batteries produced by Marathon.

*Used only on 20—cell batteries.

FIGURE 3–35 A typical battery assembly. *(Marathon Power Technologies)*

the cells are very stable even under a heavy load, and the chemicals last a long time before the battery requires replacement.

As previously explained, the active material of the negative plate of a charged nickel-cadmium cell is of metallic cadmium (Cd), and the active material of the positive plate is nickel oxyhydroxide (NiOOH). As the battery discharges, hydroxide ions (OH) from the electrolyte combine with the cadmium in the negative plates, and electrons are released to the plates. The cadmium is converted into cadmium hydroxide [$Cd(OH)_2$] during the process. At the same time, hydroxide ions from the nickel oxyhydroxide positive plates go into the electrolyte, carrying extra electrons with them. Thus electrons are removed from the positive plates and delivered to the negative plates during discharge. The composition of the electrolyte remains a solution of potassium hydroxide because hydroxide ions are added to the electrolyte as rapidly as they are removed. For this reason the specific gravity of the electrolyte remains essentially constant at any state of discharge. It is, therefore, impossible to use specific gravity as an indicator of the state of charge.

When a nickel-cadmium battery is being charged, the hydroxide ions are forced to leave the negative plate and enter the electrolyte. Thus the cadmium hydroxide of the negative plate is converted back into metallic cadmium. Hydroxide ions from the electrolyte recombine with the nickel hydroxide of the positive plates, and the active material is brought to a higher state of oxidization called nickel oxyhydroxide. This process continues until all the active material of the plates has been converted. If charging is continued, the battery will be in overcharge, and the water of the electrolyte will be decomposed by electrolysis. Hydrogen will be released at the negative plates, and oxygen will be released at the positive plates. This combination of gases is highly explosive, and care must be exercised to avoid any possibility of ignition of the gases.

Water is lost from the electrolyte during overcharge because of electrolysis. Some water is also lost by evaporation and entrapment of water particles during the venting of cell gases. By theory, 1 cubic centimeter (cm^3) of water will be lost by electrolysis for every 3 h of overcharge. In practice the loss is not this high, because there is some recombination of hydrogen and oxygen within the cell.

The separator acts as an electrical insulator and a gas barrier between the negative and positive plates. The nylon fabric provides separation to prevent contact between plates of opposite polarity. The cellophane acts as a gas barrier to prevent oxygen from reaching the negative plates. Oxygen reaching the negative plates will cause the plates to heat, with resulting plate damage, as explained earlier.

Voltage Rating

The OCV of 1.28 V is consistent for all nickel-cadmium vented cells, regardless of cell size. The OCV does vary slightly with temperature and elapsed time since the battery's last charge. Immediately after charge, the OCV may reach 1.40 V; however, it soon lowers to between 1.35 and 1.28 V. A 20-cell ni-cad battery would, therefore, have an OCV be-

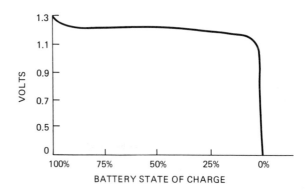

FIGURE 3–36 Typical discharge voltage curve under moderate load. *(Marathon Power Technologies)*

tween 25.6 and 27 V. The voltage obtained by a cell immediately after charging is typically slightly higher than the average OCV. A nickel-cadmium battery may reach 27.5 V immediately after charging, or 1.5 V per cell for a 19-cell battery. Near the end of the charge cycle, the same battery may reach 28.5 V if the charging current is still applied. This voltage diminishes quickly after the battery is removed from the charger and will soon reach near 25 V.

The CCV of a vented-cell nickel-cadmium battery ranges between 1.2 and 1.25 V. This voltage will vary depending on the battery temperature, the length of time since the battery's last charge, and the discharge current applied. The CCV of a nickel-cadmium cell remains nearly constant under moderate load until the cell is near the completely discharged state. Figure 3–36 illustrates the CCV of a nickel-cadmium cell.

Capacity and Internal Resistance

A nickel-cadmium battery has tremendous peak power and delivers far more power than a lead-acid battery of the same size and weight. The large amount of instantly available power produced by a nickel-cadmium battery is why it is so well suited for starting turbine engines. The **capacity of a nickel-cadmium battery** is a function of the total plate area contained inside the cells (more plate area, more capacity). Most ni-cad batteries are designed for 24-V systems with a capacity between 22 and 80 Ah. The ampere-hour rating is determined at a 5-h discharge rate unless otherwise denoted.

The capacitance of any battery is partially a function of that battery's internal resistance. The **internal resistance of most vented nickel-cadmium cells** is very low (less than 1 mΩ per cell), which allows these cells to maintain a high discharge current and still maintain acceptable voltage levels. The low internal resistance of a nickel-cadmium battery allows it to recharge very rapidly. This resistance in part results from the large surface area of active materials made available through the use of a highly porous plate.

The output of a nickel-cadmium battery is relatively constant, even in harsh operating conditions such as very cold weather. The optimum temperature range is between 60 and 90°F; above or below these values, total capacity will diminish slightly, as illustrated in Figure 3–37.

Nickel-Cadmium Storage Batteries **49**

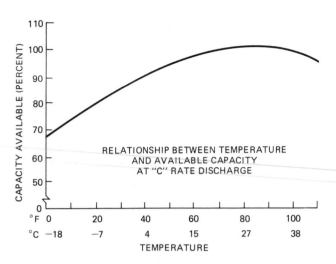

FIGURE 3–37 Relationship between temperature and available capacity. *(Marathon Power Technologies)*

6. Check the cell electrolyte level for proper amounts. If the battery is overfilled and spillage has occurred, the battery should be removed, discharged and then disassembled for repair. If a low level of electrolyte is found, add clean distilled water only after the battery has been idle for at least 2 h after charging. **Never add water to a discharged battery or a battery of unknown charge.** The electrolyte level increases significantly during charging; therefore, if water is added before charging, an overfill situation is likely.

Nickel-Cadmium Battery Reconditioning and Charging

The reconditioning of nickel-cadmium aircraft batteries is usually performed between 100 and 300 flight hours. The exact reconditioning time period depends mainly on the aircraft starting procedures, operating temperatures, and generator voltage regulator setting. These factors will also determine the frequency of water additions to the battery cells.

Reconditioning of the battery is necessary to prevent any **cell imbalance,** which may result in a temporary loss of battery capacity. Cell imbalances occur during recharging by an aircraft's constant potential charging system. This "out-of-balance" condition is caused by differences in temperature, or charge efficiency or by varying self-discharge rates in cells. Low electrolyte levels also contribute to a loss in capacity. To ensure optimum performance and battery life, any cell imbalance should be corrected through reconditioning.

Any service area used to recondition nickel-cadmium batteries should be separated from the service area for lead-acid batteries. The chemical materials of a lead-acid battery will neutralize a nickel-cadmium battery and vice versa. Two separate ventilated rooms are recommended for servicing the two types of batteries. Also, the tools used for maintenance should never be interchanged; interchanging tools may cause damage to the nickel-cadmium cell that is due to partial electrolyte neutralization.

The typical reconditioning procedures include a battery inspection as previously stated, a battery discharge, disassembly, and cleaning or repairing as needed. Finally, the battery is reassembled and recharged. **During reassembly, always observe correct cell polarity and always use the proper torque values on each cell connector bolt or nut.** An improperly torqued, dirty, or corroded cell connector is likely to cause battery failure. To correct a cell imbalance during reconditioning, the battery is typically discharged to zero capacity and then recharged. This process is often called a battery **deep cycle.** The specifics of this deep cycle vary between batteries, so always refer to the proper maintenance data.

If the battery is received in a charged condition, an **electrical leak check** should be performed. Prior to discharge, this test detects current leakage from the cells to the battery case. A leakage exceeding 50 mA measured from any positive cell connection to the case is usually excessive. If the battery arrives in a discharged state, a leak test should be performed after charging. The current flowing from the cell to the battery case (electrical leakage) is usually caused by

---

NICKEL-CADMIUM BATTERY MAINTENANCE PROCEDURES

The nickel-cadmium storage battery requires specific maintenance procedures; always follow the battery manufacturer's recommendations during service. The following general guidelines illustrate typical maintenance practices, including proper log entries. Each battery should have its own specific maintenance record. This will aid in the isolation of defects and will help to ensure optimum battery performance.

*CAUTION: During all maintenance procedures, follow the precautions previously discussed in this chapter.

Battery Inspection

Every aircraft maintenance schedule will specify a battery inspection period. This schedule should not exceed 50 flight hours for new batteries to ensure proper battery and aircraft compatibility and operation. After a few months, the inspection periods can be lengthened. Before removing the battery from the aircraft, inspect the following and repair as needed:

1. Inspect the battery case for cracks, distortion, or other damage.
2. Inspect the vent system (if installed) for proper airflow.
3. Inspect the cells and clean as needed. Often potassium carbonate will deposit as a white powder on the top of the cells. This should be removed with a nonmetallic brush or damp cloth. If excessive deposits are present, suspect overcharging or leaking cells, and remove and clean the cells individually.
4. Inspect the cell connectors for corrosion, cracks, and overheating. If these problems exist, discharge and disassemble the battery in order to repair the damage.
5. Inspect the cell caps for proper O-ring and vent sleeve condition. Wash any dirty cell caps in clean, warm water.

excessive liquid on top of or around the cells. Therefore, if an excessive electrical leakage is detected, remove, clean, and dry all cells and the battery case. During this procedure, inspect each cell for a liquid leak that may have caused the excess liquid around the cells. Any cells from which electrolyte is found to be seeping or leaking should be replaced.

Typically, to discharge the battery completely during reconditioning, discharge equipment made specifically for nickel-cadmium batteries should be used. Once the battery cells reach 0.5 V or less, shorting clips should be placed across each cell. By shorting the positive to negative terminals of each cell, the battery will become completely discharged.

Recharging the battery during reconditioning can be performed by a constant-current or constant-voltage charger. In either case, the battery will require a charge of 120 to 140 percent of its 5-h capacity rating. If a constant-current charger is used for recharging a 40-Ah battery, the applied charge should be 8 A for 7 h. This is determined as follows: 40 Ah = 8 A for 5 h (8 A × 5 h = 40-Ah) times 140 percent = 8 A for 7 h = 56 Ah.

Two common charger/analyzer systems are the Marathon PCA-131 and the Christie RF80-K. The Marathon PCA-131 charger/analyzer was designed to charge and analyze nickel-cadmium batteries automatically and simplify reconditioning procedures. This charger/analyzer features a Go, No-Go indication of battery condition. The correct charge and discharge current is preselected with the setting of the switch position. The battery can be left unattended during charge, and the system automatically adjusts for changes in line voltage. It will automatically terminate the discharge if the average battery voltage falls below a preselected voltage.

The Christie RF80-K (Figure 3–38) performs the same charger/analyzer functions as the PCA-131; however, this charger reduces maintenance time with its DigiFLEX Pre-Analysis system. The technology is similar to that used to recharge flashlight-type dry-cell ni-cad batteries. This system is designed to charge aircraft batteries in 1 h or less, increase capacity and life, and rejuvenate batteries that were rejected after conventional charging. The Christie RF80-K is also designed to overcome any cell imbalance problems during reconditioning of the battery.

Low-current charging during the last 15 percent of the ni-cad's charge is the major cause of cell imbalance. The ReFLEX system uses negative pulses during the charging cycle to remove internal gases that form on the cell plates. The gases on the cell plates increase the current density and result in the heating of the battery. The Christie charger/analyzer makes use of microcontroller technology to quickly analyze and charge ni-cad batteries.

In general, always follow the manufacturers' recommendations when charging batteries. All vented nickel-cadmium batteries will charge at approximately 140 percent of their total capacity. To achieve this value, the length of charge is typically increased by 40 percent above the total time required for 100 percent of the battery's ampere-hour rating.

Always observe precautions when charging nickel-cadmium batteries. Avoid accidental short circuits. The exposed cell connectors are very vulnerable to shorts from dropped tools or metallic jewelry.

*CAUTION: Always use tools that are insulated when servicing batteries. Always remove all metallic jewelry prior to working on or around batteries. Always wear eye protection during battery service. Before charging, always inspect the cell connectors for cleanliness, physical condition, and proper torque value. Repair any improper conditions.

FIGURE 3–38 A typical nickel-cadmium battery charger/analyzer. *(The Christie Corporation)*

Constant-Voltage Chargers

Just as with lead-acid batteries, a **constant-voltage charger** will supply a constant voltage to the battery during charging. The current supplied by this type of charger is high during the start of the charge cycle and lowers as the battery reaches a fully charged state. The exact current flows will be a function of the capacity of the charger, temperature, and the battery's state of discharge.

The correct voltage setting is very important when using a constant-voltage charger. The charging equipment must be set and regulated to ensure a complete battery charge without battery overcharge. Too low a charge voltage will not charge or will undercharge the battery. Too high a charge voltage will overcharge the battery and may damage the cells by thermal runaway.

Constant-Current Chargers

Constant-current charging of nickel-cadmium batteries is recommended to ensure better cell balance and a total battery charge, as well as to prevent the possibility of thermal runaway. However, constant-current charging typically requires a longer charging time and creates a greater water loss during overcharge than constant-voltage charging.

Once again, the manufacturer's charging data must be strictly adhered to while using constant-current chargers. Generally speaking, completely discharged batteries are charged at their ampere-hour capacity, 5-h rate for 7 h. This will apply approximately 40 percent "extra" current during charging to allow for battery charging inefficiencies.

Both the charging equipment and the battery should be monitored periodically during the charge cycle, especially during the first and last hour of charge. During the initial portion of the charge, the charge current and voltage must be monitored to ensure that they are the correct values. Until the battery reaches its maximum charge voltage, the charge current may not stabilize. During the last hour of charge, the battery should be checked for excessive boiling of the electrolyte; this is an indication that the battery has reached full charge and should be removed from the charger.

There are several variations of the two charging methods just discussed that may be used under limited applications. For the most part, however, constant-voltage or constant-current charging methods are employed. For long-term storage of nickel-cadmium batteries, a float or trickle charge can be used. These charging methods ensure that the battery will remain in a fully charged condition during storage. Both the float and trickle charge supply a very low current flow to compensate for the battery's normal self-discharge loss.

Foaming Electrolyte

If **foaming** occurs while any charging method is being used, always monitor the battery or cell in question. If the cell recently received additional water, the electrolyte may foam during charging. This typically will diminish upon further use and does indicate a defective cell. However, if foaming continues and electrolyte spills from the vented caps continuously, the cell is probably contaminated with a foreign material. In this case the cell should be replaced.

Battery Storage

Unlike lead-acid batteries, nickel-cadmium batteries can be stored for long periods of time in a charged or discharged state without damage. This assumes that the battery is cleaned properly prior to storage to prevent excessive corrosion. At room temperature a charged nickel-cadmium battery will retain most of its power for 6 months in storage. If the battery is expected to be used immediately, however, it is suggested that a trickle charge (a very low current charge) be maintained while the battery is in storage. While the battery is on trickle charge, proper ventilation should be available to it, the electrolyte level should be monitored periodically, and lost water should be replenished.

If the battery is to be stored without replenishing its charge for a long period, it is recommended that the battery be completely discharged prior to storage. This includes the use of shorting clips to bring each cell to zero capacity. During storage, the main battery positive and negative terminals should then be shorted together to help prevent any cell imbalance. Prior to storage, place a light coat of nonconductive grease, such as petroleum jelly, over the cell hardware. This will inhibit corrosion.

To return a battery to service after storage, clean the battery, remove any shorting clips (if applicable), and recharge the battery at the proper voltage and current settings. Prior to charging, the cleaning process should ensure that the cell caps are free of potassium carbonate and that the vents will function properly.

INSTALLATION OF AIRCRAFT BATTERIES

Battery Compartment

The battery compartment in an airplane should be easily accessible so that the battery can be serviced and inspected regularly; it should also be isolated from fuel, oil, and ignition systems and from any other substance or condition that could be detrimental to its operation. Any compartment used for a storage battery that emits gases at any time during operation must be provided with a ventilation system. The inside of the compartment must be coated with a paint that will prevent corrosion caused by electrolyte.

The battery must be so installed that spilled electrolyte is drained or absorbed without coming into contact with the airplane structure. The shelf or base upon which the battery rests must be strong enough to support the battery under all flying and landing conditions. The battery must be held firmly in place with bolts secured to the aircraft structure. Metal-case batteries are held down by means of bolts that extend through ears on the battery cover. Nonmetallic batteries are held down by metal clamps that hook over the handles of the battery or over the edge of the battery case.

Batteries should not be located in engine compartments unless adequate measures are taken to guard against possible fire hazards and the injurious effects on a battery of excessively high temperatures. Battery manufacturers have determined that temperatures of 110 to 115°F [43 to 46°C] and higher are likely to cause rapid deterioration of the separators and plates. The critical temperature specified by the manufacturer should not be exceeded at any time. Forced ventilation of the battery compartment may be necessary to guard against excessive battery temperatures, and this can be provided by means of a tube leading from the slip stream into the container and a suitable vent tube leading out of it.

Battery Installation

Always perform a thorough battery inspection prior to installation in any aircraft. This inspection should include an electrolyte level check (remember, add water only after charging), a cell connector inspection, and a check of general battery condition. The battery quick-disconnect plug should be inspected on both the battery and the aircraft. If any pitting, corrosion, or looseness is detected, the plug should be replaced or repaired. Be sure both the aircraft and battery connectors have the same polarity.

If a nickel-cadmium battery is installed to replace an old lead-acid battery, always neutralize the battery compartment with a soda and water solution and dry the area completely. The compartment should then be painted with an alkali-resistant paint. Also ensure that the new battery will have proper ventilation to remove any heat that may be produced during battery use. A battery that fits too tightly into its compartment may overheat. Always check that the battery charging specifications and the aircraft charging system's voltage coincide. The nickel-cadmium battery must be charged at a specific voltage for proper operation.

Finally, upon installation connect any battery temperature monitors and inspect their systems according to the aircraft manual. Make sure that all areas requiring safety wire are properly secured. If the battery output terminals are uninsulated, they may require the installation of protective covers or insulating boots. These insulators will prevent the possibility of a short circuit, which could create a serious electrical failure or fire. After completing the installation, perform a battery operational check, including a battery engine start if applicable, and a charging system verification.

Ventilating Systems

Battery compartments (battery boxes) are typically vented to remove unwanted gases and/or heat produced by the battery during charging and discharging. Figure 3–39a illustrates a typical battery vent system found on light aircraft. These systems provide a consistent ventilation airflow created by a low pressure at the output tube while the aircraft is in flight. On older aircraft a **battery sump** may be used on the outlet side of the vent system, as shown in Figure 3–39b. The sump con-

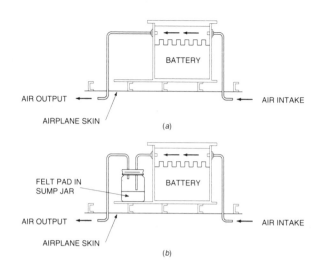

FIGURE 3–39 Battery ventilating systems.

tains a solution of bicarbonate of soda and water to neutralize the lead-acid battery fumes. The sump is then vented outside the aircraft. The acid fumes going into the sump are neutralized by the action of the soda solution; thus corrosion of the airplane's metal skin is prevented.

The ventilation systems for nickel-cadmium batteries are typically designed to remove heat from the battery compartment. The gas generated during operation is minimal, and the battery case is sealed to prevent leakage; consequently, removal of chemical gases is not typically necessary. Some nickel-cadmium battery compartments are therefore unventilated. Those systems that are vented for cooling often use ram air or even forced air regulated by a thermostatically controlled air valve. The air valve is closed if the battery temperature is below a certain level; the valve opens when the battery requires cooling.

During the inspection of any battery system, it is important to ensure that ventilation tubes remain unclogged. Compressed airflow or a water wash through the vent tubes will help ensure proper operation of the ventilation system.

Battery Cables

The electric leads to a battery in an airplane must be large enough to carry any load imposed on the battery at any time. They must be thoroughly insulated and protected from vibration or chafing and are usually attached to the airplane structure by means of rubber-lined or plastic-lined clamps or clips. Battery cables must be securely attached to the battery terminals; they are usually held in place as shown in Figure 3–40. A heavy metal lug is soldered or swaged to the end of the cable and then attached to the terminal by means of a wing nut with a flat washer and a lock washer. It must be noted that this is only one method for attaching battery terminals; others are also satisfactory.

Battery terminals must be protected from accidental shorting by means of a terminal cover. This may be a plastic or rub-

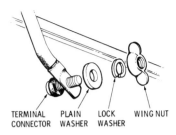

TERMINAL PLAIN LOCK WING NUT
CONNECTOR WASHER WASHER

FIGURE 3–40 Battery terminal connection

ber boot over the terminal, or the terminals may be contained within the protective battery box.

Quick-Disconnect Plugs

Quick-disconnect battery connectors are found on some lead-acid batteries and practically all nickel-cadmium batteries. The quick-disconnect consists of an adaptor secured to the battery case in place of the terminal cover and a plug to which the battery leads are attached. Two smooth contact prongs are screwed onto the battery terminals, and the plug is pulled into place on the battery by means of a large screw attached to a handwheel. This screw also pushes the plug off the terminals to disconnect the battery.

A popular battery connector is shown in Figure 3–41. The connector consists of two main assemblies: the terminal assembly, which is attached to the battery to serve as a receptacle, and the connector plug assembly, to which the battery cables are connected. The plug assembly is inserted into the receptacle on the battery and is seated firmly by means of the center screw (worm) in the plug.

The design of the contacts provides for many contact surfaces with the mating male pin, thus assuring a low-resis-

tance contact. The contacts are made of silver-plated soft copper wire and are designed to fit snugly onto the pins of the battery terminal. The pins and sockets of the connectors should be inspected at regular intervals. If loose connections, burned spots, corrosion, or pitting are noticed, the contacts should be replaced.

REVIEW QUESTIONS

1. Briefly describe a voltaic cell.
2. What is the difference between a primary cell and a secondary cell?
3. What voltage is developed by a carbon-zinc cell?
4. What is a dry cell?
5. Describe how various voltages are obtained in different cells.
6. What electrolyte material is used in an alkaline cell?
7. What are the active materials in a nickel-cadmium cell?
8. What are the active materials in a lead-acid storage cell?
9. Describe the construction of a lead-acid storage cell.
10. Describe a plate group.
11. What electrolyte is used in a lead-acid storage cell?
12. Why does the number of negative plates in a lead-acid cell exceed the number of positive plates?
13. What materials are used for separators?
14. What means is employed to prevent the sediment in the bottom of a cell container from short-circuiting the plates?
15. Explain the means used to prevent the spillage of electrolyte from an aircraft storage battery.
16. What is the purpose of antimony in the grid of a lead-acid cell?
17. How are aircraft storage batteries constructed to provide for elimination of explosive gases?
18. What determines the voltage of an aircraft storage battery?
19. What ratings are used to describe aircraft storage batteries?
20. What is the approximate open-circuit voltage of a fully charged lead-acid cell?
21. If a storage cell will deliver 20 A for 5 h, what is the ampere-hour rating?
22. Why will a lead-acid storage battery appear to be discharged after the application of a heavy load for a short time but will again deliver power after the load has been disconnected for a few minutes?
23. What occurs with respect to ampere-hour rating when the discharge rate is increased above that used to establish the rating?
24. What is the most common method for determining the state of charge of a lead-acid battery?
25. Give the specific gravity range for a lead-acid cell.
26. When must corrections be made if testing a lead-acid cell with a hydrometer?
27. Under what condition may new electrolyte be added to a lead-acid cell?

FIGURE 3–41 Elcon battery connector. *(Icore International)*

28. How does temperature affect a hydrometer reading?

29. Describe the method for testing a lead-acid cell under load by means of a voltmeter.

30. Give the principal safety precaution that must be observed in working with lead-acid storage batteries.

31. If an aircraft has a dead battery, why is it considered improper procedure to jump-start that aircraft?

32. Explain the difference between constant-voltage charging and constant-current charging.

33. What type of charging is employed in an aircraft electric system?

34. What hazards exist with respect to lead-acid batteries during charging?

35. Describe a battery compartment in an aircraft.

36. How are explosive gases from a battery eliminated from an aircraft?

37. Discuss the connection of battery cables to a battery.

38. What is a typical inspection period for a lead-acid battery?

39. Explain the detrimental effects of short circuits that may occur in careless handling of batteries.

40. How would you adjust the specific gravity for a lead-acid battery?

41. What precautions should be taken if charging a battery while it is still installed in the aircraft?

42. What electrolyte is used in nickel-cadmium batteries?

43. Describe the construction of a nickel-cadmium cell.

44. Describe the means used to separate the plates in a nickel-cadmium cell.

45. What is the function of the cellophane strip in the separator?

46. Explain the operation of a nickel-cadmium cell.

47. What are the factors affecting the performance of nickel-cadmium batteries?

48. Describe the vented cap system for expelling gases from a nickel-cadmium battery.

49. What is the specific gravity of a nickel-cadmium aircraft battery?

50. What is the danger caused by loose cell connectors in a nickel-cadmium battery?

51. What are satisfactory charging methods for nickel-cadmium batteries?

52. When a completely discharged nickel-cadmium battery is being charged, how much electric energy must be returned to it as a percentage of its ampere-hour rating for a full charge?

53. What affects the open circuit voltage of a nickel-cadmium battery?

54. What is the typical closed voltage of a nickel-cadmium battery?

55. Explain a thermal runaway.

56. What condition of charge is most suitable for nickel-cadmium batteries that are to be stored?

57. What is the internal resistance of a typical nickel-cadmium battery?

58. What is meant by an electrical leak check?

59. List precautions that should be observed in handling nickel-cadmium batteries.

60. Why should service areas for lead-acid and nickel-cadmium batteries be separated?

61. What is capacity reconditioning, and why is it performed?

62. How is a capacity check performed?

63. What conditions are observed in making an inspection of nickel-cadmium batteries?

64. How should a nickel-cadmium battery's potassium carbonate deposits be cleaned?

65. Describe the procedure for reconditioning a nickel-cadmium battery.

66. What practice should be observed in installing and tightening cell connectors?

67. What final inspections should be made when the assembly of a nickel-cadmium battery has been completed and the battery is to be installed in the aircraft?

Electric Wire and Wiring Practices 4

The electric wiring in an aircraft must be properly installed and maintained in order to ensure the safety of the aircraft's passengers and crew. On light, single-engine aircraft there is a relatively small amount of wire; on large commercial or complex military aircraft, there are literally miles of wire controlling almost every facet of flight. On any civilian aircraft operated within the United States, the electric wiring must be installed and maintained according to the FAA guidelines. This chapter describes these FAA specifications and shows their relationships to aircraft wiring practices.

CHARACTERISTICS OF ELECTRIC WIRE

There are several conditions to be considered when choosing an aircraft electric wire. The design temperature, flexibility requirements, abrasion resistance, strength, insulation, electrical resistance, weight, and applied voltage and current flow all affect the wire selection. These factors will determine the type of conductor and insulation necessary for a given installation. Most aircraft wire is made with a stranded copper conductor, either 7 or 19 strands for small wire and 19 or more for larger wire. The use of stranded, or twisted, wire increases the flexibility of the conductor, thus decreasing the chance of fatigue failure. Flexible wire is made of several small strands; less flexible wire is made of fewer, coarser strands. Solid wire (one single strand) is very inflexible and may only be used in limited areas of the aircraft.

Copper conductors are coated to prevent oxidation and to facilitate soldering. Tinned copper wire is generally used for installations where temperatures do not exceed 221°F [105°C]. Silver-coated copper wire is used for temperatures up to 392°F [200°C], and nickel-coated copper wire must be used at temperatures between 392 and 500°F [200 and 260°C]. This coating becomes quite apparent when viewing stripped wire. The copper strands are tin or silver in color (not copper). This is due to the thin coating applied to each copper strand. Under certain conditions, aluminum wire may also be used under 221°F [105°C].

Any type of single conductor surrounded by insulation is usually referred to as a **wire.** A **cable** is any group of two or more conductors separately insulated and grouped together by an outer sleeve. One cable can be routed through the aircraft and be used for several circuits. A cable's primary disadvantage is created by the inability to repair or replace a single wire. Figure 4–1 illustrates the differences between wire and cable.

Approved wire insulations include polyvinylidene fluoride (PVF), fluorinated ethylene propylene (FEP), polytetrafluoroethylene (TFE), ethylene tetrafluoroethylene copolymer (ETFE), and glass braids. The TFE-type insulations are often sold under the trade name Teflon or Tefzel. In any case, the type of insulation is governed by operating temperature, insulation resistance, abrasion resistance, chemical resistance, and strength. Design engineers determine the requirements and specify the type of wire that will meet the requirements for each circuit.

In addition to the basic insulation of electric wire, an outer sheath is often applied to protect the wire from abrasion. This

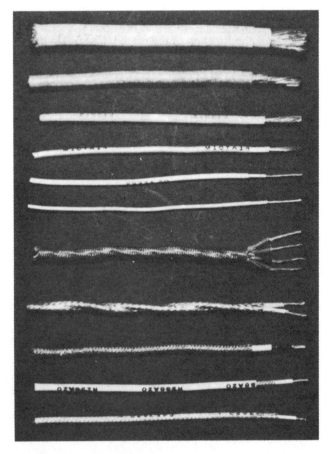

FIGURE 4–1 Typical electrical wire and cable. *(Prestolite)*

sheath may use the appropriate material reinforced with glass fiber.

Typical specification numbers for approved aircraft wire are MIL-W-5086, MIL-C-7078, MIL-W-22759, MIL-W-81381, MIL-C-27500, and MIL-W-81044. These standards specify the type of conductor, the conductor plating (if any), and the type of insulation. The standards give specific operating temperature ranges, voltage levels, and general application guidelines. All aircraft wire must meet the appropriate standard to be considered airworthy. Always install wire according to current manufacturers' data.

The specification number for aluminum aircraft wire is MIL-W-7072. When copper wire is replaced with aluminum wire, the aluminum wire should be two sizes larger than the copper wire because of the greater resistance of aluminum wire. Also, aluminum wire smaller than size 6 must not be used on aircraft.

Characteristics of some typical aircraft wires are as follows:

MIL-W-22759/1
Silver-coated copper conductor, TFE fluorocarbon and glass insulator, 600 V, 260°C
MIL-W-22759/41
Nickel-coated copper conductor, ETFE dual insulator, 600 V, 105°C
MIL-W-22759/16
Tin-coated copper conductor, ETFE insulator, 600 V, 150°C
MIL-W-22759/19
Silver-plated, high-strength copper alloy conductor; ETFE insulation; 600 V, 150°C
MIL-W-81044/9
Tin-coated copper conductor, Ployalkene-reinforced PVF insulation, 600 V, 150°C

The wires and cables described above are a few that are approved for aircraft use. Other types are also approved and are selected by engineers to meet certain specifications as required by the circuit design.

Some circuits in an airplane require the use of **shielded cable** to eliminate radio interference or to prevent undesirable voltages from being induced in the circuit. In this case the wire is manufactured with an outer metal sheath (shielding) of woven wire over the insulation. The center wire or wires are usually connected to the positive voltage of the circuit, while the outer shielding is typically grounded to the aircraft structure, or negative voltage of the circuit. The center wires are always insulated from each other and from the outer shielding. In some cases the outer shielding may be insulated with an appropriate material. A special application of shielded wire containing a solid copper conductor is called **coaxial cable.** Coaxial cable is commonly used for connecting an antenna to a radio receiver or transmitter.

Whenever possible, wires should be routed in areas of the aircraft that are not subject to extreme heat. Some electric devices, however, such as engine exhaust temperature monitors, must operate at elevated temperatures. Electric wiring or cable used in areas where high temperatures exist must have heat-resistant insulation. Fiberglass, asbestos, Teflon, and silicone are commonly used high-temperature insulators. During specific high-temperature installations, one should always follow the manufacturers' recommendations pertaining to the wire and its installation.

Electric wire used on aircraft is either white, stamped with identifying code numbers, or color-coded to allow the technician to identify specific wires. In either case, the wiring diagram of the maintenance manual will identify which wire number, or color, is connected to which circuit. This becomes very important when troubleshooting electric circuits containing several wires. Wire identification codes are discussed later in this chapter.

Wire Size

The wire used for aircraft electrical installations is sized according to the **American Wire Gage (AWG).** The size of the wire is a function of its diameter and is indicated by a unit called the **circular mil.** One circular mil is equal to the cross-sectional area of a 1-ml [0.001-in.] diameter wire, measured in thousandths of an inch. To determine the size in circular mils of a wire, simply square the wire's diameter measured in thousandths of an inch; for example, the size of a wire that is 0.025 in. in diameter is 625 circular mils. It is calculated as follows:

$$0.025 \text{ in.} = 25 \text{ thousandths of an inch}$$
$$25^2 = 625 \text{ circular mils}$$

The **square mil** is the unit of measure for rectangular conductors, such as bus bars or terminal strips. One square mil is the measure of a rectangular conductor having sides that are 0.001 in. in length. Figure 4–2 illustrates this concept. To simplify the wire size, the AWG standard has applied numbers to the various diameters of wire. Only even numbers are used. Small wires have higher numbers, typically starting at AWG 24; large wires have smaller numbers, down to AWG 4/0 (0000). A 20-gage wire is approximately 0.032 in. in diameter, and a 0 gage is approximately 0.325 in. in diameter. It should be noted that conductors of other sizes can be used on an aircraft if approved by the FAA.

To determine the size of any given wire, a wire gage tool may be used. As illustrated in Figure 4–3, the typical tool consists of a slotted piece of steel approximately 3 in. in diameter. Each slot, being a specific size, represents a given wire gage. The stripped portion of a wire is inserted into a slot, which fits snugly around the conductor. The wire size is marked adjacent to that slot.

There are two principal requirements for any wire carrying current in an aircraft electric system: The wire must be

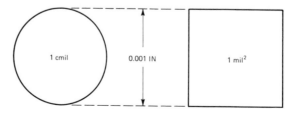

FIGURE 4–2 Circular mil and square mil dimensions.

FIGURE 4–3 A typical wire gage tool.

able to carry the required current without overheating and burning, and it must carry the required current without producing a voltage drop greater than that which is permissible for aircraft circuits. For the guidance of technicians engaged in the replacement or installation of electric wiring in civil aircraft, the FAA has prepared charts and tables setting forth the wire sizes needed to meet various conditions of installation and load. Table 4–1 gives the maximum allowable voltage drop between the bus and the electric components according to the nominal voltage of the system.

Table 4–1 establishes the maximum voltage drop that may occur between the power distribution bus and any unit of electric equipment. These voltage drops were considered when the FAA-approved electric wire charts were formed (Figure 4–4a and b). These charts are typically used when choosing a wire size for a given set of operating conditions if no manufacturers' data are available. The electric wire charts take into consideration both the maximum allowable voltage drop and the current-carrying capability of the wire for various situations. The charts are arranged for copper wire only and should not be used for the selection of aluminum wire. The following steps should be employed when using the electric wire charts:

1. Determine which chart to use for the application being considered. The chart in Figure 4–4a is used to determine wire sizes for circuits that have a **continuous** flow of current. A continuous flow is considered to be any circuit that carries current for a period longer than 2 min. The continuous flow conductor chart has two limiting curves, **curve 1** and **curve 2.**

Curve 1 is used when the wire will be installed in a conduit or bundle. Curve 2 is used when the wire will be installed as a single wire in free air. The two curves are used to distinguish between conduits or bundles and single wires because of the heat generated by the current flow. Wires in free air can expel excess heat faster and may therefore be approved for applications using a smaller wire.

The chart in Figure 4–4b is used to determine wire sizes for circuits that have an **intermittent** flow of current. Intermittent circuits carry current for intervals of 2 min or less. This chart contains the intermittent flow curve, **curve 3.**

2. After determining which chart to use, note the circuit's voltage at the top of the left-hand side of the chart. Choose the correct value for your applications, either 200, 115, 28, or 14 V.

3. Note the total wire length in the voltage column (left side of the chart). This will determine the correct horizontal line used to find the wire size. The correct horizontal line is just to the right of the wire-length value. (*Note:* The total wire length is considered to be from the circuit bus to the load. Do not assume that a single wire segment is the entire length of the circuit's wire.)

4. Locate and note the circuit's maximum current flow on the diagonal lines of the chart. (*Note:* The maximum current flow for any circuit is determined by the current rating of the circuit breaker or fuse, not the current draw of the load.)

5. Find the intersection of the diagonal (amperage) line and the horizontal (wire length) line. If the intersection is above the curve that depicts the wire installation, find the wire size using this intersection. Move straight down from this intersection, and read the wire size on an adjacent vertical line.

6. If the wire size is between two vertical lines (two wire sizes), always choose the largest wire (smallest number).

7. If the selected horizontal line intersects curve 1, 2, or 3 before intersecting the correct diagonal line, follow the *curve* to the right until it intersects the appropriate diagonal (amperage) line. At the intersection of these diagonal and *curve* lines, move straight down to find the wire size required for your circuit. As stated in step 6, if the wire size is between two vertical lines, always choose the largest wire.

8. In general, always be conservative when choosing a wire size. A wire that is too large will not adversely affect a circuit. A wire that is too small could cause overheating or circuit failure.

In order to understand the use of the wire charts, study the following examples.

Example 1. Assume that we wish to install a wire in a bundle 30 ft [8.85 m] long to carry a continuous load of 10 A in a 28-V circuit.

Solution. First, select the continuous-flow conductor chart (Figure 4–4a). Locate the 30-foot length at the left side of the chart in the 28-V column. Then follow the line horizontally to the right from 30 ft until it intersects the diagonal line for 10 A. This intersection falls between the vertical lines numbered 14 and 16, so select No. 14 cable.

TABLE 4–1 Maximum allowable voltage drop.

Nominal system voltage	Allowable voltage drop	
	Continuous operation	Intermittent operation
14	0.5	1.0
28	1.0	2.0
115	4.0	8.0
200	7.0	14.0

If a choice is to be made between two wire sizes, always pick the larger wire size, which is the smaller number; thus No. 14 gage was chosen. Note that the intersection is above curve 1. This indicates that it is safe to use this wire for a 10-A continuous load in a conduit or a bundle. Because the intersection of the horizontal and diagonal lines is above curve 2, the wire is also safe for the conditions dictated by that curve.

Example 2. If a single electric wire is to be installed in a 28-V system, 10 ft [2.95 m] from the bus bar, and fused for a maximum current of 50 A, what is the correct size for the copper wire?

Solution. Use the wire chart in Figure 4–4a as follows: First, find the 28-V column on the left side of the chart, and note the 10-ft length in that column. Normally, the technician would follow the horizontal line adjacent to the 10-ft, 28-V location until it intersected the 50-A diagonal line, but curve 2 is encountered first. Curve 2 is used because the wire is installed as a single conductor, not in a bundle or conduit. Follow curve 2 up and to the right to where it intersects the 50-A diagonal line.

At the intersection of the 50-A diagonal line and curve 2, move straight down to determine the wire size. The wire needed is between the vertical lines numbered 10 and 12. A No. 10 wire is selected because it is the larger.

Example 3. To understand the use of the intermittent-rating conductor chart, consider the installation of a 120-ft wire in a 115-V system that is fused to a maximum current of 20 A. This system operates for less than 2 min.

Solution. First, find the 115-V column on the left side of the chart, and note the 128-ft length in that column. (*Note:* Here we substitute 128 ft for 120 ft, since there is no 120-ft figure on the chart.) Follow the horizontal line adjacent to the 128-ft, 115-V location until it intersects the 20-A diagonal line. Move straight down from this intersection between the vertical lines numbered 14 and 16. A No. 14 wire is selected because it is the larger of the two. The intersection of the horizontal (wire length) line and the diagonal (current) line is above curve 3; therefore, a No. 14 wire is satisfactory for this installation.

Current-Carrying Capacity of Wire

It is often desirable to obtain more information about wire capacity and characteristics than is provided in Figure 4–4. For this purpose Table 4–2 is useful.

Note that the indicated current-carrying capacity for copper electric wire is approximately the same in Figure 4–4a and Table 4–2, although small differences show up in certain instances. Note that No. 6 wire is rated to carry 60 A safely in a conduit or bundle in both the chart and the table. In free air the No. 6 wire is rated for about 96 A in the chart and for 101 A in the table. These apparent discrepancies are not great enough to make much difference; however, for civil aircraft the chart ratings of Figure 4–4 should be used.

From Table 4–2 it is possible to compute the voltage drop for any length of copper wire with any given load. For exam-

ple, if it is desired to know the voltage drop in 100 ft [30.5 m] of No. 18 wire carrying 10 A, we use Ohm's law, but we must first determine the resistance from the figures given in the table. Note that the resistance of 1000 ft [304.8 m] of No. 18 wire is 6.44 Ω. Then for 100 ft of the same wire, the resistance would be 0.644 Ω. Then, by Ohm's law,

$$E = 10 \text{ A} \times 0.644 \ \Omega = 6.44 \text{ V}$$

Thus we see that 100 ft of No. 18 wire will produce a voltage drop of 6.44 V when carrying a current of 10 A. To find the length of this wire that will produce a voltage drop of 1 V with a 10-A load, we merely divide 100 by 6.44. The result is approximately 15.5 ft [4.57 m], which is the same length of wire for a 1-V drop as is indicated by the chart in Figure 4–4a for No. 18 wire.

Although it is permissible to use aluminum wire in aircraft installations, the size of the wire must be larger than that of a copper wire for the same load. In general, an aluminum wire two sizes larger than the copper wire will be acceptable. Table 4–3 gives capacity, resistance, size, and weight for MIL-W-7072 aluminum wire approved for aircraft use.

TABLE 4–2 Capacity, weight, and resistance for stranded copper electric wire.

Wire size, AWG	Maximum amperes		Resistance Ω/ 1000 ft (20°C)	Area, cmil	Weight, lb/ 1000 ft
	Free air	Conduit or bundled			
20	11	7.5	10.25	1 119	5.6
18	16	10	6.44	1 779	8.4
16	22	13	4.76	2 409	10.8
14	32	17	2.99	3 830	17.1
12	41	23	1.88	6 088	25.0
10	55	33	1.10	10 443	42.7
8	73	46	0.70	16 864	69.2
6	101	60	0.436	26 813	102.7
4	135	80	0.274	42 613	162.5
2	181	100	0.179	66 832	247.6
1	211	125	0.146	81 807	288.0
0	245	150	0.114	104 118	382
00	283	175	0.090	133 665	482
000	328	200	0.072	167 332	620
0000	380	225	0.057	211 954	770

TABLE 4–3 Capacity, weight, and resistance for stranded aluminum wire.

Wire or cable size, AWG	Maximum amperes		Resistance, Ω/ 1000 ft (20°)	Area, cmil	Weight. lb/ 1000 ft
	Free air	Conduit or bundled			
AL-6	83	50	0.641	28 280	
AL-4	108	66	0.427	42 420	
AL-2	152	90	0.268	67 872	
AL-0	202	123	0.169	107 464	166
AL-00	235	145	0.133	138 168	204
AL-000	266	162	0.109	168 872	250
AL-0000	303	190	0.085	214 928	303

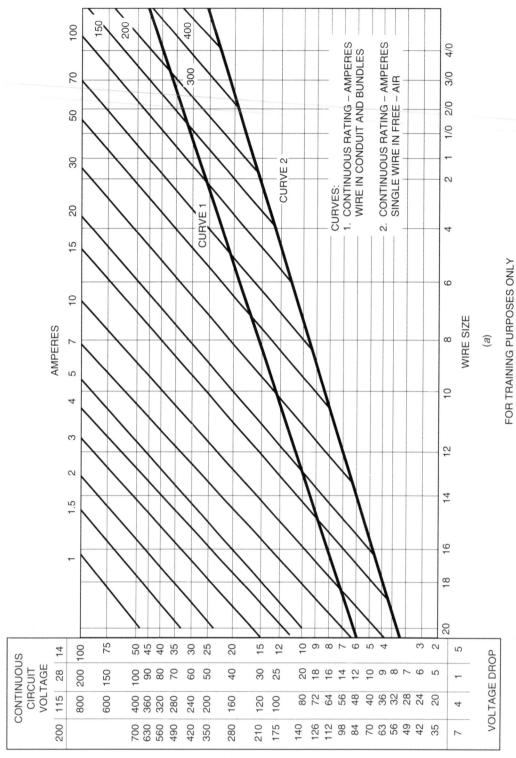

FIGURE 4–4 Wire selection chart (a) Continuous loads.

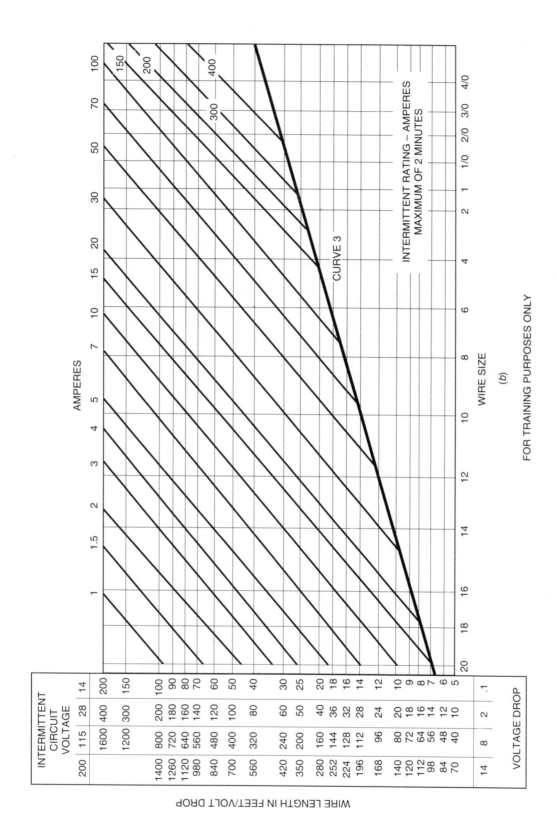

FIGURE 4–4 Wire selection chart. (b) Intermittent loads.

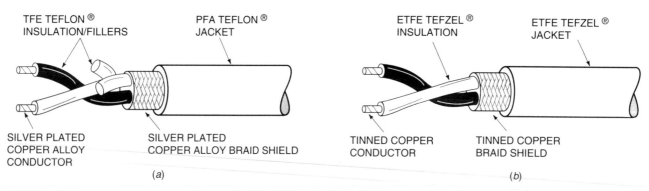

TFE TEFLON ® INSULATION/FILLERS

PFA TEFLON ® JACKET

SILVER PLATED COPPER ALLOY CONDUCTOR

SILVER PLATED COPPER ALLOY BRAID SHIELD

ETFE TEFZEL ® INSULATION

ETFE TEFZEL ® JACKET

TINNED COPPER CONDUCTOR

TINNED COPPER BRAID SHIELD

(a)

(b)

FIGURE 4–5 Two types of data bus cables. (a) MIL-STD-1553B with fillers; (b) data bus cable without filler rod.

Table 4–3 lists aluminum wire only of sizes 6 through 0000 because smaller aluminum wires are not recommended for aircraft use. It is interesting to note that aluminum wires of the larger sizes can be used advantageously to save weight, even though the aluminum wires are larger in diameter. Note that No. 00 aluminum wire has almost as much capacity as No. 0 copper wire but that in lengths of 1000 ft [304.8 m], the aluminum wire weighs only 204 lb [92.5 kg] against 382 lb [173.3 kg] for the copper wire. This is a saving of 178 lb [80.7 kg] for 1000 ft of wire. When No. 0000 aluminum wire is substituted for No. 000 copper wire, it is found that the weight of the aluminum wire is less than half the weight of the copper wire.

Data Bus Cable

One special type of cable used exclusively for various digital electronic systems is called **data bus cable.** This cable typically consists of a twisted pair of wires surrounded by an electrical shielding and insulators. (Shielding is discussed later in this chapter.) There are several different types of data bus cables; each meets a different standard and is used for specific applications. Figure 4–5 shows two different cables; one uses a filler rod to separate the inner conductors. It should also be noted that there are slight differences with regard to the inner conductors: one cable contains silver-plated copper conductors, the other has tinned copper conductors.

Data bus cables perform very specific tasks for their associated systems. Digital systems operate on different frequencies, voltages, and current levels. It is extremely important to ensure that the correct cable is used for the system installed. The cable should not be pinched or bent during installation. Data bus cable lengths may also be critical, and proper connectors must always be used. Refer to the current manufacturers' manuals for cable specifications.

REQUIREMENTS FOR OPEN WIRING

When wires or wire bundles are routed through the aircraft without the mechanical protection of conduit, it is called **open wiring.** Most aircraft use the open wiring system, and

mechanical protection is provided in critical areas by routing wires behind decorative or structural panels.

Open wiring is more vulnerable to wear, abrasion, and damage from liquids than wiring installed in conduits; hence care must be taken to see that it is installed where it is not exposed to these hazards and in a manner to prevent damage. The number of wires grouped in a bundle should be limited in order to reduce the problems of maintenance and to limit damage in case a short circuit should occur and burn one of the wires in the bundle. Shielded cable, ignition cable, and wire that is not protected by a circuit breaker or fuse should be routed separately. The bending radius of a wire bundle should not be less than 10 times the outer diameter of the bundle. This is required to avoid excessive stresses on the insulation.

Cable Lacing

The lacing of wire bundles should be performed according to accepted specifications. Approved **lacing cord** complying with specification MIL-C-5649 or twine specification JAN-T-713 may be used for wire lacing. If wire bundles will not be exposed to temperatures greater than 248°F [120° C], **cable tie straps** complying with specification MS-17821 or MS-17822 can be used. Typical tie straps are shown in Figure 4–6. Tie straps have replaced lacing cord in many aircraft installations; but always consult the current maintenance data to ensure that straps can be substituted for lacing cord. As seen in Figure 4–7, many aircraft have certain areas subject to high vibration or excessive heat where tie straps are not acceptable. To install a tie strap, simply wrap the tie around the wire bundle, being sure not to twist the strap. Insert the strap through the locking eyelet, and tighten the strap, using the proper tool. The tool is also used to cut off any excess trap, leaving a flush edge. Figure 4–8 illustrates the use of a typical tie strap installation tool.

Single-cord lacing is used for cable bundles 1 in. [2.5 cm] in diameter or less. For larger bundles, double-cord lacing should be employed. Cable bundles inside a junction box should be laced securely at frequent intervals to assure that a minimum of movement can take place. In open areas, the bundles should be laced or tied if supports for the cable are more than 12 in. [30.5 cm] apart.

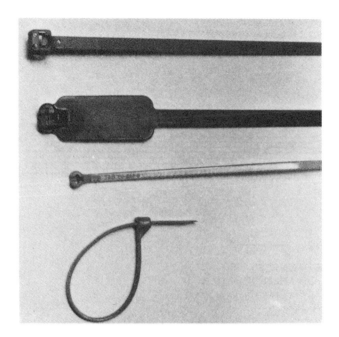

FIGURE 4–6 Tie straps for wire bundles.

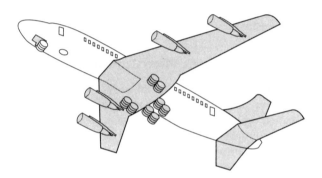

FIGURE 4–7 High vibration of typical aircraft.

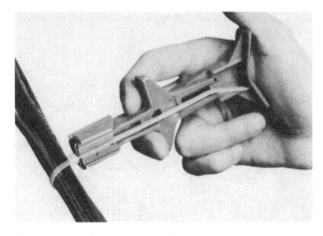

FIGURE 4–8 Typical tie strap installation tool. *(Thomas & Betts Corp.)*

Wire bundles may be laced with a continuous series of loops around the bundle as shown in Figure 4–9 or with single ties as in Figure 4–10. When the continuous lacing is applied, the first loop is a clove hitch locked with a double overhand knot as shown in Figure 4–9a. The knot is pulled

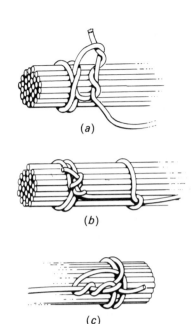

FIGURE 4–9 Lacing of wire bundles.

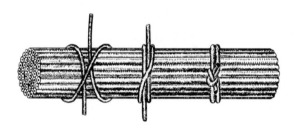

FIGURE 4–10 Single-tie lacing.

tight as shown in Figure 4–9b, and the continuing end is then looped around the wire bundle with the cord brought over and under the cord from the previous loop to form the type of loop shown in Figure 4–9b. These loops are continued at suitable intervals, and the series is then terminated with another clove hitch. The free end is wrapped twice around the cord from the previous loop and is then pulled tight to lock the loop. The terminating ends of the cord are trimmed to provide a minimum length of 3/8 in. [0.95 cm]. The method for making the terminal loop is illustrated in Figure 4–9c.

When it is desired to use single ties to secure a wire bundle, the locked clove hitch is used. The clove hitch is formed as shown, and it is then locked with a square knot. Single ties are sometimes used to separate a group of wires from a bundle for identification purposes, as shown in Figure 4–11. This helps maintenance technicians locate particular circuit wiring.

When **double-cord lacing** is required for large cable bundles, the first loop is made with a special type of slip knot similar to the "bowline-on-a-bight." This is shown in Figure 4–12. The double cord is then used to make additional loops as required in the same manner as the single cord is used. The terminal lock knot is made by forming two single loops around the bundle and then tying the two ends with a square knot.

Requirements for Open Wiring **63**

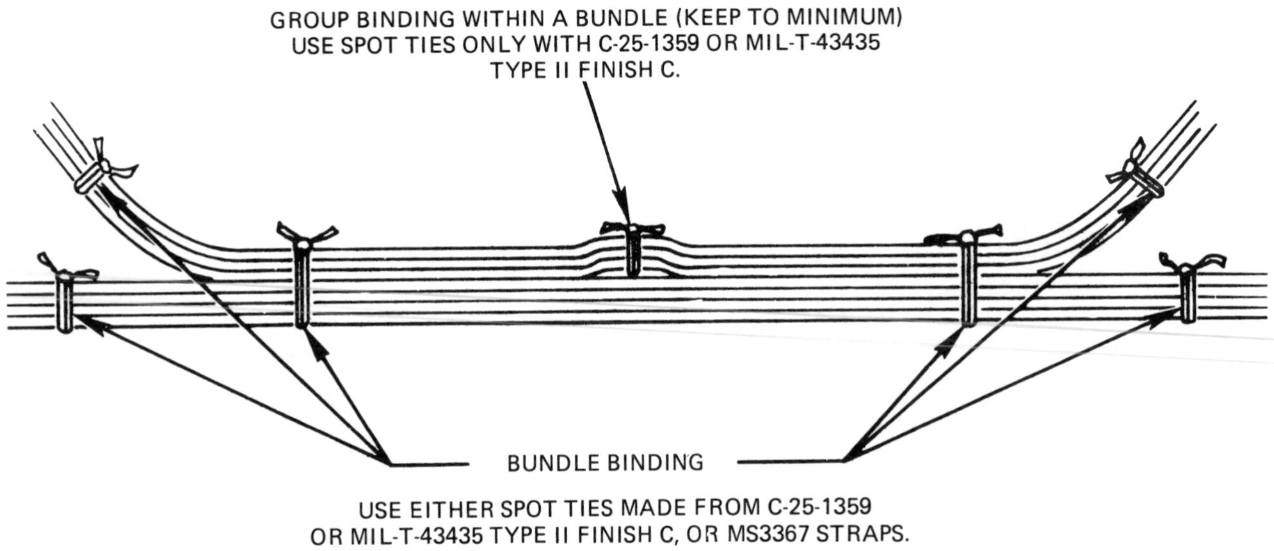

GROUP BINDING WITHIN A BUNDLE (KEEP TO MINIMUM)
USE SPOT TIES ONLY WITH C-25-1359 OR MIL-T-43435
TYPE II FINISH C.

BUNDLE BINDING

USE EITHER SPOT TIES MADE FROM C-25-1359
OR MIL-T-43435 TYPE II FINISH C, OR MS3367 STRAPS.

FIGURE 4–11 Separation of wire groups for identification. *(Lockheed)*

FIGURE 4–12 Beginning loop for double-cord lacing.

Wire and Cable Clamping

Electric cables or wire bundles are secured to the aircraft structure by means of metal clamps lined with synthetic rubber or a similar material. Specification MS-21919 cable clamp meets the requirement for civil aircraft use. Such a clamp is illustrated in Figure 4–13.

In the installation of cable clamps, care must be taken to assure that the stress applied by the cable to the clamp is not in a direction that will tend to bend the clamp. When a clamp is mounted on a vertical member, the loop of the clamp should always be at the bottom. Correct methods for installing clamps are shown in Figure 4–14.

When a wire bundle is routed through a clamp, the bundle must be held within the rubber lining of the clamp, and no wires must be pinched between the flanges of the clamp. Pinching of the wire could cause the insulation to be damaged, and a short circuit could result.

In installing electric wiring in a particular make and model of aircraft, it is the best practice to make the installation in accordance with the manufacturer's original design unless a specific change has been ordered. The clamps, wiring, and connectors should be of the same types specified and used by the manufacturer.

Routing of Electric Wire Bundles

The routing of electric wire should be done in a manner that will provide the protection previously mentioned, namely protection against heat, liquids, abrasion, and wear. Clamps should be installed in such a manner that the wires do not come in contact with other parts of the aircraft when subjected to vibration. Sufficient slack should be left between the last clamp and the electric equipment to prevent strain at the wire terminals and to minimize adverse effects on shock-mounted equipment. Where wire bundles pass through bulkheads or other structural members, a grommet or suitable

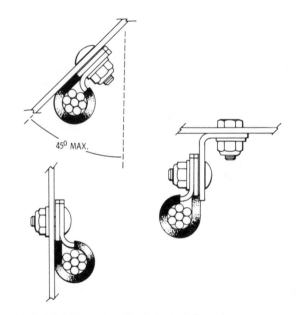

FIGURE 4–14 Correct methods for installing clamps.

RUBBER LINER

FIGURE 4–13 Clamp for electric cable.

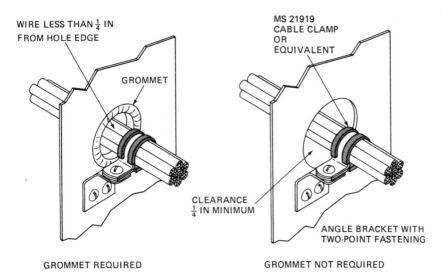

WIRE LESS THAN ¼ IN FROM HOLE EDGE

GROMMET

MS 21919 CABLE CLAMP OR EQUIVALENT

CLEARANCE ¼ IN MINIMUM

ANGLE BRACKET WITH TWO-POINT FASTENING

GROMMET REQUIRED

GROMMET NOT REQUIRED

FIGURE 4–15 Wire bundle passing through bulkhead.

clamping device should be provided to prevent abrasion as shown in Figure 4–15. If a wire bundle is held by a clamp in the center of a hole through a bulkhead, and the clearance between the edge of the hole and the bundle is more than ¼ in. [0.64 cm], a grommet is not required.

At points in an installation where electric wire may be exposed to oil, hydraulic fluid, battery acid, or some other liquid, the cable should be enclosed in a plastic sleeve. At the lowest point in the sleeve, a hole ⅛ in. [0.32 cm] in diameter should be cut to provide for drainage. The sleeve can be held in place by clamps or by lacing.

If a hot wire terminal should come into contact with a metal line carrying a flammable fluid, the line might be punctured and the fluid ignited. This, of course, would result in a serious fire and probable loss of the airplane. Consequently, every effort should be made to avoid this hazard by physical separation of the cables from lines carrying oil, fuel, hydraulic fluid, or alcohol. When separation is impractical, the electric wire should be placed above the flammable-fluid line and securely clamped to the structure.

Particular care must be used in installing electric wire on and in the vicinity of landing gear, flaps, and other moving structures. Slack must be allowed for required movement, but the wire must not be too loose. Routing of the wire must be such that it is not rubbed or pinched by moving parts during operation of the mechanism. An examination of the wire during a ground check of the operation of the mechanism will usually reveal any hazards.

Electric wiring must be protected from excessive heat. As noted previously, electric wiring is insulated and protected with various types of materials, some of which can withstand temperatures as high as 392°F [200°C]. In areas where a wire must be subjected to high temperatures, it is necessary to use wiring with insulation made of asbestos or some other heat-resistant material. Wires should not be routed near exhaust pipes, resistors, or other devices that produce high temperatures except as required for special purposes and then only if the wires are protected with adequate heat-resistant insulation.

ELECTRICAL CONDUIT

Electrical **conduit** consists of thin-walled aluminum tubing, braided metal tubing called **flexible conduit,** and nonmetallic tubing. The purpose of conduit is to provide mechanical protection, and metal conduit is often used as a means of shielding electric wiring to prevent radio interference.

Approved flexible conduit is covered by specification MIL-C-6136 for aluminum and specification MIL-C-7931 for brass. The aluminum conduit is made in two types. Type I is bare, and type II is rubber-covered.

The size of conduit should be such that the inside diameter is about 25 percent larger than the largest diameter of the cable bundle. To obtain the correct inside diameter of a conduit, subtract twice the wall thickness from the outside diameter. Typically, conduits are specified according to their outside diameter.

The inside of the conduit should be clean and free of burrs, sharp edges, or obstructions. When conduit is being cut and prepared, all edges and holes should be deburred to assure a smooth surface that will not damage the cable. The conduit should be inspected carefully after the end fittings are installed to assure that the interior is clean and smooth. If a fitting is not installed on the end of a conduit section, the end should be flared to prevent the edge of the tubing from rubbing and wearing the insulation of the cable.

Installation of conduit should be such that it is protected from damage of all types. It should be securely attached to the structure with metal clamps so there can be no movement or vibration. A clean metal-to-metal contact will assure good bonding to aid in shielding. The installed conduit should not be under appreciable stress and should not be located where it may be stepped upon or used as a hand support by a member of the crew. Drain holes must be provided at the lowest point in any conduit run.

Rigid conduit that is cut or has appreciable dents should be replaced to prevent damage to the electric cable. Bends in the conduit must not be wrinkled and must not be flattened to the extent that the minor diameter is less than 75 percent of the

Nominal tube outside diameter		Minimum bend radii	
in	cm	in	cm
$\frac{1}{8}$	0.32	$\frac{3}{8}$	0.96
$\frac{3}{16}$	0.48	$\frac{7}{16}$	1.11
$\frac{1}{4}$	0.64	$\frac{9}{16}$	1.43
$\frac{3}{8}$	0.96	$\frac{15}{16}$	2.38
$\frac{1}{2}$	1.27	$1\frac{1}{4}$	3.18
$\frac{5}{8}$	1.60	$1\frac{1}{2}$	3.81
$\frac{3}{4}$	1.92	$1\frac{3}{4}$	4.46
1	2.54	3	7.62
$1\frac{1}{4}$	3.18	$3\frac{3}{4}$	9.53
$1\frac{1}{2}$	3.81	5	12.7
$1\frac{3}{4}$	4.46	7	17.8
2	5.08	8	20.3

TABLE 4–4 Minimum bend radii for rigid conduit.

Nominal internal diameter of conduit		Minimum bending radius inside	
in	cm	in	cm
$\frac{3}{16}$	0.48	$2\frac{1}{4}$	5.72
$\frac{1}{4}$	0.64	$2\frac{3}{4}$	6.99
$\frac{3}{8}$	0.96	$3\frac{3}{4}$	9.53
$\frac{1}{2}$	1.28	$3\frac{3}{4}$	9.53
$\frac{5}{8}$	1.60	$3\frac{3}{4}$	9.53
$\frac{3}{4}$	1.92	$4\frac{1}{4}$	10.80
1	1.54	$5\frac{3}{4}$	14.61
$1\frac{1}{4}$	3.18	8	20.32
$1\frac{1}{2}$	3.82	$8\frac{1}{4}$	20.96
$1\frac{3}{4}$	4.46	9	20.86
2	5.08	$9\frac{3}{4}$	24.77
$2\frac{1}{2}$	6.35	10	25.40

TABLE 4–5 Minimum bend radii for flexible conduit.

nominal tubing diameter. Table 4–4 shows the minimum tubing-bend radii for rigid conduit.

Flexible conduit cannot be bent as sharply as rigid conduit. This is indicated by Table 4–5, which gives the minimum bending radii for flexible aluminum or brass conduit.

When sections of flexible conduit are being replaced and it is necessary to cut the conduit, the operation can be greatly improved by wrapping the area of the cut with transparent adhesive tape. Fraying of the end will be greatly reduced because the tape will hold the fine wires in place as the cut is made with a hacksaw. Before a wire or cable bundle is placed in a conduit, the bundle should be liberally sprinkled with talc.

CONNECTING DEVICES

Wire and Cable Terminals

Since aircraft electric wires are seldom solid but are usually strands of small-gage, soft-drawn tinned copper or bare aluminum twisted together to provide flexibility, the separate strands must be held together and fastened to connectors. These connectors are commonly called **terminals** or terminal lugs and are required to connect the wires to terminal posts on electric equipment or on terminal strips.

Compatibility between the wire's material and the terminal's material is very important. Copper wires must be used with terminals suitable for copper wires, and aluminum wires must be used with terminals compatible with aluminum wires. If the incorrect terminal-wire combination is used, dissimilar-metal corrosion may create an insufficient electrical connection. Always verify the type of wire and terminal to be used before installation. Compatibility between the terminal and the terminal post is also necessary to reduce corrosion and possible electrical failure in this area.

Approved terminals of the **swaged** or **crimped** type are available from several manufacturers. They are designed according to wire size and the size of the terminal stud to which the terminal is to be connected. One size of terminal will usually fit two or three different sizes of wire; for example, one size of terminal will fit wire from No. 18 to No. 22. The terminal is attached to the wire by means of a special crimping tool. First, the insulation is removed with a wire stripper as shown in Figure 4–16. Care must be taken that the stripper is of the correct size and type for the wire being stripped. This will help to ensure that the wire strands are not damaged during stripping. For each type of terminal, the length of insulation to be removed from the wire is specified. The bare wire is then inserted in the end of the terminal, and the terminal is crimped with the proper tool. The result must be such that the terminal attachment has a tensile strength at least equivalent to the tensile strength of the wire.

In the stripping of electric wire, the technician should see that the tool is sharp and is correctly adjusted. It is also important to ensure that the correct tool and cutting blades are used. Some insulations, such as Teflon, require a different cutting edge than conventional wire insulation. It is important to check the manufacturers' recommendations for

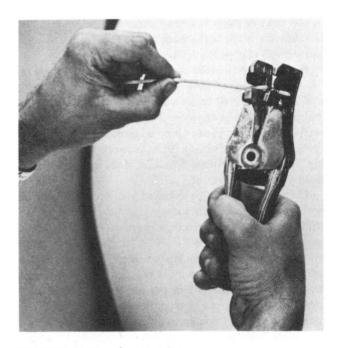

FIGURE 4–16 Use of a wire stripper.

Wire Size #		Number of strands per conductor	Total allowable nicked and broken strands
Copper Wire	24-14	19	2 nicked, none broken
	12-10	37	4 nicked, none broken
	8-4	133	6 nicked, none broken
	2-1	665-817	6 nicked, 2 broken
	0-00	1,045-1330	6 nicked, 3 broken
	000	1,665-	6 nicked, 4 broken
	0000	2,109-	6 nicked, 5 broken
Aluminum Wire	6-000	All numbers of strands	None, None

FIGURE 4–17 Allowable nicked or broken strands.

compatibility of tools and wire types. It is also very important to use the proper stripping procedures to avoid damaging conductor strands. Damage can occur in the form of **nicked, broken,** or **scraped wire strands.** Minor longitudinal scrapes are acceptable; however, the allowable number of nicked or broken strands is regulated by the FAA. The data on allowable nicked or broken strands for a copper or copper alloy conductor are shown in Figure 4–17. Nicked and broken strands are *not* acceptable for any size of aluminum conductor. Whenever stripping wires, the technician should always refer to the latest version of FAA Advisory Circular 43.13-1A, or other pertinent data, to determine the exact number of permissible nicked or broken strands.

Figure 4–18 shows a group of crimp-type terminals properly attached to electric wire. Note that the terminal sleeves are crimped on both bottom and top (once on the conductor of the wire, once on the insulation). Figure 4–19 shows the construction of a typical crimp-type terminal for aircraft. The terminal is equipped with a plastic, copper-reinforced insulating sleeve, which makes it unnecessary to install insulation after the terminal is attached to the wire. It is important to note that after the terminal is installed, the strands of the wire extend approximately $\frac{1}{32}$ in. [0.079 cm] beyond the terminal sleeve. This condition is required to make sure that the terminal has sufficient grip on the wire.

The crimping tool used with solderless-type terminals is designed so that it will not release until the terminal has been sufficiently crimped. This feature is provided by a ratchet installed between the handles of the tool. Care must be taken to ensure that the technician using the tool is well informed about its proper operation. Manufacturers of terminals and installation tools supply instructions and specifications that give all the necessary information and data for proper installation. Figure 4–20 shows a typical installation tool. Many tools are color-coded to match the color of the terminal sleeve or insulation. This coding assures that the tool of the proper size will be used for each terminal. The various colors are used to designate the different sizes of terminals and the wire gages that fit those terminals.

The crimp-type terminals used on aircraft are constructed with two metal sleeves, as illustrated in Figure 4–19. One sleeve is part of the electric terminal and is crimped to the

copper conductor. The second is a thin metal sleeve surrounded by the terminal's insulation cover. This sleeve is used to crimp the wire's insulation. When the wire's insulation is secured with a second crimped sleeve, the vibration stress is transmitted into the wire's insulation. This reduces the stress and the likelihood of fatigue failure of the wire's conductor. It is very important to ensure that the terminal is crimped twice: once to secure the conductor, once to secure the insulation.

In addition to the manual crimping tools, manufacturers provide power crimpers that are driven by either hydraulic or pneumatic power. When large numbers of crimps are to be

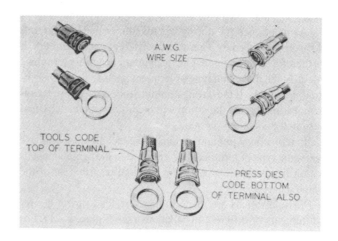

FIGURE 4–18 Crimp-type terminals. *(AMP Specialties)*

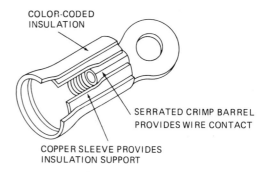

FIGURE 4–19 Construction of crimp-type terminal.

FIGURE 4–20 Typical installation tool.

made, the power tools save time and effort. Special tensile test machines are also available that can be used to test the security of the attachment of a terminal to the wire. The tensile test ensures equipment accuracy and proper installation by the technician. A tensile test should also be done in the field after a technician installs a crimp-type terminal. In the aircraft it is nearly impossible to use a tensile test machine on crimped terminals; however, a simple "pull" test will help to ensure connector reliability. A finished crimp should always be tested by applying a moderate pull to the terminal and the wire.

As mentioned previously, approved terminals for aircraft wire are produced by a number of different manufacturers. It is therefore important that the technician installing terminals identify the make and type of terminal and use the proper installation tools. If the wrong crimping tool is used on a terminal, it is likely that the crimp will be faulty, and the wire and terminal may fail in service.

In replacing wires and terminals, the technician should use the same type of terminal that was used in the original installation if possible; however, the terminal need not be of the same make.

Wire terminals are made in many styles to meet the requirements of different installations (see Figure 4–21). For most aircraft applications, ring terminals, rather than slotted or hook-type terminals, must be used. This helps eliminate any circuit failure due to terminal disconnection. On aircraft, it is always the best practice to replace terminals with those of similar design and always to use terminals approved for aircraft installations. Not all terminals are produced alike; technicians should be sure that those they install are of aircraft quality.

Soldered terminals are typically considered unsatisfactory for general electrical use in aircraft electric systems, even though soldering is considered a good practice in electronic units such as radio receivers, radar equipment, and auto pilot controlling circuits. Electric wires in the main electric system of an aircraft are of the flexible type. When a terminal is soldered to such a wire, the solder tends to pene-

trate the wire and make it rigid in the vicinity of the terminal. This makes the wire and terminal less resistant to vibration, with the result that the wire may become crystallized by fatigue and break off at the terminal. Another disadvantage of the soldered terminal is that the flux used for soldering may be of a corrosive type, thus bringing about failure through corrosion.

Because of the necessary unsoldering and resoldering, the maintenance of soldered-terminal systems is more difficult than that of systems with swaged terminals. A technician well skilled in soldering techniques is required, because a poorly soldered terminal is a hazard in itself. An unskilled operator may burn the insulation, may fail to make sure that the solder is thoroughly sweated into the terminal, or may use a corrosive flux.

In the event that a joint must be soldered in an aircraft electric system, there are certain conditions that have to be observed. The flux should be of a noncorrosive type such as rosin. Rosin-core wire solder is most commonly used because the flux is automatically applied as the solder is melted

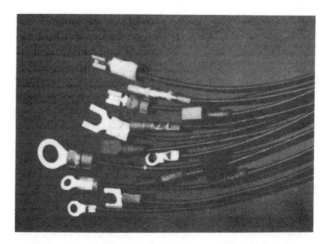

FIGURE 4–21 Assorted wire terminals. *(Thomas & Betts Corp.)*

on the joint. The two metal parts being joined by the solder must be brought up to the melting temperature of the solder so that the solder will flow smoothly into the joint and form a solid bond with the metal. Care must be exercised so that adjacent insulation or electric units are not damaged by the heat. After completion of the solder connection, the flux should be removed from the terminal and the wire. **Denatured alcohol** or a commercially available **rosin remover** can typically be used for this purpose. On some solder installations (particularly printed circuit boards), the manufacturer may also recommend that the connection be covered with an environmentally protective coating.

Splicing

The splicing of electric wires may be done if approved for a particular installation. Typically, the splice is made with an approved crimp-type splice connector. The **splice connector** is a metal tube with a plastic insulator on the outside or a plain metal tube that is covered with a plastic tube after the splice is made. The stripped wire is inserted into the end of the tube in the same manner in which wire is inserted into a terminal sleeve. The tube is then crimped with a terminal crimping tool. A typical crimp-type splice connector is shown in Figure 4–22.

When splices are made in wires that are in a wire bundle, the spliced wires are placed on the outside of the bundle. The bundle ties or straps are located where there are no splices. If several splices are to be made in any wire bundle, the splices should be staggered to reduce the bundle's diameter, as illustrated in Figure 4–23. The FAA regulations recommend that electric wire splices be kept to a minimum and be avoided entirely in locations of extreme vibrations.

Soldered splices are permitted; however, they are particularly brittle and are *not recommended* by the FAA. If a soldered splice must be accomplished, it should be a lap-type splice where the stripped portions of the wires are aligned side by side, twisted gently, and soldered. Butt-type soldered splices are relatively weak and should not be used. All soldered splices should be covered with at least one layer of insulation tubing to protect the connection against accidental shorts.

Insulation Tubing

Many electrical connections require the addition of insulation after the connection has been made. Typically, an insulation tube is slipped onto the wire before the connection is made and then slid over the exposed connection and secured in place. There are two basic types of insulation tubing. One is secured in place with lacing cord drawn tight and tied at each end of the tube. The other type of insulation tubing is called **heat shrink tubing,** since it is held in place on the connection through a shrinking process. The heat shrink insulation is more reliable when installed properly and can also provide some resistance to vibration stress of the connection.

Heat shrink insulation should be of the 200 to 300°C type and should always overlap the connection at least $\frac{1}{2}$ in. [1.27 cm.] after shrinking. It is important to choose a tube of the correct diameter that will shrink and form a tight fit around the conductor. If additional vibration or abrasion resistance is needed for an application, two or more layers of tubing should be installed. The innermost tube should be installed and shrunk first; the second tube, approximately $\frac{1}{2}$ in. [1.27 cm] longer than the first, should be installed next and a third added if needed. The heating process should be performed with the proper heat gun, adjusted for the correct temperature range. Reflective heat shields are often used to protect adjacent wiring or other components.

In some cases it is very important to seal electrical connections from the corrosive effects of the environment. Special heat shrink tubing is available with a sealant that is inside the tube. As the tube is heated, the sealant becomes soft and melts around the conductor, hence sealing out the environment. Some insulated crimp-type terminals and splices also contain this sealant and must be heated after the crimping process is completed.

Special **insulation tapes** are also approved for certain applications on aircraft terminals and splices. The tape is typically a high-temperature Teflon or silicon-type material and requires application in a spiral wrap over the splice. Tape may be used over insulation tubing as a second layer of protection. In all cases the tape is overlapped 50 percent on each wrap and should extend past the splice by a minimum of $\frac{1}{2}$ in. [1.27 cm] on each side.

Electric Terminal Strips

On vintage aircraft the joining of separate sections of electric wire is often accomplished by means of terminal strips like those shown in Figure 4–24. A terminal strip is made of a strong insulating material with metal studs molded into the material or inserted through it. The studs are anchored so that they cannot turn and are of sufficient length to accommodate four terminals. Between each pair of studs are barriers to prevent wire terminals attached to different studs from coming into contact with each other.

When it is necessary to join more than four terminals at a terminal strip, two or more of the studs are connected with a metal bus or jumper wire, and the terminals are then connected to the studs with no more than four terminals on any one stud. The stud sizes in terminal strips must be adequate to withstand the stresses imposed during installation and tightening of the nut. For this reason it is common practice to use No. 10, or $\frac{3}{16}$-in. [0.48-cm], studs for aircraft electric systems.

A stud in a terminal strip to which wire terminals are to be connected is usually mounted in the insulating strip with two flat washers, two lock washers, and two nuts as shown in Figure 4–25. The stud is secured in the strip with a flat washer, a lock washer, and a plain nut. The terminal to be

FIGURE 4–22 A crimp-type splice connector.

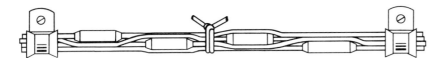

FIGURE 4–23 Staggered splice connections.

FIGURE 4–24 Terminal strips.

connected to the stud is placed directly upon the bottom nut, and a flat washer is placed over the terminal. A lock washer is placed upon the flat washer, and a plain nut is tightened against it. The nut should be tightened with a nut driver-type wrench or a socket on an extension. The torque applied to the nut must not exceed the safe limit, because excess stress can easily crack the material of the terminal strip. It is important to note that lock washers, either star-type or split-type, are never placed directly against a terminal or against the aircraft structure. A plain washer is always used under the lock washer to prevent damage to the terminal or structure.

Terminal strips must be mounted in a manner and position such that loose objects cannot fall on the terminals. This may be accomplished by installing the strips on vertical bulkheads or overhead and providing them with suitable covers. A most effective method for protecting the terminal strips is to install them in a box made of metal or some other strong material. A junction box is shown in Figure 4–26.

Electric wiring inside a junction box should be laced or clamped in such a manner that terminals are not hidden, the operation of equipment is not hampered, and motion between wires and equipment is prevented. The openings through which the wire bundles enter the box must be provided with clamps or grommets so that the insulation on the wires cannot become worn or otherwise damaged.

Aluminum Wires and Terminals

The installation of aluminum wiring and terminals requires exceptional care to ensure satisfactory operation. Aluminum wire hardens as a result of vibration more quickly than copper wire. For this reason aluminum wire should not be used where there is appreciable movement of the wire during operation of the aircraft. Also, aluminum wire smaller than

AWG size 6 is not acceptable. Aluminum wire is typically found only on large aircraft for power distribution cables requiring large-diameter wire.

Terminals, nuts, bolts, and washers used with aluminum wiring must be compatible with aluminum to avoid the **electrolytic corrosion** that takes place between dissimilar metals when they are in contact. Such hardware should be made of aluminum or an aluminum alloy or should be plated with cadmium or some other compatible metal. It is extremely important that all components of any connection be made of compatible materials, even when the terminals are connected to both aluminum wire and copper terminal studs.

Since aluminum always has an oxide coating on the surface, special procedures must be used to eliminate the coating when joining aluminum parts for electrical contact. In installing terminal lugs on aluminum wire, a means must be provided to destroy the oxide coating on the wire and inside the terminal sleeve. One method for doing this is to procure aluminum lugs with a petrolatum and zinc powder compound inside the sleeve. The abrasive action of the zinc dust in the

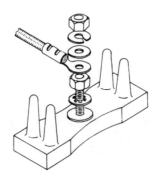

FIGURE 4–25 Attachment of a terminal to a stud.

FIGURE 4–26 A junction box.

compound removes the oxide film from the aluminum wire and the terminal. In most cases special terminal lugs are available for the "dry" connection of aluminum wire. The material and construction are such that the crimping process automatically destroys the oxide film and produces a good contact. It should also be noted that in many cases aluminum wire requires a special wire-stripping tool to ensure that the strands are not nicked or broken.

Connectors

Electric connectors are designed in many sizes and shapes to facilitate the installation and maintenance of electric circuits and equipment in all types of flying vehicles. For example, it may be necessary to replace a damaged section of electric harness in an aircraft. If the section of harness is connected to other sections by means of connectors, it is a simple matter to unplug the damaged section at both ends and remove it. A completely new section may then be quickly installed. If the damaged section were connected to other sections by terminal strips, it would be necessary to disconnect each wire from the terminal studs separately; this operation would consume considerable time, especially if the harness contained many wires. Connectors are also used to connect electric and electronic assemblies, or **line replaceable units (LRUs),** such as voltage regulators, flight computers, inverters, and radio equipment. When it is necessary to replace such an assembly, the connector makes it possible to disconnect the unit quickly and to reconnect the new unit with no danger of connecting any of the leads incorrectly.

A **connector assembly** actually consists of two principal parts. These parts are often called the **plug** and the **receptacle.** The plug section generally contains the **sockets,** and the receptacle contains the **pins.** The pins and sockets are connected to the individual wires that make up the circuit. When the plug and receptacle are assembled together, the pins slip inside the sockets and form the electrical connection.

When a connector assembly is installed, the "hot," or voltage-positive, side of the circuit should be connected to the socket section, and the ground side of the circuit should be connected to the pin section. This arrangement will reduce the possibility of shorting the circuit when the connector is separated. For this reason LRUs typically contain the pins, and the wiring harness is connected to the socket section of the plug.

Problems experienced with connectors are often due to corrosion caused by moisture condensation inside the shell of the connector. If a connector is to be installed where corrosion is a problem, the assembly should be coated with a chemically inert, waterproof jelly, or a special waterproof connector should be installed. In all cases, any unused contact hole should be filled with a wire or plug to prevent the entrance of moisture or other foreign matter. The free end of a stub wire should be covered with potting compound or some other material to prevent electric contact.

In working on large aircraft electric systems, a technician will encounter many different types and makes of connectors. Among the brand names that may be encountered are Cannon, Bendix, Burndy, AMP, Deutsch, and Amphenol. The earlier connectors were designed for the wires to be soldered to the pin and socket contacts; however, most connector assemblies are now designed with crimp-type pins and sockets. The pins and sockets are first crimped to the wires and then are installed in the connectors by means of special tools.

Because of the almost infinite variety of possible electric circuits and installations, it is readily understandable that there must also be a wide variety of connectors and other connecting devices. Various connectors may have from just one contact pin and socket to more than one hundred. Several different types of connectors are shown in Figure 4–27. For the installation of any particular connector assembly, the specification of the manufacturer or the appropriate governing agency must be followed.

FIGURE 4–27 Different types of connectors.

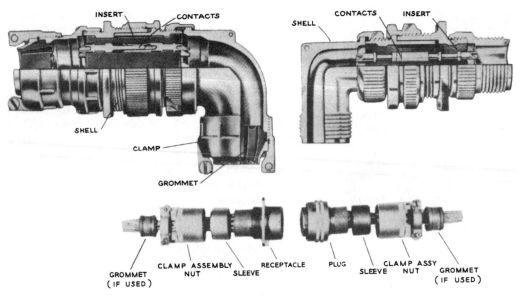

INSERT   CONTACTS   CONTACTS   INSERT
                    SHELL

SHELL

CLAMP

GROMMET

GROMMET   CLAMP ASSEMBLY   RECEPTACLE   PLUG   CLAMP ASSY   GROMMET
(IF USED)   NUT   SLEEVE   SLEEVE   NUT   (IF USED)

FIGURE 4–28 Bendix MS-type connectors.   *(Bendix Corporation)*

All connectors must be labeled as to the pin or socket locations within the connector's insulation housing. This identification label is necessary to ensure that the wire is connected to the correct pin or socket during installation. Labeling also facilitates troubleshooting by allowing the technician to easily identify a particular circuit connection. Most aircraft connectors are labeled by stamping a letter on the connector's insulation housing adjacent to the correct pin or socket hole. The technician should be sure to correctly identify the electrical connection of a connector prior to performing maintenance.

Connectors currently being manufactured for aircraft use are often required to meet military specifications and are called MS electric connectors. **Military specifications (MIL specs)** are revised from time to time to incorporate performance requirements as dictated by design advances and more stringent operating requirements of equipment. Many older connectors adhere to the AN (Air Force-Navy or Army-Navy) specifications. These standards have been replaced with MS or MIL specifications.

The general specification **MIL-C-5015** provides for several designations of connectors to meet different requirements. These connectors carry MS numbers such as MS3100A-20-27S. In this designation, the number *3100* indicates a wall-mounted unit, the letter *A* indicates general utility usage, the number *20* indicates shell size, and *27S* shows that the plug contains 27 socket-type connections. Typical MS connectors are shown in Figure 4–28.

Connector assemblies are manufactured in many shapes and sizes to meet the requirements of modern electric and electronic equipment. The round connector is popular because it lends itself to easy joining and securing by means of a threaded collar. Many connectors are made in a rectangular shape, however, and these are often used when a harness is connected to an electronic unit.

The construction of the pins and sockets in an MIL or other type of connectors may be designed for solder connections to the electric wires, or the connector may be designed

with crimp pins and sockets. At the end of the pin or socket in a solder-type connector is a small solder pocket. A short section of insulation is removed from the wire, and the bare stranded wire is then inserted in the pocket. Enough insulation should be removed from the wire so that none extends into the solder pocket. With the wire in the pocket, solder is applied with a small-pointed soldering iron or soldering gun. The solder should be of the rosin-core wire type and should be applied to the pocket as it is heated with the soldering iron. As soon as the solder flows smoothly into the pocket and penetrates the wire, the soldering iron should be removed to avoid the possibility of burning the insulation of either the wire being soldered or the adjacent wires. Only enough solder should be applied to fill the pocket, and all small drops of solder should be removed from between the pins. After each pin is soldered, a plastic sleeve insulator should be pushed down over the soldered joint and metal pin to prevent the possibility of short-circuiting. The insulating sleeves should be tied or

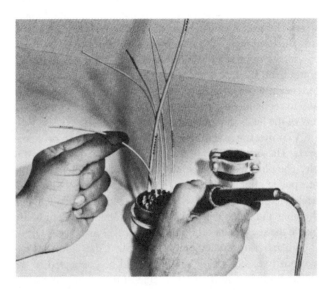

FIGURE 4–29 Soldering a wire to a connector.

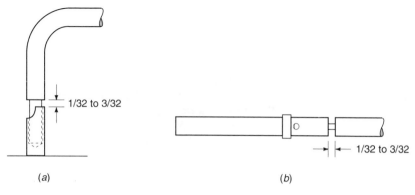

(a)   (b)

FIGURE 4–30 Wire strip lengths. (a) In the case of a solder pot connection, the wire should be long enough to reach to the bottom of a solder cup, with $\frac{1}{32}$ to $\frac{3}{32}$ in. of wire exposed beyond the cup. (b) The wire in a crimp connection should be visible in the inspection hole of the crimp contact, with $\frac{1}{32}$ to $\frac{3}{32}$ in. of wire exposed beyond the barrel.   *(The Deutsch Co.)*

clamped to prevent them from slipping off the pins. Figure 4–29 shows a method for soldering a wire to a connector.

As mentioned earlier, crimp type solderless contacts are used in a variety of connectors. Because of their relatively simple connection means, crimp-type connectors are selected for most modern aircraft electrical systems. The process of crimping the wire to the connectior relies on a mechanical tool used to compress the pin around the stripped portion of the wire. Some of the characteristics of soldered and crimp-type connectors are as follows:

Soldered Connections

a. The flux used for soldering is corrosive and can weaken the connection over time.

b. Errors such as too much heat, too much solder, not enough heat, and lack of connection cleanliness are difficult to eliminate.

c. Gold-plated contacts can be destroyed by the soldering process.

d. Solder wicking into the wire strands can create additional stress in the wire.

Crimped Connections

a. The use of the appropriate tools eliminates the chance of human errors.

b. No corrosive fluxes are used during crimping.

c. Gold-plated contacts are completely compatible with the crimping process.

d. The connection is easily inspected prior to pin installation.

e. Field repair can be performed more easily and with less error than repair of soldered connections.

When preparing a wire for installation into a pin or socket, the technician must first remove the proper length of insulation from the end of the wire. During this process it is very important not to damage the conductor through broken or nicked strands beyond acceptable limits. The proper length of insulation to remove depends on the conductor size and the type of pin or socket to be used. The manufacturer's installa-

tion data should be consulted for strip length specifications. A properly stripped wire for both a solder connection and a crimp-type connection is shown in Figure 4–30.

There are three basic procedures that must be performed to install a wire for all crimp-type connectors. (1) The wire must be stripped, (2) the wire must be crimped to the pin or socket, and (3) the pin or socket must be installed in the connector housing. Each of these procedures should be followed by a visual inspection of the work performed, and defects must be eliminated. After this process has been successfully completed, the connector should be reassembled and installed in the aircraft. Figure 4–31 gives detailed instructions for the installation of a contact into a typical connector assembly.

Wire installation instructions are as follows:

Step 1 Install the pin or socket into the crimp tool with the wire barrel facing up. During this process be sure to use the correct tool and any necessary adapters and/or the appropriate setting for the pin or socket desired.

Step 2 Remove the correct length of insulation from the wire, using appropriate methods.

Step 3 Install the stripped portion of the wire into the connector barrel, and compress the connector by squeezing the tool handles together. The tool will release the contact when the crimping process has been completed. If the contact will not release, the crimp cycle has not been completed and the contact must be compressed further.

Step 4 Inspect the finished crimp through the inspection hole in the contact. The wire must be visible; if it is not, the crimp must be redone using a new contact.

Step 5 Install the contact into the connector housing, using the appropriate tool. During this process it is very important to ensure that the contact is installed completely and reaches a firm stop inside the connector housing. Often a small click is heard as the contacts reach their stopping point.

There are two methods commonly used to install a contact into a connector housing: **front release** and **rear release.** The front-release contact is held in place by a relatively complex combination of retainers molded into the connector housing.

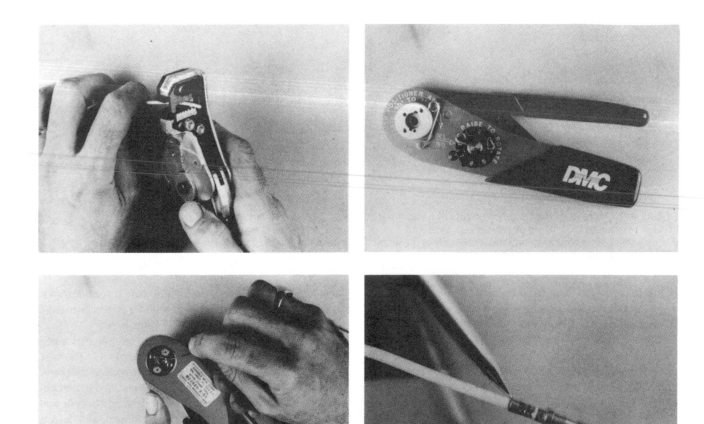

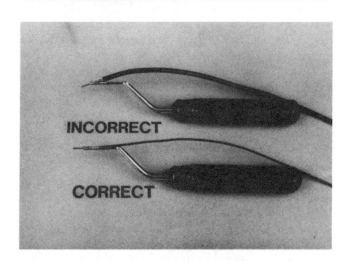

FIGURE 4–31 Steps to install wire and pin. (a) Strip wire approximate length. (b) Select correct crimping tool and proper crimp setting. (c) Crimp wire to pin. (d) Inspect wire crimp installation. (e) Insert tool used to install pin/wire into the connector housing. *(The Daniels Mfg. Corp.)*

A special tool is used to install a front-release contact in the front of the connector housing. A rear-release contact is installed in the rear of the connector housing and is secured with two or more small tines as shown in Figure 4–32. The rear-release method provides better front-end support for the contact; therefore, the contact is less likely to bend during reassembly of the connector.

Contact Removal. To remove a contact from a connector assembly, the technician must first remove any outer-shell components to expose the wire and contact to be removed. A special removal tool is slid gently into the connector housing; it releases the locking tabs holding the contact in place. In many cases a double-ended tool is used for both contact installation and removal. This helps to eliminate confusion

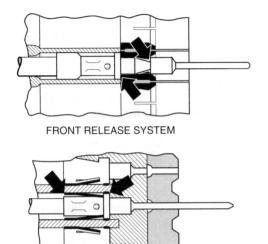

FRONT RELEASE SYSTEM

REAR RELEASE SYSTEM

FIGURE 4–32 Pin retention systems. *(The Deutsch Co.)*

when repairing defective connections. To remove a rear-release contact, simply slide the tool over the wire and onto the contact. Once the tool has reached the end of its travel, the locking tabs have been depressed and the contact can be removed. A similar procedure is used for the removal of front-release contacts; however, extra care must be taken not to damage front-release components. Figure 4–33 shows the removal of front- and rear-release pins.

Potting. The process of encapsulating electric wires and components in a plastic or similar material is called **potting.** Potting is typically used for the purpose of reducing vibration stress or inhibiting moisture transfer. The process is sometimes recommended for certain components and should be accomplished as directed by the appropriate instructions.

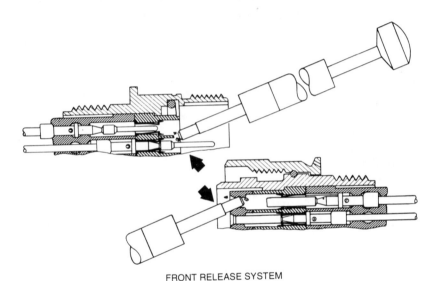

FRONT RELEASE SYSTEM

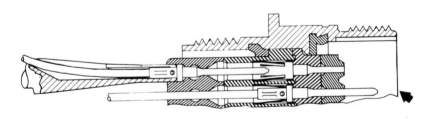

REAR RELEASE SYSTEM

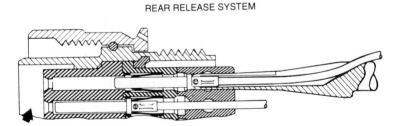

FIGURE 4–33 Removing contact pins. Releasing the retention system from the rear assures that the intermating seal on either the socket or the pin insert is not damaged by the contact removal tool. *(The Deutsch Co.)*

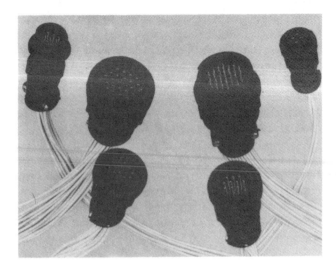

FIGURE 4–34 Typical light-duty connectors. *(AMP Products Corporation)*

With the invention of a variety of environmentally sealed connectors, potting is used only in limited situations on modern aircraft.

If potting is to be performed, the first task is to thoroughly clean all components. Petroleum solvent, methylene chloride, or some other specified cleaner may be employed. Potting is accomplished by forming a cavity with nylon or a plastic membrane around the item to be encapsulated. An approved potting compound is then mixed and poured into the cavity of the potting boot. The potting compound should cure for the appropriate time set forth by the manufacturer's data.

Many potting compounds in the unmixed state consist of the base compound and an accelerator or catalyst. After the two materials are mixed, they will harden beyond use within about 30 min. It is therefore important that work be planned so the materials will be used within the working time of the compound.

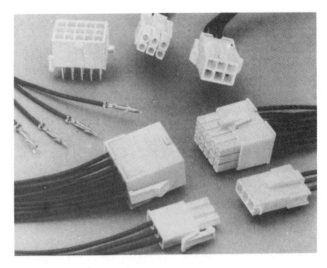

FIGURE 4–35 Various pins and housings of typical connectors. *(AMP Products Corporation)*

There are several means of potting an electrical connection. In each case, always follow the manufacturer's instructions carefully to ensure a safe electric circuit.

Light-Duty Connectors. Recently, many connectors using nylon or plastic housing have been employed on aircraft. These connectors are used in applications where a high stress resistance or waterproofing is not required. Several manufacturers produce these connectors; therefore, it is vital to ensure complete compatibility prior to installing pins, sockets, or connectors. Figure 4–34 shows a variety of light-duty connectors. Typically these connectors are limited to use in light aircraft, in areas not critical to flight safety.

The housings of light-duty connectors are produced by molding a nylon or plastic material into a single unit. Two mating housings produce one connector. The connectors may be designed to fasten to an electrical unit or installed on the end of a wire or a wire bundle. The pins and sockets are connected to their respective wires by means of a special crimping tool designed by the connector manufacturer. Pins and sockets are installed by pushing them into the housing until

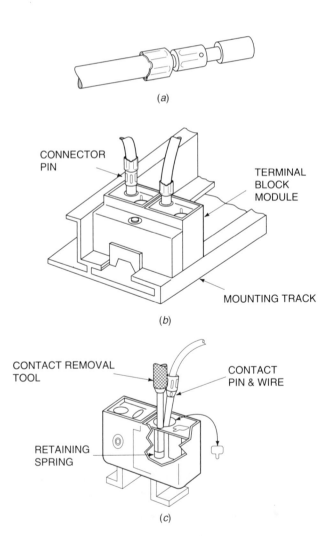

FIGURE 4–36 Terminal block module. (*a*) Connector pin; (*b*) module installed on track; (*c*) contact removed.

they are locked in place. To remove a pin or socket from its housing, a special tool is needed to depress the locking tabs that hold the pin or socket in the housing. Once the tabs are depressed, the connection may be easily removed. Various housings and pins are shown in Figure 4–35. There are several variations of each connector available; be sure to choose any replacement using current manufacturer's data.

Terminal Blocks

Another type of quick connect/disconnect system used for individual wires is known as the **terminal block.** This system is typically foun d on large aircraft in areas not critical to flight safety, such as passenger entertainment or reading light systems. Terminal blocks are typically made up of individual modules that connect two or more wires. The modules are designed to fit into a mounting track. The track is then mounted to the aircraft's structure. Tracks come in various lengths to accommodate one or several modules. As seen in Figure 4–36, the wire is attached to a special crimped-on pin; the pin then fits into the module. To insert the pin, simply apply light pressure using the installation tool. To remove the wire and pin, simply slide the removal tool into the module alongside the pin. This will relax the retaining spring and release the wire.

BONDING AND SHIELDING

Bonding

Bonding is the process of electrically connecting the various metallic parts of an aircraft or some other flight vehicle so that they will collectively form an integral electric unit. That is, there will be a very low resistance path for current from any one part of the structure to any other part. The specific purposes of bonding are to provide a low-resistance path for electric equipment, thereby eliminating ground wires; to reduce radio interference; to decrease the probability of lightning damage to such aircraft elements as control hinges; and to prevent the buildup of static charges between parts of the structure, thus reducing the fire hazard that could result from spark discharges between these parts.

Bonding should take place between any components that do not have a consistent connection to the airframe.

A **bonding jumper** is a short length of metal braid or metal strip with a terminal at each end for attaching to the structure. These jumpers should be as short as practicable and installed in such a manner that the resistance of each connection does not exceed 0.003 Ω. They should also be installed in locations that provide reasonably easy access for inspection and maintenance. Care must be taken so that bonding jumpers do not interfere with the operation of any movable parts of the aircraft and so that the normal movement of such parts does not result in damage to the bonding jumpers.

When bonding jumpers are installed, it is important that all insulating coatings, such as anodizing, paint, oxides, and grease, be removed so that clean, bare metal surfaces come into contact. After the bonding is secured, it is good practice to coat the junctions with a sealing coating to prevent the entrance of moisture, which could produce corrosion. Electrolytic corrosion may occur quickly at a bonding connection if adequate precautions are not taken. Aluminum-alloy jumpers are recommended in most cases, but copper jumpers are used to bond together parts made of stainless steel, cadmium-plated steel, copper, brass, and bronze. Where contact between dissimilar metals cannot be avoided, the choice of jumper and hardware should be such that corrosion is minimized and the part likely to corrode is the jumper or hardware.

A guide to the selection of metals that can be joined without the danger of corrosion is given in the following grouping of metals. Metals in any one group may be joined with a minimum likelihood of corrosion.

Group 1. Magnesium alloys
Group 2. Zinc, cadmium, lead, tin, steel
Group 3. Copper and its alloy, nickel and its alloys, chromium, stainless steel
Group 4. All aluminum alloys

The screws, washers, nuts, bolts, or other fasteners used for securing bonding jumpers must be of a material that is compatible with the metals being joined. For example, where aluminum jumpers are attached to aluminum-alloy structures, the fasteners should be made of aluminum.

The use of solder to attach bonding jumpers should be avoided for the same reasons that it is not recommended for electric wire connections. Tubular members should be bonded by means of clamps or clamp blocks as illustrated in Figure 4–37. In this installation a thin aluminum strip lines both inner surfaces of the clamp block, and the ends of the metal strips are carried around the ends of the clamp block so that they make contact with the aircraft structure. Bonding braid may be connected between electrically separate parts of a structure as shown in Figure 4–38. Each terminal of the bonding braid is securely attached to the sheet metal structure by means of machine screws and nuts. A typical attachment of a bonding jumper to an aluminum alloy structure is shown in Figure 4–39.

When a bonding jumper is installed in a location where it will be required to carry a **ground load** for a unit of electric equipment, care must be taken to assure that the jumper has

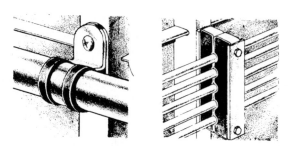

FIGURE 4–37 Bonding for tubular members.

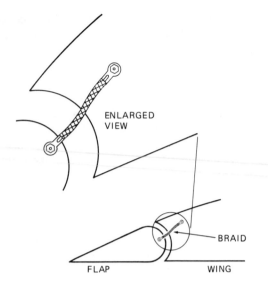

FIGURE 4–38 Installation of bonding braid.

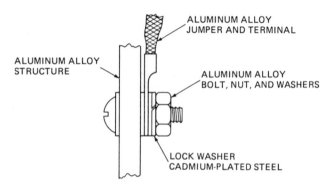

FIGURE 4–39 Attachment of a bonding jumper to an aluminum structure.

sufficient capacity to carry the load. This could occur when an electronic unit is mounted on shock mountings. If the equipment is grounded through the case, the ground current must be carried through the mounting structure and then through a bonding braid to the main structure. A bonding braid too small to carry the load could overload and melt, thus creating a fire hazard and also causing the equipment to fail.

Shielding

With the increased number of highly sophisticated and extremely sensitive electronic devices found on modern aircraft, it has become very important to ensure proper shielding for many electric circuits. **Shielding** is the process of applying a metallic covering to wiring and equipment to eliminate interference caused by stray electromagnetic energy. Shielded wire (or cable) is typically connected to the aircraft's ground at both ends of the wire, or at connectors in the cable. The word *shielding* is also applied to the process of enclosing wires or electric units with metal.

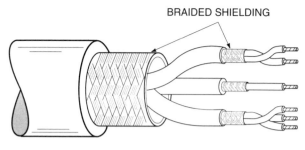

FIGURE 4–40 Shielded cable.

As noted above, the purpose of shielding is to reduce the adverse effects of **electromagnetic interference (EMI).** EMI is caused when electromagnetic fields (radio waves) induce high-frequency (HF) voltages in a wire or component. The induced voltage can cause system inaccuracies or even failure, hence putting the aircraft and passengers at risk. Shielding helps to eliminate EMI by protecting the primary conductor with an outer conductor, called the *shield* (see Figure 4–40). The induced voltage that would normally reach the primary conductor is induced in the shield and sent directly to the aircraft's electrical ground. Shielding is used when one unit is to be protected from the effects of a high-frequency (HF) current in an adjacent unit, or when a wire is to be protected from high-energy electromagnetic fields generated by the airborne or ground equipment. Protecting one unit from the interference of another is called **electromagnetic compatibility.** The interference caused by **high-energy radiated (electromagnetic) fields** is known as **HERF.** Shielding also eliminates the ability of any primary conductor or electrical unit to generate its own interference.

Shielding of a wire is typically done by surrounding the primary conductor(s) with a finely braided copper wire, as shown in Figure 4–40. This technique provides adequate protection for most circuits; however, some very sensitive equipment may require the use of a second shield, or two braided conductors. In some cases a thin metallic foil is wrapped around the primary conductor and used for the shield. Unfortunately, this technique increases the wire's rigidity. Another type of EMI shield is made of a composition of ferrite and polymers. Cable with this shield is lighter in weight and has good vibration resistance.

WIRE IDENTIFICATION

To facilitate installation and maintenance, all wiring should be indelibly marked with **wire identification numbers.** Any consistent numbering system is considered adequate if the numbers are placed at each end of each section of wire and also at intervals along the wire. To accomplish this marking during aircraft assembly, the wire is usually run through a numbering machine, which stamps the numbers along the wire at specified intervals. The identification numbers and letters should clearly show the circuit in which the wire is installed, the particular wire in the circuit, the wire gage, and other pertinent information. Care must be taken when marking coaxial cable with a machine. If too much pressure is applied to the cable, it may be flattened, and this will change the electrical characteristics of the cable.

Electric wires or cables may be identified by both numbers and letters, especially on large aircraft. For example, on a typical airliner the following letter system is used to identify specific circuits:

AC power	X
Deicing and anti-icing	D
Engine control	K
Engine instrument	E
Flight control	C
Flight instrument	F
Fuel and oil	Q
Ground network	N
Heating and ventilating	H
Ignition	J
Inverter control	V
Lighting	L
Miscellaneous	M
Power	P
Radio navigation and communication	R
Warning devices	W

Numbers used with the letters for identification also have a specific purpose. In the identification number 2P281C-20, there are two letters and three separate numbers. The number *2* indicates that the wire is associated with the No. 2 engine, and the letter *P* means that the wire is a part of the electric power system. The number *281* is the basic wire number and remains unchanged between the electric units of any particular system regardless of the number of junctions the wire may have. The letter *C* identifies the particular section of wire in the circuit, and the number *20* indicates the gage of the wire.

On large, complex aircraft, the wire-numbering system may also include a **wire bundle code** to identify which bundle contains a specific wire. This system helps the technician to find the desired wire quickly without searching individual bundles. The above system is only an example of a typical wiring code; many different systems have been devised by manufacturers. It is therefore very important to be familiar with the wiring code of the aircraft currently being serviced. Most manufacturers include a wire code explanation in the maintenance manual for each aircraft.

Identification markings are usually stamped directly on the insulation of the wire or cable. In an aircraft factory, the markings are stamped on the wires by means of a marking machine. The markings are placed at each end of a wire or cable section and at intervals of 12 to 15 in. [30.5 to 38.1 cm] along the wire. Sections of wire that are less than 3 in. [7.6 cm] long need not be marked. Wires that are 3 to 7 in. [17.8 cm] long should be marked midway between the ends.

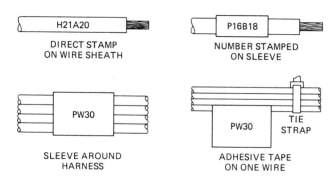

FIGURE 4–41 Methods of marking wires and harnesses.

If the outer coating or surface of a wire sheath is such that it cannot be easily marked, sleeving or tubing can be marked and placed over the wire. High-temperature wires, shielded wiring, multiconductor cable, and thermocouple wires usually require special sleeves to carry identification marks. Metallic sleeves or bands must not be used on electric wires.

Wire harnesses are often identified by number to indicate the particular section installed in a system. These harnesses are identified by means of a marked sleeve or pressure-sensitive tape. Methods of marking wires and harnesses are shown in Figure 4–41.

When wires are installed in an aircraft for an additional circuit or when existing wires are replaced, the technician should label each of the new wires with the appropriate code. A special type of heat shrink tubing is often used for this purpose. The tubing is designed to accept writing prior to being shrunk to fit the wire or cable. The original wire-marking identification is to be retained whenever possible to facilitate maintenance and service. In most maintenance situations, a wire-marking machine is not available; therefore, an appropriate stamped sleeve should be used for wire identification.

Wiring and Schematic Diagrams

During the design, manufacture, and repair of electrical systems, it is imperative to understand the various current paths and types of wire being used in each system. For this purpose electrical **diagrams,** or **schematics,** are included in the maintenance and installation data for the aircraft's electrical systems. The schematics use various symbols to represent different types of wire and connections within a circuit. A schematic can be thought of as a "road map" that helps technicians to find their way around an electric circuit. In some cases wire identification numbers are also included on the electrical schematic.

If two wires are shown to intersect on a schematic, it does not always mean that they are electrically connected. On a wiring diagram there are two common ways to show electric wires that intersect. Two wires that intersect on the diagram and *do not* connect in the actual circuit are shown in Figure 4–42a. Two wires that intersect on the wiring diagram and *do* form an electrical connection are shown in Figure 4–42b. To

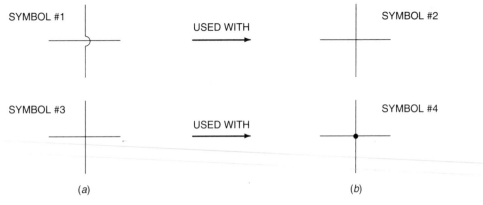

SYMBOL #1   USED WITH →   SYMBOL #2

SYMBOL #3   USED WITH →   SYMBOL #4

(a)   (b)

FIGURE 4–42 Wiring diagram symbols for electric wire. (a) Wires without an electrical connection; (b) wires with an electrical connection.

eliminate confusion, symbols 1 and 2 are always used together on a specific diagram, and symbols 3 and 4 are used together. To further your understanding of wiring diagrams, review the set of schematic symbols found in the Appendix of this text.

For most diagrams a single wire is represented by a solid black line. Shielded cable is typically shown with a dotted line surrounding a solid line; this represents one conductor inside another. An electrical ground symbol means the wire is terminated at the metal structure of the airframe. On composite aircraft, the ground symbol may mean that the wire is terminated at a central location of the aircraft, which is then connected to the negative terminal of the battery. Symbols for plugs, wire connections, and other items can vary from the examples shown in the Appendix, so be sure to refer to the manufacturer's data prior to maintenance on any electrical system. Refer to Chapter 13, "Design and Maintenance of Aircraft Electrical Systems," for more information on schematics.

REVIEW QUESTIONS

1. List the various approved aircraft electric wires.
2. Why is aircraft wire stranded?
3. What types of materials are used for wire insulation?
4. Describe shielded wire or cable.
5. What is the maximum allowable temperature for silver-coated copper wire?
6. What is the maximum allowable temperature for aluminum wire?
7. What is a coaxial cable?
8. How is the cross-sectional area of a wire indicated?
9. What is the circular-mil dimension of a wire 0.050 in. in diameter?
10. Describe the method used to determine a wire size using a wire gage tool.
11. What voltage drop is allowed for a 115-V system in a continuously operating circuit?

12. What size of copper wire should be used in a 28-V circuit for a 10-A continuous load when the wire is in a conduit and the distance from the bus to the load is 20 ft [6.1 m]?
13. What size of copper cable should be installed for a starter that draws 200 A?
14. What size of aluminum wire should be used for a 200-A continuous load?
15. What is the minimum bend radius for a wire bundle?
16. What can be done to protect wire and wire bundles from abrasion?
17. Describe a satisfactory method for wire-bundle lacing.
18. What method is employed to prevent cable or wire bundles from vibrating, swinging, or otherwise moving?
19. Describe an approved clamp for wire bundles.
20. Describe how wire bundles should be routed in an aircraft.
21. Describe electrical conduit.
22. When it is impractical to separate an electric wire from a line carrying flammable fluids, where should the wire be located with respect to the fluid line?
23. Describe typical terminals used with aircraft electric wire or cable.
24. Describe the use of wire strippers.
25. Why is it important to use specific types of crimping tools when installing wire terminals?
26. Why are soldered terminals not recommended for aircraft electric systems?
27. Describe the proper method of soldering if a soldered terminal is used.
28. How may aircraft wires be spliced?
29. Describe a terminal strip.
30. What is the maximum number of terminals that should be attached to a single stud?
31. What is the position requirement for mounting a terminal strip?
32. Describe the use of washers and lock washers in the installation of studs with terminals.
33. Describe the installation wire in a junction box.

34. When electric wires pass through holes in bulkheads or the side of a junction box, what protection must be provided?

35. Discuss the precautions necessary in the installation of aluminum wire in an aircraft system.

36. What is the advantage provided by the use of connectors in electric systems?

37. Describe the procedure for soldering wires to connectors.

38. Describe the installation of pins and sockets in crimp-type connectors.

39. Describe the process of potting the wire connections to an electric connector.

40. Explain the purposes of bonding in an aircraft.

41. What is the maximum resistance permitted for a bonding connection?

42. Describe the procedure for installing a bonding jumper.

43. What is the requirement for a bonding jumper that must carry a ground load for a unit of electric equipment?

44. Explain the importance of noting the types of metals joined by bonding.

45. What is the purpose of shielding?

46. How is shielding accomplished?

47. Explain how electric wires are identified in a system.

48. Where should wire identification marks be placed on a wire or bundle?

49. What methods are used for attaching identification numbers and letters to wires and harnesses?

Alternating Current 5

A thorough understanding of alternating current is becoming increasingly important to aviation maintenance technicians, because modern aircraft utilize this type of power for both flight and ground operations. Those who wish to be classified as master maintenance technicians, those who expect to become electrical and electronics specialists, and others who seek supervisory positions in the aircraft industries should study this chapter most carefully. Every principle discussed should be completely understood before the next is studied.

In recent years, alternating current (ac) has become increasingly popular for powering aircraft systems. The advances in modern electronics have made it possible for even light, single-engine aircraft to maintain small ac power systems. Larger ac power systems are now employed on virtually all modern transport-category aircraft. Most of the electrical systems found on these aircraft operate on ac power, although a direct current (dc) emergency system is still used. The dc emergency system must be maintained because technology has yet to produce an ac storage battery, and a battery may be the sole power source during emergency situations. In the future, however, one might find aircraft powered totally by ac systems, including emergency backups.

Some of the units operated by alternating current in airplanes are instruments, fluorescent lights, radio equipment, electric motors, navigation equipment, and automatic pilots. This list does not include all the devices that are or may be operated by alternating current, nor is it intended to indicate that all types of the above-named devices require alternating current. Airplanes not carrying the equipment listed or any other equipment requiring alternating current are very common. Some small airplanes have practically no electric equipment, but their utility is limited to daylight flight.

A good knowledge of the principles of alternating current is essential for the understanding of various electric devices. This is especially true with regard to ac electric motors, alternators, and transformers. This chapter explains the nature of alternating current and many of its characteristics and uses.

DEFINITION AND CHARACTERISTICS

Alternating current is defined as current that periodically changes direction and continuously changes in magnitude. The current starts at zero and builds up to a maximum in one direction, then falls back to zero, builds up to a maxi-

mum in the opposite direction, and returns to zero. In like manner, the voltage attains a maximum in one direction, drops to zero, rises to a maximum in the opposite direction, and then returns to zero. Voltage (electrical pressure) does not actually flow; therefore, when voltage changes direction, the positive and negative values simply reverse. That is, polarity of the circuit reverses.

It is difficult for some students to visualize the nature of alternating current, but there are many common devices that can be used to illustrate this principle. First, consider reciprocating (moving back and forth) devices such as a carpenter's saw, a connecting rod in an engine, or the pendulum in a clock. Each of these devices performs useful work with a reciprocating motion. Figure 5–1 shows a hydraulic analogy of an ac circuit performing work. The pump forces the fluid back and forth in the pipes and causes the working piston to move back and forth. This piston is connected to a crankshaft, which converts the reciprocating motion of the piston into the rotary motion of the flywheel.

Values of alternating current and voltage are indicated by a **sine curve** or **sine wave**. In Figure 5–2, this curve represents a definite voltage or current value for a certain degree of

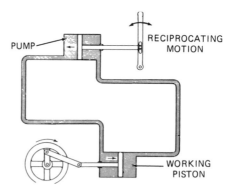

FIGURE 5–1 Hydraulic analogy of alternating current.

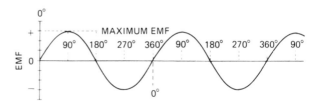

FIGURE 5–2 A sine curve.

rotation through the alternating cycle. One cycle begins at 0° and ends at 360°. The value of the alternating current is zero at 0°, maximum in one direction at 90°, zero at 180°, maximum in the opposite direction at 270°, and zero at 360°, as shown in the sine curve. The values of 360° and 0° along the horizontal axis of the sine wave are virtually identical. At 360° (0°), one cycle ends and the next cycle begins.

For practical purposes, the values of an alternating current can be considered to follow the sine curve. This can be understood by considering the generation of alternating current by a simple generator (see Figure 5–3). A single loop of wire is placed so that it can be rotated in a magnetic field. As the loop is turned, the sides of the loop cut through the lines of force, and an electromagnetic force (emf) is induced in the sides of the loop. At position 1 the conductor moves parallel

to the flux lines as the loop rotates in a clockwise direction. The voltage increases as the loop moves from position 1 to position 2. At position 2 the induced voltage is at a maximum value. This occurs because the loop is moving perpendicular to the flux lines. As the conductor moves beyond position 2, the voltage decreases. The voltage once again reaches zero at position 3. One-half of a revolution has been completed at position 3. As the conductor passes this position, the induced voltage reverses. This reversal occurs because side B of the conductor loop now moves down and side A moves up. Originally, side A moved down and side B moved up; hence the induced voltage reverses polarity because the flux lines are being cut in the opposite direction. The negative voltage increases until position 4 and then decreases as the conductor loop travels back to its original position. At position 5 the

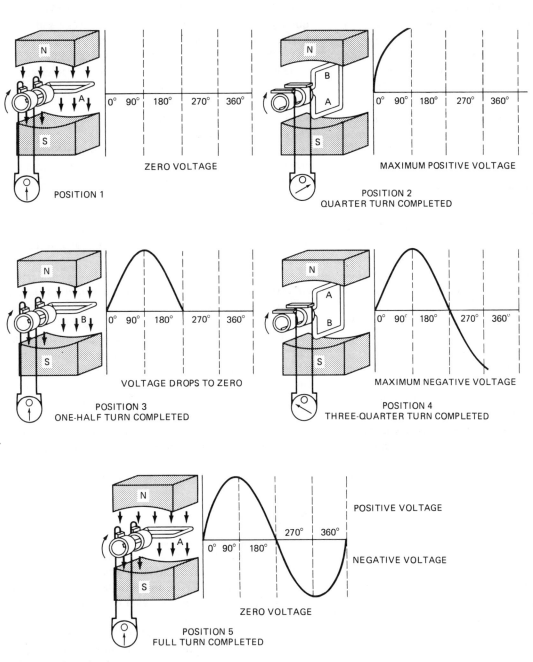

FIGURE 5–3 A simple ac generator.

loop has made one complete cycle, and the process repeats. At position 5 it can be seen that the rotating loop has produced a sine wave of voltage-values.

When the current is carried to an external circuit by means of slip rings, it travels in one direction while the loop moves from 0 to 180°, and in the other direction while the loop moves from 180 to 360°. When it is horizontal, the loop is in either the 0 or 180° position, and no voltage is induced. When the loop is in a vertical position, the maximum voltage is induced because at this time the sides are cutting the greatest number of lines of force.

It has been found that the instantaneous value of the voltage induced in a loop as it rotates in a magnetic field is proportional to the sine of the angle through which the loop has rotated from 0°. Hence we use the sine curve to represent the values from 0 to 360°. The value of either the voltage or the amperage can be represented in this manner.

As illustrated in Figure 5–3, the sine wave is above the horizontal axis as current travels in one direction and below the horizontal axis as current travels in the other direction. Often the values above the curve are considered positive and values below the curve, negative. The assignment of positive and negative is completely arbitrary; however, they do represent a change in the direction of current flow. This concept of changing direction can also be applied to the voltage of an ac circuit.

RMS, or Effective, Values

In order to determine the amount of power available from an alternating current, we must arrive at its effective value. It is obvious that effective value does not equal maximum value, because maximum value is attained only twice in the cycle. Even though the current during one half-cycle is equal and opposite in direction to that during the other half-cycle, the currents do not cancel each other; work is done whether the current is moving in one direction or the other. Therefore, the effective value must lie somewhere between the zero value and the maximum value.

The effective value of an alternating current is calculated by comparing it with direct current. The comparison is based on the amount of heat produced by each current under identical conditions. Since the heat produced by a current is proportional to the square of the current ($P = I^2R$), it is necessary to find the square root of the mean square of a number of instantaneous values. The resultant value is called the **root-mean-square (rms)** current. In other words, all the instantaneous values of the sine wave are squared, the results are added together, and the mean is determined. The square root of the mean is equal to the effective voltage. For all general-purpose applications, the effective voltage or current can be determined using the following equations:

$$E_{eff} = 0.707 \times E_{max}$$
$$I_{eff} = 0.707 \times I_{max}$$

Conversely,

$$E_{max} = 1.414 \times E_{eff}$$
$$I_{max} = 1.414 \times I_{eff}$$

In all practical applications of alternating current, the values of voltage, or current, are stated according to their effective values rather than the maximum values. For example, when the voltage is given as 110 V, the maximum value of the voltage is 1.414×110 V $= 155.6$ V. Keeping this in mind, technicians should always make certain that any instrument or equipment that they use with a nominal voltage rating in alternating current has a safety factor sufficient to handle the maximum voltage.

Frequency

It has been explained that one cycle of alternating current covers a period in which the current value increases from zero to maximum in one direction, returns to zero, increases to maximum in the opposite direction, and then returns to zero. The number of cycles occurring per second is the **frequency** of the current and is measured in a unit called the **hertz,** named for Heinrich Rudolph Hertz, a German physicist of the late nineteenth century who made a number of important discoveries and valuable contributions to electrical science. One hertz (Hz) is equal to 1 cycle per second [1 cps]. The terms **kilohertz** and **megahertz** are often used when describing radio frequencies. One kilohertz is equal to 1000 Hz, and one megahertz is equal to 1,000,000 Hz.

City lighting and power systems in the United States generally operate at a frequency of 60 Hz. Alternating currents in airplane circuits usually have a frequency of 400 Hz. This frequency is commonly used for modern aircraft as well as for a number of other applications. The word **alternation** is frequently used in discussing alternating current, and it means one-half cycle. It is apparent, therefore, that there are 120 alternations in a 60-Hz current.

The frequency of an alternating current has a considerable effect on the operation of a circuit, for many units of electric equipment operate only on current of a certain frequency. Wherever such equipment is used, it is important to make sure it is designed for the frequency of the current in the circuit in which it is to be used. Units such as synchronous motors operate at speeds proportional to the frequency of the current, even though the voltage is somewhat lower or higher than the rated voltage of the machine. It is also important to remember that a circuit designed for a given frequency may be easily overloaded by using a current of a different frequency, even though the voltage may remain the same. This is because of effects of inductive and capacitive reactance, which will be explained later in this chapter.

Phase

The phase of an alternating current or a voltage is the angular distance it has moved from 0° in a positive direction. The phase angle in electrical equations is usually represented by the Greek letter theta (θ). The **phase angle** is the difference in degrees of rotation between two alternating currents or voltages, or between a voltage and a current. For example, when one voltage reaches maximum value 120° later than another, there is a phase angle of 120° between the two voltages. Figure 5–4a shows a 120° phase difference between three

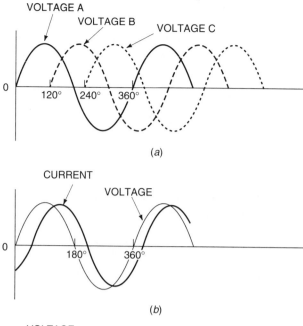

(a)

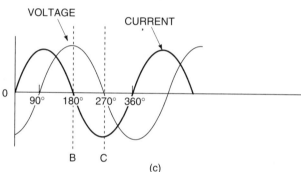

(b)

(c)

FIGURE 5–4 Out-of-phase voltage and current curves.
(a) Voltage curves of a three-phase circuit; (b) voltage leads
current; (c) current leads voltage.

different voltage curves. This type of phase relationship is very common in aircraft circuits that employ a three-phase ac electrical system. Three-phase systems are known as **polyphase circuits** and will be discussed later in this chapter.

In most ac circuits a phase shift exists between voltage and current. Figure 5–4b shows sine curves representing a current lagging the voltage, or voltage leading the current. In circuits where the current and voltage do not reach maximum at the same time, they are said to be *out of phase.* In Figure 5–4b, notice that the heavy current line crosses the zero line after the light voltage line has crossed it. This means that the current reaches zero after the voltage. In like manner, the peak value of current occurs after the peak value of voltage. For this reason we know that the current is lagging the voltage by several degrees. In Figure 5–4c, it will be seen that the voltage is approximately 90° out of phase with the current. That is, the voltage follows the current by approximately 90°.

Capacitance in AC Circuits

Capacitance can be defined as the ability to store an electric charge. Most capacitance in a circuit is created by a device called a **capacitor.** Capacitor theory is discussed in

detail in Chapter 6; that section should be studied carefully to gain a full understanding of how capacitors react in an ac circuit. In short, capacitors oppose the change of current flow in a circuit. In an ac circuit, since the current is constantly changing in magnitude and direction, a capacitor will create a constant opposition to the applied current. This opposition to current is similar to a resistance; however, it also creates a phase shift within the circuit.

When a capacitor is connected in series in an ac circuit, it appears that the alternating current is passing through the capacitor. In reality, electrons are stored first on one side of the capacitor and then on the other, thus permitting the alternating current to flow back and forth in the circuit without actually passing through the capacitor.

A hydraulic analogy can be used to explain the operation of a capacitor in a circuit (see Figure 5–5a). The capacitor is represented by a chamber separated into two sections by an elastic diaphragm. The ac generator is represented by the piston-type pump. As the piston moves in one direction, it forces fluid into one section of the chamber and draws it out of the other section. The fluid flow represents the flow of electrons in an electric circuit. Thus it can be seen that there is an alternating flow of fluid in the lines and that work is done as the fluid moves back and forth, first filling one side of the chamber and then the other.

As seen in Figure 5–5b, the operation of a capacitor in an ac circuit is for all practical purposes identical to the operation of the chamber just described. The electrons build up on one plate of the capacitor, and this negative charge forces the electrons to flow away from the other plate. As the ac current reverses direction, the capacitor is charged with the negative charge moving to the opposite plate on the capacitor. With each ac cycle, as voltage from the source begins to drop, the current starts to flow out of one plate of the capacitor and into the other; and the cycle repeats as long as the current is flowing.

This constant charging and discharging of the capacitor creates an electrostatic field and dielectric stress within the capacitor. A **dielectric** is an insulating material used to separate the plates of a capacitor. The dielectric stress is similar to

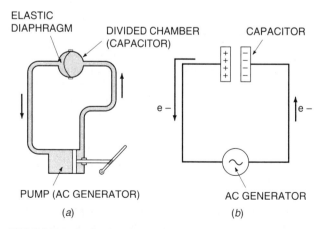

(a)   (b)

FIGURE 5–5 Hydraulic analogy of a capacitor. (a) The hydraulic pump forces fluid into the right side of the divided chamber; (b) the ac generator forces electrons into the right side of the capacitor.

the stress of the elastic diaphragm of the hydraulic circuit (Figure 5–5a). The dielectric stress creates a force that opposes the applied current. In other words, the capacitor will create a current flow in the opposite direction of the applied current. This current flow has two effects on the circuit: (1) it opposes, or "resists," the applied current, and (2) it creates a **phase shift** between voltage and current.

The phase shift in capacitive circuits causes the current to lead the voltage. This phase shift causes current to reach its maximum and minimum values before the voltage of the circuit. If it were possible to have a circuit with only capacitance and no resistance, the current would lead the voltage by 90° (Figure 5–4c). Studying Figure 5–4c, it can be seen that as the voltage rises, the current begins to drop because of the dielectric stress in the capacitor. This, of course, means that opposition to the flow of current is developing. By the time the voltage has reached its maximum value, the capacitor is completely charged; hence no current can flow. At this point B the current has a value of zero. As the voltage begins to drop, the current flows out of the capacitor in the opposite direction because the potential of the capacitor is higher than the potential on the applied voltage. By the time the voltage has dropped to zero, the current is flowing at a maximum rate because there is no opposition. This point on the curve is represented by the letter C.

It must be remembered that the above action takes place only when there is no resistance in the circuit. Since this is impossible, a circuit in which the current leads the voltage by as much as 90° does not exist. However, the study of such a circuit gives the student a clear understanding of the effect of capacitance. In an ac circuit where both capacitance and resistance are present, the phase shift will be between 0 and 90°.

The effects of capacitance in ac circuits are most pronounced at higher frequencies. Modern electronic circuits often produce frequencies of many millions of cycles per second (Hz). For this reason special types of electronic and electric devices and equipment have been designed to reduce the effects of capacitance where these effects are detrimental to the operation of the circuit.

Capacitive Reactance

If capacitance is considered the *ability* to oppose changes in current flow, then **capacitive reactance** is the *actual* opposition to current flow in a given ac circuit. Since capacitive reactance opposes the flow of current in ac circuits, it is measured in ohms. It should be noted that capacitive reactance also creates a phase shift in the circuit and therefore cannot be thought of as resistance. Capacitive reactance is represented by X_C and is a function of both the circuit's ac frequency and the total capacitance.

The capacitive reactance in a circuit is inversely proportional to the capacitance and the ac frequency. This is because a large-capacity capacitor will take a greater charge than a low-capacity capacitor; hence it will allow more current to flow in the circuit. If the frequency increases, the capacitor charges and discharges more times per second; hence

more current flows in the circuit. From the following equation for capacitive reactance, it can be seen that reactance will decrease as capacitance or frequency increases.

The formula for capacitive reactance is

$$X_C = \frac{1}{2\pi f C}$$

where X_C = capacitive reactance, Ω

f = frequency, Hz

C = capacitance, F

To determine the capacitive reactance in a circuit in which the frequency is 60 Hz and the capacitance is 100 μF, substitute the known values in the formula. Then

$$X_C = \frac{1}{2\pi \times 60 \times 100/1,000,000}$$

Remember that 1 μF is one-millionth of a farad; hence 100 μF is equal to 100/1,000,000 F. Therefore,

$$X_C = \frac{1}{6.283 \times 0.006} = \frac{1}{0.037698} = 26.5 \ \Omega$$

Inductance in AC Circuits

The effect of **inductance** in ac circuits is exactly opposite to that of capacitance. Capacitance causes the current to lead the voltage, and inductance causes the current to lag. Figure 5–6 shows the voltage and current curves for a purely inductive circuit. In order to completely understand the effects of inductive reactance, one should first study the section discussing the inductance coil in Chapter 6.

According to **Lenz's law,** whenever a current change takes place in an inductance coil, an emf (voltage) is induced that opposes the change in current. The induced voltage will then be maximum when the rate of current change is the greatest. Since the current change is most rapid in an ac circuit when the current is passing through the zero point, the induced voltage will be maximum at this same time, marked A in Figure 5–6. When the current reaches maximum, there is momentarily no current change, and hence the induced voltage is zero at point B. Remember, to induce a voltage in any circuit, there must be a current change; thus a "moving" magnetic field is created around the inductor coil. So at point B, where there is no current change, there will be no induced voltage. This effect causes the current to lag the voltage by 90° in a purely inductive circuit. But since a purely inductive

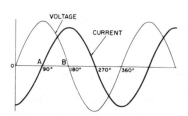

FIGURE 5–6 Current lagging voltage.

circuit is impossible because there is always resistance present, the current lag of 90° is purely theoretical. In an ac circuit where both resistance and inductance are present, the current will lag the voltage somewhere between 0 and 90°.

Inductive Reactance

The effect of inductance in an ac circuit is called **inductive reactance** and is measured in ohms because it "resists" the flow of current in the circuit. **Inductive reactance (X_L) is the *actual* opposition to current flow created by inductors in an ac circuit.** Inductance L is the ability of a coil to oppose changes in current flow. The inductive reactance of any given circuit is a function of the ac frequency and the inductance of that circuit.

The inductive reactance in a circuit is proportional to the inductance of the circuit and the frequency of the alternating current. As the inductance is increased, the induced voltage (which opposes the applied voltage) is increased; hence the current flow is reduced. Likewise, when the frequency of the circuit is increased, the rate of current change in the inductance coil is also increased; hence the induced (opposing) voltage is higher and the inductive reactance is increased. As inductive reactance increases, current in the circuit flow is reduced.

We can clearly see that the effects of capacitance and inductance are opposite, since inductive reactance increases as the frequency increases and capacitive reactance decreases as the frequency increases. The formula for inductive reactance is

$$X_L = 2\pi f L$$

where X_L = inductive reactance, Ω

f = frequency, Hz

L = inductance, H

Let us assume that an inductance coil of 7 H is connected in a 60-Hz circuit and it is necessary to find the inductive reactance. By substituting the known values in the formula,

$$X_L = 2 \times 3.1416 \times 60 \times 7 = 2638.94 \ \Omega$$

Combining Resistance, Capacitance, and Inductance

In practical applications found on a typical aircraft, there are various components that have resistance, capacitance, and inductance. In that case the circuit is known as an **RCL** circuit. A circuit containing only resistance is called a **resistive circuit (R).** Other circuits are known as **resistive inductive (RL),** and **resistive capacitive (RC).** Each name describes the types of elements that are contained in the circuit. For example, an RC circuit contains both resistive units and capacitive units. For any circuit that is not purely resistive, the total opposition to current flow is called *impedance*. As noted earlier, all circuits contain some resistance, inductance, and capacitance; however, in some cases the inductance and capacitance effects are considered negligible.

IMPEDANCE

In the study of Ohm's law for dc circuits, it was found that the current in a circuit was equal to the voltage divided by the resistance. In an ac circuit it is necessary to consider capacitive reactance and inductive reactance before the net current in such a circuit can be determined. The combination of resistance, capacitive reactance, and inductive reactance is called **impedance,** and the formula symbol is Z.

It might appear that we could add the capacitive reactance, inductive reactance, and resistance to find the impedance, but this is not true. Remember that capacitive reactance and inductive reactance have opposite effects in an ac circuit. For this reason, to find the total reactance we use the difference in the reactances. If we consider inductive reactance as positive, because inductance causes the voltage to lead the current, and capacitive reactance as negative, because it causes the voltage to lag, then we can add the two algebraically; that is,

$$X_L + (-X_C) = X_t \quad \text{or total reactance}$$

Now it might appear that we could add this result to the resistance to find the impedance, but again we must consider the effect of resistance in the circuit. We know that resistance in a circuit does not cause the current to lead or lag, and for this reason its effect is 90° ahead of inductance and 90° behind capacitance. Therefore, it is necessary to add resistance and reactance vectorially.

A **vector** is a quantity having both magnitude and direction. Vectors are often represented graphically by a line pointing in a given direction. Vectors may be used to represent a given force. The strength of the force is indicated by the length of the line representing the vector. The values for X_L, X_C, and R can be represented using a vector diagram as illustrated in Figure 5–7*a*. Resistance is always shown on the horizontal axis, inductive reactance on the vertical axis pointing up, and capacitive reactance on the vertical axis pointing down. As illustrated, it is easy to see that the effects of X_L and X_C cancel each other, and the effects of resistance are 90° from either reactance. As demonstrated in Figure 5–7*b*, using vector addition, the three vectors X_L, X_C, and R can be combined into one resultant vector called *impedance* (Z). The length of the impedance vector can be determined graphically or algebraically. The Pythagorean theorem, $A^2 + B^2 = C^2$, can easily be applied to solve for Z; that is,

$$X_t^2 + R^2 = Z^2$$

or

$$Z = \sqrt{R^2 + X_t^2}$$

Substituting for X_t,

$$Z = \sqrt{R^2 + (X_L - X_C)^2}$$

This formula is typically used to solve for Z or an unknown R or X value. It should be noted that this formula can be applied to determine the impedance in *series* circuits only. That is, the total values of resistance, capacitance, and inductance must be in series with each other, as shown in Figure 5–8. The equations for parallel circuits will be discussed later.

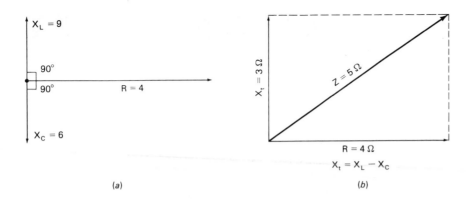

(a)

(b)

NOTE: X_t IS CONSIDERED TO BE INDUCTIVE
BECAUSE X_L IS LARGER THAN X_C.

FIGURE 5–7 Vector diagrams of resistance, reactance, and impedance.

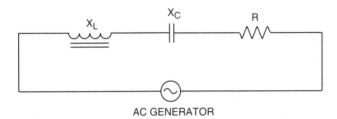

AC GENERATOR

FIGURE 5–8 A simple series ac circuit.

After the impedance is found in an ac circuit, the other values can be found by Ohm's law for alternating current. In this formula we merely substitute the symbol Z, meaning impedance, for the normal symbol R, meaning resistance. The formula then reads

$$I = \frac{E}{Z}$$

Sample Problem. If a series ac circuit contains an inductor with inductive reactance X_L equal to 12 Ω, a capacitive reactance X_C of 18 Ω, a resistance R of 5 Ω, and an applied voltage of 120 V, what is the current flow through that circuit?

Solution. In an ac circuit, $I = E/Z$; therefore, the value for Z must be determined.

$$Z = \sqrt{R^2 + (X_L - X_C)^2}$$
$$= \sqrt{5^2 + (12 - 18)^2}$$
$$= \sqrt{25 + 36}$$
$$= 7.8 \ \Omega$$

Note: Always subtract X_C from X_L, then square the result; $X_L - X_C$ may produce a negative number. This negative number will become positive when it is squared.

To find I, use $I = E/Z$:

$$I = \frac{120 \text{ V}}{7.8 \ \Omega}$$
$$= 15.4 \text{ A}$$

Current will lead voltage in this circuit because X_C is larger than X_L and therefore has a greater effect on the phase shift.

Phase Angle

As stated earlier, a *phase angle* is the angular distance between current and voltage in an ac circuit. The phase angle is designated by the Greek letter theta (θ). To better understand the phase shift created in an ac circuit, study the vector diagrams in Figure 5–9. As illustrated, θ is always measured between the horizontal line and the resultant Z vector. For a simple R, C, or L circuit, the resultant vector equals the R, C, or L vector. For a pure resistive (R) circuit, the phase shift is 0°. For a pure inductive (L) or capacitive (C) circuit, the phase shift angle is always 90°. Voltage leads current in the inductive circuit, and current leads voltage in the capacitive circuit.

To determine the exact value of a phase shift angle, the trigonometry function of sine, cosine, or tangent is used. Figure 5–10 demonstrates the calculations to find θ in an RL circuit. The sine of θ is employed in this case to find the angle's value. The sine of θ is determined to be 0.446; a calculator is then used to determine the actual angle of 26.5°. If the tangent was used to find θ, the calculations would be as follows:

$$\tan \theta = \frac{\text{opposite side}}{\text{adjacent side}}$$

$$\tan \theta = \frac{10}{20} = 0.5$$

$$\theta = \tan^{-1} 0.5 = 26.5°$$

The cosine function could be used in a similar manner to find the angle θ.

The value of the phase angle has great importance when an ac circuit is designed. A small phase angle means the resistance vector is large compared with the reactance vectors. A large phase angle occurs when the inductive or capacitive reactance vector is much larger than the resistance vector. If

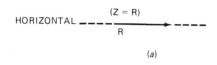

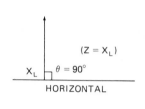

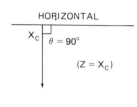

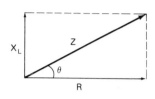

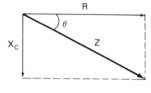

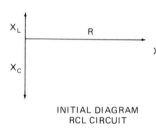

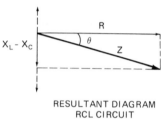

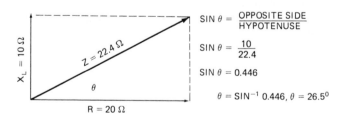

FIGURE 5–9 Various phase shift angles. (a) Pure resistive (R) circuit; θ = 0°. (b) Pure inductive (L) or pure capacitive (C) circuit; θ = 90°. (c) Resistive inductive (RL) or resistive capacitive (RC) circuit; θ is greater than 0° but less than 90°. (d) Resistive capacitive inductive (RCL) circuit; θ is greater than 0° but less than 90°.

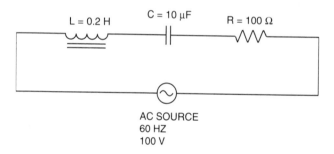

$$X_L = 2\pi f L$$
$$= 2\pi 60(0.2)$$
$$= 75.4 \ \Omega$$
$$X_C = \frac{1}{2\pi f C}$$
$$= \frac{1}{2\pi 60(0.00001)}$$
$$= \frac{1}{0.0038}$$
$$= 265.26 \ \Omega$$

The inductive reactance is 75.4 Ω; rounding gives 75 Ω. The capacitive reactance is 265.26 Ω; rounding gives 265 Ω.

Step 2. Find the circuit's impedance.

$$Z = \sqrt{R^2 + (X_L - X_C)^2}$$
$$= \sqrt{100^2 + (75.4 - 265.26)^2}$$
$$= \sqrt{46,046}$$
$$= 214.58 \ \Omega$$

The impedance is 214.58 Ω; rounding gives 214 Ω.

Step 3. Determine the phase shift angle (Figure 5–12). (*Note:* Current will lead the voltage, since X_C is larger than X_L.)

FIGURE 5–11 The sample series ac circuit for sample problem.

SIN θ = OPPOSITE SIDE / HYPOTENUSE

SIN θ = 10/22.4

SIN θ = 0.446

θ = SIN⁻¹ 0.446, θ = 26.5°

FIGURE 5–10 Phase angle calculations.

this situation occurs, the ac circuit becomes very inefficient and may overload the power source. The aircraft technician must be aware of this potential problem because of the large number of ac electrical systems found on modern aircraft.

Sample Problem. In a series ac circuit (see Figure 5–11), determine the total impedance Z, the total current flow I_t, and the phase angle θ.

Step 1. Find the inductive and capacitive reactance. (*Note:* Remember to convert microfarads to farads.)

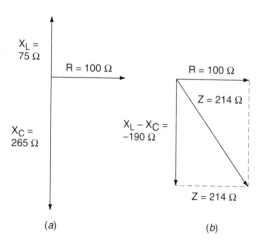

FIGURE 5–12 Vector diagram of the series ac circuit for sample problem.

$$\sin \theta = \frac{190}{214}$$

$$\sin \theta = 0.8879$$

$$\theta = 62.6°$$

The phase shift angle is 62.6°.

Step 4. Find the total current flow.

$$I_t = \frac{E_t}{Z}$$

$$= \frac{100 \text{ V}}{214 \text{ }\Omega}$$

$$= 0.47 \text{ A}$$

The total current flow equals 0.47 A.

Power Factor Calculations for AC Circuits

There are typically two types of power used to describe the work performed by an ac circuit. **True power** is the power consumed by the resistance of an ac circuit. **Apparent power** is the power consumed by the entire ac circuit. Thus apparent power takes into consideration the power consumed because of the resistance and the inductive and capacitive reactances. True power considers only resistance. Just as impedance is found vectorially, so is apparent power, as illustrated in Figure 5–13. Reactive power Q is placed on the vertical axis. **Reactive power** is a function of the total reactance of a circuit. True power P is placed on the horizontal axis, and its magnitude is found by $P = I^2R$. Apparent power U is the resultant vector and can be found using

$$U = \sqrt{P^2 + Q^2}$$

To distinguish between the types of power, true power is measured in **watts (W)**, apparent power is measured in **volt-amps (VA)** and reactive power is measured in **volt-amps-reactive (VAR)**.

Power factor (PF) is the ratio of true power to apparent power; a circuit's efficiency will determine the power factor. That is, the more inductive or capacitive reactance, the lower the circuit's efficiency and the smaller the power factor. Power factor can be calculated by

$$PF = \frac{\text{true power}}{\text{apparent power}}$$

$$= \frac{P}{U}$$

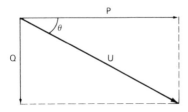

Q = REACTIVE POWER
P = TRUE POWER
U = APPARENT POWER

FIGURE 5–13 Apparent power vector diagram.

Since power factor is a ratio, it has no units of measure. If $P = 130$ W and $U = 140$ VA, the power factor can be calculated as follows:

$$PF = \frac{P}{U}$$

$$= \frac{130}{140}$$

$$= 0.928$$

A power factor of 1.0 indicates a purely resistive circuit. A power factor of zero indicates a purely reactive circuit.

Power factor can also be calculated by using the cosine of the phase shift angle. That is,

$$PF = \text{cosine } \theta$$

Therefore, if the power factor is known, the phase shift angle (not direction) is also known. The greater the reactance in a circuit for a given resistance, the larger the phase shift angle, the lower the power factor, and the lower the circuit's efficiency. Once you understand this relationship, it is easy to see why large quantities of inductive or capacitive reactance create inefficiencies and are undesirable in an ac circuit.

Most aircraft ac alternators contain critical specifications for both apparent power and a power factor range. Apparent power is used because it describes the power used by the entire circuit—that is, both resistance and reactive power. Apparent power would be measured in **kilovolt-amps (kVA),** and the power factor would be given as a range. For example, a particular aircraft alternator may be rated as follows: apparent power maximum of 200 kVA and a power factor range of 0.90 to 1.0. These specifications must never be exceeded; otherwise, the alternator may be internally damaged.

Parallel AC Circuits

As mentioned earlier, the calculations used to find impedance for parallel circuits are different from those used with series circuits. The letter Z is still used to represent impedance; however, for parallel circuits $Z_1 = 1/Y$. To solve for Y, one must use the following equation:

$$Y = \sqrt{G^2 + (B_L - B_C)^2}$$

$$\text{where } G_1 = 1/R$$

$$B_L = 1/X_L$$

$$B_C = 1/X_C$$

Here we can see that G, B_L, and B_C are simply the inverse of the resistance R, the inductive reactance X_L, and the capacitive reactance X_C for a given circuit. G, B_L, and B_C have no practical value in electrical terms; however, they must be used as an interim step whenever the total impedance of a parallel circuit is determined. Once you have calculated these values, the equation

$$Y = \sqrt{G^2 + (B_L - B_C)^2}$$

can be used to find Y. The value for Y must then be inverted to find Z, as shown next.

$$Z = \frac{1}{Y}$$

For parallel ac circuits, voltage leads current if B_L is greater than B_C. Voltage will lag current if B_C is greater than B_L (Figure 5–14a). The phase shift angle (θ) is determined by the angle between the resultant vector Y and the horizontal vector G as seen in Figure 5–14b.

As seen in the following example, the calculation of impedance Z for parallel circuits is relatively easy if you always remember to invert the values for R, X_L, and X_C before calculating for Y. Then simply invert Y to find total impedance Z.

Sample Parallel Problem. To find the total impedance and the phase shift angle for the circuit in Figure 5–15, take the following steps:

Step 1. Convert R, X_L, and X_C to G, B_L, and B_C.

$$G = \frac{1}{R} \qquad G = \frac{1}{5} \qquad G = 0.20 \ \Omega$$

$$B_L = \frac{1}{X_L} \qquad B_L = \frac{1}{12} \qquad B_L = 0.083 \ \Omega$$

$$B_C = \frac{1}{X_C} \qquad B_C = \frac{1}{4} \qquad B_C = 0.25 \ \Omega$$

Step 2. Solve for Y.

$$Z = \sqrt{G^2 + (B_L - B_C)^2}$$
$$= \sqrt{0.02^2 + (0.083 - 0.25)^2}$$
$$= \sqrt{0.0283}$$
$$= 0.2606 \ \Omega$$

Step 3. Invert Y to determine Z.

$$Z = \frac{1}{Y}$$
$$= \frac{1}{0.2606}$$
$$= 3.84 \ \Omega$$

The total impedance Z is equal to 3.84 Ω.

Step 4. Calculate the phase shift angle for the vector diagram in Figure 5–16.

$$\sin \theta = \frac{\text{opposite side}}{\text{hypotenuse}}$$
$$\sin \theta = \frac{0.167}{0.2606}$$
$$\sin \theta = 0.641$$
$$\theta = 39.8°$$

This circuit has a phase shift angle of 39.8°, and voltage lags current.

To determine the total current for the circuit, divide the total voltage by the total impedance as shown below.

$$I_t = \frac{E_t}{Z}$$
$$= \frac{120 \ V}{3.84 \ \Omega}$$
$$= 31.25 \ A$$

Total current is 31.25 A.

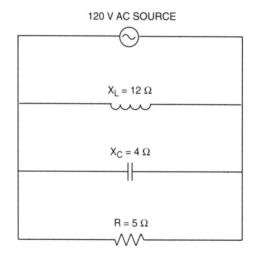

FIGURE 5–15 Circuit for the sample parallel problem.

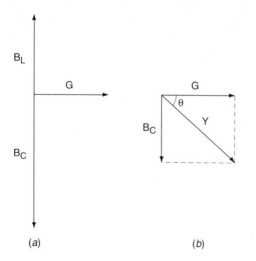

FIGURE 5–14 Vector diagrams of parallel ac circuits; θ is measured between the horizontal and resultant vectors.

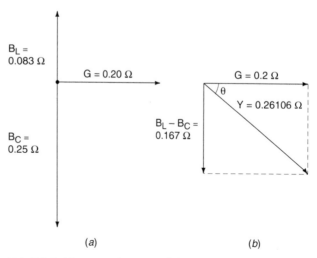

FIGURE 5–16 Vector diagrams of the parallel ac circuit.

Vector Addition of Voltage and Current

In a series dc circuit, total voltage is equal to the sum of the individual voltage drops. In a series ac circuit, however, the voltage total is equal to the voltage drops summed vectorially. That is, the voltage drops created by the resistances are summed, the voltage drops across the capacitances are summed, and the voltage drops across the inductances are also summed. These three totals are then added vectorially, with resistance voltage drops placed on the horizontal axis, inductive voltage drops placed vertically pointing up, and capacitive voltage drops placed vertically pointing down. This is illustrated in Figure 5–17a. The total voltage drop is then calculated using the equation

$$E_t = \sqrt{E_R^2 + (E_{X_L} - E_{X_C})^2}$$

The total current in a parallel dc circuit can be found by summing the current flows through each individual path. Total current in a parallel ac circuit is found by adding currents vectorially. The vector diagrams in Figure 5–17b demonstrate the summation of ac current flows. The equation

$$I_t = \sqrt{I_R^2 + (I_{X_L} - I_{X_C})^2}$$

is used to find I, where I_t is the total current in the ac circuit, I_R is the sum of the resistive current flows, I_{X_L} is the sum of the inductive currents, and I_{X_C} is the sum of the capacitive currents.

The following is a sample problem: If a parallel ac circuit contains two inductors carrying 2 A each, a capacitor carrying 1 A, and three resistors carrying 1.5 A, 0.5 A, and 3 A, what is the circuit's total current flow?

First, add the individual current flows for resistance, inductance, and capacitance.

$$I_R = 1.5\,\text{A} + 0.5\,\text{A} + 3.0\,\text{A}$$
$$= 5\,\text{A}$$
$$I_{X_L} = 2\,\text{A} + 2\,\text{A}$$
$$= 4\,\text{A}$$
$$I_{X_C} = 1\,\text{A}$$

To find I_t, use

$$I_t = \sqrt{I_R^2 + (I_{X_L} - I_{X_C})^2}$$
$$= \sqrt{(5\,\text{A})^2 + (4\,\text{A} - 1\,\text{A})^2}$$
$$= \sqrt{34}\,\text{A}$$
$$= 5.8\,\text{A}$$

POLYPHASE AC CIRCUITS

A **polyphase** ac circuit consists of two or more circuits that are usually interconnected and so energized that the currents through the separate conductors and the voltages between them have exactly equal frequencies but differ in phase. A difference in phase means that the voltages do not reach peak positive or peak negative values at the same time. Also, the corresponding values of current are usually separated by an equal number of degrees. For example, in a **three-phase** ac system, no. 1 phase will reach a peak voltage 120° before the no. 2 phase, the no. 2 phase will reach the maximum positive voltage 120° before no. 3 phase, and so on. Thus the three phases are separated by an angle of 120° (Figure 5–18).

Modern, large, transport-category aircraft of all types employ a **three-phase** ac electrical system. This system is considerably more efficient than a comparable single-phase ac system or a dc electrical system. Because of the great electric power requirements on large aircraft, a dc power system would add hundreds of pounds of weight in comparison with a three-phase ac system. The three-phase system found on these aircraft uses a polyphase ac generator that produces three ac voltages 120° apart.

Figure 5–19 is the schematic diagram of a delta-connected alternator stator. This alternator, which also can be called an ac generator, supplies three separate voltages spaced 120° apart. It is called a **delta-connected** alternator because the diagram is in the form of the Greek letter delta (Δ). With a delta

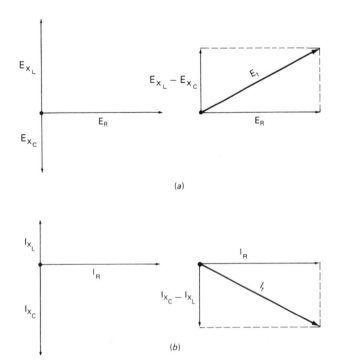

(a)

(b)

FIGURE 5–17 Vector addition in an ac circuit. (a) Voltage; (b) current.

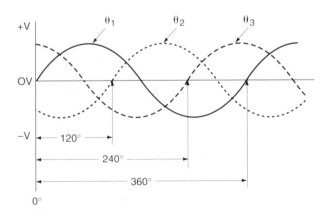

FIGURE 5–18 A typical sine wave for a three-phase circuit.

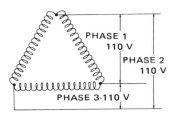

FIGURE 5–19 Schematic diagram of a delta alternator stator winding.

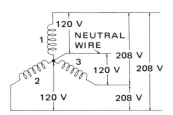

FIGURE 5–20 Schematic diagram of a Y-wound alternator stator.

connection, the voltage between any two terminals of an alternator is equal to the voltage across one phase winding.

Another method for connecting the phase windings of a three-phase system is illustrated in Figure 5–20. This is known as a **Y connection.** An alternator of this type can have three or four terminals. When there are three terminals, the voltages between any two of the terminals are equal but 120° apart in phase. To operate single-phase equipment, any two of the terminals are used. When the alternator has four terminals, the fourth is common to all windings and is called the **neutral wire.** This makes it possible to obtain two different voltages from the same alternator. In the ac power system of a transport-category aircraft, the neutral wire is grounded, and the three phase connections, which may be A, B, and C or 1, 2, and 3, are distributed to the power system of the airplane. In all cases, the separate phase terminals must be properly identified.

The voltage between any two of the three phase windings of the Y system is equal to the vectorial sum of the voltages of two of the phase windings. For example, if the voltage across one winding is 120 V, the voltage between two of the three phase terminals is 1.73 times 120, or 208 V. An arrangement of this kind is convenient because the 120-V circuit may be used for operating lights and other small loads and the three-phase 208-V circuit used to operate larger power equipment. On an airplane in which the neutral wire is grounded, a single-wire system can be used for all single-phase 120-V circuits. The 208-V three-phase power can be directed to a three-phase motor or some other device requiring this type of power. Where direct current must be obtained for certain power needs, the three-phase current can be directed through a three-phase, full-wave rectifier. Rectifiers are explained in Chapter 6.

Single-phase equipment and such items as lights are operated by connecting one terminal of the unit to one of the three-phase conductors and the other terminal to the metal structure of the airplane (ground). Thus the single-phase circuits make use of the voltage from one phase winding, and

three-phase equipment is connected to the three separate phase windings. The voltage from ground to any one of the phase terminals is 120 V (single-phase voltage); the voltage from one phase terminal to another phase terminal is 208 V; and the voltage for the three-phase system is 208 V.

ALTERNATING CURRENT AND THE AIRPLANE

You might be wondering, why use alternating current to power aircraft electrical systems? Simply stated, ac power is much more flexible than dc power. Alternating current is produced by all aircraft generators and alternators. This power must be converted to dc if such power is desired. Since converting ac to dc does require some power itself, it only makes sense to convert as little as possible and use mainly ac electrical systems.

There are three principal advantages in the use of alternating current for electric power systems. (1) The voltage of ac power can be changed easily by means of transformers. This makes it possible to transmit power at a high voltage with low current, thus reducing the size and weight of wire required. (2) Alternating current can be produced in a three-phase system, thus making it possible to use motors of less weight for the same amount of power developed. (3) AC machinery, such as alternators and motors, does not require the use of commutators; hence service and upkeep are greatly reduced.

A simple example will demonstrate the advantage of using high voltages for power transmission. We shall assume that we have a 1-hp [746-W] motor that must be driven at a distance of 100 ft [30.5 m] from the source of electric power. With a dc source of 10 V, the motor will require approximately 125 A, assuming that the motor is 60 percent efficient. Now when we consider the current-carrying capacity of copper wire, we find that a No. 1 cable is required to carry the current for the motor. One hundred feet of this wire weighs approximately 25 lb [11 kg]. If we substitute a 1-hp, 200-V ac motor for the 10-V dc motor, the current required is only about 5 or 6 A, depending on the efficiency of the motor. This will require a No. 18 wire, which weighs about 1 lb [0.5 kg] for 100 ft. This comparison clearly demonstrates the advantage of higher voltages for power transmission.

On large transport-category aircraft, three-phase alternating current is produced by the engine-driven generator (alternator). Three-phase current is used to power most of the large motors found on this type of aircraft. Three-phase ac motors are much lighter and smaller than if they were produced as single-phase ac or a direct-current motor. Three-phase motors are typically used to power hydraulic pumps, equipment-cooling blower fans, and other systems that require large amounts of mechanical energy. Single-phase ac is used to power light-duty motors, such as those used to operate a valve assembly. Single-phase ac is also used to power a variety of other systems, such as lighting.

Alternating current can be converted to different voltages much more easily than direct current. Through the principle of electromagnetic induction, ac voltage can easily be increased or decreased to virtually any desired level. A device called a **transformer** is used for this purpose. Most aircraft alternators produce power at 115 V and 400 cycles per second (Hz). However, various voltages are often desirable for specific electrical equipment. For example, fluorescent lamps operate on a relatively high voltage obtained from the output of a ballast transformer. If the same amount of light was to be produced using dc, a much greater amount of power would be consumed. Just as voltage can be increased, it can also be decreased to a relatively low level for charging a battery or operating other systems requiring only 28 V.

Even in large aircraft, some dc power is required for specific systems. Alternating current can easily be converted to dc when this becomes necessary. Usually, the dc power required is only a very small percentage of the total electric power consumed in the aircraft.

In some cases, light aircraft use ac power systems for the operation of certain equipment. Since light aircraft generate electric power using *dc* generators or alternators, they require an **inverter** to produce *ac*. An inverter is a device that changes dc voltage to ac voltage. An inverter is used only when a small amount of ac is required. Inverters can be designed to produce virtually any voltage value. Typically, 26- or 115-V inverters are used for light-aircraft systems.

Alternating-current systems do have certain disadvantages, such as the radiation of an electromagnetic field around each conductor. This field can interfere with communication or navigation systems if not properly controlled. For the most part, however, the advantages of ac far outweigh the disadvantages. For this reason, the majority of transport-category aircraft contain ac electrical systems.

REVIEW QUESTIONS

1. Define *alternating current*.
2. What are the advantages of alternating current in large transport aircraft?
3. Explain the sine curve.
4. What is meant by *rms*, or *effective*, *values* of alternating current?
5. Explain frequency.
6. What is meant by *phase* in speaking of alternating current?
7. Why does alternating current appear to flow through a capacitor?
8. What is the effect of capacitance in an ac circuit?
9. Give the formula for capacitive reactance.
10. What is the capacitive reactance in a circuit when the capacitance is 1 μF and the frequency is 60 Hz?
11. Define *capacitive reactance*.
12. What unit is normally used to indicate the capacitive reactance?
13. What unit is used to measure inductive reactance?
14. Give the formula for inductive reactance.
15. Compute the inductive reactance in a circuit where the frequency is 1000 kHz and the inductance is 20 mH.
16. Explain impedance.
17. Describe a diagram used to show the combination of 15-Ω inductive reactance, 10-Ω capacitive reactance, and 4-Ω resistance.
18. Compute the impedance in an ac circuit that has the following values: $f = 1400$ kHz, $L = 5$ mH, $C = 2$ μF, $R = 600$ Ω.
19. Define *phase angle*.
20. What is the symbol for phase angle?
21. Define *power factor*.
22. What are the formulas for true power and apparent power?
23. Define the relationships between power factor and phase angle.
24. How are individual current flows added in an ac circuit?
25. How are individual voltage drops added in an ac circuit?
26. What is a polyphase ac circuit?
27. On transport-category aircraft, why are three-phase motors used on much of the equipment?
28. What voltage and frequency alternating current is produced by most aircraft alternators?

✈

Electrical Control Devices 6

There are numerous means of controlling the voltage and current of any particular circuit. The use of switches, resistors, transistors, and other electrical devices has become commonplace in all modern aircraft. These devices are necessary in order to assure the correct operation and control of any electrical load. This chapter will discuss the theory of operation of and installation practices for the various electrical control devices used on modern aircraft.

SWITCHES

A **switch** can be defined as a device for closing or opening (making or breaking) an electric circuit. It usually consists of one or more pairs of contacts, made of metal or a metal alloy, through which an electric current can flow when the contacts are closed. Switches of many types have been designed for a wide variety of applications. The switches can be manually operated, electrically operated, or electronically operated. The manual switch is usually operated by either a lever or a push button. Electrically operated switches are generally called **relays** or **solenoids.** An electronically operated switch utilizes a transistor or integrated circuit to control the current flow through a circuit. The "switch" is turned on or off by means of an electric signal applied to the transistor or integrated circuit.

To be suitable for continued use, a switch must have contacts that are capable of withstanding thousands of cycles of operation without appreciable deterioration due to arcing or wear. The contacts are usually made of special alloys that are resistant to burning or corrosion. The operating mechanism of a switch must be ruggedly constructed so it will not fail owing to wear or load stresses. For aircraft use, a switch must be of a type and design approved by appropriate governmental agencies and by the manufacturer of the aircraft.

The type of electrical load that a switch is required to control will determine to some extent the type and capacity of switch to be employed in a circuit. Some electric circuits will have a high surge of current when first connected, and then the current flow will decrease to the normal operating level. This is typical of circuits for incandescent lamps or electric motors. An incandescent lamp will draw a high current while the filament of the lamp is cold. The resistance of the filament increases severalfold as the temperature reaches maximum; hence the current is reduced at this time. The switch for an in-candescent-lamp circuit must be able to carry the high starting current without damage.

An electric motor will draw a high current during starting because of the extra torque required for initial rotation. The countervoltage of the armature is also weak during initial motor starting. When the motor reaches normal operating speed, the countervoltage increases and opposes the applied voltage, substantially reducing current flow.

Inductive circuits, those which include electromagnetic coils of various types, have a momentary high voltage at the time the circuit is broken. This high voltage causes a strong arc to occur at the switch contacts.

It is apparent from the foregoing discussion that a switch must be able to carry a greater load than the nominal running load of the circuit in which it is installed. Accordingly, **derating factors** are applied in determining the capacity of a switch for a particular installation. The derating factor is a multiplier that is used to establish the capacity a switch should have in order to control a particular type of circuit without damage. For example, if an incandescent-lamp circuit operates continuously at 5 A in a 24-V system, the capacity of the switch should be 40 A because the derating factor is 8. That is, the surge current for the lamp circuit can be almost eight times the steady operating current. Table 6–1 gives the derating factors for aircraft switches in various types of dc circuits.

The installation of switches should be in accordance with a standard practice so the operator will always tend to move the switch lever in the correct direction for any particular operation. Switches should always be installed in panels so the lever will be moved up or forward to turn the circuit on. Switches that operate movable parts of the aircraft should be installed so the switch lever is moved in the same direction that the aircraft part will be moved. The landing gear switch

TABLE 6–1 Derating factors.

Nominal system voltage	Type of load	Derating factor
24	Lamp	8
24	Inductive	4
24	Resistive	2
24	Motor	3
12	Lamp	5
12	Inductive	2
12	Resistive	1
12	Motor	2

should be installed so the switch lever will be moved down to lower the landing gear and up to raise the gear. The same principle should apply for wing flap operation.

Switches are designed with varying numbers of contacts to make them suitable for controlling one or more electric circuits. The switch used to open and close a single circuit is called a single-pole single-throw (SPST) switch. A switch designed to turn two circuits on and off with a single lever is called a two-pole, or double-pole, single-throw (DPST) switch. A switch designed to route current to either of two separate circuits is called a double-throw switch. Schematic diagrams of several different types of switches are shown in Figure 6–1. Double-throw switches can be designed with or without a center OFF position. The switch's OFF position disconnects the pole from both throws. A three-position switch (one containing a center OFF position) would be used when it is necessary to connect a wire to a choice of two circuits or disconnect it from both. A two-position switch would be used when the circuit must always be connected to either of the two throws. Two-position DPDT switches contain no OFF position. When installing any switch, be sure it is capable of controlling the circuit properly. The schematic symbols for a switch are not always consistent among manufacturers. As illustrated in Figure 6–1, more than one type of symbol may be used to represent a given switch configuration.

Switches are available in several different configurations. Toggle, rotary, micro, rocker, and electromagnetic switches are each designed for a specific application. Figure 6–2 illustrates several switch types. Toggle or rocker switches are used to control most of the aircraft's electrical components. In situations where one contact must be connected to a choice of more than two circuits, a rotary switch is usually employed. Rotary switches are commonly found on radio control panels.

Microswitches require very little pressure applied to the actuator in order to move the switch's internal contacts. All microswitches are spring-loaded; therefore, once the external pressure is removed from the actuator, the electrical contacts will return to their normal position. The normal position of any spring-loaded switch is defined by the position of the contact points when there is no external force acting upon the switch actuator. Spring-loaded switches can be either **normally open** or **normally closed.** The contact points of a normally open switch are disconnected (open) until pressure is applied to the switch-actuating mechanism. If pressure is applied to the switch's actuator, the contact points connect (close). A normally closed switch contains closed contact points when there is no force applied to the switch actuator, and open points when a force is applied.

Microswitches are often SPDT or DPDT. This allows the switch to be used in several different configurations. As illustrated in Figure 6–3, the pole of a microswitch is labeled "C" for common, and the throws are labeled "NC" for normally closed and "NO" for normally open. For example, a circuit that is needed to turn on a light when pressure is applied to the switch would be connected to the C and NO terminals. If the light must turn off when pressure is applied to the switch, the C and NC terminals would be used. Microswitches are used chiefly for detecting the position or limit of a moving component; therefore, they are often referred to as limit switches. Landing gear, flaps, speed breaks, spoilers, and other moving components may all contain some type of microswitch (limit switch) to ensure proper positioning.

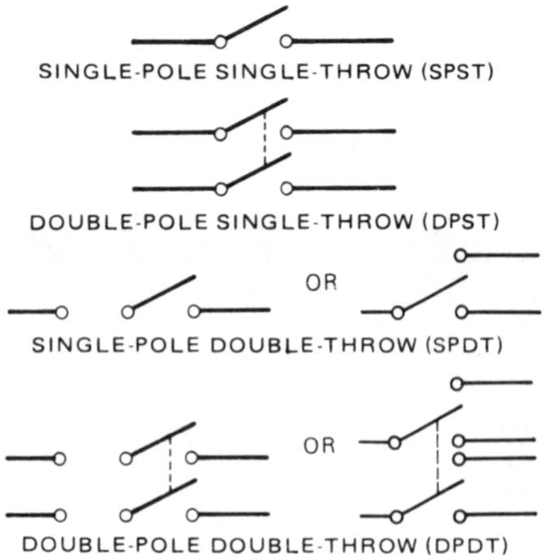

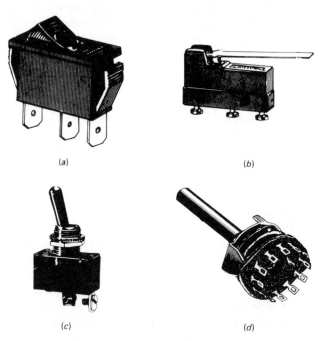

FIGURE 6–2 Typical switch designs. (a) Rocker switch; (b) microswitch; (c) toggle switch; (d) rotary switch. (GC Electronics)

SINGLE-POLE SINGLE-THROW (SPST)

DOUBLE-POLE SINGLE-THROW (DPST)

OR

SINGLE-POLE DOUBLE-THROW (SPDT)

OR

DOUBLE-POLE DOUBLE-THROW (DPDT)

FIGURE 6–1 Schematic diagrams for different types of switches.

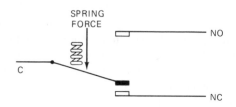

FIGURE 6–3 A schematic diagram of an SPDT microswitch.

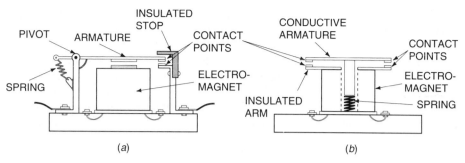

FIGURE 6–4 Electromagnetic switches. *(a)* Relay; *(b)* solenoid.

FIGURE 6–5 A typical lighted push-button switch.
(StacoSwitch, Aero Products Group)

Electromagnetic switches, as discussed in Chapter 1, are called **relays** or **solenoids.** These switches use an electromagnet to move one or more sets of switch contacts. The power to engage the electromagnet is controlled through a separate switch or an electronic control unit typically located in a different part of the aircraft. Relays and solenoids are spring-loaded switches; therefore, their contacts are designated as normally open, normally closed, and common, as seen in Figure 6–4.

Solenoids and relays may also be designated by their **duty cycle** (*continuous* or *intermittent*). A solenoid designed to operate for 2 min or less is considered intermittent-duty. A solenoid designed to be left in the activated position for longer than 2 min is a continuous-duty solenoid. If an intermittent-duty solenoid or relay is left in the activated position for too long, it will most likely overheat and fail.

Proximity sensors are a type of electronic switch with no moving contact points. They are used in conjunction with electronic circuitry to detect the position of various moving components on the aircraft, such as flaps and landing gear. On many high-tech aircraft, proximity sensors have replaced microswitches, since they are considered more reliable. Proximity sensors are discussed in Chapter 13.

Lighted push-button switches are found on many modern aircraft instrument panels. Each of these switches displays a lighted description (*legend*) of the circuit it controls (see Figure 6–5). The flight crew can easily identify switches

FIGURE 6–6 A lighted push-button switch assembly.   *(StacoSwitch, Aero Products Group)*

and determine the status of a circuit by the description on the front of the switch. Typically, the legends can be lit in two different configurations. This allows the aircraft designer to choose different colors for various operating modes of a circuit.

As shown in Figure 6–6, these switches are constructed of two basic units: the switch assembly and the lighted push button. The switch assembly comes in one of various configurations such as momentary contact or continuous contact. The lighted push button contains up to four lightbulbs to provide redundancy for the legends. This type of switch is typically designed to work in conjunction with computerized equipment; therefore, the contacts carry relatively small current flows. The electrical connections on the rear of the switch are typically soldered to their associated conductors.

CIRCUIT-PROTECTION DEVICES

A common cause of circuit failure is called a **short circuit.** A short circuit exists when an accidental contact between conductors allows the current to return to the battery through a short, low-resistance path, as shown in Figure 6–7. This failure is prevented by making sure that all insulation on the wires is in good condition and strong enough to withstand the voltage of the power source. Furthermore, all wiring should be properly secured with insulated clamps or other devices so that it cannot rub against any structure and wear through the insulation.

The danger in a short circuit is that an excessive amount of current may flow through limited portions of the circuit, causing wires to overheat and burn off the insulation. If the short circuit is not discovered immediately, the wiring is likely to become red hot and may melt. Many fires are caused by short circuits, but the danger is largely overcome by the installation of protective devices, such as fuses or circuit breakers.

Fuses

A **fuse** is a strip of metal having a very low melting point. It is placed in a circuit in series with the load so that all load current must flow through it. The metal strip is made of lead, lead and tin, tin and bismuth, or some other low-melting-temperature alloy.

When the current flowing through a fuse exceeds the capacity of the fuse, the metal strip melts and breaks the circuit. The strip must have low resistance, and yet it must melt at a comparatively low temperature. When the strip melts, it should not give off a vapor or gas that will serve as a good

conductor, because this would create an arc between the melted ends of the strip. The metal or alloy used must be of a type that reduces the tendency toward arcing.

Fuses are generally enclosed in glass or some other heat-resistant insulating material to prevent an arc from causing damage to the electric equipment or other parts of the airplane. Fuses used in aircraft are classified mechanically as cartridge type, plug-in type, or clip type, although others are manufactured. All these types are easily inspected, removed, and replaced. Typical fuses are shown in Figure 6–8.

During flight, spare fuses must be accessible to the pilot so that he or she can replace any fuses that may have accidentally failed. The FAA flight regulations stipulate that for each type of fuse found on an aircraft, the number of spare fuses must be 50 percent of the number of that type or one (whichever is greater). Therefore, if an aircraft contains four 20-A fuses, five 10-A fuses, and one 30-A fuse for circuit operations, there must be at least two 20-A and three 10-A spare fuses and one 30-A spare fuse. These spare fuses are required only for flight under Federal Air Regulation part 91; however, during any aircraft inspection, it is advisable to ensure the proper number and location of spare fuses.

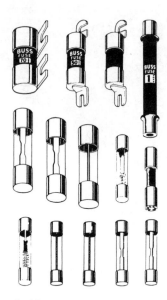

FIGURE 6–8 Typical fuses.

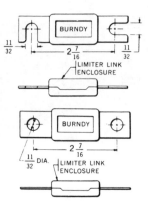

FIGURE 6–9 Typical current limiters.

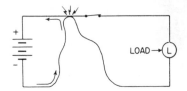

FIGURE 6–7 A schematic diagram of a short circuit.

A **current limiter** is essentially a **slow-blow fuse.** That is, when the circuit becomes overloaded, there is a short delay before the metal link melts and disconnects the circuit. This is because the link is made of copper, which has a higher melting point than the alloys used in other types of fuses. The current limiter will carry more than its rated capacity and will also carry a heavy overload for a short time. It is designed to be used in heavy-power circuits where loads may occur of such short duration that they will not damage the circuit or equipment. The capacity of a current limiter for any circuit is so selected that the current limiter will always interrupt the circuit before an overload has had time to cause damage. Current limiters are shown in Figure 6–9.

Circuit Breakers

A circuit breaker serves a purpose similar to that of a fuse; however, the circuit breaker can usually be reset after the circuit fault has been removed. A typical aircraft circuit breaker, as shown in Figure 6–10, can be described as a manually operated switch that has an automatic tripping device. This tripping device breaks the circuit when the current reaches any predetermined value. The switch-type circuit breaker shown in Figure 6–11 serves as both a fuse and a switch. A single circuit breaker may be used to control several circuits if they are not critical to flight safety.

One type of circuit breaker is known as the magnetic or electromagnetic type. The latter term is technically more correct, since the device operates through an electromagnet that pulls on a small armature and trips the breaker when energized with an overload current.

When a circuit breaker has opened a circuit, it must be reset after the circuit fault has been removed. To do this, the switch lever is moved to the FULL OFF position and then returned to the ON position. If there is still too much current flowing in the circuit—that is, if the overload still exists—the circuit breaker will trip again without damaging the circuit. It is common practice to attempt to reset a circuit breaker immediately after it has "kicked out" because the reason for its having been tripped may have been a transient overload. In this case the circuit breaker will reset easily and the circuit will again be operative. If there is a fault in the circuit, however, it must be removed before the breaker can be reset.

Thermal circuit breakers, that is, those tripped by excessive temperature acting on a bimetallic strip, cannot be reset

FIGURE 6–10 Typical circuit breakers.

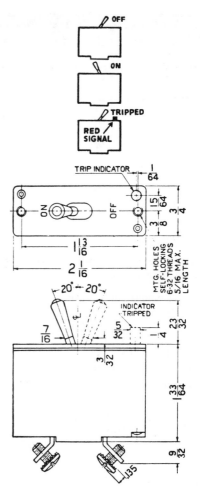

FIGURE 6–11 A switch-type circuit breaker.

until the temperature has returned to normal. Since many circuit breakers are of this type, it is often necessary to wait for a short time before attempting to reset.

Requirements for Circuit-Protection Devices

Circuit breakers and fuses should in all cases protect the wire in the circuit from overload and should be located as close as possible to the source bus. Remember that a bus is a metal strip to which a power supply is connected and from which other circuits receive power for operation. A bus is fitted with connection points to which the wire terminals are secured.

A circuit breaker or fuse should open the circuit before the wire becomes heated sufficiently to emit smoke. The time-current characteristic of the protective device should therefore be below that of the associated wire, with the result, of course, that the circuit protector will open the circuit before the wire is damaged. The term **time-current** refers to the product of multiplying the amount of current by the time during which it flows. In order to obtain maximum protection of the connected equipment, the characteristics of the circuit protector should match as closely as possible those of the connected wire.

TABLE 6–2 Wire and cable protection chart.

Wire AN gage: copper	Circuit breaker, A	Fuse, A
22	5	5
20	7.5	5
18	10	10
16	15	20
14	20	15
12	25(30)*	20
10	35(40)	30
8	50	50
6	80	70
4	100	70
2	125	100
1		150
0		150

*Figures in parentheses may be substituted where protectors of the indicated ratings are not available.

Table 6–2 is a guide to the selection of circuit-breaker and fuse ratings to protect copper cable. The conditions for the figures given in the table are as follows:

1. Wire bundles in 135°F [57.2°C] ambient temperature and altitudes up to 30,000 ft [9144 m].

2. Wire bundles of 15 or more wires (cables), with wires carrying no more than 20 percent of the total current-carrying capacity of the bundle as given in specification MIL-W-5088 [Aeronautical Standards Group (ASG)].

3. Protectors in 75 to 85°F [23.9 to 29.4°C] ambient temperatures.

4. Copper wire specification MIL-W-5088 (ASG) or equivalent.

5. Circuit breakers to specification MIL-C-5809 or equivalent.

6. Fuses to specification MIL-F-15160 or equivalent.

If the actual conditions of an installation deviate materially from those stated for Table 6–2, a rating above or below the value recommended may be justified. For example, a wire run individually in the open air may possibly be protected by a circuit breaker of the next higher rating rather than that shown on the chart. In general, the chart is conservative for all ordinary aircraft electric installations.

All resettable circuit breakers should be designed to open the circuit regardless of the position of the operating control when an overload or circuit fault exists. Such circuit breakers are described as **trip-free.** One cannot manually override a trip-free circuit breaker if the circuit fault still exists. Automatic-reset circuit breakers, which reset themselves periodically, should not be used as aircraft circuit protectors.

RESISTORS

A resistor is a circuit element designed to insert resistance in the circuit. A resistor may be of low value or of extremely high value. The nature of resistance and its effect in electric circuits were discussed in Chapter 1; the construction of resistors and their use in circuits is described briefly here.

Resistors in electronic circuits are made in a variety of sizes and shapes. They are generally classed as fixed, adjustable, or variable, depending on their construction and use. A typical fixed resistor is illustrated in Figure 6–12. This type of resistor is constructed of a small rod of a carbon compound. The value of the resistance for each resistor is determined by the makeup and size of the carbon compound and may vary from only a few ohms up to several million ohms. The two important values associated with resistors are the value in **ohms** of resistance and the value in **watts,** which represents the capacity of the resistor to dissipate power.

The resistance value of small fixed resistors is indicated by a code color. The numerical values of the colors used in this coding are as follows:

Black	0	Yellow	4	Violet	7
Brown	1	Green	5	Gray	8
Red	2	Blue	6	White	9
Orange	3				

On most resistors there are four **color bands.** The band at the end of the resistor is called band A and represents the first digit of the resistance value. The next, band B, represents the second digit of the resistance value; the third, band C, represents the number of zeros to be placed after the first two digits; and the fourth, band D, indicates the degree of accuracy or tolerance of the resistor.

If a resistor's four bands are colored with green (band A), blue (band B), orange (band C), and silver (band D), then the value of the resistor is determined as follows. The green band indicates the figure 5 and the blue band the figure 6; the orange band shows that three zeros are to follow the 5 and 6. Therefore, the resistor has a value of 56,000 Ω. The silver, or D, band represents a tolerance of 10 percent. If this band were gold, the tolerance would be 5 percent.

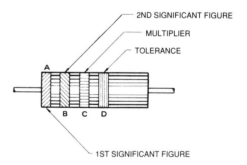

FIGURE 6–12 A typical fixed resistor.

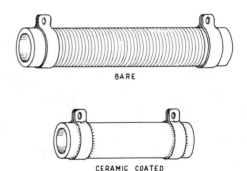

FIGURE 6–13 Wire-wound resistors.

Resistors required to carry a comparatively high current and dissipate high power are usually of the wire-wound ceramic type (see Figure 6–13). A wire-wound resistor consists of a ceramic tube wound with fine resistance wire, which is then covered with a ceramic coating or glaze. The terminals for the resistance wire extend out at each end of the resistor as shown. The value of the wire-wound resistor is usually printed on the ceramic coating.

Adjustable and Variable Resistors

An **adjustable resistor,** shown in Figure 6–14, is usually of the wire-wound type with a metal collar that can be moved along the resistance wire to vary the value of the resistance placed in the circuit. In order to change the resistance, the contact band must be loosened and moved to the desired position and then tightened so that it will not slip. In this way the

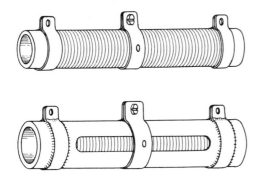

FIGURE 6–14 Adjustable resistors.

resistor becomes, for all practical purposes, a fixed resistor during operation.

A **variable resistor** is arranged so that it can be changed in value at any time by the operator of the electronic circuit. This change is usually accomplished by rotating a small adjustment knob or by turning a screw adjustment. Variable resistors are commonly known as **rheostats** or **potentiometers** (see Figure 6–15). A rheostat is typically connected in a circuit merely to change the current flow and has a comparatively low resistance value (usually below 500 Ω). Its circuit connections are as shown in Figure 6–16. Note that the rheostat has two terminals, one connected to the wire-wound resistor and the other connected to the sliding contact arm, which moves along the resistor.

A potentiometer normally is connected with three terminals. One terminal is connected at each end of the resistor, and the third terminal is connected to the sliding contact arm. The value of the resistance of a potentiometer is comparatively high, and the resistor is normally made of a material such as a carbon or graphite compound. The purpose of a potentiometer is to vary the value of the voltage in a circuit.

A diagram of a potentiometer in a transistorized dimming circuit is shown in Figure 6–17. The voltage applied between the base and emitter of the transistor is controlled by the potentiometer; subsequently, the voltage to the lamp is controlled by the transistor emitter-collector circuit. Transistors are discussed in greater detail later in this chapter.

It must be pointed out that the use of a resistor of any type must be very carefully considered. The capacity of a fixed resistor, rheostat, or potentiometer must be such that the resistor can handle the current through the circuit without damage. It is always necessary to compute the current

FIGURE 6–15 An assortment of variable resistors.   *(Clarostat Mfg. Co.)*

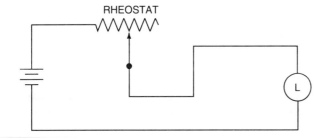

FIGURE 6-16 A rheostat circuit.

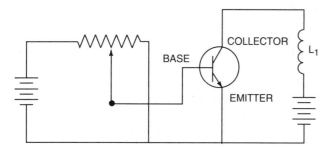

FIGURE 6-17 A potentiometer circuit.

through the resistor by means of Ohm's law before placing the circuit in operation.

Voltage Dividers

Resistors are often arranged in such a manner as to create a **voltage divider** circuit. A voltage divider is simply two resistors placed in series with each other and in parallel with a voltage source. The voltage divider creates a voltage drop across each resistor that is proportional to the resistance of the individual resistor (see Figure 6-18). When a high resistance load is placed across R_1 of the voltage divider, the load receives a voltage equal to (or close to) the original drop over R_1.

The actual voltage applied to the load is a function of the total parallel resistance of R_1 and R_{load}. In many cases, however, if the load resistance is large enough, the change in the voltage drop over R_1 is negligible with the addition of the load.

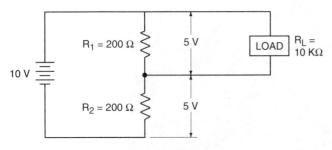

FIGURE 6-18 Example of a voltage divider circuit.

Capacitor Theory

A capacitor consists of two conductors that are capable of holding an electric charge and which are separated by an insulating medium. A simple capacitor, consisting of two metal plates separated by air, is shown in Figure 6-19. The air, or some other insulating material, between the plates of a capacitor is called the **dielectric.** When the plates of a capacitor are connected to a voltage source, the capacitor becomes *charged.* This charge consists of an excess of electrons on the negative plate and a corresponding deficiency of electrons on the positive plate. If the capacitor is disconnected from the voltage source, the charge will remain in the capacitor for a length of time depending on the nature of the dielectric (Figure 6-20).

Unless there is a complete vacuum between the plates, the dielectric material between the plates of a capacitor consists of a large number of atoms. This holds true whether the dielectric is gaseous, liquid, or solid. Since the dielectric is an insulator, it takes a very high voltage to cause the free electrons to break away from the dielectric's atoms and move through the material. When the capacitor is charged, a voltage exists between the plates and acts upon the dielectric. Although the voltage is not great enough to cause the electrons in the dielectric to break away from the atoms, it does

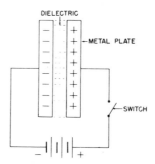

FIGURE 6-19 A simple capacitor circuit.

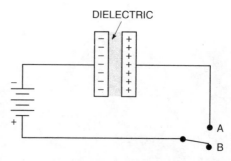

FIGURE 6-20 A capacitor circuit. When the switch is placed in position A, the capacitor will become charged. With the switch in position B, the capacitor will become discharged through the dielectric.

cause them to shift a small distance in their orbits. This shifting of the electrons toward the positive plate of the capacitor creates what is known as a **dielectric stress,** which can be compared to a stretched rubber band. When the plates of a charged capacitor are connected by a conductor, the electrons flow from the negative plate to the positive plate, thus neutralizing the charge and relieving the dielectric stress. Then the capacitor is said to be *discharged.*

A clear understanding of the operation of a capacitor can be had by studying the hydraulic analogy shown in Figure 6–21. The capacitor is represented by a chamber separated into two equal sections by an elastic diaphragm representing the dielectric. These chambers are connected to a centrifugal pump by means of pipes. The pump represents the generator in an electric circuit, and the valve in one of the pipes represents a switch. When the pump rotates, it forces water into one of the chambers and causes the diaphragm to stretch. Water from the other chamber then flows out toward the pump. One of the chambers contains more water than the other, and the diaphragm, being stretched, maintains a pressure differential between the chambers. When the diaphragm pressure is equal to the pump pressure, the water will stop flowing, and the chamber will be "charged." If the valve is then closed, the diaphragm will maintain the differential of pressure between the sections of the chamber.

In the corresponding electric circuit, when the generator is running, electrons are forced into one plate of the capacitor and withdrawn from the other plate. When the potential difference between the plates is equal to the voltage of the generator, the current flow will stop, and the capacitor will be charged. Now the switch can be opened, and the charge will remain in the capacitor.

In the hydraulic "circuit," when the pump is stopped and the valve opened, the water will immediately flow from the section that has the higher pressure to the section that has the lower pressure. As soon as the pressures are equal, the flow will stop. In like manner, if the generator in the electric circuit is stopped and the switch is closed, the electrons will flow from the negatively charged plate, through the generator, and back to the positive plate.

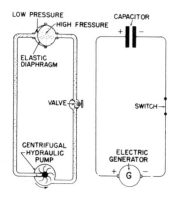

FIGURE 6–21 Hydraulic analogy of a capacitor.

Capacitance

The effect of a capacitor, that is, its ability to store an electric charge, is called **capacitance** (*C*). The unit of capacitance is the **farad** (F), which is the capacitance present when one volt will store one coulomb of electric energy in the capacitor. The farad is much too large a unit for practical purposes, and so a smaller unit called the **microfarad** (μF) is generally used. One microfarad is one-millionth of a farad, and μ is the Greek letter mu. It is of interest to note that a capacitor with a capacitance of 1 F would be so large that it would probably weigh several thousand pounds.

Some capacitors have such small capacitance that even the microfarad is too large a unit for convenient expression of the value. In such cases the **picofarad** (pF) is used. One picofarad is equal to one-trillionth of a farad. This value can also be expressed as $1 pF = 10^{-12} F$.

The capacitance of a capacitor depends on three principal factors: the area of the plates, the thickness of the dielectric, and the material of which the dielectric is composed. It will be readily apparent that two capacitors of the same size can differ considerably in capacitance because of a difference in the dielectric material.

To measure the dielectric characteristics of a material, a **dielectric constant** is used. Air is given a dielectric constant of 1 and is used as a reference for establishing the dielectric constants of other materials. Mica, which is commonly used as a dielectric in capacitors, has a dielectric constant of 5.5. This means that a capacitor having mica as the dielectric has 5.5 times the capacitance of a similar capacitor having air as the dielectric.

In addition to the dielectric constant of a material, its insulating quality must be considered. The insulating quality of a material is called its **dielectric strength** and is measured in terms of the voltage required to rupture (break down) a given thickness of the material. In selecting a capacitor for any purpose, it is important that the capacitance be correct and that the breakdown voltage of the capacitor be greater than the voltage to which the capacitor will be subjected when in use.

Types of Capacitors

There are two general types of capacitors: **fixed** and **variable.** The fixed capacitor is constructed with the plates and dielectric placed firmly together and covered with a protecting material such as waxed paper, plastic, ceramic material, or an insulated metal case. Because of the construction of a fixed capacitor, its capacitance cannot be changed.

Variable capacitors normally have fixed plates and movable plates arranged in such a manner that the dielectric effect between the plates can be changed by varying the distance between the plates or by moving one set of plates into or out of the other set. The construction of a typical variable tuning capacitor is shown in Figure 6–22. Variable capacitors are used in radios and other electronic devices where it is necessary to change the capacitance to meet the requirements of a given circuit. The dielectric material in a variable capacitor is usually air.

FIGURE 6–22 A variable capacitor.

FIGURE 6–24 Capacitors.

Although the conducting elements of a capacitor are called plates, in a fixed capacitor they frequently consist of long strips of foil insulated with waxed paper and rolled together. The rolled plates are then covered with an insulating material and may be placed in a protective case. The leads from the plates may be brought out at one end or both ends of the case, depending on the design of the capacitor. Fixed capacitors of the mica type are often constructed as shown in Figure 6–23. The plates are connected to form two groups with mica sheets separating the alternate plates.

When a relatively high capacitance is desired in a small physical size, an **electrolytic** capacitor is used. In a capacitor of this type, the dielectric is a liquid or paste known as an *electrolyte*. The electrolyte forms an oxide on one of the plates, which effectively insulates it from the other plate. The dielectric constant of the electrolyte is much greater than that of the commonly used dry materials; hence the capacitor has a considerably higher capacitance than the capacitors using dry materials. An electrolytic capacitor must be connected in a circuit with the correct polarity, because such a capacitor will allow current to flow through it in one direction. If the current flows through the plate of an electrolytic capacitor, the capacitance will be lost and the plates will decompose. Precautions must be taken to ensure that electrolytic capacitors are not connected in reverse and that they are not overloaded. Often these capacitors will overheat and burst if they are not connected and used properly. Incorrect use may thus create a safety hazard. Fixed capacitors of both the dry and the electrolytic types are manufactured in a wide variety of shapes, as shown in Figure 6–24.

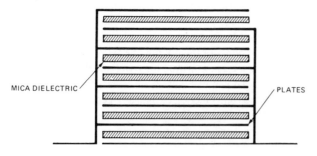

FIGURE 6–23 Arrangement of a fixed capacitor.

Multiple Capacitor Circuits

When capacitors are connected in parallel (Figure 6–25a), **the combined capacitance is equal to the sum of the capacitances.** The effect is the same as if one capacitor were used having a plate area equal to the total plate area of all the capacitors in the parallel circuit. Any multiple-plate capacitor is actually a group of capacitors connected in parallel. Since the capacitance varies directly as the area of the plates, it is apparent that two capacitors having the same plate area and connected in parallel have twice the capacitance of one, because the two capacitors have twice the plate area of the one.

The formula for capacitors connected in parallel is

$$C_t = C_1 + C_2 + C_3 \cdots$$

For capacitors in series, the formula is similar to that used for resistances in parallel. **When capacitors are connected in series, the total capacitance is equal to the reciprocal of the sum of the reciprocals of the capacitances.** The formula is

$$C_t = \frac{1}{1/C_1 + 1/C_2 + 1/C_3 \cdots}$$

From the foregoing formula, it will be found that the total capacitance, when capacitors are connected in series, is less than the capacitance of the lowest-rated capacitor in the series. The reason for this can be understood by observing a circuit where two capacitors of equal rating are connected in series (Figure 6–25b). The two center plates will not contribute to the capacitance, because their charges are opposite and will neutralize each other. The effect is that of two outside plates acting through a dielectric that has twice the thickness of the dielectric of one of the capacitors. Therefore, the total capacitance of the two capacitors is equal to one-half the capacitance of one of the capacitors. Remember that the capacitance of a capacitor varies inversely as the thickness of the dielectric.

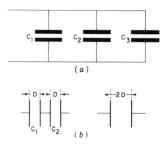

FIGURE 6–25 Capacitors connected (a) in parallel and (b) in series.

Time Constant

When a capacitor is connected to a voltage source, it takes a certain length of time for the capacitor to become fully charged. If a high resistance is connected in series with the capacitor, the time for charging is increased. For any given circuit containing capacitance and resistance only, the time in seconds required to charge the capacitor to 63.2 percent of its full charge is called the **time constant** for that circuit. This same time constant applies when the capacitor is discharged through the same resistance and is the time required for the capacitor to lose 63.2 percent of its charge.

The charging and discharging of a capacitor in terms of time constants is illustrated in the graph of Figure 6–26. It will be noted that it takes six time constants to charge the capacitor to 99.8 percent of full charge. The discharge curve is the exact reverse of the charge curve. When the capacitor is short-circuited, it will lose 63.2 percent of its charge in one time constant and almost 99.8 percent of its charge in six time constants.

To determine the length of a time constant in seconds for any particular capacitor-resistance circuit, it is necessary to multiply the capacitance (in microfarads) by the resistance (in megohms, or MΩ); that is,

$$T = CR$$

As an example of how the time constant can be used in determining the performance of a capacitor-resistance circuit, we shall assume that a 20-μF capacitor is connected in series with a 10,000-Ω resistor and that 110 V is applied to the circuit at intervals of 0.5 s.

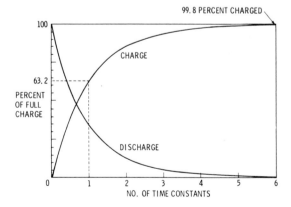

FIGURE 6–26 Curves showing the charging and discharging of a capacitor according to the time constant.

The time constant is equal to 20×0.01, or $T = 0.2$ s. (Note that 10,000 Ω is equal to 0.01 MΩ.) The time interval is given as 0.5 s; hence the number of time constants is 2.5. If we examine a time-constant chart or graph, we find that the voltage at 2.5 time constants will be approximately 92 percent of full voltage. Applying this to our problem shows us that 92 percent of 110 V is approximately 101 V. Thus we find that the capacitor in this problem will charge to approximately 101 V.

Effects and Uses of Capacitors in Electric Circuits

When a capacitor is connected in *series* in a dc circuit, no current can flow through the capacitor because of the insulating quality of the dielectric. When a voltage is applied to such a circuit, there is a momentary flow of electrons into the negative plate of the capacitor and a corresponding flow out of the positive plate. As soon as the dielectric stress is equal to the applied voltage, the flow of electrons stops. If the voltage is removed, the charge remains in the capacitor until a path is provided through which the electrons can flow from the negative plate back to the positive plate.

When a capacitor is connected in *parallel* in a dc circuit, it opposes any change in the circuit voltage. As voltage from the source rises, current flows into the capacitor and thus slows the voltage rise. If the voltage of the source remains at a higher level, the capacitor will charge to that level and will have no further effect on the circuit as long as the voltage remains constant. If the voltage from the source drops, the capacitor discharges into the circuit and holds the circuit voltage above that of the source for a short time. The property of capacitors to oppose changes in voltage is utilized in dc circuits to reduce or eliminate voltage pulsations. The voltage from a dc generator pulsates; that is, it varies slightly above and below the average value. When a capacitor of sufficient capacitance is connected in parallel with the generator, voltage pulsations are largely eliminated, and a more steady direct current is delivered. This is discussed further in Chapter 10.

Another use for capacitors is to reduce the arcing at breaker points or switch contacts. When a switch opens and stops the current flowing through a circuit, a spark jumps across the switch's contact points. If this spark is uncontrolled, the contact points will soon become pitted and burned. This damage creates a high resistance at the switch contacts and can result in poor circuit efficiency or complete circuit failure. When a capacitor is connected in parallel with the contact points, the spark is absorbed by the capacitor; thus the capacitor prevents burning of the points. When the switch is again closed, the capacitor discharges back into the circuit.

Fluctuating voltages and currents in electric circuits cause the emanation of electromagnetic waves. These waves induce currents in radio circuits and other sensitive circuits and interfere with their normal operation. Capacitors are connected in the electric circuits at points where they will be most effective in absorbing the momentary fluctuations of voltage; in this way they reduce the emanation of electromagnetic waves.

In an ac circuit a capacitor is often used to block direct current but permit the flow of the alternating current. A capacitor can also be used in combination with an inductor and/or a resistor to allow certain ac frequencies to pass through the circuit; other frequencies are blocked. This technique is known as *filtering*.

INDUCTORS

Any electric conductor possesses the property of inductance; however, most **inductors** are specifically designed coils of wire. **Inductance is the ability of a conductor to induce a voltage into itself when a change in current is applied to the inductor.** An inductor can be a straight piece of wire or a coil. Figure 6–27 shows a variety of inductance coils.

The inductance of a single straight wire is usually negligible. However, if the wire is wound into a coil, the inductance value increases significantly. This is due to the relatively strong magnetic field produced by the current flowing through a coil of wire. It is the increase or decrease (a change) of this magnetic field that produces the coil's inductance. As discussed in Chapter 1, if there is relative motion between a conductor and a magnetic field, a voltage will be induced into that conductor. Figure 6–28 illustrates the formation of a magnetic field around an inductor coil with respect to time. As illustrated, the magnetic field strength increases for a short time period, from point A to point B. This occurs at the

instant current flow begins. Immediately after the switch opens, the current flow drops to zero. At this time, the magnetic field strength decreases as illustrated between points C and D. This increase and decrease of the magnetic field strength creates a relative motion between the conductor and the magnetic field. That is, if the conductor is stationary and the magnetic field "grows" or "shrinks" around it, a voltage is induced into that conductor. This induced voltage is opposite in polarity to the source voltage. Since the induced voltage is opposite in polarity to the source voltage, it opposes the source voltage. This is illustrated in Figure 6–29. The induced voltage is always much less than the source voltage; therefore, the induced voltage only weakens the source voltage. Current reacts in a similar manner. The induced current always opposes the applied current. The induced voltage exists only during changes in current flow (see Figure 6–28). Therefore, inductance can be thought of as the ability to oppose changes in current flow.

Inductance can be compared to inertia of an object. Inertia tends to keep objects at rest that are at rest and keep objects moving that already have motion, just as induced currents tend to keep electron flow zero if it is already zero and keep electrons flowing if they are already flowing. That is, induced currents tend to oppose **changes** in electron motion, just as inertia tends to oppose changes in an object's motion.

It was stated in Chapter 1 that the field strength of an electromagnet depends on the number of turns of wire in the coil, the current flowing in the coil, and the material in the core.

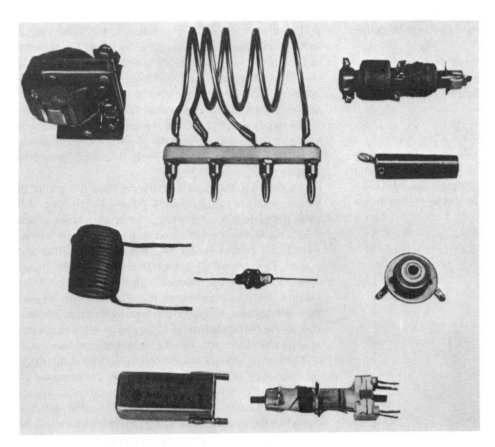

FIGURE 6–27 A variety of inductance coils.

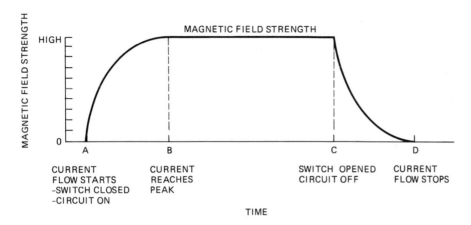

FIGURE 6–28 An inductance coil's magnetic field strength plotted with respect to time.

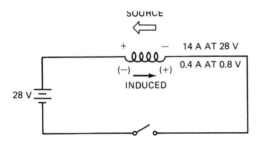

FIGURE 6–29 Relationships of applied voltage to induced voltage.

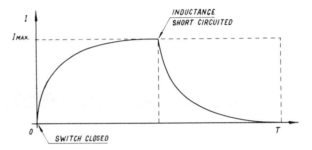

FIGURE 6–30 Curves showing the effects of inductance on the rise and fall of current.

Actually, an electromagnet and an inductance coil are essentially the same; hence the effect of an inductance coil in a circuit also depends on the number of turns of wire in the coil, the current flowing in the coil, and the material used in the core. Inductance coils are made with soft-iron cores when a high inductive effect is desired. When a low inductive effect is desired, the inductance coil has no core; that is, the core is made of air.

The inductance of a coil is measured in a unit called the **henry** (H), named for Joseph Henry (1797–1878), an American physicist. **One henry is the inductance of a coil when a change of current of one ampere per second will induce an emf of one volt.** The symbol for inductance is the letter L. The henry is too large a unit for most applications, and so a smaller unit called the **millihenry** (mH) is used. One millihenry is one-thousandth of a henry.

As in the case with capacitance in a circuit with resistance, a time constant is applied in a circuit containing inductance in series with a resistance. In Figure 6–26, the curves shown apply to an inductive circuit as well as a capacitive circuit. The curves for a circuit with inductance and resistance only are also shown in Figure 6–30.

In the case of inductance, the maximum current flow in a circuit is delayed for a short time after the inductance coil is connected to a power source. The time constant is the time in seconds that is required for the current flow to reach 63.2 percent of maximum after the circuit is connected to the power source. The time constant for a decaying current is the time in seconds required for the current flow to fall to 36.8 percent of

maximum. This is the same time as is required for the increase to 63.2 percent of maximum.

To determine the time constant for a circuit containing only inductance and resistance, it is necessary to divide the inductance (L) in henrys by the resistance (R) in ohms. Hence,

$$T = \frac{L}{R}$$

If a 10-H inductance coil is connected in series with a 200-Ω resistance, the time constant is 10/200, or 0.05 s.

Multiple Inductor Circuits

In some cases two or more inductors are combined in a series or parallel arrangement. In either case the total inductance of the circuit changes owing to this combination. When inductors are wired in series with each other, the total inductance is increased. The resultant is the sum of the inductor values within the circuit, as shown by the following equation.

$$L_t = L_1 + L_2 + L_3 \cdots$$

When inductors are placed in parallel with respect to each other, the total inductance is decreased. The following equation is used to find the total inductance of two or more inductors in parallel.

$$L_t = \frac{1}{1/L_1 + 1/L_2 + 1/L_3 \cdots}$$

Inductors   **107**

Uses of Inductors

As explained in Chapter 5, the opposition to current flow in an ac circuit created by an inductor is called **inductive reactance** and is measured in ohms. Since radio signals are transmitted using a rapidly changing (high-frequency) electromagnetic energy, inductors are often used in combination with capacitors to provide tuned circuits. These tuned circuits are most valuable in radio and television for filtering out unwanted frequencies and passing the desired frequencies.

In many electronic circuits it is desirable to use inductors that are variable in inductance. This means that devices must be provided that enable the operator to change the inductance of the inductance coil. A common method for changing the inductance is to use a powdered-iron core in the inductor and provide a means whereby this core can be moved in and out of the coil. An inductance coil that contains a movable core for tuning purposes is often called a *slug-tuned* inductor. Inductors of this type are often found in small radio receivers.

TRANSFORMERS

A **transformer** is a device used to increase or decrease the voltage in an ac circuit. In fact, one of the chief advantages of alternating current is that it can be transmitted at a high voltage with a low power loss; the voltage can then be reduced to any desired value by means of a transformer. We therefore frequently find transformers in ac systems.

A schematic diagram of a transformer is shown in Figure 6–31. It was explained in Chapter 1, in the section on electromagnetic induction, that every conductor of an electric current has a magnetic field. If alternating current is flowing in a conductor, the magnetic field around the conductor expands and collapses rapidly as the current changes in magnitude and direction. This makes it possible to change the voltage of an alternating current by means of a mutual induction coil (transformer).

A transformer consists of a **primary winding** and a **secondary winding** on either a laminated soft-iron core or an annealed sheet steel core. The secondary coil may be wound on the primary coil or on a separate section of the same core. This is illustrated in Figure 6–32. The laminated core reduces the effect of **eddy currents,** which otherwise would cause considerable heat and a loss of power.

The transformer theory of operation is similar to that of an induction coil. As an ac current flow is fed through the pri-

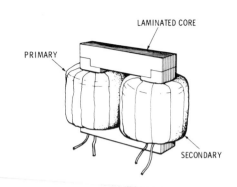

FIGURE 6–32 A transformer.

mary winding, a magnetic field expands and contracts around that winding. If another inductance coil, the secondary, is placed near the primary, it will receive an induced voltage from the constantly changing magnetic field of the primary. If the second coil is connected to a circuit, the induced voltage will produce a current flow. The greater the applied ac frequency (within limits), the better the voltage transformation between the primary and secondary. A relatively higher-frequency ac, therefore, allows for the use of smaller transformers. Because of the inductive reactance of the primary and secondary coils, the induced voltage in the secondary is nearly 180° out of phase with the primary voltage. This is because the primary current is nearly 90° out of phase with the primary emf, owing to the inductance of the primary winding, and the emf of the second coil is 90° out of phase with that of the primary. In theory, the secondary emf of a circuit with no resistance would be exactly 180° out of phase with the emf of the primary, but since no circuit can be free of resistance, the two voltages cannot be 180° out of phase but will be somewhat less than 180°, depending on the resistance of the circuit.

A study of Figure 6–33 will help the student to understand the phase relationships in a transformer circuit. The curve E_p represents the emf applied to the primary coil of the transformer. I_p is the current in the primary, which lags behind the primary emf by almost 90° because of the inductance of the primary winding. Since the current change is greatest as the current reverses direction, a maximum emf (E_s) is induced in the secondary at this point. When the current reaches a maximum value at 180° on the curve, there is an instant when there is no current change; hence at this point there is no induced emf in the secondary. As the current value decreases, the rate of change increases, and the secondary emf increases to oppose this change.

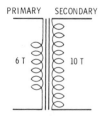

FIGURE 6–31 A schematic diagram of a transformer.

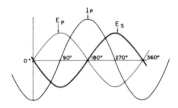

FIGURE 6–33 Voltage and current in the coils of a transformer.

One of the most important features of a transformer is that the primary coil can be left connected to the line and will consume very little power unless the secondary circuit is closed. This is because of the inductive reactance of the primary winding. The primary current sets up a field that induces an opposing emf in the primary coil. This opposing emf is called **counter emf** and is almost equal to the emf applied to the coil; hence only a very small current will flow in the coil.

We can consider the field as a reservoir of power, and when the secondary circuit is closed, power is being drawn from the reservoir. Then current will flow in the primary circuit sufficient to maintain the field flux at a maximum value. If the secondary circuit is disconnected, no more power will be drawn from the field; hence very little current will be necessary to maintain the field strength. From this we can see that the strength of the field remains almost constant as long as the load does not exceed the ultimate capacity of the transformer.

When the primary and secondary coils of a transformer are wound upon the same core, they are both affected by the same magnetic field. It will be remembered that the emf induced in a coil depends on the lines of force being cut per second. Since both the primary and secondary coils are being cut by the same magnetic field, the ratio of the primary emf to the secondary emf is proportional to the ratio of the number of turns of wire in the primary to the number of turns in the secondary. For example, if the primary coil has 100 turns of wire and the secondary has 200 turns, then the emf of the secondary will have twice the value of the emf in the primary. The formula for these values is

$$\frac{E_p}{E_s} = \frac{N_p}{N_s}$$

where E_p = voltage in primary

E_s = voltage induced in secondary

N_p = number of turns in primary winding

N_s = number of turns in secondary winding

It is obvious that the power output of a transformer cannot be greater than the power input. Since the power in a transformer is approximately equal to the voltage times the amperage, we can see that if the voltage in the secondary is higher than the voltage in the primary, then the amperage in the secondary must be lower than the amperage in the primary. In a transformer that is 100 percent efficient, the ratio of the amperage in the primary to the amperage in the secondary is inversely proportional to the ratio of the voltages. The formula for this relationship is

$$\frac{E_p}{E_s} = \frac{I_s}{I_p} \quad \text{or} \quad E_p I_p = E_s I_s$$

When the secondary of a transformer has more turns of wire than the primary and is used to increase voltage, the transformer is called a **step-up** transformer. When the transformer is used to reduce voltage, it is called a **step-down** transformer. In many cases the same transformer can be used as either a step-up or a step-down transformer. The coil connected to the input voltage is called the primary, and the coil connected to the load is called the secondary. When a transformer is in use, the voltage capacity of its primary winding, which can usually be ascertained from the name or data plate, must not be exceeded.

If it becomes necessary to use more than one transformer in a circuit, with the transformers connected either in series or in parallel, it is most important that they be properly *phased*. Figure 6–34 illustrates a simplified circuit for two transformers connected in series. Note that the primary terminals P_1 of the first transformer and P_1 of the second transformer are connected to the same line of the power supply and that the P_2 terminals of the transformers are likewise connected to the same line of the power supply. With the primary circuits connected in this manner, the secondary terminals S_1 will be positive at the same time and negative at the same time. Therefore, to connect the two secondary circuits in series to obtain maximum voltage, S_2 of one transformer should be connected to S_1 of the other transformer and the opposite terminals S_1 and S_2 then used as output terminals. With this arrangement the voltages are additive, and the total output will be 220 V if the individual secondary windings produce 110 V each. If the two secondary windings in this series were connected so that S_2 of one transformer was connected to S_2 of the other, then there would be no output from the two S_1 terminals, because the voltages would be working in opposite directions.

In Figure 6–35, transformers are shown connected in parallel. The primary windings are connected in the same manner as those in the circuit of Figure 6–34. To connect the secondary windings in parallel, the two terminals S_1 and the two terminals S_2 are connected to the same line. The output between these lines will then have the same voltage as each individual winding, and the available amperage will be additive. If the connection for one of the secondary windings is reversed, a short circuit will be created between the two secondary windings, and the transformers will be burned out, or the circuit breaker in the power supply will be opened.

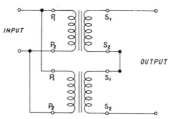

FIGURE 6–34 Transformers connected in series.

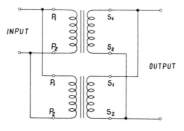

FIGURE 6–35 Transformers connected in parallel.

A **rectifier** is a device that allows current to flow in one direction but will oppose, or stop, current flow in the opposite direction. A rectifier can be compared to a check valve in a hydraulic system. A check valve is a one-way gate for fluids; rectifiers are one-way gates for electrons.

There are several types of solid-state rectifiers currently in use. The term **solid-state** refers to a device in which a solid material is used to control electric currents through the manipulation of electrons. Solid-state devices have proved to be a very reliable and efficient means of electron control for a wide range of applications.

Semiconductor Rectifiers

To understand the principles of rectification as performed by a semiconductor rectifier, it is necessary to gain a concept of what actually takes place in the material of the rectifier. We shall, therefore, give a brief description of the structure of semiconductor materials and the electronic activity within such materials. Semiconductors are commonly called solid-state devices because they are solid and contain no loose or moving parts.

The principal semiconductor materials used for rectifiers are **silicon** and **germanium.** It was explained in Chapter 1 that a semiconductor element has four electrons in the outer orbit, or shell, of each atom. Silicon has a total of 14 electrons in the atom, 4 of these being in the outer shell. Germanium atoms have 32 electrons, with 4 in the outer shell. In the pure state, neither of these materials will conduct an electric current easily. This is because the atoms have a strong **valence bond** formed as the electrons in the outer shell of each atom pair with the electrons in adjacent atoms. This is shown in Figure 6–36. The illustration is a two-dimensional concept of the **crystal lattice** for germanium. Actually, the electrons are in spherical shells rather than rings, and they rotate about the nuclei of the atoms. However, they still form energy bonds in the outer shells, and they are not easily moved from one atom to another. The only way this can happen is when a very high voltage is applied across the material and the valence bonds are broken. It can be stated that pure germanium and silicon do not have free electrons to serve as current carriers.

To make germanium or silicon capable of carrying a current, a small amount of another element (impurity) is added.

This is called **doping.** The element **antimony,** having the chemical symbol Sb, has five electrons in the outer shell of each atom. When this material is added to germanium, the germanium becomes conductive. The reason for this is that the fifth electron from the Sb atom cannot bond with the germanium electrons and is left free in the material. This is shown in Figure 6–37. Remember that the germanium atoms have four electrons in the outer shell of each atom; hence only four of the Sb electrons can become paired in the valence bonds.

When germanium is treated with antimony, the resulting material is called **n-type** germanium because it contains extra electrons, which constitute negative charges. It must be remembered, however, that the material is still electrically neutral because the total number of electrons in the material is balanced by the same number of protons. The Sb atom has 51 protons, and their positive charge balances the negative charge of the 51 electrons in each atom. One of 51 electrons is forced out of the outer shell of the Sb atom, and this becomes a free electron. The Sb is called a **donor** because it *donates* electrons to the material.

When the element indium (In) is added to germanium, vacant spaces are left in the valence bonds because indium atoms have only three electrons in the outer shell. The vacant spaces are called **holes.** The holes can be filled by electrons that break away from the valence bonds. When this occurs, another hole is left where the electron previously was situated. Thus the holes appear to move through the material.

The hole represents a net positive charge because a balanced condition requires that a pair of electrons occupy each bond. When one of the electrons is missing, the bond lacks the normal negative charge; hence it is positive and attracts electrons. An illustration of **p-type** germanium is shown in Figure 6–38. The holes can be seen adjacent to the indium

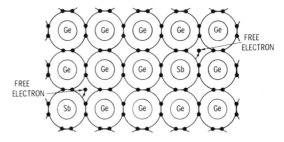

FIGURE 6–37 Effect of adding antimony to germanium to form *n*-type material.

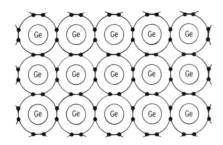

FIGURE 6–36 Germanium valence bonds.

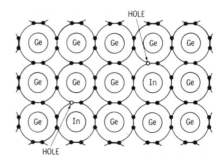

FIGURE 6–38 Addition of indium to form *p*-type material.

atoms. Indium added to germanium is called an **acceptor** because it *accepts* electrons from other atoms.

The *n*-type material of a rectifier is known as the **cathode;** the *p*-type material is the **anode.** The cathode is the electron emitter, or negative connection. The anode is the electron acceptor, or positive connection. These polarities must be observed if the diode is to conduct electricity.

When a piece of *n*-type germanium forms a **junction** with a piece of *p*-type germanium, an interesting phenomenon takes place. Since there are holes (positive charges) in the *p*-type germanium and electrons (negative charges) in the *n*-type germanium, there is a drift of holes and electrons toward the junction. The holes are attracted by the negative charge of the electrons in the *n*-type material, and the electrons are attracted by the positive charge of the holes in the *p*-type material. Some of the electrons diffuse across the junction to fill holes on the positive side. This movement of charges leaves a large number of negative ions in the *p*-type material farthest from the junction and a large number of positive ions in the *n*-type material farthest from the junction. Remember that the material is electrically neutral, as a whole, before the junction is made, because the number of electrons is balanced by the number of protons. The material is still electrically neutral as a whole after the junction is made, but some portions have negative charges, and other portions have positive charges.

The stationary ions on each side of the junction provide charges that stop the movement of electrons across the junction. These charges result in a **potential barrier** with a voltage of approximately 0.3 for germanium and 0.6 for silicon. Figure 6–39 illustrates the condition that exists when a junction of two different types of germanium is made. Note that holes move toward the junction from the *p*-type material, and electrons move toward the junction from the *n*-type material until the charges are balanced.

When two types of germanium or silicon are joined as described in the preceding paragraphs, a **diode** is formed. The word *diode* means *two electrodes*. The two parts of the material, *p*-type and *n*-type, constitute the two electrodes. If we connect a battery or some other power source to the *pn* diode, we find that current will flow through it in one direction but not in the other direction. The diode therefore becomes a rectifier. This is explained in Figures 6–40 and 6–41.

In Figure 6–40 the battery is connected such that its negative terminal is joined to the *n* side of the diode. In this way the electrons flowing from the negative side of the battery neutralize the effect of the positive ions, which would otherwise affect the current flow. This makes it possible for the

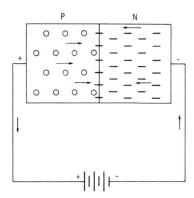

FIGURE 6–40 A forward-biased diode.

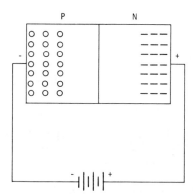

FIGURE 6–41 A reverse-biased diode.

electrons to flow across the barrier (junction) to occupy the holes and flow on toward the positive battery terminal. Thus the diode has become a good conductor in one direction, that is, from *n* to *p*. The diode in its conducting mode is said to be **forward-biased.**

In Figure 6–41 we observe the condition when the battery is connected in the opposite direction, which is called **reverse bias.** Here the positive terminal of the battery is connected to the *n* side of the diode. The free electrons are drawn toward the positive charge until the potential balances. The holes in the *p* side of the diode move toward the negative charge so there can be no movement of electrons across the junction. Under this condition no current can flow.

From the foregoing explanations it can be seen that a single *pn* diode (crystal diode) can serve as a half-wave rectifier, and four diodes connected in a bridge circuit can serve as a full-wave rectifier (see Figure 6–42).

Figure 6–43 is a photograph of the components of a silicon diode. As explained previously, the word *diode* means that the device has two electrodes, or terminals. In the illustration the components of the diode are arranged as follows, from left to right: terminal wire ("pigtail"), base, special soldering alloy, silicon wafer, aluminum connector, case with insulator and anode terminal, and terminal wire. During the assembly of this diode, the base terminal wire is first welded to the base. This assembly is then placed in a jig with the terminal wire pointing downward. A small disk of solder is then placed in the circular depression in the base, and the silicon wafer is placed on the solder. Finally, the pure aluminum wire

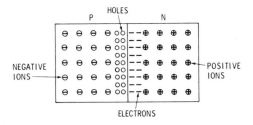

FIGURE 6–39 Junction of *p*- and *n*-type materials to form a potential barrier.

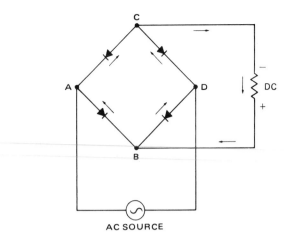

FIGURE 6–42 A full-wave rectifier.

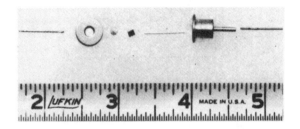

FIGURE 6–43 Components of a silicon diode.

is placed in a vertical position with its end bearing against the center of the silicon wafer. Still held in the jig, the unit is then passed through a furnace with carefully controlled temperatures. The heat fuses all parts together and creates an alloy junction at the point where the aluminum wire is joined to the silicon. This junction establishes the barrier that makes the unit a rectifier. Upon completion of the initial assembly and testing for performance, a case is placed over the unit, and the flange of the case is resistance-welded to the base. Thereafter, the unit is placed in a heated vacuum chamber to remove all air and moisture and is then sealed by compressing the terminal stem. Several additional tests and inspections are performed before the unit is ready for the customer.

Diodes are available in several different sizes and shapes. The current and voltage ratings of a diode will determine its physical size. Power diodes that are relatively large have been developed and are capable of carrying over 2500 A.

A principal consideration in the installation of power diodes is to ensure that the diode is firmly attached to the mounting, which serves as a **heat sink.** Diodes that carry substantial current will become overheated and damaged or destroyed unless the heat developed is conducted away by the mounting structure. Many power diodes are provided with cooling fins by which heat is dissipated.

Large diodes are constructed with heavy metal bases to be mounted securely to a metal structure heavy enough to act as a heat sink. A mounting stud is provided on the base and is integral with it. Before a diode is mounted, the base of the diode should be inspected for cleanliness and smoothness, and the mount to which it is to be attached should be similarly in-

spected. This is to assure that there will be maximum metal-to-metal contact between the base of the diode and the mounting. In some cases a heat-conducting jelly is placed between the diode base and the mounting to fill any gaps caused by irregularities in the surfaces to be joined. This assures maximum heat conductance from the diode to the mounting.

Heat sinks can also be used for other solid-state devices that generate large amounts of heat. Often cooling fans are used in conjunction with heat sinks to help dissipate heat from the component. A typical heat sink assembly is illustrated in Figure 6–44.

Diode Testing. A diode is a one-way gate for electrons; therefore, it can be tested by applying a voltage to the diode and measuring the current flow. The voltage polarity is then reversed, and the current is measured again. An ohmmeter can be used as a source of power for this test. An ohmmeter applies a positive voltage to its red probe and a negative voltage to its black probe. To test a diode, place the red probe on one diode connection, the black probe on the other. Set the ohmmeter on a low-resistance scale, and note the meter's indication. Now reverse the meter's probes, and note the indication again. When testing a good device, the ohmmeter will indicate a high resistance with the probes connected one way and a low resistance with the probes reversed. When testing a defective diode, the ohmmeter will indicate like readings, both high or both low, with the probes connected in either direction. A good diode will block current flow from the anode to the cathode and pass current from the cathode to the anode. An ohmmeter connected to the diode can detect the presence or absence of current flow as indicated in Figure 6–45.

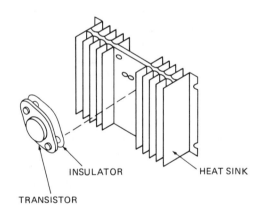

FIGURE 6–44 A transistor heat sink assembly.

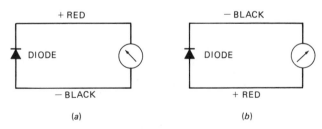

FIGURE 6–45 An ohmmeter test of a diode. (a) The ohmmeter indicates high resistance; (b) the ohmmeter indicates low resistance.

Several test instruments contain a separate diode (semiconductor) test function. A diode tester indicates the actual voltage drop across the semiconductor junction. An ohmmeter indicates the resistance of the diode. A diode tester is the preferred instrument if available, since it will give the actual voltage drop present when current flows through the diode. When a diode is tested with a semiconductor test instrument, the instrument's negative (black) lead is connected to the cathode of the diode, and the instrument's positive (red) lead is connected to the anode. The test should indicate the appropriate voltage drop for the type of diode being tested. The correct testing procedures are shown in Figure 6–46. The correct voltage drop should be approximately 0.3 V over germanium diodes and 0.6 V over silicon diodes. If the diode is shorted, the reading will be 0 V drop; if the diode is open, the reading will be well over 0.6 V. Be sure to consult your instrument's instruction guide for exact testing instructions.

Half-Wave Rectifiers

When a single rectifier (diode) is placed in series with an ac circuit, the result is called **half-wave rectification.** Only one-half of the alternating current can pass through the rectifier into the circuit. As stated earlier, ac current is constantly changing direction and polarity. Therefore, any diode in a series ac circuit will receive a current flow in two different directions. In one direction, negative voltage is applied to the cathode and positive is applied to the anode; the diode will conduct and current will flow. In the opposite direction, the diode will offer high resistance and current will not flow. The result is a pulsating direct current. Figure 6–47a illustrates a half-wave rectifier circuit; its associated current curve is illustrated in Figure 6–47b. In this circuit the dc motor will receive a pulsating direct current 60 times per second. The solid

line of the current curve represents the pulsating direct current; the dotted line represents the ac wave that is blocked by the half-wave rectifier.

Full-Wave Rectifiers

Half-wave rectifiers make use of only one-half of the ac current available; therefore, they are used in limited applications. Whenever a smoother ripple direct current or a more efficient use of power is required, a **full-wave rectifier** is used. In order to convert both halves of the ac wave to direct current, the diodes must be arranged into a bridge circuit. A full-wave bridge circuit can be made using four individual diodes or a single solid-state rectifier assembly. The rectifier assembly simply contains four diodes combined into a single compact package. The bridge rectifier assembly allows for easy installation to printed circuit boards or other circuits.

A typical bridge rectifier circuit is shown in Figure 6–48. The ac source is connected to points A and D (the rectifier input). The rectifier's output, points C and B, is connected to the dc load. Figure 6–48a illustrates the current flow during the first half of the ac wave. At this time the negative side of the ac voltage is connected to point A, the positive to point D. The current flows through the forward-biased diode D_1 but is blocked by D_4 and D_2 because they are reverse-biased. The current travels through the dc load and returns to point B. At this point the current will take the path of least resistance back to the positive side of the ac source; thus the current travels through D_3 and the cycle is complete.

During the second half of the ac wave (see Figure 6–48b), the current polarity is reversed, and the negative voltage is now connected to point D and the positive to point A. The current flows through D_2 and is blocked by the high resistance of D_3 and D_1. Current then travels through the load from

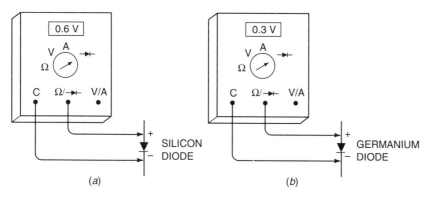

FIGURE 6–46 Testing a diode. *(a)* Silicon diodes should indicate 0.6 V; *(b)* germanium diodes should indicate 0.3 V.

FIGURE 6–47 *(a)* A half-wave rectifier circuit; *(b)* the associated current curve.

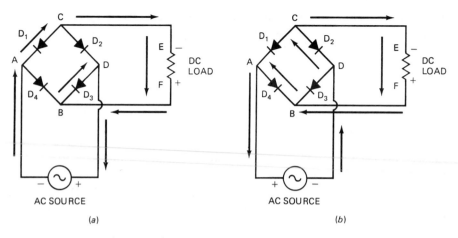

FIGURE 6–48 A full-wave rectifier circuit.

point *E* to *F*; hence the polarity of the dc load remains constant. Point *E* remains negative, and point *F* remains positive. The current reaches point *B* and takes the most direct path, through D_4, back to the positive side of the source.

The ac wave is converted into a ripple direct current through the full-wave rectifier. As illustrated in Figure 6–49, the bottom half of the ac voltage is reversed in polarity to produce a ripple direct current. The ripple dc voltage is designated by the solid line, and the ac voltage is designated by the dotted line. A full-wave rectifier makes total use of the applied ac voltage and produces a relatively smooth dc voltage; therefore, full-wave rectifiers are more common than half-wave rectifiers.

Three-Phase Rectifier

It is often necessary to obtain direct current from three-phase power systems in aircraft; hence three-phase rectifier units are employed. It would be possible to use a single-phase full-wave rectifier in one leg of a three-phase system; however, it is more efficient to use a rectifier system that utilizes the power from all three legs of the three-phase circuit. The output of a three-phase alternator is indicated in Figure 6–50. It will be noted in the diagram that the voltages reach maximum 120° apart. A rectifier consisting of six diodes is connected in such a manner as to provide one-way paths for the ac output as shown in Figure 6–51. It can be seen in the diagram that

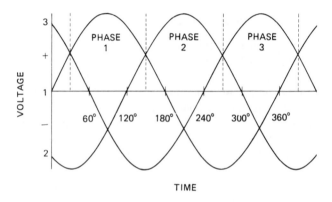

FIGURE 6–50 Output of a three-phase alternator.

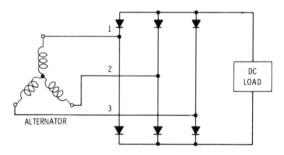

FIGURE 6–51 Full-wave rectifier for a three-phase alternator.

the current flowing in each section of the three-phase circuit will always be flowing in the same direction in the output side of the rectifier, even though it reverses direction in the ac side of the circuit.

In Figure 6–52, the current flow is illustrated by the arrows through each phase of the three-phase system. The solid line represents current flow through phase 1, the long dashed line represents phase 2 current, and the short dotted line represents phase 3 current. Each phase of the current is directed through the rectifier in such a manner as to apply a dc voltage to the load.

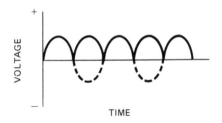

FIGURE 6–49 Ripple dc voltage produced by a full-wave rectifier.

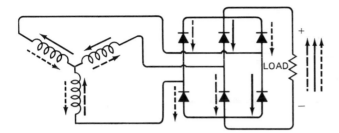

FIGURE 6–52 Current flow through a three-phase rectifier.

TRANSISTORS

A **transistor** is a solid-state device that can be used for controlling electric signals. Transistors are typically mass produced and relatively low in cost. During operation, transistors produce heat that may damage the transistor if the heat becomes excessive. Since even small amounts of heat can damage many transistors, it is often essential to provide heat sinks and adequate air circulation to ensure proper transistor operation.

It was explained earlier, in the discussion of diodes, that *n*-type and *p*-type materials are semiconductors. When formed into a junction, these semiconductors will allow current flow in one direction and block current flow in the opposite direction. When *n*- or *p*-type materials are joined in the correct combination containing two junctions, a junction transistor is formed.

Junction Transistor

The **junction transistor** is only one of several types of transistors currently available. Essentially, they all require junctions established between *n*-type and *p*-type germanium or silicon, so we shall examine the junction transistor here.

A junction transistor consists of three principal sections and will be manufactured as one piece. In an ***npn*** transistor the crystal consists of a section of *n*-type germanium, then a very thin section of *p*-type germanium, and another larger section of *n*-type germanium. This is shown in Figure 6–53. One end of this transistor is called the **emitter,** the small *p*-type section is called the **base,** and the other end is called the **collector.**

A ***pnp*** transistor also contains two *np* junctions; however, in this case the emitter is a *p*-type material, the base is an *n*-type material, and the collector is a *p*-type material. A *pnp* transistor is illustrated in Figure 6–54. For all practical purposes, *pnp* and *npn* transistors are functionally the same, except their connections have opposite polarities. For example,

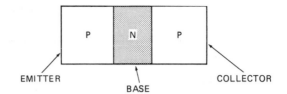

FIGURE 6–54 A *pnp* transistor diagram.

the emitter of an *npn* transistor is negative because it is an *n*-type material; the emitter of a *pnp* transistor is positive, because it is a *p*-type material. The transistor's schematic symbols make it easy to identify which type of transistor is being used. Figure 6–55 shows the symbols for a *pnp* transistor and an *npn* transistor. It can be seen from this illustration that the arrow, which represents the emitter lead of the transistor, points in opposite directions for the two different types of transistors. The arrow points in for *pnp* transistors and points out for *npn* transistors. This is easily remembered by noting that the arrow of an *npn* transistor symbol does *not* point in (*npn*).

The direction of the arrow also helps to define the direction of current flow through the transistor. As with diodes, current always flows against the arrow. Therefore, to allow correct current flow through the transistor, any connection at the point of the arrow must be negative. The connection at the back of the arrow must be positive. The *npn* and *pnp* transistors are not interchangeable because of their opposite polarities.

Transistors can be used for switching or amplification. A **switching transistor** is similar to an electrical solenoid or relay; that is, a transistor can act as a remote-control switch. By connecting the correct voltage to the transistor's base connection, the resistance between the emitter and collector is lowered to near zero. The solenoid analogy of a switching transistor is illustrated in Figure 6–56. The solenoid's coil connection performs a function similar to that of the transistor's base connection; if either is connected to ground, the lamp will illuminate because the emitter-collector or the solenoid (switch) connection lowers to near zero resistance.

Amplification is another function commonly performed by transistors. **Amplification is defined as an increase in a signal's power.** A weak signal can be fed into a transistor and a stronger output signal produced. The transistor does require an added power source for the amplified signal. Figure 6–57 demonstrates the principle of amplification.

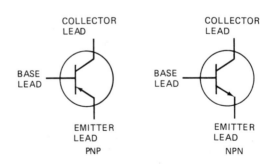

FIGURE 6–55 Schematic symbols of *npn* and *pnp* transistors.

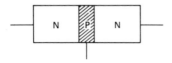

FIGURE 6–53 An *npn* transistor diagram.

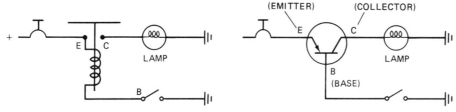

FIGURE 6–56 Solenoid analogy of a switching transistor.

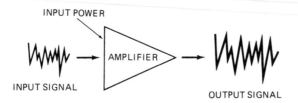

FIGURE 6–57 Principle of amplification.

Amplification in a transistor is called **gain.** True amplification of a signal must increase the total power of the signal without changing the characteristics of that signal. If the voltage of an input signal increases and the current decreases, as in a step-up transformer, no amplification takes place. To amplify, there must be a total power gain. Transistor gain is symbolized by the Greek letter β (beta).

The gain of a transistor is often defined as the ratio of collector current to base current. That is,

$$\text{Gain} = \frac{\text{collector current}}{\text{base current}}$$

or

$$\beta = \frac{I_c}{I_b}$$

Power transistors may have a gain as low as 20, and signal transistors are available with a gain as high as 400. Gain is not always a constant number for any given transistor. The gain of a transistor may vary with the collector current or voltage and is often affected by temperature changes.

Transistor Operation

Transistors contain at least two junctions between an *n*-type and a *p*-type material. Both junctions must be biased correctly to allow the transistor to conduct. **For the transistor to**

be operating, the emitter-base junction of a *pnp* or an *npn* transistor must be forward-biased, and the base-collector junction must be reverse-biased (Figure 6–58). As demonstrated in Figure 6–58, the emitter-base junction of an *npn* transistor is forward-biased when the emitter is negative with respect to the base. The base-collector junction is reverse-biased when the collector is positive with respect to the base. "More positive" in this illustration represents a greater positive voltage at the collector than at the base, labeled "positive."

The junctions of a *pnp* transistor are biased in a conducting mode when the emitter is positive with respect to the base and the collector is negative with respect to the base. It becomes confusing when dealing with transistors if voltage is thought of as only an absolute positive and absolute negative. Voltage of a point must be defined with respect to a second point. Voltage measured at any given point may be negative with respect to one reference and positive with respect to a second. Figure 6–59 demonstrates this concept. Point *B* is negative if measured from point *A*. Point *B* becomes positive if measured from point *C*. This concept of voltage must be employed when studying the operation of transistors.

A picture of an *npn* transistor circuit is shown in Figure 6–60. As indicated, the electron flow produced by the circuit's dc batteries travels through two paths, the emitter-base path and the emitter-collector path. The emitter-base current is a relatively weak "control" signal. Approximately 1 percent of the total current travels through the emitter-base circuit. This is the signal to be amplified. The current that travels through the emitter-collector circuit is the amplified signal. The majority of the transistor's current, approximately 99 percent, is sent through this path. The current in this path is controlled by the current through the base circuit.

During normal operation of an amplifying transistor, when the base current increases, the collector current will increase proportionally. If the base current decreases, the col-

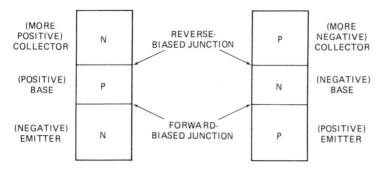

FIGURE 6–58 Transistor junction bias.

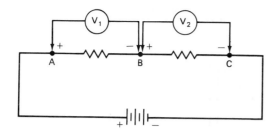

FIGURE 6–59 Voltage measurement. Voltage at point *B* is positive with respect to point *C*; point *B* is negative with respect to point *A*.

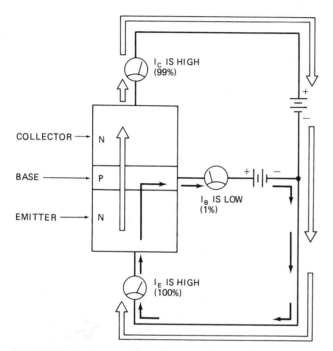

FIGURE 6–60 A current flow diagram of an *npn* transistor.

lector current will decrease. A change in voltage applied to the emitter-base circuit or the emitter-collector circuit will have very little effect on the current of the emitter-collector.

It should be considered that any *np* junction will have a **breakdown threshold.** This threshold is equal to the voltage required to overcome the resistance of the transistor junction in a reverse-biased condition. The breakdown threshold is usually a relatively high voltage applied with reversed polarity. When the transistor is subjected to this type of voltage, it is often damaged and must be replaced.

As indicated by Figure 6–61, the *pnp* transistor can have a current ratio similar to that of an *npn* transistor (99 to 1). The emitter-base current is the controlling, weaker signal, while the emitter-collector signal is the stronger signal being controlled. The voltage and currents of a *pnp* transistor are reversed from those of an *npn* transistor; therefore, the two types of transistors are not interchangeable. In general, the *pnp* transistor is less popular than the *npn* transistor because the *npn* type responds more quickly to base current changes. This is particularly important when amplifying high-frequency signals.

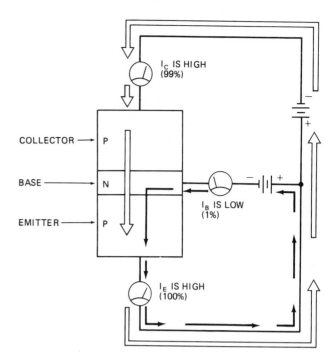

FIGURE 6–61 A current flow diagram of a *pnp* transistor.

Transistor Characteristics

One of the best ways to define the operating characteristics of any transistor is by plotting its current flows and voltages. The **collector current curves,** as illustrated in Figure 6–62, are the most common means of plotting a transistor's characteristics. The horizontal axis is measured in volts and represents emitter-collector bias voltage (V_{e-c}). The vertical axis is calibrated in milliamperes and represents collector current (I_c).

The curves in Figure 6–62 show the output data (I_B) for any given input situation (V_{e-c} and I_C). From these curves it is easy to see that most of the variance in output (collector current) is caused by changes in base current, not changes in the emitter-collector voltage. The initial voltage, between 0 and 0.5 V, applied to the emitter-collector circuit does change emitter-collector current by approximately 3 mA. However, a 0.5-V change between 10 and 10.5 V affects emitter-collector current very little. This change in voltage without a proportional change in current shows that transistors do not react similarly to resistors; that is, their current flow is not directly proportional to the voltage applied.

The major factor controlling collector current is base current. Base current (I_B) is measured in microamperes and represented on the right-hand side of the graph. A base current of 10 μA at an emitter-collector voltage of 20 V will allow 3.2 mA of collector current flow (see point *A* in Figure 6–62). At a base current of 20 μA, the same emitter-collector voltage of 20 V will allow 5.7 mA of collector current flow (see point *B* in Figure 6–62). It can easily be seen from examining these two points of the collector curves that a slight change in base current of 10 μA will create a significantly greater change in collector current, 2.5 mA. The weak base signal does control the stronger collector signal, and amplification is present.

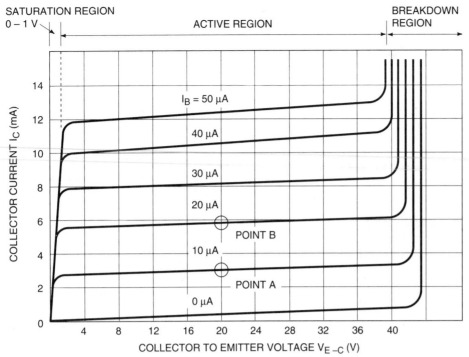

FIGURE 6–62 Collector current curves.

We have already seen from Figure 6–62 how base current controls collector current; it is also true that emitter-base voltage controls collector current. This is because the voltage between the base and the emitter controls the amount of emitter-base current. Figure 6–63 illustrates the relationship between emitter-base voltage and collector current for both germanium and silicon transistors. As the voltage to the emitter-base circuit of either a germanium or silicon transistor increases, the collector current increases. For a germanium transistor, a 0.1-V change (0.2 to 0.3 V) in the emitter-base circuit creates a 12-mA change (2 to 14 mA) in collector current.

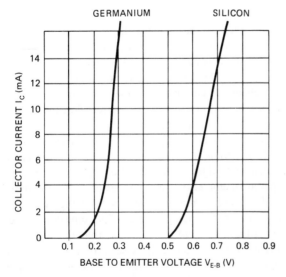

FIGURE 6–63 Voltage curves for germanium and silicon transistors.

The emitter-base voltage required to turn on a germanium transistor is approximately 0.2 to 0.3 V; a silicon transistor requires approximately 0.5 to 0.6 V. Studying these bias voltages may be helpful when troubleshooting transistors and their related circuits. For example, if the bias voltage applied to the emitter-base circuit of a silicon transistor was less than 0.6 V, the technician would not expect the transistor to turn on. If the voltage is above 0.6 V, the transistor should turn on; if it does not, the transistor is probably defective.

Transistor Regions of Operation

There are three regions of operation for transistors: the **active region,** the **saturation region,** and the **breakdown region.** The different regions are defined by the change in collector current due to different voltage levels applied to the transistor's emitter-collector. The **active region** is the flat area of the collector current curves (Figure 6–62). In this region, the voltage of the emitter-collector must be between 1 and 40 V approximately. The transistor is typically operated in this region for amplifying purposes. In the active region, the changes in the collector current are controlled by changes in the base current.

The **saturation region** is the vertical area of the curves (Figure 6–62). Here small changes in collector voltage will result in substantial changes in the collector current. Typically in this region, emitter-collector voltage will be less than 1 V. Transistors are often operated in this region when used for switching purposes.

The **breakdown region** of a transistor is the area of the collector curves above approximately 40 V. The transistor should not be operated in this region, since it will most likely damage the semiconductor material.

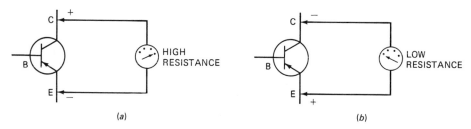

FIGURE 6–64 Ohmmeter test of a transistor emitter-collector. *(a)* First test, positive probe to the collector, negative probe to the emitter; *(b)* second test, negative probe to the collector, positive probe to the emitter.

Transistor Testing

Transistors can be tested with an ohmmeter, provided that the transistor is removed from its circuit. This test should be limited to transistors that are not extremely delicate and to ohmmeters with range settings below $R \times 100 \ \Omega$.

The first part of the test is to measure the resistance between the emitter and collector. As illustrated in Figure 6–64, this resistance should be high with the ohmmeter connected in both directions. If the emitter-collector resistance measures less than a few thousand ohms, then the transistor is defective.

The next step is to check the emitter-base junction in both forward and reverse polarities. As illustrated in Figure 6–65, the junction should measure high resistance in one direction (forward bias) and low resistance in the opposite direction (reverse bias). If the resistance of the emitter-base junction is the same in both directions, the transistor is defective.

The last step is to check the collector-base junction in forward and reversed polarities. Once again the resistance should be high in one direction and low in the opposite direction. If the readings are equal in both directions, the transistor is defective. This test is illustrated in Figure 6–66.

There are several transistor testers currently available. These testers will perform a more thorough test than an ohmmeter. However, transistors seldom partially fail, and any complete failure is detectable with an ohmmeter. One advantage of most transistor checkers is that they often have "in-circuit" test capabilities. This is very important when dealing with transistors soldered in place. Often the removal of such

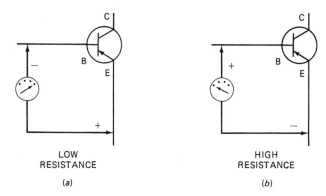

FIGURE 6–65 Ohmmeter test of a transistor emitter-base. *(a)* First test, negative probe to the base, positive probe to the emitter; *(b)* second test, positive probe to the base, negative probe to the emitter.

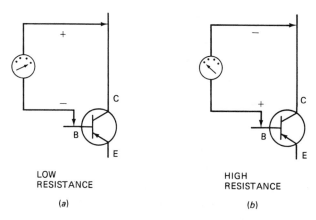

FIGURE 6–66 Ohmmeter test of a transistor collector-base. *(a)* First test, negative probe to the base, positive probe to the collector; *(b)* second test, positive probe to the base, negative probe to the collector.

a transistor will damage the component; therefore, in-circuit testing is the preferred method. Voltmeters and oscilloscopes can also be used to test transistors if the circuit is complete and normal operating power is available.

OTHER SOLID-STATE DEVICES

There are several types of hybrid solid-state devices currently available. Many of these have been designed to function under specific conditions and perform specific operations. Eight specialized semiconductors are commonly known as JFETs, MOSFETs, thyristors, zener diodes, light-emitting diodes, photodiodes, LASCRs, and LCDs.

The **junction field-effect transistor** (JFET) is very similar to a junction transistor; however, a JFET is considered to be voltage-sensitive; a junction transistor is current-sensitive. Figure 6–67 shows the symbol of a JFET. The connections of a JFET are the **gate,** which is the control lead, and the **drain** and the **source,** which are the leads being controlled. If the voltage applied between the gate and the source is increased, the current flow between the drain and source will increase. This relationship is illustrated in Figure 6–68. The JFET is used in circuits where the voltage input must control the current output. One such circuit is found in digital voltmeters and oscilloscopes.

Another advantage of the JFET is that it will produce a very low noise output. This makes the JFET a perfect device

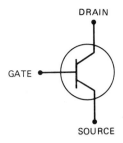

FIGURE 6–67 JFET symbol.

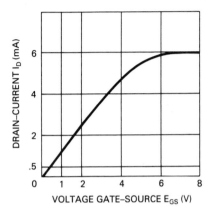

FIGURE 6–68 The relationship between gate voltage and drain-source current for a JFET.

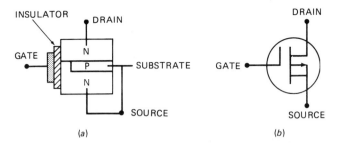

FIGURE 6–69 A MOSFET. *(a)* Structure of an *n*-type MOSFET; *(b)* symbol for an *n*-type MOSFET.

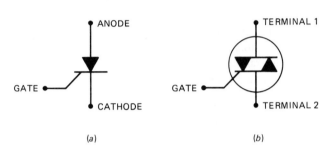

FIGURE 6–70 Thyristor symbols. *(a)* An SCR; *(b)* a triac.

for amplifiers with weak input signals. The major disadvantage of a JFET is that it is less sensitive to voltage changes than a junction transistor.

The **metal-oxide semiconductor field-effect transistor** (MOSFET) has a source, gate, and drain similar to the connections of a JFET. The major difference is that the gate is insulated from the current channel of the drain and source. The structure and the schematic symbol of a MOSFET are illustrated in Figure 6–69. The major advantage of a MOSFET is that either a positive or negative voltage can be applied to the gate to produce a drain-source current flow.

A **thyristor** is a semiconductor that is used for switching purposes. In many cases a thyristor may replace a solenoid or relay for controlling the load current to motors. Electrical load switching through thyristors is advantageous because there are no moving parts; thus problems with wear, corrosion, and arcing are eliminated.

There are two common types of thyristors, the **silicon controlled rectifier** (SCR) and the **triode ac semiconductor** (triac). The SCR symbol is shown in Figure 6–70*a*, and the triac symbol is shown in Figure 6–70*b*. Either the SCR or triac will allow current to flow once a certain level of gate signal is achieved. If the gate signal is then removed, the current will continue to flow through the anode-cathode circuit until that signal is interrupted. This characteristic of a thyristor makes it ideal for switching warning circuits on aircraft. For example, if an excessive temperature is reached in a turbine engine for only a split second and then decreases, an engine warning light will illuminate. The light will continue to glow

until the pilot interrupts the light circuit and turns off the thyristor, thus turning off the light. The thyristor has made it possible for the pilot to receive a continuous indication of a warning condition that existed for only a very short time period. Thyristors are also used for controlling large amounts of current flow to motors, heaters, or lighting circuits.

The **zener diode** is also a popular device on modern aircraft. The zener diode, as illustrated in Figure 6–71, will conduct electricity only under certain voltage conditions; hence it is ideal for use in voltage regulator circuits. A zener diode is designed to operate at or above its breakdown voltage. The breakdown voltage is a given voltage at which the zener diode will conduct; below this level the zener will not conduct. The **avalanche effect** is caused when a zener reaches its breakdown voltage in a reverse-bias mode. If the correct value of reverse voltage (negative anode, positive cathode) is applied to the zener, it will act as a very low resistance. Below this voltage the zener will offer high resistance. This is known as the avalanche effect because the zener offers nearly infinite resistance until the breakdown voltage is reached. At that point the resistance falls dramatically to nearly zero (an avalanche). As with conventional diodes, if the polarity is reversed, the zener diode will not conduct.

Bidirectional zener diodes are also found on many aircraft. These diodes are often connected in parallel with

FIGURE 6–71 Schematic symbol of a zener diode.

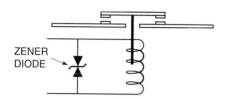

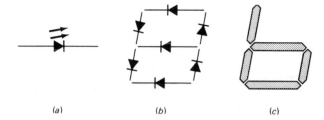

FIGURE 6–72 Bidirectional zener diode installed on a solenoid coil.

FIGURE 6–73 Light-emitting diodes. (a) A schematic symbol; (b) an LED arrangement to display a digit; (c) an LED display of the digit 6.

the magnetic coil of a relay or solenoid (Figure 6–72). Bidirectional zener diodes are used to eliminate voltage spikes (transients) that are created during the expansion or contraction of the magnetic field of the relay (solenoid) coil. Similar to zener diodes, the bidirectional zener will conduct current above a certain voltage level. However, the bidirectional zener will conduct current in either direction or polarity.

The **light-emitting diode** (LED) is widely used on aircraft instruments and test equipment. Single LEDs are used as indicator lights. An arrangement of LEDs, such as in Figure 6–73, is used to display letters and numbers. LEDs require 1.5 to 2.5 V and 10 to 20 mA to produce adequate light for most applications. For an LED to conduct, the applied voltage must be connected in the forward-biased condition.

The light of an LED comes from the energy given off when the diode is forward-biased. At this time free electrons travel from a high to a low energy level and produce light and heat. Those diodes which do not emit light will expend all their "extra" energy in heat. LEDs expend most of their extra energy in light. The various colors available from LEDs are determined by their active elements, such as gallium, phosphorus, and arsenic. LEDs that produce red, green, yellow, blue, orange, and infrared light are all currently available.

Photodiodes are semiconductors that respond to light. Photovoltaic diodes, or solar cells, as they are commonly called, produce a dc voltage when they are exposed to light. A relatively small amount of power is produced by photovoltaic cells; however, through the use of modern electronics, several calculators and other low-power devices can operate using the current produced by photovoltaic cells. Outside the earth's atmosphere, the sun's rays are much stronger and help create a greater amount of electric power through photodiodes. Several modern satellites operate solely on electric power generated by photovoltaic cells.

A **light-activated SCR** (LASCR) is a device that is turned on by light rays. In this device, when the light is strong enough, the valence electrons of the gate become free electrons and allow current to flow from the cathode to the anode. Figure 6–74 shows the symbol of an LASCR. The arrows indicate the light needed to trigger the LASCR. As with SCRs, LASCRs will continue to conduct after the trigger source is removed. In other words, when the light source has diminished, the LASCR will continue to offer very low resistance (a closed circuit).

Liquid crystal displays (LCDs) are found in many state-of-the-art aircraft instruments. The display can be configured to form letter and number patterns, or it can form a full pic-

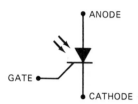

FIGURE 6–74 Schematic symbol of an LASCR.

ture. Commonly, liquid crystal displays are gray; however, some full-color systems are available. Color LCDs show great promise in replacing heavier displays used in electronic flight instruments owing to their weight savings and electrical efficiencies.

Liquid crystal displays get their name from the *liquid crystal* that is used to arrange light patterns within the unit. Liquid crystals are fluid materials that contain molecules arranged in crystal forms. The molecules are typically twisted and therefore "bend" the light that passes through the crystal (see Figure 6–75a). If a voltage is applied to the liquid crystal, the molecules align and the light passes "straight" through the material (see Figure 6–75b). LCDs use this phenomenon to align light waves with polarized filters. The polarized filters block, or pass, the light to form specific patterns for the display.

Figure 6–76 shows a typical 7-segment liquid crystal display. In this example, voltage is applied to the individual segments to form the number 5. The segments that do not have voltage applied are *light gray*. The light waves pass through these segments and reflect off a mirror mounted behind the rear polarizer. The segments that form the number 5 are dark gray, since they reflect the light. The liquid crystals of these displays are aligned by an applied voltage.

Thermistors are heat-sensitive devices used on some aircraft to monitor the temperature of certain electric equipment. For example, nickel-cadmium battery sensors may use thermistors to monitor the battery temperature. As the name implies, *thermistor* comes from the words *therm*al and res*is*tor. Thermistors are semiconductor devices that change resistance as their temperature changes. Thermistors are formed from metal oxides and coated with an epoxy, glass, or similar material. There are a variety of thermistor styles, such as rods, discs, and washers, which are mounted within a temperature probe.

Other Solid-State Devices   **121**

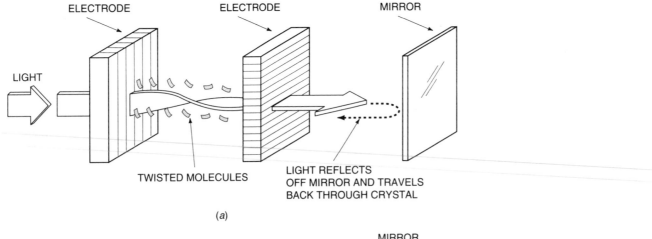

FIGURE 6–75 Liquid crystal theory. *(a)* With no current applied to the electrodes, light passes through the polarizer. Light reflects off the mirror and passes through the polarizer again. This segment is light. *(b)* With current applied to the electrodes, light does not travel through the polarizer. This segment is dark.

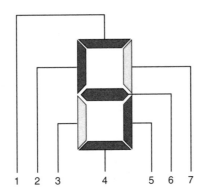

FIGURE 6–76 A 7-segment liquid crystal display. Electrodes 1, 2, 4, 5, and 6 are energized to form the number 5.

PRINTED CIRCUIT BOARDS

Modern electronic equipment uses **printed circuit boards** (PCBs), which provide a mounting surface and the electric current paths for the individual components of a system. Printed circuit boards are typically constructed of a rigid insulator material approximately $\frac{1}{16}$ in. thick. The surface of the board is covered with a copper foil, and holes are drilled through the materials where component connections are made (see Figure 6–77). Printed circuit boards allow for compact installation of hundreds of individual components

into electronic equipment. On some equipment the PCBs, known as *cards,* can be easily removed for repair.

During the construction of PCBs, the copper foil covers the entire surface of the insulator board. The foil is then etched with chemical solvents to form the specific current paths needed for the circuits. The components are installed on the board and soldered to the copper foil. In many cases the complete assembly (components and PCB) is covered with a protective coating to seal out moisture. This protective coating must be removed prior to replacement of components on a PCB. Be sure to follow the manufacturer's recommendations for component replacement.

Surface mounted components are designed to be mounted to both sides of a printed circuit board. As seen in Figure 6–78, the component leads do not extend through the PCB. The leads are bent at a 90° angle and sit flat on the surface of the PCB. The leads are then soldered to the copper foil conductor and secured in place. This arrangement allows for installation of components on both sides of the PCB; hence a more compact unit can be designed. Virtually all types of electrical components, from resistors to integrated circuits, can be designed for surface mounting. In most cases components designed for surface mounting are extremely small and difficult to handle. In many cases component identification is virtually impossible since the parts are too small to print identification numbers or resistor color codes on their surface.

It is difficult to install and remove surface mounted components owing to their compact design. Special soldering and

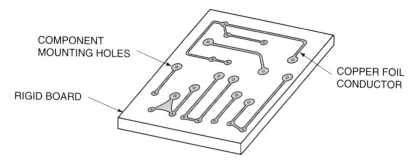

FIGURE 6–77 A printed circuit board.

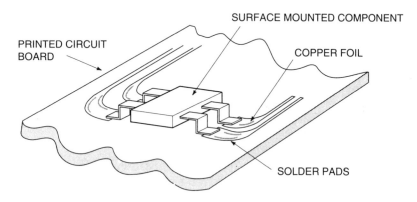

FIGURE 6–78 An example of a surface mounted component.

desoldering equipment is vital to the success of a component replacement. Surface mounted components should be removed and installed only by technicians well trained in PCB and surface mounted repair techniques.

CATHODE-RAY TUBE

For many years the **cathode-ray tube** (CRT) has been familiar to millions of persons as the picture tube in a television set. The CRT in a television is designed to reproduce an undistorted picture on a screen. The picture is developed from a series of pulses and varying voltages applied to the elements of the tube.

Fundamentally, the CRT consists of an electron "gun," a phosphorescent screen, and deflecting devices to control the movement of the electron beam "shot" from the gun. A diagram of a CRT is shown in Figure 6–79. As in any other thermoemitting tube, the heated cathode supplies the electron emission, and these electrons are accelerated toward the screen by the positive charges on the anodes. The intensity of the electron beam is regulated by means of the control grid charge. After the electron beam is accelerated and focused by the anodes, it is controlled in direction by the **deflection plates.** When the electrons strike the phosphor-coated screen, they cause a bright spot to appear. If an alternating voltage is applied to the vertical deflection plates, the spot will move up and down and form a straight line, as shown in Figure 6–80a. In like manner, if an alternating voltage is ap-

plied to the horizontal deflection plates, a horizontal straight line will appear on the screen, as in Figure 6–80b.

The deflection of the electron beam can be accomplished by either electrostatic or electromagnetic means. Electrostatic deflection positions the electron beam by producing electric fields that change the direction in which the beam travels. **Deflection plates,** which create the electric field, are typically used in CRTs for oscilloscopes and many aircraft

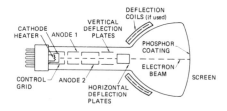

FIGURE 6–79 A diagram of a typical CRT.

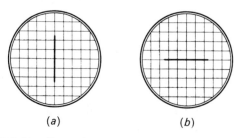

(a)   (b)

FIGURE 6–80 Effects when ac voltage is applied to the deflection plates.

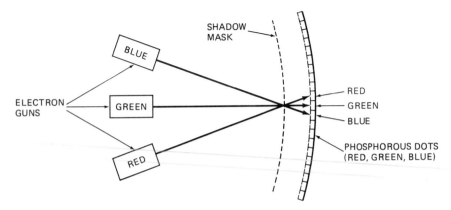

FIGURE 6–81 A diagram of a color CRT.

instrument displays. Electromagnetic fields are used to direct the electron beam in many television receivers and similar video displays. **Deflection coils** create magnetic fields to direct the electron beam in both the horizontal and vertical direction.

The CRT screen is coated on the inside with a phosphorous material that glows after being hit by an electron beam. The time required for the phosphorous material to glow is determined by the sweep time of the electron beam. Any point of the CRT screen must glow until the electron beam returns to that spot and reactivates the phosphorous material.

A color CRT uses three electron beam guns in order to achieve the three primary colors. A separate gun is needed for each color: red, blue, and green. The screen of a color CRT is made of three separate phosphorous materials; the type of phosphorous material determines the color achieved when the electron beam strikes the CRT screen. The CRT screen is divided into hundreds of groups of three small dots of phosphorus, one each for red, blue, and green. Each electron beam must pass through a **shadow mask** before it reaches its respective phosphorous dot. The shadow mask helps to prevent overlap of the electron beams into any incorrect color dot. A diagram of a color CRT is shown in Figure 6–81. When one electron beam reaches its color dot on the CRT screen, that color appears on the face of the CRT. If two or three electrons reach their corresponding phosphorous dots in any given area, the colors blend together to create a new color. For example, if the red, blue, and green guns are emitting electrons directed to the same area of the CRT screen, white is produced. If the red and blue guns emit electrons to their respective phosphorous dots, the color mauve (bluish-purple) will be produced. Color CRTs are becoming very popular in aircraft weather radar systems and instrument display panels.

Every CRT requires some external form of control for the electron beam(s). These controls must turn on or off and control the intensity of the electron beam(s) at the correct time intervals. The direction of the electron beam(s) must also be controlled in order to ensure that the beam(s) travels to the correct portion of the CRT screen in the correct sequence. These controls are usually provided through a variety of complex circuits. Although either digital or analog signals can be used for this purpose, most modern airborne equipment employs digital techniques.

REVIEW QUESTIONS

1. Describe the principles of switch installation.
2. Why should switches be derated for some circuits?
3. Describe a DPST switch.
4. Describe a rotary switch.
5. When are microswitches typically used?
6. What is meant by the normally closed contact of a switch?
7. What is meant by the common contact of a microswitch?
8. Describe a fuse.
9. What must be done before resetting a circuit breaker that is tripped?
10. Describe a thermal circuit breaker.
11. What should be the capacity of a circuit breaker used with No. 16 wire?
12. What is a trip-free circuit breaker?
13. When may more than one circuit be protected by the same circuit breaker?
14. Define *resistor*.
15. Name the general types of resistors.
16. Give the color code for resistor values.
17. If the first color band of a fixed resistor is orange, the second is green, and the third is blue, what is its resistance value?
18. What is the principal difference between adjustable and variable resistors?
19. By what other names is a variable resistor commonly called?
20. What is the principal difference between a potentiometer and a rheostat?
21. Define *capacitor*.
22. What is a dielectric?
23. Describe the two general types of capacitors.
24. What is the formula for capacitors connected in parallel? in series?
25. Describe the time constant for capacitors.
26. What is the function of an inductor in an electronic circuit?
27. What is an electrical transient?
28. Compare inductance with capacitance.

29. What precaution must be taken in connecting an electrolytic capacitor in a circuit?

30. What is meant by the wattage of a resistor?

31. Define a capacitance of 1 F.

32. What units are usually used to indicate capacitance in an electronic circuit?

33. What is a step-up transformer? a step-down transformer?

34. Why is it that a transformer will not draw an appreciable amount of power when no load is connected to the secondary, even though the primary is connected?

35. Compare the number of windings in the primary coil of a step-up transformer with the number of windings in the secondary coil.

36. Give the formula for expressing the voltage values in the circuits of a transformer with respect to the number of turns in the primary and secondary windings.

37. Give the formula for maximum current values in the primary and secondary windings of a transformer.

38. What is a rectifier?

39. What are the two principle materials used in a semiconductor?

40. What materials are used in crystal diodes?

41. What is the difference between an n-type material and a p-type material in a semiconductor unit?

42. What is a hole in a semiconductor?

43. Explain the operation of a rectifier.

44. What is the difference between a half-wave rectifier and a full-wave rectifier?

45. How is full-wave rectification accomplished?

46. What procedures are used to test a diode with an ohmmeter?

47. Describe the purpose of a heat sink.

48. Describe the characteristics of a zener diode.

49. Describe a thyristor.

50. What are two common types of thyristors?

51. What is the difference between an npn transistor and a pnp transistor?

52. Name the three working parts of a transistor.

53. Discuss the importance of temperature with respect to the operation of a transistor.

54. Explain the difference between a junction field-effect transistor and a junction transistor.

55. Which lead carries the majority of the current in a junction transistor?

56. What is amplification?

57. What is transistor gain?

58. Which circuit is the control circuit of a transistor?

59. Which circuit is the controlled circuit of a transistor?

60. What emitter-base voltage is required to turn on a germanium transistor? a silicon transistor?

61. Describe the process used to test a transistor with an ohmmeter.

62. What is an LED?

63. What is a cathode-ray tube?

64. How is the direction of the electron beam(s) changed within a CRT?

65. What are the three colors used in a color CRT?

66. What is the purpose of the shadow mask of a CRT?

7 Digital Electronics

Digital systems became practical with the invention of the integrated circuit. However, there have been several examples of digital circuits throughout the evolution of electronics. The first means of transmitting an information signal, the telegraph, relied on basic digital principles. The telegraph used a code system of voltage on and voltage off combinations to produce letters, words, and, therefore, information. This concept of voltage on and voltage off is the heart of the modern digital circuit.

Currently, integrated circuits are capable of providing thousands of combinations of voltage on and voltage off signals. This large number of voltage combinations per component allows modern digital circuits to perform a seemingly infinite number of tasks. Digital circuits are used extensively on modern aircraft computer systems. Computers operate virtually every system on a state-of-the-art airplane. The concept of digital electronics has changed the way we design, fly, and maintain aircraft. Some modern aircraft, such as the Airbus A-320 or the Boeing 777, employ a complete fly-by-wire system; that is, the cockpit controls are linked to the control surfaces only through electrical wiring and computer systems. There are several computers used on modern commercial and military aircraft that control a variety of functions. Some common examples are flight management computers, thrust management computers, and bus power control computers. There are also numerous peripheral devices used to send information to, or receive information from, the different computers. One of the most common peripheral devices is the cathode-ray tube (CRT). CRTs are used on modern aircraft to display flight parameters and aircraft system data. This display system replaces many of the conventional electromechanical instruments, which have relatively high maintenance costs. Solid-state CRT instrument display systems are known as electronic flight instrument systems, or EFIS.

This chapter will introduce the language used by computers and the basic functions of digital circuits, computers, and various peripheral devices. The electronic structure of computers will also be presented, including integrated circuits and microprocessors.

THE DIGITAL SIGNAL

A **digital signal** is one that contains two distinct values. These values are often considered to be on and off, or 1 and 0. An analog signal, on the other hand, is one that contains an infinite number of voltage values. A digital signal is represented in Figure 7–1. As illustrated, if zero voltage is present at point A, it is considered a digital 0; 5 V positive at point A is considered a digital 1. In this circuit the digital 0 is created when the switch is open. The digital 1 is created when the switch is closed, thus connecting point A to the 5-V positive source.

An analog signal created by a variable resistor is shown in Figure 7–2. In this circuit +5 V, or digital 1, is present at point A of the circuit, when the potentiometer is set to position 1. Zero voltage, or digital 0, is present when the potentiometer is moved to position 2. As illustrated by the graph of Figure 7–2b, the analog signal does not produce a distinct value of +5 or 0 V. As the potentiometer moves from position 1 to position 2, it provides an infinitely variable voltage, ranging from 0 to +5 V.

Logic circuits, the fundamental components of all computers, utilize digital signals. Through the use of integrated circuits, thousands of combinations of 1s and 0s can be produced and manipulated to perform a countless number of functions. A modern aircraft flight computer receives millions of digital input signals during a typical flight. These signals are compared and processed by the computer circuits. The computer then produces output signals that are also composed of a combination of 1s and 0s. The computer's output

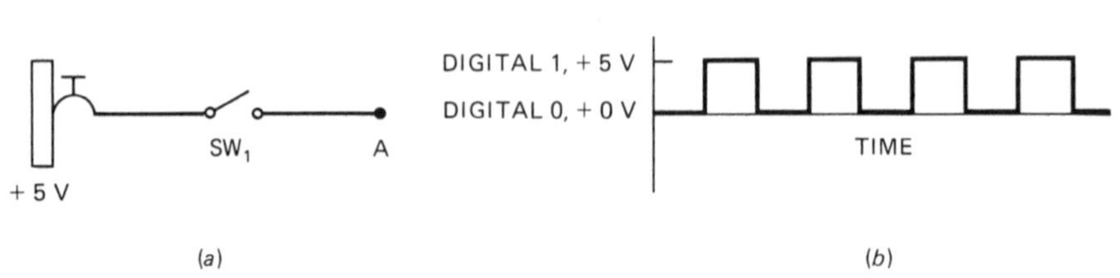

FIGURE 7–1 A digital signal. (a) A simple digital circuit; (b) a digital waveform.

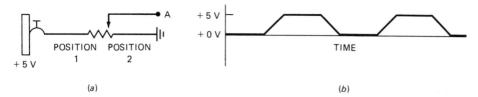

FIGURE 7–2 An analog signal. (a) A simple analog circuit; (b) an analog waveform.

signals may be used to display data on flight instruments, control flight altitudes, maintain engine power, and provide fault information on system malfunctions.

DIGITAL NUMEROLOGY

In order for a computer to understand a variety of information, a code system must be used that is composed of digital 1s and 0s. The **binary number system,** as the name implies, is composed of two components, 1 and 0. There are several varieties of binary systems used in modern computers. These code systems provide the "language" for communication between computers and their related components. For example, a digital combination of 1011 might represent the number eleven. If a binary code is used to represent or label a particular component, the digital combination of 1001 might represent temperature probe no. 1. As illustrated in Figure 7–3, an input code of 1001 1011 might represent 11° measured at temperature probe no. 1. Virtually any binary code system may be used as a computer's language, as long as all components of the system are programmed for the same code. Digital 1s and 0s are often used to represent other characteristics of aircraft systems. For example, a light being on is often represented by a digital 1; the light off equals digital 0. An item being present as stated, such as hydraulic pressure, is represented by a digital 1. A digital 0 would be used to represent the absence of hydraulic pressure. The list below presents common designations of the binary digits 1 and 0:

1	2
Positive voltage	Negative voltage (positive logic)
Negative voltage	Positive voltage (negative logic)
On	Off
Yes	No
Present as stated	Not present as stated
True	False
Conduction	Nonconduction

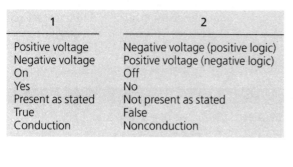

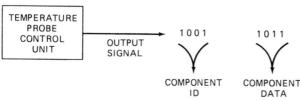

FIGURE 7–3 Representation of digital to binary conversions.

Decimal System

In any number system a symbol is given to each quantity to be represented. The **decimal number system** uses 10 figures to represent the quantities 0 through 9. Each quantity of dots shown in Table 7–1 is represented by a symbol of the decimal system. This example of the decimal number system may seem quite obvious; however, this is only because we use this system during everyday activities. We already recognize what quantity each symbol represents. Unfortunately, the decimal number system is not suited for use as a computer language, because the digital system contains only two symbols, 1 and 0.

Binary Numbers

The **binary number system** represents different quantities using only two symbols, 1 and 0. If a quantity larger than 1 must be represented by binary numbers, the symbols systematically repeat. As in the decimal number system, the binary system repeats by adding to the left of the first digit. The column just to the left of the first digit is considered a **higher-order column.**

In Table 7–2 each quantity of dots is represented by a symbol of the binary number system. Each digit of a binary number has a specific value. As illustrated, the first digit of a binary number is represented by 2 to the power 0, or 1; the second digit equals 2 to the power 1, or 2; the third digit equals 2 to the power 2, or 4; and so on. The value of 2^0 is 1, the value of 2^1 is 2, the value of 2^2 is 4, and so on.

To determine the total value of any binary number, add the individual values (2^0, 2^1, 2^2, etc.) of the digits containing a bi-

TABLE 7–1 Representation of the decimal number system.

Decimal number symbol	Quantity of dots
0	none
1	•
2	• •
3	• • •
4	• • • •
5	• • • • •
6	• • • • • •
7	• • • • • • •
8	• • • • • • • •
9	• • • • • • • • •

TABLE 7–2 The binary number system. (*a*) The value of binary numbers; (*b*) the value of binary digits.

Binary number symbol	Quantity of dots
0	none
1	•
10	• •
11	• • •
100	• • • •
101	• • • • •
110	• • • • • •
111	• • • • • • •
1000	• • • • • • • •
1001	• • • • • • • • •

(a)

Binary digit	10th	9th	8th	7th	6th	5th	4th	3rd	2nd	1st
Power	2^9	2^8	2^7	2^6	2^5	2^4	2^3	2^2	2^1	2^0
Decimal value	512	256	128	64	32	16	8	4	2	1

(b)

Conversion of decimal number 96 to its binary equivalent:

	Quotient	Remainder
$\frac{96}{2} =$	48	0 (last binary digit)
$\frac{48}{2} =$	24	0
$\frac{24}{2} =$	12	0
$\frac{12}{2} =$	6	0
$\frac{6}{2} =$	3	0
$\frac{3}{2} =$	1	1
$\frac{1}{2} =$	0	1 (first binary digit)

Decimal number 96 = binary number 1 1 0 0 0 0 0

As illustrated, 96 is divided by 2 and the remainder, 0, is recorded. The quotient, 48, is divided by 2 and the remainder recorded. Twenty-four is divided by 2, and so on. The binary number is found by listing the calculated remainders. To determine the binary number, simply record each remainder, starting from the bottom of the list. The bottom digit of the remainder list becomes the left-most digit of the binary number. The top digit of the list becomes the right-most digit of the binary number. Two more examples of the conversion from decimal numbers to binary numbers are shown in Figure 7–4.

The binary number code is the language of logic circuits and computers. The decimal code system is the language of humans. If we desire to communicate with computers, we must understand their language. Conversely, computers and their related components must be capable of converting information to the decimal system for the display of human-compatible information. Comprehension of the binary code system is essential for a basic understanding of computer operations.

The term **bit** may be used when referring to binary digits. One bit is equal to one binary digit. A bit will always be expressed as a high (1) or low (0) logic level. Bits handled as a group are referred to as a **byte.** Therefore, an eight-digit binary number is a byte containing eight bits. Another category of data is called the **word.** A word is a grouping of bits that the computer uses as a standard information format. For example, many systems communicate using a 16- or 32-bit word. Each word for a particular system will conform to a specific format that enables the computer to decode the message.

nary 1. Any digit with a binary 0 has a value of zero; therefore, it need not be added. For example, the binary number 101 is equal to the decimal number 5. This is found by adding the individual values: the first digit, 2^0 (1), plus the third digit, 2^2 (4), or (1 + 4 = 5). Note that the second digit is not added because it is a binary zero. The total decimal value of the binary number 111011 is determined as follows: $2^5 + 2^4 + 2^3 + 2^1 + 2^0 = 32 + 16 + 8 + 2 + 1 = 59$. Studying Table 7–3 will help you understand the concept of binary number values.

A decimal number may be converted to its binary equivalent by sequentially dividing the number by 2 and recording each remainder. The quotient from the first division must be divided by 2 and its remainder recorded. This process is repeated until the quotient is 0. The remainders make up the binary number. The following example indicates this conversion system.

TABLE 7–3 Conversion of binary numbers to their decimal equivalents (101 binary = 5 decimal; 111011 binary = 59 decimal).

Binary digit	6th	5th	4th	3rd	2nd	1st	
power	2^5	2^4	2^3	2^2	2^1	2^0	
Decimal value	32	16	8	4	2	1	
				1	0	1	4 + 1 = 5
	1	1	1	0	1	1	32 + 16 + 8 + 2 + 1 = 59
		binary number					decimal equivalent

	Quotient	Remainder
$\frac{18}{2} =$	9	0
$\frac{9}{2} =$	4	1
$\frac{4}{2} =$	2	0
$\frac{2}{2} =$	1	0
$\frac{1}{2} =$	0	1

Decimal number = 18
Binary number = 10010
(a)

	Quotient	Remainder
$\frac{69}{2} =$	34	1
$\frac{34}{2} =$	17	0
$\frac{17}{2} =$	8	1
$\frac{8}{2} =$	4	0
$\frac{4}{2} =$	2	0
$\frac{2}{2} =$	1	0
$\frac{1}{2} =$	0	1

Decimal number = 69
Binary number = 1000101
(b)

FIGURE 7–4 Examples of decimal to binary conversions.

Addition of Binary Numbers

Addition using any number system is a means of combining quantities. For example, in the decimal system $6 + 1 = 7$ may represent a quantity of X's or X X X X X X plus X equals X X X X X X. In the binary system, quantities may be added in a similar manner. In binary, $110 + 1 = 111$ would represent $6 + 1 = 7$ of the decimal system.

There are four basic rules to remember when adding binary numbers:

Rule no. 1: $0 + 0 = 0$
Rule no. 2: $0 + 1 = 1$
Rule no. 3: $1 + 0 = 1$
Rule no. 4: $1 + 1 = 10$

In the binary system only two symbols are used, 1 and 0; therefore, when 1 and 1 are added, we must record zero and carry the 1 into a higher-order column. This is identical with the process used in the decimal system, as illustrated in Figure 7–5. To add the binary 100 and 100, the process would be as follows:

$$\begin{array}{r} 100 \\ + 100 \\ \hline 1000 \end{array}$$

That is, first, $0 + 0 = 0$; second, $0 + 0 = 0$; and third, $1 + 1 = 10$. If the binary numbers 11 and 11 were added, the process would be as follows:

$$\begin{array}{r} 11 \\ + 11 \\ \hline 110 \end{array}$$

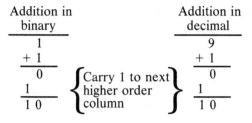

FIGURE 7–5 Comparison of addition of binary numbers and addition of decimal numbers.

Addition of 11011 + 110		Addition of 11000 + 1101		Addition of 11101 + 11101	
Binary	Decimal	Binary	Decimal	Binary	Decimal
11011	27	11000	24	11101	29
+ 110	+ 6	+1101	+ 13	+11101	+ 29
100001	33	100101	37	111010	58

FIGURE 7–6 Examples of binary addition showing decimal equivalents.

That is, first, $1 + 1 = 10$; here it becomes necessary to record the 0 and carry the 1 to the next higher-order column. Second, $1 + 1 + 1 = 11$; therefore, the result is 110. Figure 7–6 shows three more examples of binary addition.

Subtraction of Binary Numbers

There are four that apply to binary number subtraction; they are as follows:

Rule no. 1: $0 - 0 = 0$
Rule no. 2: $1 - 0 = 1$
Rule no. 3: $1 - 1 = 0$
Rule no. 4: $10 - 1 = 1$

When subtracting binary numbers, start with subtraction in the right-hand column and proceed to the left. Borrow from the adjacent higher-order column when necessary. That is, if the right-hand column is $0 - 1$, borrow from the adjacent column in order to subtract 1 from 10 ($10 - 1$). Next, subtract in the adjacent column to the left; once again borrow if necessary. Repeat this procedure until a subtraction has been done in each column. For example, in subtracting 110 from 1101, proceed as follows:

$$\begin{array}{r} 1101 \\ - 110 \\ \hline 111 \end{array}$$

First, subtract in the right column ($1 - 0 = 1$); second, subtract in the adjacent column ($0 - 1$; with a carried digit, $10 - 1 = 1$); third, subtract in the next column ($0 - 1$; with a carried digit, $10 - 1 = 1$). The fourth column no longer exists because we borrowed the 1 for subtraction in the third column. In decimal form the same subtraction would be $13 - 6 = 7$.

Three more examples of binary number subtraction are given in Figure 7–7. The respective decimal equivalents are also shown.

Subtraction of 1011 + 100		Subtraction of 1110011 − 101001		Subtraction of 111000 − 101	
Binary	Decimal	Binary	Decimal	Binary	Decimal
1011	11	1110011	115	111000	56
− 100	− 4	−101001	− 41	− 101	− 5
111	7	1001010	74	110011	51

FIGURE 7–7 Examples of binary subtraction showing decimal equivalents.

```
  11010      26        1011      11         110001
×   110     × 6       ×   11    × 3        × 1000       49
  00000      156        1011     33        000000      × 8
 11010                 1011                000000       392
11010                100001                000000
10011100                                  110001
                                          110001000
```
(a)

```
      1001
1100)1101100            9              111          7
     1100           12)108          11)10101       3)21
       11              108             11            21
       00                0             100            0
      110                               11
      000                               11
     1100                               11
     1100                               11
        0                                0
```
(b)

FIGURE 7–8 (a) Examples of binary multiplication with decimal equivalents; (b) examples of binary division with decimal equivalents.

Binary Multiplication and Division

In order to multiply or divide binary numbers, four rules must be considered; they are as follows:

Rule no. 1: $0 \times 0 = 0$
Rule no. 2: $0 \times 1 = 0$
Rule no. 3: $1 \times 0 = 0$
Rule no. 4: $1 \times 1 = 1$

To multiply a binary number, use the same procedures that are used for multiplying decimal numbers; that is, multiply to form partial products and add. For example, to multiply 10 by 11, proceed as follows:

```
   10
 × 11
   10    (partial product)
   10    (partial product)
  110    (addition of partial products)
```

Division of a binary number is performed in a manner similar to division of a decimal number. To divide 1110 by 10, proceed as follows:

```
       111
   10)1110
       10
       11
       10
       10
       10
        0
```

Always divide the left-most digits first and progress to the right; the procedure is complete when the remainder is zero. Several examples of multiplication and division problems are given in Figure 7–8. The decimal equivalents are also given.

BINARY CODE SYSTEMS

General Theory

There are several varieties of code systems used to convert information bits into letters or decimal numbers. The pure binary number system is often too clumsy for a computer to manipulate data quickly. Three of the more common systems used to produce a faster computer are called the **binary-coded decimal, octal notation,** and **hexadecimal** systems. All of these systems utilize the binary digits 1 and 0 to represent decimal, or base 10, numbers.

Binary-Coded Decimal System

The **binary-coded decimal** (BCD) system uses a group of four bits to represent each digit of a decimal number. Each digit of the decimal system (1 to 9) can be represented by four binary digits. For example, 9 in the decimal system equals 1001 in the binary system; 2 in decimal equals 0010 in binary. This system is extremely useful when dealing with large quantities of data interchanged between the inputs and outputs of a computer system.

Another advantage of the BCD is that there are "extra" bits unused by the four-digit system. Since 9 (1001) is the highest decimal number to be represented by the four-bit binary code, there are six unused digit combinations for each byte. These additional digits are often assigned various alphabetic symbols, thus expanding the information carried by each four-bit byte.

Octal Notation System

The **octal notation** system is a binary representation of an octal number. **Octal numbers** are composed of eight different symbols. Since computers can handle only two symbols, 1 and 0, octal notation was developed. Octal notation is comprised of a series of three-bit groups. Since the largest decimal number represented by three binary digits is 7 (111), this is a base 8, or octal, system. In short, octal notation is a means to represent octal (base 8) numbers in a binary language.

Three-digit group (triad)	5th	4th	3rd	2nd	1st
Power of eight	8^4	8^3	8^2	8^1	8^0
Decimal value of base eight number	4096	512	64	8	1

(a)

Triad group	4th	3rd	2nd	1st
3-digit octal notation	001	010	100	001
Decimal equivalent of triad	1	2	4	1
Power of eight	8^3	8^2	8^1	8^0
Decimal value of base eight number	512	64	8	1
Decimal equivalent of octal groups	(1×512) 512	(2×64) 128	(4×8) 32	(1×1) 1

Sum the decimal equivalents of each octal group
512 + 128 + 32 + 1 = 673
Octal notation 001 010 100 001 = decimal 673

(b)

Octal notation is useful for certain programming techniques where large quantities of binary numbers must be manipulated. Octal notation is often used for the transmission of data by aircraft computers and their related peripherals.

The octal notation system has a base of 8; that is, each three-digit binary group (**triad**) can be represented as in Table 7–4*a*. Table 7–4*b* shows the octal notation 001 010 100 001 converted to its decimal equivalent, 673. To determine the decimal value of an octal notation, find the decimal value of each triad and sum those values. To find the value of a triad, multiply the decimal equivalent of the base 8 number (8^0 or 1, 8^1 or 8, 8^2 or 64, etc.) by the decimal equivalent of the triad (001 = 1, 010 = 2, 011 = 3, etc.). For example, to convert 001 010 100 001 to decimal form, proceed as follows. First, determine the decimal value of each triad. The decimal value of the first triad equals 001×8^0, or $1 \times 1 = 1$. The second triad group is 100×8^1, or $4 \times 8 = 32$. The third triad group is 010×8^2, or $2 \times 64 = 128$. The fourth triad group is 001×8^3, or $1 \times 512 = 512$. Next, sum the values of the individual triad groups: $1 + 32 + 128 + 512 = 673$. The decimal value of the octal notation 001 010 100 001 is 673.

To determine the octal notation code of a decimal number, proceed as follows. First, divide the decimal number by 8, and record the quotient and the remainder. Second, divide the previous quotient by 8, and record the quotient and remainder. Repeat this procedure until the quotient is zero, and then convert each remainder into a three-digit binary number (triad). Last, record each triad, starting from the bottom and moving upward to determine the octal notation code.

To determine the octal code for the decimal number 741, proceed as follows:

Division	Quotient	Remainder	Triad Equivalent of Remainder in Binary
$\frac{741}{8}$	92	5	101 (first binary triad)
$\frac{92}{8}$	11	4	100
$\frac{11}{8}$	1	3	011
$\frac{1}{8}$	0	1	001 (last binary triad)

The triad list is 001 011 100 101, which is the octal notation code of the decimal number 741. Figure 7–9 shows several examples of conversions from octal notation to decimal codes and from decimal to octal codes. In the triad groups presented in the preceding examples, each triad is separated by a blank space. In a typical computer language, the blank spaces would be removed, and the triad groups would be bunched together. For example, the octal notation number 100 011 010 111 would be represented as 100011010111.

Hexadecimal Number System

The **hexadecimal number system** (Hex) uses base 16. The main purpose of this number system is to represent the very large numbers of memory locations utilized by microcontrollers and microprocessors. The instructions that make up

Conversion from octal notation to decimal

101 010 octal

equals	$(101 \times 8^1) + (010 \times 8^0)$
equals	$(5 \times 8) + (2 \times 1)$
equals	$40 + 2$
equals	42 decimal

100 010 111 100 octal

equals	$(100 \times 8^3) + (010 \times 8^2) + (111 \times 8^1)$ $+ (100 \times 8^0)$
equals	$(4 \times 512) + (2 \times 64) + (7 \times 8) + (4 \times 1)$
equals	$2048 + 128 + 56 + 4$
	equals 2, 236 decimal

(a)

Conversion of 2460 decimal to octal notation

Division	Quotient	Remainder	Triad equivalent of remainder	
$\frac{2460}{8}$	307	4	100	First triad
$\frac{307}{8}$	38	3	011	
$\frac{38}{8}$	4	6	110	
$\frac{4}{8}$	0	4	100	Last triad

Octal code 100 110 011 100 equals decimal 2460

Conversion of 137 decimal to octal notation

Division	Quotient	Remainder	Triad equivalent of remainder	
$\frac{137}{8}$	17	1	001	First triad
$\frac{17}{8}$	2	1	001	
$\frac{2}{8}$	0	2	010	Last triad

Octal code 010 001 001 equals decimal 137

(b)

FIGURE 7–9 (a) Conversion of octal notation numbers to decimal numbers; (b) conversion of decimal numbers to octal notation numbers.

TABLE 7–5 The relationship between hexadecimal, decimal, and binary numbers.

Hexadecimal	Decimal	Binary
0	0	0000
1	1	0001
2	2	0010
3	3	0011
4	4	0100
5	5	0101
6	6	0110
7	7	0111
8	8	1000
9	9	1001
A	10	1010
B	11	1011
C	12	1100
D	13	1101
E	14	1110
F	15	1111

the assembly language used to program microcontrollers and microprocessors also make use of the hexadecimal number system. The hexadecimal number system uses the numbers 0 through 9, along with the letters A, B, C, D, E, and F, to make up the 16 symbols. The relationship between the hexadecimal, decimal, and binary numbers is shown in Table 7–5.

An important point to recognize about the hexadecimal number system is that 4 binary digits represent a single hexadecimal digit. This is significant because a 16-bit binary code can be represented with a 4-digit hexadecimal number.

Hex to Decimal Conversion

Each digit in a base 16 number possesses a magnitude that corresponds to that digit's position. The right-most digit in a sequence, or the least significant digit (LSD), has a magnitude of 16^0, or 1. The left-most digit in the sequence, or the most significant digit (MSD), has a magnitude of 16^{n-1}, where n is the number of digits in the sequence. To convert the number 653_{16} to its decimal equivalent:

$$653_{16} = 6 \times 16^2 + 5 \times 16^1 + 3 \times 16^0$$
$$= 1536 + 80 + 3$$
$$= 1619_{10}$$
$$FA2_{16} = 15 \times 16^2 + 10 \times 16^1 + 2 \times 16^0$$
$$= 3840 + 160 + 2$$
$$= 4002_{10}$$

Note that 15 was substituted for F and 10 was substituted for A in the second example.

Decimal to Hex Conversion

As with decimal to binary conversion, repeated division is used to calculate the equivalent base 16 number.

To convert 324_{10} to its hex equivalent:

$$\frac{324}{16} = 20 + \text{remainder of } 4$$
$$\frac{20}{16} = 1 + \text{remainder of } 4$$
$$\frac{1}{16} = 0 + \text{remainder of } 1$$

Therefore, $324_{10} = 144_{16}$.

To convert 412_{10} to its hex equivalent:

$$\frac{412}{16} = 25 + \text{remainder of } 12$$
$$\frac{25}{16} = 1 + \text{remainder of } 9$$
$$\frac{1}{16} = 0 + \text{remainder of } 1$$

Therefore, $412_{10} = 19C_{16}$.

Note how the remainders of the division processes form the digits of the hex number.

Hex to Binary Conversion

Hexadecimal notation is a "shorthand" method for representing binary numbers. Each hexadecimal digit is converted to its 4-bit binary equivalent.

To convert $2F9_{16}$ to its binary equivalent:

$$
\begin{array}{ccc}
2 & F & 9 \\
\downarrow & \downarrow & \downarrow \\
0010 & 1111 & 1001
\end{array}
$$

Therefore, $2F9_{16} = 001011111001_2$.

Binary to Hex Conversion

This is simply a matter of converting each group of 4 binary digits to its hexadecimal equivalent. Starting from the right of the binary number, count every 4 bits and convert.

To convert 10110001101_2 to its hexadecimal equivalent:

$$
10110001101_2 = \underbrace{0101}_{5} \quad \underbrace{1000}_{8} \quad \underbrace{1101}_{D}
$$

Therefore, $10110001101_2 = 58D_{16}$.

LOGIC GATES

An Introduction

Logic gates, or **gates,** are fundamental functions performed by computers and related equipment. A single integrated circuit (IC) within a computer contains several gate circuits. Each gate may have several inputs and must have only one output. There are six commonly used logic gates: the **AND,** the **OR**, the **INVERT**, the **NOR**, the **NAND**, and the exclusive **OR**. The name of each gate represents the function it performs.

Truth tables are a systematic means of displaying binary data. Truth tables illustrate the relationship between a logic gate's inputs and output. This type of data display can be used to describe the operation of a gate or an IC. For troubleshooting purposes, the truth table data for a specific IC is often reviewed in order to determine the correct output signal for a given set of inputs.

Each logic gate has a symbol of a specific shape. The symbols are designated to "point" in a given direction. That is, the inputs are always listed on the left of the symbol and the output on the right. Since logic gates operate using digital data, all input and output signals will be composed of 1s or 0s. Typically, the symbol 1 represents "on," or voltage positive. The symbol 0 represents "off," or voltage negative. Voltage negative is often referred to as zero voltage or the circuit's ground.

The AND Gate

The **AND gate** is used to represent a situation where all inputs to the gate must be 1 (on) to produce a 1 (on) output. For an AND gate, input no. 1, input no. 2, input no. 3, etc., must be 1 to produce a 1 output. If any input is a 0 (off), the output will be 0 (off). The symbol and the truth table for a two-input AND gate are illustrated in Figure 7–10.

A simple AND circuit may be represented by two switches in series used to turn on a light. If both switches (inputs) are on (1), the light will turn on (1). If either switch is off (0), the light will be off (0). Figure 7–11 illustrates this simple AND circuit.

A different AND circuit is shown in Figure 7–12. This circuit uses solid-state components to produce the gate; therefore, it is more consistent with those circuits found within an

INPUTS		OUTPUT
A	B	C
0	0	0
0	1	0
1	0	0
1	1	1

FIGURE 7–10 An AND gate. (*a*) Logic symbol; (*b*) truth table.

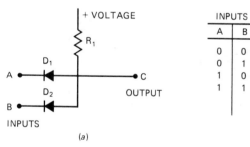

INPUTS		OUTPUT
A	B	C
0	0	0
0	1	0
1	0	0
1	1	1

FIGURE 7–12 An AND gate. (*a*) Solid-state AND circuit; (*b*) truth table.

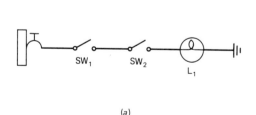

INPUTS		OUTPUT
SW_1	SW_2	L_1
0	0	0
0	1	0
1	0	0
1	1	1

FIGURE 7–11 An AND circuit. (*a*) Series circuit; (*b*) truth table.

Logic Gates **133**

IC. In this circuit, each diode is reverse-biased by a positive signal at inputs A and B, respectively. If both inputs are positive (1), the diodes are both reverse-biased and no current flows through R_1; therefore, there is no voltage drop over R_1, and point C is positive. If either input is negative (0), current will flow through R_1, and point C will be negative (0) owing to the voltage drop across R_1. In this circuit the only way to produce a positive at point C (a 1 output) is to provide both diodes with a positive voltage (both inputs are 1); therefore, the circuit performs an AND function.

The OR Gate

The **OR gate** is used to represent a situation where any input being on (1) will produce an on (1) output. For an OR gate, input no. 1 or input no. 2 or input no. 3, etc., must be 1 to produce a 1 output. Only if all inputs become 0 will the OR gate produce a 0 output. If any input is a 1, regardless of the other input values, the OR gate will produce a 1 output. A two-input OR gate symbol and the corresponding truth table are illustrated in Figure 7–13.

A simple OR circuit may be composed of two switches in parallel controlling one light, as illustrated in Figure 7–14. If either switch is on (1), the light will turn on (1).

Figure 7–15 shows a solid-state OR circuit and its corresponding truth table. In this circuit, if a positive voltage (1) is applied to either input, the corresponding diode will be forward-biased and current will flow through R_1.

Point C will be positive owing to the voltage drop over R_1. Point C will be negative only if there is no current flow through R_1. This occurs when both inputs are negative. If either or both inputs become positive, the output is positive; therefore, this is an OR circuit.

The INVERT Gate

The **INVERT gate** is used to reverse the condition of the input signal. The INVERT gate contains only one input and one output, and it is most often used in conjunction with other gates. The INVERT gate is sometimes referred to as a **NOT** gate. The symbol and truth table for an INVERT gate are shown in Figure 7–16.

An INVERT circuit might be composed of a switch controlling a normally closed relay that turns on or off a light. As illustrated in Figure 7–17, if the switch is turned on (1), the light is off (0).

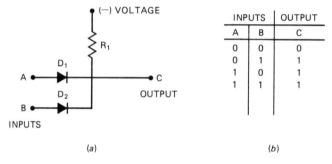

INPUTS		OUTPUT
A	B	C
0	0	0
0	1	1
1	0	1
1	1	1

(a)   (b)

FIGURE 7–15 An OR gate. (a) Solid-state OR circuit; (b) truth table.

INPUT	OUTPUT
A	B
0	1
1	0

(a)   (b)

FIGURE 7–16 An INVERT gate. (a) Logic symbol; (b) truth table.

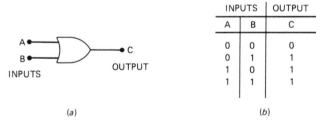

INPUTS		OUTPUT
A	B	C
0	0	0
0	1	1
1	0	1
1	1	1

(a)   (b)

FIGURE 7–13 An OR gate. (a) Logic symbol; (b) truth table.

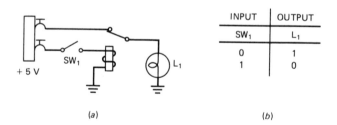

INPUT	OUTPUT
SW_1	L_1
0	1
1	0

(a)   (b)

FIGURE 7–17 An INVERT gate. (a) Simple INVERT circuit; (b) truth table.

INPUTS		OUTPUT
SW_1	SW_2	L_1
0	0	0
0	1	1
1	0	1
1	1	1

(a)   (b)

FIGURE 7–14 An OR circuit. (a) Parallel circuit; (b) truth table.

A basic solid-state inverter circuit containing a transistor is illustrated in Figure 7–18. As shown by the circuit's truth table, if the input is negative voltage (0), the transistor is reverse-biased and point B is positive (1). Point B becomes positive owing to the lack of current through R_1. If no current flows in R_1, there will be no voltage drop across R_1; therefore, point B is positive. If point A becomes positive, the transistor is forward-biased and point B is connected to ground (0) through the emitter-collector circuit of the transistor; therefore, point B is negative (0).

The NOR Gate

The **NOR gate** is an OR gate with an inverted output. This results in a gate where any input being 1 will create a 0 output. The NOR symbol and truth table are given in Figure 7–19.

The electronic circuit used to represent a NOR symbol is illustrated in Figure 7–20. If the inputs of this circuit are both negative (0), the transistor is reverse-biased and the output is positive (1). If either input is positive (1), the corresponding diode conducts and the transistor is forward-biased. This, in turn, connects point C to ground (0), and the output becomes negative (0).

The NAND Gate

The **NAND gate** is an AND gate with an inverted output. The output of this gate will be 1 if any input is 0. This, of course, is the exact opposite of the situation with an AND gate. The symbol and truth table of a NAND gate are shown in Figure 7–21. The circuit for a NAND gate is shown in Figure 7–22. If either input is connected to a negative voltage (0), the cor-

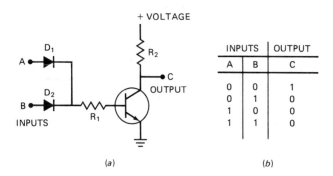

FIGURE 7–20 A NOR gate. (a) Solid-state NOR circuit; (b) truth table.

INPUTS		OUTPUT
A	B	C
0	0	1
0	1	0
1	0	0
1	1	0

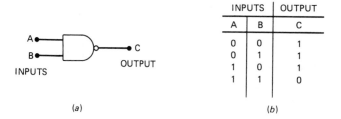

FIGURE 7–21 A NAND gate. (a) Logic symbol; (b) truth table.

INPUTS		OUTPUT
A	B	C
0	0	1
0	1	1
1	0	1
1	1	0

responding diode is forward-biased. Current then flows from the negative input voltage through R_1 to the positive source. Current takes this path instead of flowing from ground (0) through the emitter-base circuit and through R_1 and R_2 to the positive source. Since no current flows in the emitter-base circuit, the transistor is reverse-biased and point C is positive (1).

If both inputs are connected to a positive (1) source, both diodes are reverse-biased, the transistor is forward-biased, and point C is connected to ground (0); therefore, the output is 0.

The Exclusive OR Gate

The **exclusive OR gate** is designed to produce a 1 output whenever its input signals are dissimilar. An illustration of the exclusive OR symbol is shown in Figure 7–23a. This gate compares a maximum of two input signals to determine its output. The exclusive OR gate is often referred to as a digital comparator. As shown in Figure 7–23b, if the input signals are like values, the output is 0; if the input signals are unlike values, the output is 1.

Variation of Basic Gates

The basic logic gates can be used in an infinite number of combinations. This variety of gate combinations allows a computer to perform a multitude of functions. One typical combination of logic gates adds an INVERT gate to the input of an AND or OR gate. A symbol and truth table for an AND

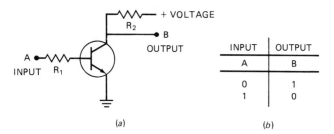

FIGURE 7–18 An INVERT gate. (a) Solid-state INVERT circuit; (b) truth table.

INPUT	OUTPUT
A	B
0	1
1	0

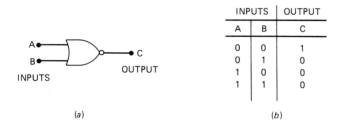

INPUTS		OUTPUT
A	B	C
0	0	1
0	1	0
1	0	0
1	1	0

FIGURE 7–19 A NOR gate. (a) Logic symbol; (b) truth table.

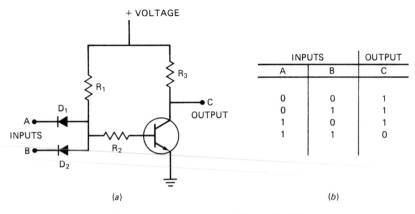

INPUTS		OUTPUT
A	B	C
0	0	1
0	1	1
1	0	1
1	1	0

(a) (b)

FIGURE 7–22 A NAND gate. (a) Solid-state NAND circuit; (b) truth table.

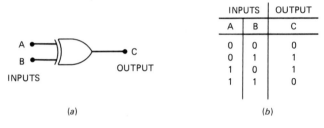

INPUTS		OUTPUT
A	B	C
0	0	0
0	1	1
1	0	1
1	1	0

(a) (b)

FIGURE 7–23 An exclusive OR gate. (a) Logic symbol; (b) truth table.

INPUTS		OUTPUT
A	B	C
0	0	0
0	1	1
1	0	0
1	1	0

(a) (b)

FIGURE 7–24 An AND gate with an inverted input. (a) Logic symbol; (b) truth table.

gate with an inverted input are shown in Figure 7–24. An inverted input is often referred to as a NOT input.

The circuit for an AND gate with an inverted input is shown in Figure 7–25a. As illustrated, this circuit is simply a combination of an INVERT circuit and an AND circuit. INVERT gates are often added to the inputs, or outputs, of the basic gates previously discussed. Three other basic gate combinations are illustrated in Figure 7–26.

Positive and Negative Logic

As stated earlier, logic circuit input and output signals consist of two distinct levels. These levels are often referred to as binary 1 and binary 0. The actual voltage levels required to achieve a binary 1 or 0 may vary between circuits.

If **positive logic** is used in the digital circuit, a binary 1 equals a high voltage level and a binary 0 equals a low voltage level. The actual voltage values may be both positive, both negative, or one positive and one negative. The only stipulation for positive logic is that a binary 1 be created by a greater positive voltage than a binary zero. Four examples of

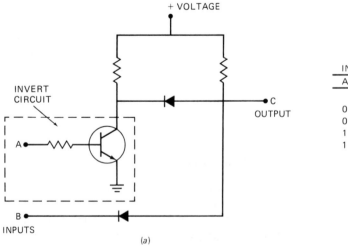

INPUTS		OUTPUT
A	B	C
0	0	0
0	1	1
1	0	0
1	1	0

(a) (b)

FIGURE 7–25 An AND gate with an inverted input. (a) Solid-state circuit; (b) truth table.

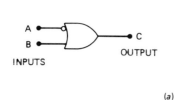

INPUTS		OUTPUT
A	B	C
0	0	1
0	1	1
1	0	0
1	1	1

(a)

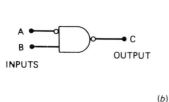

INPUTS		OUTPUT
A	B	C
0	0	1
0	1	0
1	0	1
1	1	1

(b)

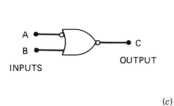

INPUTS		OUTPUT
A	B	C
0	0	0
0	1	0
1	0	1
1	1	0

(c)

FIGURE 7–26 Variations of basic logic gates. (a) An OR gate with an inverted input; (b) a NAND gate with an inverted input; (c) a NOR gate with an inverted input.

a digital signal are shown in Figure 7–27. Each signal represents the greater positive voltage value as a binary 1; therefore, each example employs the positive logic concept. Most digital systems employ positive logic throughout the entire computer and related component circuitry.

The voltage levels of any digital system are designed into the circuit by the limitations of the individual components. Since countless varieties of ICs are available, the circuit's design engineer must decide which voltage values best suit the design parameters. The most common voltage levels are zero (or circuit ground) for binary 0 and +5 V for binary 1. These values, of course, do have some tolerance. A signal of +4.95 V may be sufficient to produce a binary 1. The tolerance of each IC must be known to determine the exact voltage values required.

The **negative logic** concept defines binary 1 as the lower voltage value and binary 0 as the higher voltage value (more positive). Although less popular, negative logic is used in some systems in order to meet certain design parameters. Examples of negative logic digital signals are shown in Figure 7–28. As illustrated, the binary 0 is the higher positive voltage value, and the binary 1 is the lower voltage value.

Display of Digital Data

The display of digital data can take several forms. The **truth table** and the **voltage waveform graph** are the two most common types of digital data displays. Truth tables have been discussed previously in this chapter; voltage waveform graphs will be presented in the following paragraphs.

The digital waveform may be used to describe the operation of any digital circuit by displaying its input and output data. Figure 7–29 shows a typical digital waveform. As illustrated, the vertical line represents voltage values, and the horizontal line represents time. The manipulation of most digital signals requires an extremely short time interval; therefore, the horizontal axis is often scaled in milliseconds. One millisecond equals one-thousandth of a second.

The vertical axis on some waveform graphs is labeled according to the binary code, as shown in Figure 7–30a. This illustration demonstrates the waveform of an AND gate. The horizontal and vertical axes have been eliminated to help

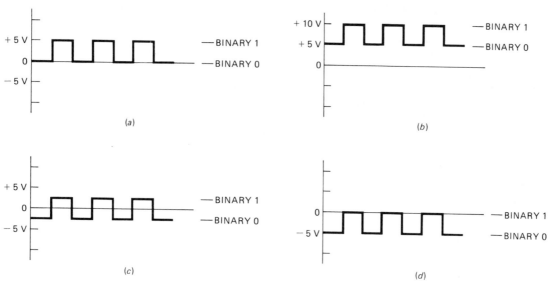

FIGURE 7–27 Four examples of positive logic digital signals. (a) Binary 1 = +5 V, binary 0 = +0 V; (b) binary 1 = +10 V, binary 0 = +5 V; (c) binary 1 = +2.5 V, binary 0 = −2.5 V; (d) binary 1 = +0 V, binary 0 = −5 V.

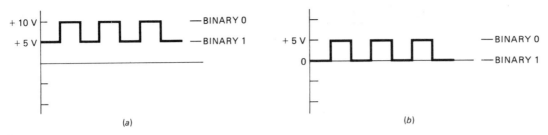

FIGURE 7–28 Examples of negative logic digital signals. (a) Binary 1 = +5 V, binary 0 = +10 V; (b) binary 1 = +0 V, binary 0 = +5 V.

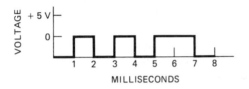

FIGURE 7–29 A typical digital waveform.

clarify the data. The corresponding logic symbol and truth table are shown in Figure 7–30b and c. From this waveform diagram it can be seen that the AND gate produces a 1 output when both inputs are a digital 1. Waveform displays of digital data are commonly used during the troubleshooting of computer circuits. An extremely high speed oscilloscope can display a digital waveform with a pulse time of less than 1 ms. The waveform displayed by the scope can be compared with known operational data in order to evaluate the circuit's performance.

INTEGRATED CIRCUITS

An **integrated circuit** IC is simply an assembly of diodes, transistors, and/or other circuit elements combined into an extremely small package. Since a computer "thinks" by moving electrons through a variety of circuits, it is important to keep circuits as close together as possible to reduce the electron travel time. The greater the number of circuits contained in a small area, the faster the computer. In order to produce an extremely small circuit, most manufacturers use a process called **photolithography**. This process has been used to produce ICs capable of handling thousands of bits (1s and 0s). Recall that a *bit* is equal to one piece of digital information.

Photolithography imprints a circuit on a silicon wafer by focusing a pattern of light into a concentrated area. This process is similar to that used in a darkroom to imprint a negative's image on photographic paper. Chemical solvents are then used to etch the circuit's design into the silicon. By adding other materials to specific areas of the silicon wafer, "doping" of the circuit is accomplished. A second layer of silicon may be added and another combination of circuits produced within the IC. The silicon wafers are then cut to size and assembled into the IC package. Figure 7–31 shows various silicon wafers ready to be cut and assembled into ICs. Extremely small gold wires are used to connect the silicon to the IC pin terminals. The silicon chip is usually housed in a ceramic or plastic package. Figure 7–32 shows three **40-pin microprocessors.** From the top down, they are a ceramic piggyback package microprocessor, a standard ceramic package microprocessor, and a standard plastic package microprocessor. A microprocessor is a complex or very large scale IC and will be discussed later. These microprocessors each contain 40 pin terminals, which are used to transmit the electrical signals between the circuit board and the silicon wafer.

Integrated circuits are produced in a wide range of circuit complexities. Some ICs are simple adders or subtracters of binary digits, while others contain an entire digital system. There are four categories that define the complexity of an IC: **small-scale, medium-scale, large-scale,** and **very large scale integration** circuits. Small-scale integration (SSI) ICs may contain as few as 10 logic gates and eight pin terminals, while very large scale ICs may have over 1000 gates contained within the silicon wafer and up to 40 external pin terminals.

Integrated circuits are divided into classes called **logic families.** A family contains ICs that operate at similar power levels and speeds. Two common logic families are the **TTL** and **CMOS** families.

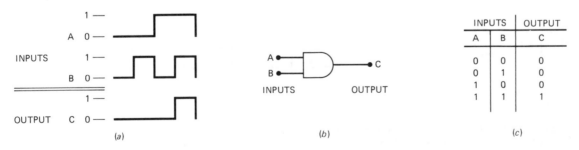

FIGURE 7–30 Digital data. (a) Waveform graph; (b) logic gate symbol; (c) truth table.

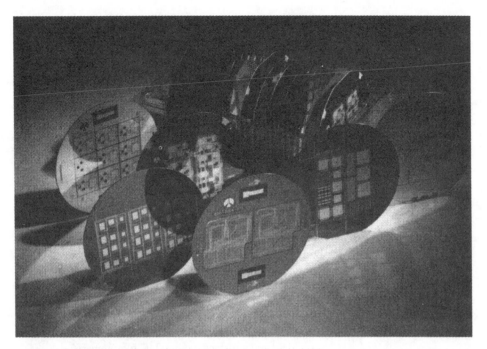

FIGURE 7–31 Silicon wafers ready to be cut and assembled into ICs. *(Collins Divisions, Rockwell International)*

FIGURE 7–32 Three 40-pin microprocessors. *(Texas Instruments, Inc.)*

The TTL Logic Family

A **transistor-transistor logic,** or TTL, circuit contains bipolar transistors as its primary elements. All TTL ICs operate with a +5-V power source and positive logic; therefore, binary 1 equals +5 V, and binary 0 equals ground or +0 V. Most TTL circuits will accept voltage values within a tolerance of plus or minus 5 percent. There are five common TTL circuits; they are the **standard TTL,** the **low-power TTL,** the **high-power TTL,** the **Schottky TTL,** and the **low-power Schottky TTL.** The members of the TTL family employ slightly different circuit components to perform various logic functions. The use of different components often changes the power requirement of the circuit. When the power requirement of the circuit changes, typically the speed of the circuit changes. As a general rule, faster circuits consume more power.

An example of a basic TTL inverter circuit is shown in Figure 7–33. It can be seen from this illustration that even a simple TTL inverter contains four bipolar transistors.

The CMOS Logic Family

CMOS is the acronym for **complementary metal-oxide semiconductor.** A CMOS is a metal-oxide semiconductor field-effect transistor (MOSFET) using both p and n channel inputs. The schematic in Figure 7–34 is a CMOS inverter. This circuit replaces the bipolar junction transistors of the TTL gates with MOSFET devices. The major advantage of

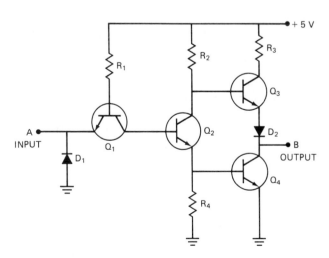

FIGURE 7–33 A TTL inverter circuit.

Integrated Circuits **139**

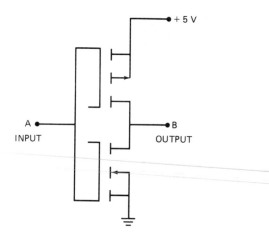

FIGURE 7–34 A CMOS inverter.

the CMOS logic family devices is that they require less power than TTL devices. CMOS devices will also operate over a wider range of input voltage levels, typically between +3 and +18 V.

Integrated Circuit Standards

The most obvious standard for ICs is the connecting pin arrangement. As illustrated in Figure 7–35a, all ICs adhere to the **dual in-line package (DIP) standard.** This means there is an equal number of connections (pins) on both sides of the IC, spaced within specific dimensions. Integrated circuits are available in 8-, 14-, 16-, 18-, 24-, 28-, and 40-pin configurations. Generally speaking, an IC with a greater number of pins is capable of performing a greater number of logic functions.

The pin numbering arrangement is also part of the DIP standard. If the IC is viewed vertically, as in Figure 7–35b, the top left pin is number 1. The pin numbers progress in a counterclockwise direction, down the left side of the IC and then up the right side. To identify the upper left corner of an

IC, manufacturers place a notch or dot on each component. The notch, or dot, should be on top when the IC is viewed vertically with the terminal pins facing away.

A typical IC pin diagram is shown in Figure 7–36. This IC is a type 7408 TTL unit containing four 2-input AND gates. This IC is known as a quadruple 2-input AND gate. The power connections of this IC are pins 7 and 14, as labeled. Diagrams such as this one become very useful during the design or troubleshooting of electric devices containing integrated circuits.

In order to build faster and more powerful computers, IC manufacturers are constantly looking for ways to shrink circuitry. Recently a new line of electronic parts has been introduced; these parts are known as **surface mounted components.** As seen in Figure 7–37, the electrical connections on these components are bent at a 90° angle, which allows the components to sit flat on the printed circuit board (PCB). The components are then attached by soldering their leads to the surface of the PCB. Traditional components contained leads that were mounted through the PCB and soldered on the side opposite the component. Figure 7–38 shows both a standard and a surface mounted IC for comparison.

Surface mounted components enable manufacturers to install electronic parts on both sides of a printed circuit board. This technique can, in effect, double the computing capacity of any given PCB. There are some problems associated with surface mounted components; for example, (1) compact installations require more cooling air, (2) the parts are made smaller than conventional ICs, resistors, capacitors, etc., and so removal and replacement are extremely difficult, and (3) identification of these subminiature components is difficult, since in many cases there is no room to print identification numbers on the part.

The DIP Switch

A **DIP switch** is a common switching device used in computers and logic circuits. All DIP switches adhere to the dual

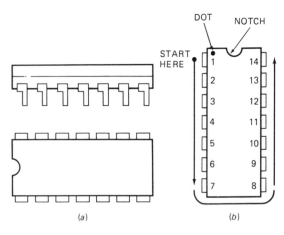

FIGURE 7–35 IC connection arrangement. (a) Typical IC pins; (b) pin numbering configuration.

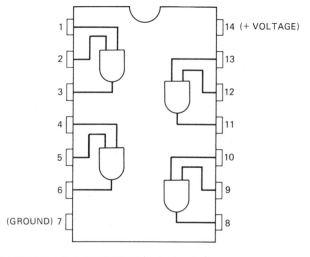

FIGURE 7–36 A 7408 TTL IC logic circuit diagram.

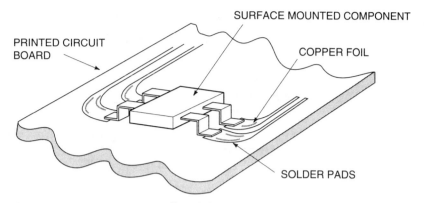

FIGURE 7–37 An example of a surface mounted component.

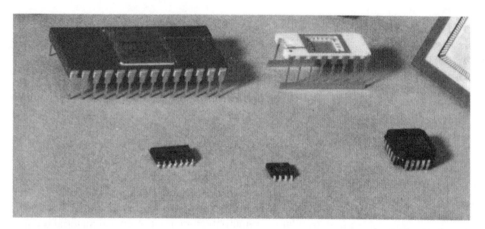

FIGURE 7–38 Examples of common integrated circuits, (top) standard ICs, (bottom) surface mounted ICs. *(Collins Divisions, Rockwell International)*

in-line package format; therefore, they are compatible with standard circuit board configurations. Figure 7–39 shows several DIP switches. Most DIP switches are double-throw, two-position switches; thus their inputs must be connected to

one of two different outputs, as illustrated in Figure 7–40. The input of a DIP switch is connected to ground (0) or voltage positive (1) for most logic circuits. This schematic shows four independent switches contained in one housing. Each pole is shown connected to the 5-V positive throw.

Most ICs require that each input be connected to a binary 1 or 0 signal. For example, a TTL device will assume a binary 1 for an unconnected input; a CMOS device will increase its power consumption and may overheat if its inputs are disconnected. Considering these problems, it is easy to understand why single-pole, double-throw DIP switches are used when controlling logic circuit inputs.

FIGURE 7–39 Typical DIP switches. *(AMP Products Corporation)*

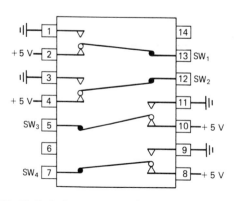

FIGURE 7–40 Typical connections of a DIP switch.

Integrated Circuits   **141**

COMMON LOGIC CIRCUIT FUNCTIONS

There are several basic logic circuits that are common to almost every computer or related peripheral device. These circuits use simple combinations of the AND, OR, INVERT, and exclusive OR gates. Five of the most common logic circuits are **adders, subtracters, clocks, latches,** and **flip-flops.**

Adders and Subtracters

Adder and **subtracter** circuits are used to perform basic calculations in computer systems. **Adders,** as their name implies, add binary digits. Since binary numbers consist of only two digits, 1 and 0, it is almost always necessary to carry a digit to the next higher-order column when adding. For example, in binary numbers $1 + 1 = 10$; two single digits were added to form a two-digit result. **Half-adder** circuits are capable of adding two binary digits, but they cannot carry a digit to the next higher-order column. **Full-adder** circuits use a combination of two half-adders in order to carry any necessary digits.

A full-adder symbol and logic diagram are shown in Figure 7–41. Points A, B, and CI are inputs into the logic circuits. A and B are the two bits to be added; CI is the digit carried from the adjacent lower-order column (if applicable). The outputs of the full-adder are CO and S; S represents the sum of the digits added, and CO is the bit to be carried over to the next higher-order column. Each full-adder is capable of adding only two binary digits and a carried digit; therefore, one full-adder must be used for every two digits to be added in any binary calculation. A typical IC used for adding binary digits contains several full-adder subcircuits. This provides the IC with the capability to sum several binary digits.

Subtracter circuits are a combination of basic gates, as shown in Figure 7–42. The inputs to a subtracter are A, B, and BRI; A and B are the digits for the subtraction, and BRI is the digit borrowed from the subtraction in the adjacent

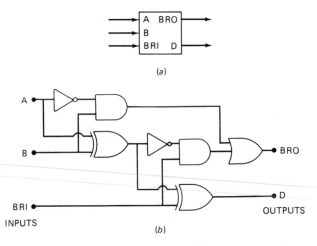

FIGURE 7–42 Subtracter. (a) Logic symbol; (b) logic circuit diagram.

lower-order column (if applicable). The outputs are D, the difference between the digits in the subtraction, and BRO, the digit to be borrowed from the adjacent higher-order column (if applicable).

Digital Clock Circuits

Certain functions of digital circuits require a consistently timed binary signal. A **digital clock** provides a stable frequency of binary 1s and 0s. The heart of a digital clock is an oscillator, or multivibrator circuit. A crystal material is commonly used to control the pulse time of a logic circuit to produce a consistent binary 1 and 0 waveform. This type of waveform is known as a **square wave.** Figure 7–43 shows a typical crystal-controlled multivibrator and its corresponding waveform.

Latches and Flip-Flop Circuits

Latches and **flip-flop** circuits are combinations of logic gates that perform basic memory functions for computers and peripherals. Both types of circuits retain their output signal even after the input signal has been removed; therefore, these circuits "remember" the input data. There are two basic latch

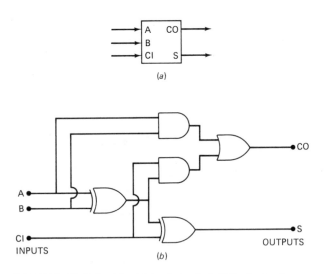

FIGURE 7–41 Full-adder. (a) Logic symbol; (b) logic circuit diagram.

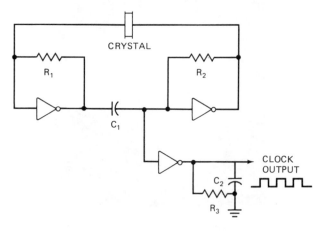

FIGURE 7–43 A crystal-controlled multivibrator.

circuits, the **RS latch** and the **data-type latch.** As illustrated in Figure 7–44, the RS latch contains two input signals, *set* S and *reset* R, and two output signals, Q and $\bar{Q}$. A binary 1 at the S input will set the latch memory, and Q will equal 1, while $\bar{Q}$ will be 0. A binary 1 at the R input will reset the latch, and Q will equal 0, while $\bar{Q}$ equals 1.

The data-type (d-type) latch contains only one input signal, as shown in Figure 7–45. In this circuit an input, D, of binary 1 will set the latch (Q = 1, $\bar{Q}$ = 0). Binary 0 at D will reset the latch (Q = 0, $\bar{Q}$ = 1).

Flip-flop circuits are similar to latch circuits; however, flip-flops change their output upon the presence of a trigger pulse. As shown in Figure 7–46, a flip-flop circuit contains three inputs. The set S and reset R signals are identical with those of a latch circuit. The **clock pulse** (CP) is an input that controls the circuit switch time. That is, the flip-flop will change its output signals (Q and $\bar{Q}$) only at given time intervals. The switching time intervals are determined by the clock pulse and the set or reset signals.

The advantage of using a clock input for a memory circuit is that all flip-flop output signals change at the same time. This becomes very important when several memory circuits are used simultaneously. If one circuit was to change its output signal out of sequence, the entire memory might become invalid.

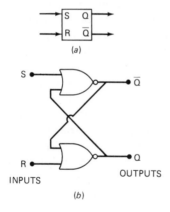

FIGURE 7–44 RS latch. (*a*) Logic symbol; (*b*) logic diagram.

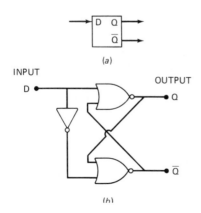

FIGURE 7–45 Data-type latch. (*a*) Logic symbol; (*b*) logic diagram.

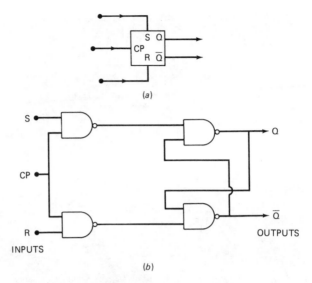

FIGURE 7–46 A digital flip-flop. (*a*) Logic symbol; (*b*) logic diagram.

MICROPROCESSORS

Microprocessors are simply complex digital circuits that can be thought of as miniature computers. Microprocessors are typically very large scale ICs that contain thousands of gates arranged to perform specific functions. A common example of a microprocessor is found in every handheld calculator. The calculator's microprocessor performs all the necessary functions for the calculator's operation. In effect, the microprocessor is a computer designed for one function, to calculate numbers.

To improve the speed of modern computers, microprocessors have been designed to be as compact as current technology will allow. Many state-of-the-art computer systems employ **multichip modules** (MCMs). Multichip modules are miniature circuit boards that contain a multitude of microprocessors, ICs, and other electrical components assembled by machine into one unit. The MCM shown in the foreground of Figure 7–47 replaces the two circuit boards in the background. The MCM in this photograph is approximately 3.5 in. [8.9 mm] square. The MCMs are then assembled onto circuit cards, which are part of the entire electronic system.

All microprocessors are made up of at least three basic elements: the central processing unit (CPU), the arithmetic logic unit (ALU), and a memory. The CPU is the primary control element of the microprocessor. The CPU processes and directs data according to requests made by the operator or another circuit in the system. The CPU coordinates the activities of the ALU, which performs the various calculations of the binary numbers to perform a specific function. The ALU performs its functions through the use of a variety of logic gates.

The memory of a microprocessor may be one of two types, permanent or temporary. The permanent memory provides information for the basic operations of the microprocessor. The temporary memory is used as a "notepad," or for short-

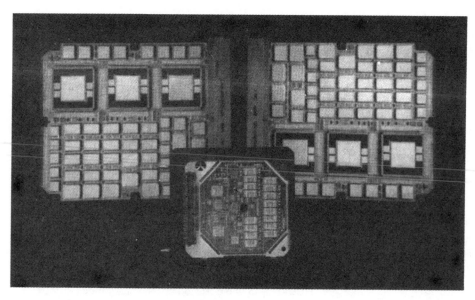

FIGURE 7–47 A multi-chip module (front), this module replaces the printed circuit cards (rear). *(Collins Divisions, Rockwell International)*

term storage of data needed during the manipulation of numbers.

As seen in Figure 7–48, all microprocessors require some means to communicate with the rest of the system. This communication link is called the **data bus.** The microprocessor also needs some type of timing device that is used to coordinate the activities of the system. A **synchronizer** or **clock** circuit is used for this function. Some microprocessors contain an internal clock, and some are accessed through the data bus.

Microprocessor Operation

The specific operation of a microprocessor is determined by the program contained in the memory and information received from data inputs. However, virtually all microprocessors follow a standard operating protocol. At power-up a microprocessor always starts at its **initialization routine.** The program functions of the microprocessor are divided into several subroutines. **Subroutines** are basically small programs that operate when called for by the CPU. The initialization routine is an example of a subroutine. When performing a routine, the CPU retrieves the routine's operating instructions from memory and performs the required op-

erations; then it moves on to the next routine. This process is repeated until all functions have been completed or until the process is interrupted by a higher-order command.

During operation, the microprocessor must have a means of communication with the rest of the computer system. As mentioned above, the communication is accomplished through a data bus. A **data bus** typically consists of a shielded pair of wires used to transmit binary data. There are several different types of digital buses; some carry data, some carry address information, and some supply control information. **Address** information is needed because many elements of the computer may be connected to the same bus. The address tells the individual elements when to send or receive messages on the data bus. The **control bus** is needed to coordinate the data transmission activities. These buses are capable of moving binary data at an extremely fast rate. The speed of these buses is measured in megahertz (MHz) and has a dramatic effect on the overall speed of a computer.

COMPUTER OPERATIONS

Every computer contains five fundamental sections: input, control, memory, processing, and output. Each of these sections manipulates binary digits through the use of logic circuits. The entire computer is composed of a power supply, microprocessor-based circuitry, and related peripherals. **Peripherals** are typically devices that allow the computer to communicate with humans or other electronic devices. For example, a keyboard is an input peripheral; a printer or CRT display is used as an output peripheral. Figure 7–49 shows the flow diagram of a typical computer. As illustrated, the **central control unit** must connect to each section of the computer in order to coordinate the activities of the entire system. Strict coordination is essential for the proper operation of any computer. Each system must operate only upon command by the central control unit. To coordinate the com-

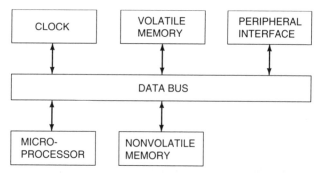

FIGURE 7–48 A typical microprocessor data link.

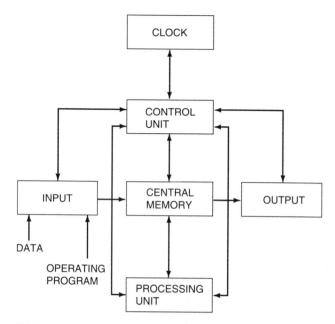

FIGURE 7–49 Computer block diagram.

puter activities, the central control unit must have access to an accurate timing source. The computer's **clock** is used to perform this synchronizing function. The clock is a crystal oscillator used to generate a constant-frequency digital pulse. This pulse is used as a time base for all computer operations. The clock speed on a typical computer is between 1 and 60 MHz. The clock can be an integral part of the control unit, or it can be a separate element.

The Central Processing Unit

Similar to a microprocessor, the **central processing unit** (CPU) of any computer performs the actual addition and subtraction and other logic functions. That is, the CPU receives input data, manipulates that data in accordance with specific instructions, and responds with the respective output data. Figure 7–50 shows a block diagram of a typical CPU. The input to a CPU usually consists of a program, which is needed to "run" the specific jobs of the computer, and the input data, which is the information to be processed. The output data of the CPU is created by the manipulation of the input data in accordance with the computer program. For example, the input

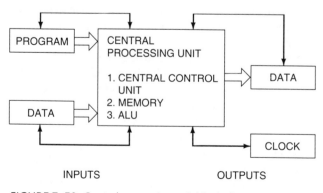

FIGURE 7–50 Central processing unit block diagram.

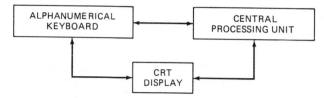

FIGURE 7–51 Data link to CPU.

data to a typical CPU might be wind velocity and direction, airspeed, temperature, and distance to a destination. The output data might be the display of true airspeed and the time to that destination. Of course, the input would also include the program (such as mathematical equations) necessary to manipulate the data. In many cases the program is contained within the CPU memory and need not be added to the CPU input for each operation.

The input keyboard, the CPU, and the output display of a typical airborne flight computer are linked as shown in Figure 7–51. The lines connecting the keyboard, the CPU, and the CRT represent the data link between these components. The CPU of every computer must be capable of communicating with its input and output peripherals. A computer's central processing unit can be subdivided into three essential subsystems: the **central control unit,** the **memory,** and the **arithmetic logic unit** (ALU). Depending on the computer, these subsystems can be combined in one microprocessor, or they can be completely separate elements.

The Central Control Unit

A **control system** is used to coordinate the functions performed by each section of the computer. To do this, a communication link must be provided between the central control unit (CCU) and the various computer sections. A **data transfer bus** is typically used to provide this communication link. A data transfer bus is a digital connection, or link, between two or more digital devices. On some systems this is a two-way communication bus; on other systems the data bus is a one-way link. In Figure 7–50 the two-way data transfer is shown as a thin dark line pointing in both directions; the one-way data bus is shown as a wide line pointing in only one direction. There are a variety of different data bus systems currently being used. Data transmission will be discussed later in this chapter.

Since the CCU performs the main coordination of the functions of each peripheral device, it must be connected to each unit via a data transfer bus. The concept of a CCU and data bus is illustrated in Figure 7–52. Here each section of the computer is linked to the data bus, which is linked to the CCU.

Memory

The memory of a CPU is often divided into two basic categories: **volatile memory** and **nonvolatile memory.** Data in a nonvolatile memory will *not* be destroyed when the computer is turned off. Memory of this type may be contained on

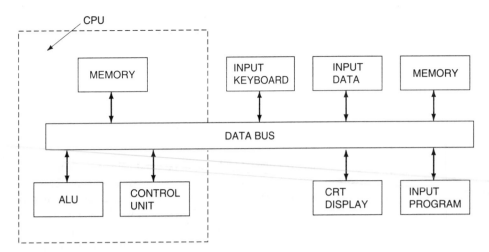

FIGURE 7–52 Data bus block diagram.

magnetic disks, on magnetic tape, or within special ICs. The data in a volatile memory are lost whenever the computer loses electric power. Data of this type are simply stored by ICs containing latch or flip-flop circuits. When the power is removed from an IC, the memory returns to a neutral state, and the data are erased.

Data stored in any memory must first be converted into the binary language. The memory unit then "remembers" the appropriate byte combinations of 1s and 0s. These combinations are labeled in order to facilitate future access. Often a computer will have several memory units, some volatile, some nonvolatile. Most aircraft computers contain a relatively large nonvolatile memory in order to store the information necessary to process the flight data. The processing data are often referred to as the **operating program.**

Semiconductor memory circuits are divided into two categories: **random-access memory** (RAM) and **read-only memory** (ROM). Random-access memory is often considered a write-and-read memory. That is, you must record (write) information into the RAM before someone can access (read) the information. A RAM may be changed at any time by using the correct procedures; therefore, it is considered to be a volatile semiconductor memory.

A read-only memory is a nonvolatile semiconductor memory. The logic pattern of 1s and 0s is permanently programmed into the semiconductor material of an IC. If the power is removed from that IC, the memory remains intact. Read-only memories are used often in aircraft computer systems to remember the operational program and subroutines for each system.

There is a special type of ROM that can be altered by the user, but only under special conditions. **EAROM (electronically alterable read-only memory)** can be changed if desired. To change an EAROM, the operator must follow the correct procedures and "reprogram" the memory chip(s). A typical example of this type of ROM is used to store flight path information on an aircraft's flight management computer. Periodically the information must be updated. To do the updating, a floppy disk containing the new information is transferred to the EAROM. During this update, the original data are lost and the new data stored in a read-only memory.

The Arithmetic Logic Unit

The **arithmetic logic unit** (ALU) is the "thinking" section of the CPU. That is, the ALU performs all the calculations and/or comparisons necessary to process the input data. The ALU received its name from the fact that it mainly performs arithmetic calculations using logic circuits. Of course, the ALU receives its coordination signals from the CCU. The resultant data from the ALU calculations are sent to the system's various output devices. The ALU is often thought of as the combination of logic circuits needed to process the computer program.

Data Transmission

An aircraft computer system typically contains peripherals in various locations throughout the aircraft. In order for these remote devices to communicate with the CPU, there must be a means of data transmission between each component of the system and the CPU and/or between the CPU and the individual components. Most digital communication data are transmitted in a **serial** form, that is, only one binary digit at a time. Transmission of data in serial form means that each binary digit is transmitted for only a very short time period. In most systems, the data transmission requires less than 1 ms. After one bit of information is sent, the next bit follows; this process continues until all the desired information has been transmitted. This type of system is often referred to as **time sharing,** because each transmitted signal shares the wires for a short time interval. **Parallel data transmission** is a continuous-type transmission requiring two wires (or one wire and ground) for each signal to be sent. Parallel transmission is so named because each circuit is wired in parallel with respect to the next circuit. One pair of transmitting wires may be used to handle enormous amounts of serial data. If the information signals were transmitted in parallel form, hundreds of wires might be required to perform a similar task.

Serial data transmission requires less wire than a parallel system; however, an interpretation circuit is needed to convert all parallel data to serial-type information prior to transmission. The device for sending serial data is called a

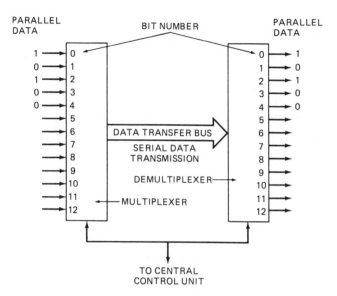

FIGURE 7–53 Data transfer system.

multiplexer (MUX), and the device for receiving serial data is called a **demultiplexer** (DEMUX). As illustrated in Figure 7–53, parallel data are sent to a multiplexer, where they are converted into serial data and sent to the data transfer bus. The **data transfer bus** is a two-wire connection between the multiplexer and the demultiplexer. The demultiplexer receives the serial data and reassembles them into parallel form. In this example, the byte 10100 is being received by the multiplexer in parallel form. Starting at the top and working down, the multiplexer transmits each digit individually. Bit number 0 is the first to be transmitted. Bit number 1 is the next digit transmitted, bit number 2 is next, and so on. This system is repeated until all the parallel data have been converted into serial form and have been individually connected to the data transfer bus. The demultiplexer receives a serial data input from the transmission bus and reassembles it into a parallel form. The output of the demultiplexer is identical to the input of the multiplexer (10100).

Some means of control must be used to coordinate the MUX/DEMUX arrangement. As shown in Figure 7–53, the CCU is typically used to coordinate the transmission and reception of data. This control is essential to ensure that all serial data are transmitted and received at the proper time intervals. This system of serial data transmission may seem somewhat complex; however, the alternative, parallel data transmission, would require one wire for each data bit to be transmitted. Since thousands of bits of information are transferred among various airborne systems, serial data transmission techniques are the obvious choice.

The use of multiplexers and demultiplexers is necessary only when a change from serial to parallel (or vice versa) is required. In many cases serial data are transmitted to another component that can "read" serial data. In those cases no change in format is required.

ARINC Specifications

Aeronautical Radio Incorporated (ARINC) is a corporation established by foreign and domestic airlines, aircraft manufacturers, and transport companies. The purpose of this organization is to aid in the standardization of aircraft systems. ARINC specifications have been established for digital flight data recorders (ARINC 573), inertial navigation systems (ARINC 561), digital information transfer systems (ARINC 429 and 629), and various other aircraft communication and navigation systems.

ARINC 429 sets specifications for the transfer of digital data between aircraft electronic system components. An ARINC 429 data bus is a one-way communication link between a single transmitter and multiple receivers. The ARINC 429 system provides for the transmission of up to 32 bits of information in each byte or word. One of three languages must be used to conform to the ARINC 429 standards: **binary, binary-coded decimal** (BCD), or **discrete.** A synchronizing clock pulse is accomplished by a four-bit null between each word. A **null** is a signal that is equal to neither binary 1 nor binary 0. As shown in Figure 7–54, ARINC 429 assigns the first 8 digits of a byte as the word label; digits 9 and 10 are a **source-destination indicator** (SDI), digits 11 through 28 provide the data information; digits 29 through 31 are the **sign-status matrix** (SSM), and bit number 32 is a **parity bit.**

There are 256 combinations of word labels in the ARINC 429 code. Each word is coded in an octal notation language and written in reverse order. The source-destination indicator serves as the address of the 32-bit word. That is, the SDI identifies the source or destination of the word. All information sent to a common serial bus is received by any receiver connected to that bus. Each receiver accepts only that information labeled with its particular address; the receiver ignores all other information.

The information data of an ARINC 429 coded transmission must be contained within the bits numbered 11 through 28. These data are the actual message that is to be transmitted. For example, an airspeed indicator may transmit the binary message 0110101001. Translated into decimal form, this means 425, or an airspeed of 425 knots. The sign-status matrix (SSM) provides information that might be common to several peripherals. Examples of common information include north, south, plus, minus, right, left, east, and west.

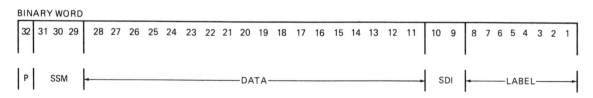

FIGURE 7–54 ARINC 429 format. *(Boeing Company)*

Using the SSM bits, fewer binary characters are required to represent this type of information than if each item were "spelled out" in the data segment of the word.

The parity bit of the ARINC 429 code is included to permit error checks by the ARINC receiver. The receiver also performs a **reasonableness check,** which deletes any unreasonable information. This ensures that if a momentary defect occurs in the transmission system resulting in unreasonable data, the receiver will ignore that signal and wait for the next transmission. The parity bit check is used to detect any recurring defect. To identify properly transmitted data, the parity bit must be a binary 1 if there is an even number of 1 bits in a transmitted word (digits 1 through 31). The parity bit must be 0 if there is an odd number of 1 bits in the word. The parity bit is assigned to the ARINC code prior to transmission. If an error occurs during data transfer, the parity bit will most likely be incorrect. The data receiver monitors the parity bit and identifies any errors. If an error is detected, the system's CPU will make any necessary adjustments and record the defect in a nonvolatile memory.

Figure 7–55 represents an ARINC 429 code for an airspeed indicator. The octal notation code of the first eight bits is read in reverse to achieve the word label. The bits 011, 000, 01 represent 206 (602 reversed). According to the ARINC 429 code, 206 represents computed airspeed. The SDI label 00 indicates transmission of these data to all receivers connected to the serial bus. The data segment of the word is also read from left to right. The binary number 0110101001 represents the sum of 0×512 (2^9), 1×256 (2^8), 1×128 (2^7), 0×64 (2^6), 1×32 (2^5), 0×16 (2^4), 1×8 (2^3), 0×4 (2^2), 0×2 (2^1), and 1×1 (2^0). In decimal form, this number is 425. In other words, the ARINC message being transmitted is "Calibrated airspeed equals 425 knots."

The sign-status matrix digits—29, 30, and 31—are binary 0, 1, and 1, respectively. This represents a normal operation of plus value data; that is, the airspeed data are a positive value. The parity bit (number 32) is a 1, which denotes an even number of binary 1s in the transmitted word. No error is present according to the parity bit.

ARINC 629 is a new digital data bus format that offers more flexibility and greater speed than the 429 system. ARINC 629 permits up to 120 devices to share a **bidirectional serial data bus,** which can be up to 100 m long. The bus can be either a twisted wire pair or a fiber-optic cable. The Boeing Company, in developing the new B-777, is anticipating the use of ARINC 629 in a two-wire format. ARINC 629 has two major improvements over the 429 system. First, there is a substantial weight saving. The 429 system requires a separate wire pair for each data transmitter. With the increased number of digital systems on modern aircraft, the 629 system will save hundreds of pounds by using one data bus for *all* transmitters. Second, the 629 bus operates at speeds up to 2 Mbits/s; the 429 is capable of only 100 kbits/s. Figure 7–56 shows simplified diagrams of the 429 and 629 bus structures. Here it can be seen that the 629 system requires much less data cable.

The ARINC 629 system can be thought of as a party line for the various electronic systems on the aircraft. Any particular unit can transmit on the bus or "listen" for information. At any given time, only one user can transmit, and one or more units can receive data. This "open bus" scenario poses some interesting problems for the 629 system: (1) how to en-

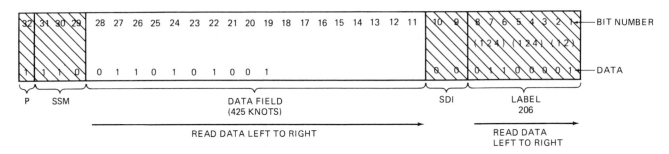

FIGURE 7–55 ARINC 429 code for computed airspeed; the message indicates 425 knots. *(Boeing Company)*

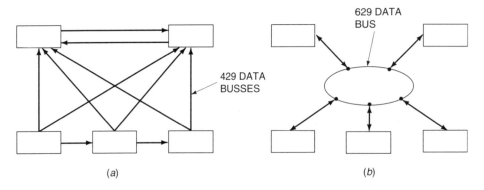

FIGURE 7–56 Typical data bus structures. (a) One-way ARINC 429; (b) Two-way ARINC 629.

sure that no single transmitter dominates the use of the bus, (2) how to ensure that the higher-priority systems have a chance to talk first, and (3) how to make the bus compatible with a variety of systems.

The answer is found in a system called the **periodic aperiodic multitransmitter bus.** To understand this system, study the examples using four receivers/transmitters in Figure 7–57. Here each transmitter can use the bus if it meets a certain set of conditions. First, any transmitter can make only one transmission per **terminal interval.** Second, each transmitter is inactive until the **terminal gap** time for that transmitter has ended. Third, each transmitter can make only one transmission; then it must wait until the **synchronization gap** has occurred before it can make a second transmission.

The **terminal interval** (TI) is a time period common to all transmitters. The TI begins immediately after any user starts a transmission. The TI inhibits another transmission from that same user until after the TI time period. A **periodic interval** occurs when all users complete their desired transmission prior to the completion of the TI. If the TI is exceeded **(aperiodic interval),** one or more users have transmitted a longer than average message (Figure 7–57b).

The **terminal gap** (TG) is a unique time period for each user. The terminal gap time determines the priority for user transmissions. Users with a high priority have a short TG. Users with a lesser need to communicate (lower priority) have a longer TG. No two terminals can ever have the same terminal gap. The TG priority is flexible and can be determined through software changes in the receivers/transmitters.

The **synchronization gap** (SG) is a time period common to all users. This gap can be thought of as the reset signal for the transmitters. Since the synchronization gap is longer than the terminal gap, the SG will occur on the bus only after each user has had a chance to transmit. If a user chooses not to transmit for a time equal to or longer than the SG, the bus is open to all transmitters once again.

Any message transmitted by a bus user has a limited time in which it is allowed to be transmitted. The **message** transmitted is composed of a maximum of 16 word strings. The word strings contain a label word and up to 256 data words. Each word is limited to 16 bits of data and a parity bit.

Another unique feature of the ARINC 629 bus is the **inductive coupling** technique used to connect the bus to receivers/transmitters. As shown in Figure 7–58, the bus wires are fed through an inductive pickup, which uses electromagnetic induction to transfer current from the bus to the user, or from the user to the bus. This system improves reliability, since no break in the bus wiring is needed to form connections.

Other Data Bus Systems

There are several other bus systems currently in use on modern aircraft. The military has a unique bus architecture that resembles both the ARINC 429 and the ARINC 629 structures.

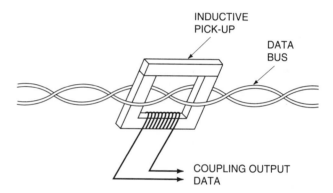

FIGURE 7–58 An example of inductive coupling in a bus system.

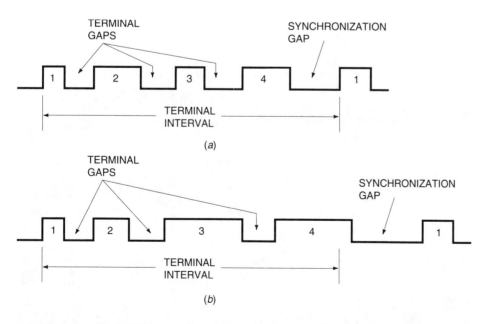

FIGURE 7–57 ARINC 629 bus structure. (*a*) A periodic interval; (*b*) an aperiodic interval caused by the extended message of user 3.

There are also two common data bus systems found on general aviation aircraft. The **CSDB (commercial standard digital bus)** is used extensively in Collins general aviation electronic equipment. The **ASCB (avionics standard communication bus)** is used for communications on much of the Sperry flight control equipment. The specifications for the CSDB and the ASCB were standardized by the General Aviation Manufacturers Association (GAMA).

The CSDB is a one-way communications link from one transmitter to one or more receivers. The system operates through a shielded pair of twisted wires at 50 kbits for high speed or 12.5 kbits for low speed. The major advantage of the CSDB is that there are a number of commercially available aviation products using this system. The ASCB is a bidirectional data bus operating at 2 MHz. This system uses a bus controller, which coordinates bus transmission activities.

DIGITAL AIRCRAFT SYSTEMS

Digital electronics provide for greater reliability, faster response, smaller components, lighter equipment, and lower operating costs than can be provided by analog systems. It is no wonder that modern commercial, corporate, and military aircraft contain countless digital circuits. Smaller civilian aircraft also contain a limited amount of digital equipment; however, this type of aircraft traditionally lags several years behind in utilizing state-of-the-art technologies. Digital systems increase the mean time between failures and reduce the subsequent repair time for failed equipment. The built-in test equipment (BITE) found in most digital systems provides rapid fault isolation. The majority of the digital aircraft systems contain several line replaceable units (LRUs). Defective LRUs may be quickly identified by the BITE system and exchanged during ground maintenance. Use of the LRU and BITE concepts greatly reduces aircraft maintenance downtime.

Another concept of digital aircraft technologies is to remove as many moving parts from the electrical system as possible. Throughout the aircraft, switches are replaced with proximity indicators, relays are replaced with transistors, and instruments are replaced with digital displays. In the flight compartment, the CRT replaces conventional analog instruments, thus eliminating thousands of moving parts. Fewer moving parts means greater system reliability. A modern digital instrument panel is shown in Figure 7–59. The CRT displays shown in this instrument panel employ high-efficiency phosphors and multiband pass optical filters in order to achieve enough contrast, brightness, and resolution to make them sunlight-readable.

TROUBLESHOOTING DIGITAL CIRCUITS

With the advent of digital logic circuits came the introduction of **logic troubleshooting techniques.** This troubleshooting system can be applied to both digital and analog circuits, as well as hydraulic, pneumatic, and other mechanical systems. A logic troubleshooting sequence simply employs a flowchart of "logical" faults and repairs for a given system.

Figure 7–60 shows a typical flowchart for troubleshooting an avionics system. This system "asks" a yes or no question

FIGURE 7–59 A Beechcraft Starship 1 instrument panel. *(Collins Divisions, Rockwell International)*

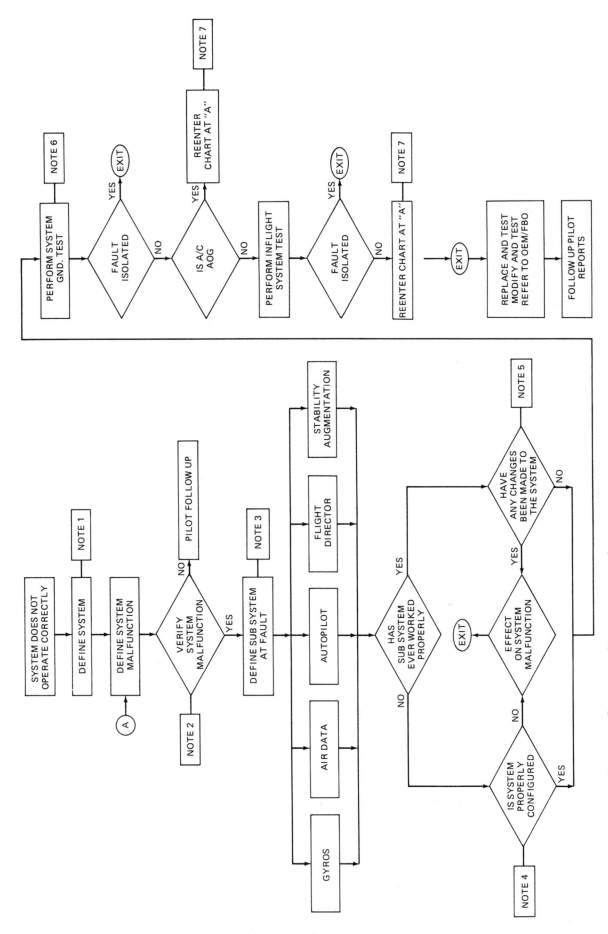

FIGURE 7–60 A typical troubleshooting flowchart. ° *(Sperry Corporation)*

151

and directs the technician to the correct means of repair. The increased use of LRUs has made this troubleshooting method feasible. Since any given system contains only a limited number of replaceable parts, a relatively simple flowchart can identify most defective components.

Logic Diagrams

A **logic diagram** is a simplified schematic of a complex digital circuit. Logic diagrams of digital circuits will typically show only the logic gates and not the actual transistors, diodes, and resistors performing the logic functions. For example, Figure 7–61 represents a landing gear down indicator and warning horn system. A comparable analog circuit is also shown. In order for the down light to illuminate, all three gears must be down and locked; therefore, an AND gate is used for this function. All three gears must not be down and locked and the throttle must be closed for the warning horn to sound. A NOR and AND gate combination is used to perform this task. This type of logic diagram is quite common for describing modern aircraft electrical systems. As in this circuit representation, most logic diagrams do not show ground connections.

Logic diagrams are not limited to describing electric circuits. Many complex mechanical systems can be simplified by using logic diagrams. Figure 7–62 shows a logic diagram used to represent the pneumatic sources available for starting the no. 2 turbine engine of a three-engine aircraft. On this airplane there are four means to start engine no. 2. To start engine no. 2 using engine no. 3 compressor bleed air, the pneumatic overpressure valve must be closed (off) and valve 2-3 must be open (on). This means of starting engine no. 2 is represented by the logic symbols within the lower dashed lines (see Figure 7–62b). To start engine no. 2 using the auxiliary power unit (APU), the APU must be on and the APU load control valve must be open (on). This portion of the logic diagram is within the upper dashed lines (see Figure 7–62b). The rest of the logic diagram is used to represent alternative means of starting engine no. 2.

Built-In Test Equipment

Built-in test equipment (BITE) systems are used in conjunction with many digital circuits. BITE systems are designed to provide **fault detection, fault isolation,** and **operational verification after defect repair.** Fault detection is performed continuously during system operation. If a defect is sensed, the BITE initiates an appropriate control signal to isolate any defective component(s). In order to repair the defective system, the line technician can utilize the BITE to identify faulty components or wiring. After the appropriate repairs have been made, the system should be run through a complete operational check. The BITE will once again monitor the system and verify correct operation if the system has been properly repaired.

A typical commercial airliner may contain several BITE units used to monitor a variety of systems. A Boeing 757 or 767 aircraft, for example, utilizes built-in test equipment systems in order to monitor electric power, environmental control, auxiliary power, and flight control systems. Seven separate BITE units located in the aircraft's electric equipment bay or aft equipment center are used to accomplish this task. Each of these BITE boxes receives inputs from several individual components of the system being tested. Other individual systems also contain their own dedicated built-in

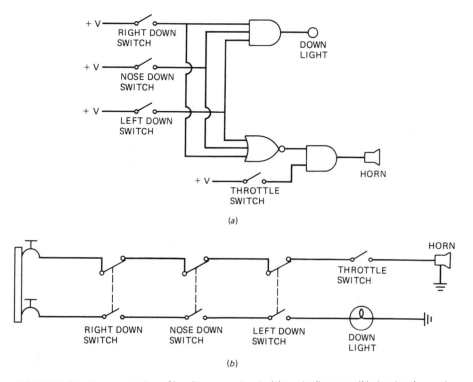

FIGURE 7–61 Representation of landing gear circuit. (*a*) Logic diagram; (*b*) circuit schematic.

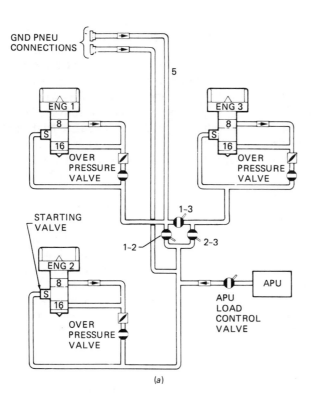

(a)

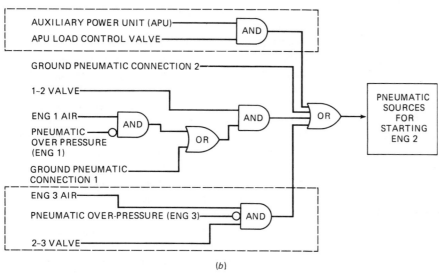

(b)

FIGURE 7–62 Various means of starting engine no. 2. (a) Pictorial representation; (b) logic diagram. (McDonnell Douglas Corp.)

test equipment. These BITE systems are relatively simple, and each is usually contained within a line replaceable unit of the system being monitored. Systems that employ dedicated built-in test equipment include the following:

Engine indicating and crew alerting

VHF communication radios

HF communication radios

ARINC communication addressing and reporting

Selective calling

Passenger address

Weather radar

ATC transponder

Radio altimeter

Automatic direction finder

Inertial reference

Air data computer

Electronic flight instruments

Flight management computer

Radio distance magnetic indicator

Lighting

Fuel quantity

Fire and overheat

Antiskid-autobrake

Instrument landing

VHF omnirange receiver

Distance measuring equipment

Window heat

Proximity switch electronic unit

Hydraulic management

A complex BITE system is capable of testing thousands of input parameters from several different LRUs. The system performs two types of test programs: an **operational test** and a **maintenance test.** Normal operational checks start with initialization upon acquisition of system power (see Figure 7–63). The operational BITE program is designed to check input signals, protection circuitry, control circuitry, output signals, and the operational BITE circuitry. During normal system operation, the built-in test equipment monitors a watchdog signal initiated by the BITE program. The watchdog routine detects any hardware failure or excessive signal distortion that may create an operational fault. If the BITE program detects either of these conditions, it automatically provides isolation of the faulty component(s); initiates warning, caution, or advisory data; and records the fault in a nonvolatile memory.

The maintenance program of the built-in test equipment is entered into the system only when the aircraft is on the ground and the maintenance test routine is requested. When requested, the maintenance BITE will exercise all input circuitry and software routines of the system being checked. The corresponding output data are then monitored, and faults are recorded and displayed by the BITE system. A flow diagram of the bus power control unit (BPCU) maintenance BITE routine is shown in Figure 7–64. This routine checks

the input circuitry, voltage regulator circuitry, protection software, logic software, and operational BITE system. The test results are returned to the BPCU for storage and display. The software, or operating programs, of the system are tested through utilization. That is, input data are initiated by the BITE and manipulated by the software program. The corresponding output data are evaluated by the BITE program in order to determine the system's performance. If a discrepancy in the output data is detected, the BITE system considers the operational software faulty and provides the appropriate indication.

Some newer aircraft like the Boeing 747-400 incorporate a **central maintenance computer system** (CMCS). This system is designed to monitor the individual BITE systems of various components and record the data in a central location. The CMCS is also used in conjunction with any flight deck displays that indicate a fault during flight. This advanced built-in troubleshooting system is more comprehensive than older systems, and it displays fault information in a more user-friendly manner.

Data Bus Analyzers

The **data bus analyzer** is a common carry-on piece of test equipment used to troubleshoot digital systems. There are many types of data bus analyzers, but their basic purposes are quite similar. Bus analyzers are used (1) to receive and review transmitted data or (2) to transmit data to a bus user. Before using any analyzer, one must first be sure that the bus language is compatible with the bus analyzer. For example, the **DATATRAC 200** shown in Figure 7–65 can monitor, simulate, and record data transmissions for avionics equipment using ARINC 429 or CSDB standard data buses.

When monitoring a system, the analyzer captures a stream of data being transmitted between digital devices. The recorded data can then be displayed by the analyzer for evaluation. If inconsistencies are detected, the transmitter or the data bus system is faulty. Some data bus analyzers are capable of reading several transmission lines, or **channels,** at one time. This allows for comparisons to be made that might expedite troubleshooting.

For the DATATRAC 200 there are three basic modes of operation:

Receive This allows the technician to select specific labels to evaluate or to receive all incoming data. Hexadecimal, decimal, or binary format may be selected for the data display.

Transmit The data bus analyzer is capable of sending digital data in order to simulate communications between avionics equipment or sending analog data using a D/A converter and driver.

Record A particular data label and record rate are selected, and the avionics bus analyzer collects information sent to this address. All the data can then be displayed in a numerical format. Some avionics bus analyzers have the capability of displaying the information in a graphic mode.

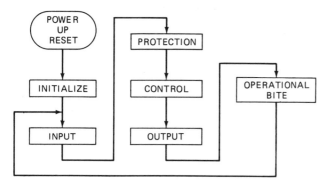

FIGURE 7–63 BITE flow diagram.

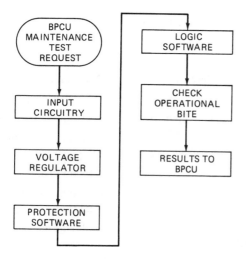

FIGURE 7–64 Bus power control unit BITE flow diagram.

FIGURE 7–65 A typical data bus analyzer. *(Atlantic Instruments, Inc.)*

In most cases the data analyzer must be connected to the system at the connector plug of an LRU. For example, if you plan to receive data transmitted to the generator control unit (GCU), you must unplug the GCU from the bus and connect the analyzer to the data bus cable. If the correct connections are made and the test equipment is properly adjusted, the an-

alyzer will display all messages sent to the GCU. On some aircraft a specific connector is available to allow for direct connection into the data bus system. Figure 7–66 shows a carry-on bus analyzer connected to the data bus. In this case the technician is monitoring the radar altimeter (RA). *Note:* The selector is set for RA, and the analyzer indicates –12 ft.

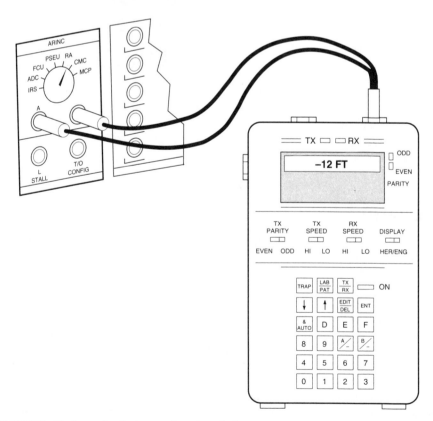

FIGURE 7–66 Example of a bus analyzer installed on a transport-category aircraft.

Measurement of Logic Levels

There are two common instruments used to measure logic levels: the **logic probe** and the **logic monitor.** A logic probe measures one point in a circuit to determine its logic level (high or low). A logic monitor is capable of measuring the logic levels of an entire integrated circuit. That is, every pin of an IC may be simultaneously tested for its logic level through the use of a logic monitor.

A typical logic probe is shown in Figure 7–67. The tip of the probe is touched to any connection in the logic circuit in order to detect its logic state. For example, a signal of + 5 V or greater activates the HIGH logic LED indicator, and a signal of less than + 5 V activates the LOW indicator. The actual voltage levels for a response of HIGH or LOW may vary between logic probes. Most logic probes will respond to a digital pulse of 50 ns or longer. This quick response rate is well beyond the capabilities of most voltmeters. The logic probe is therefore a vital instrument needed to test any rapidly changing digital signal.

Appropriate switch adjustments allow many logic probes to test both diode-transistor logic (DTL) and transistor-transistor logic (TTL) ICs. All logic probes must be connected to a reference signal; therefore, each contains a separate lead that must be connected to the logic circuit's power source.

Logic monitors use a special test clip that connects to each lead of an integrated circuit. A typical logic monitor and various test clips are shown in Figure 7–68. The logic monitor pictured contains 16 LEDs. Each LED represents one of the connection pins of the IC being tested. When a connection pin of the IC is at a HIGH logic level, its corresponding LED will illuminate. If the IC connection is at a LOW logic level, its LED will not illuminate. Both TTL and DTL integrated circuits can be tested by most logic monitors. Many monitors also contain an adjustment that allows the operator to set the desired voltage thresholds. This adjustment determines the voltage level required to indicate a logic HIGH on the monitor's LEDs.

Many variations of logic probes and logic monitors are currently available. Be sure to familiarize yourself with any equipment used to test digital circuits. Most ICs are very delicate; any unnecessary voltage may damage the circuit being tested.

High-Energy Radiated Electromagnetic Fields (HERF)

Today's digital aircraft systems communicate through the movement of thousands of bits of digital data. These data streams operate at extremely low power levels and can easily be overridden by a more powerful signal. This is the underlying concept that makes radiated energy from electromagnetic fields so troublesome. **High-energy radiated electromagnetic fields** (HERF) are emitted by virtually every radio broadcast tower in the world. Every FM transmitter, every TV station antenna, and every radar station emits electromagnetic energy. This radiated energy can and will induce a current into a nearby conductor. If the conductor happens to be a data bus and if the induced current is strong enough, the

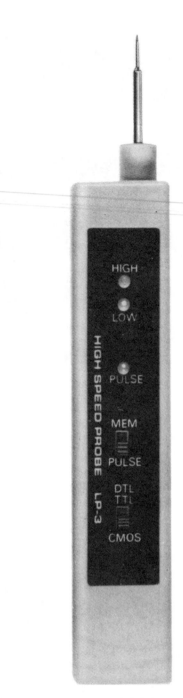

FIGURE 7–67 A logic probe. *(Global Specialties, an Interplex Electronic Company)*

possibility exists that the data on the bus can be lost. For an aircraft, this loss of data can be catastrophic.

Recent FAA standards require all new equipment to undergo HERF testing. These tests are used to ensure that the electronic components used on modern digital systems will not fail when subjected to high-energy electromagnetic fields. An aircraft might be subjected to such high-energy fields if it flies too close to a powerful radio transmitter or perhaps if it is struck by lightning. There is concern that even indirect lightning (lightning that comes close to but does not hit the aircraft) could cause digital system failure.

Currently many questions need to be answered about the vulnerability of the digital aircraft. Composite structures,

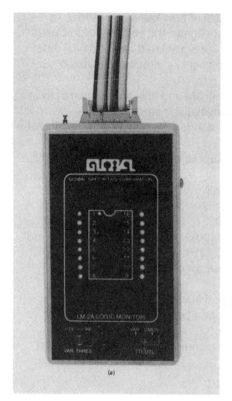 

FIGURE 7–68 A logic monitor. (a) The indicator; (b) the IC test clips. *(Global Specialties, an Interplex Electronic Company)*

which block less magnetic energy than conventional aluminum aircraft, also complicate the issue. Today's standards for HERF testing may be adequate or may be increased; but if aircraft continue to rely on low-energy digital circuits, the threat of HERF will remain.

Electrostatic Discharge–Sensitive (ESDS) Components

Many digital electronic devices are susceptible to damage from the discharge of static electricity. These components are known as **electrostatic discharge–sensitive** (ESDS) parts. Since digital circuits are manufactured from extremely small silicon chips, they can be damaged by the discharge of static voltages as low as 100 V. During movement for everyday activities, a technician can easily generate well over 1000 V. If this voltage is discharged into an ESDS component, the part will be damaged. There are several simple steps a technician can take to avoid damaging ESDS parts. Each step ensures that the sensitive components are never subjected to high levels of static electricity. The damage-prevention techniques are discussed in Chapter 13 of this book.

REVIEW QUESTIONS

1. Describe a digital circuit.
2. Give an example of a simple digital circuit.
3. What device made digital circuits practical?
4. What are computer peripheral devices?
5. Explain the difference between a digital signal and an analog signal.
6. Explain the binary number system.
7. What two symbols make up the binary number system?
8. How is the decimal number 12 represented in the binary number system?
9. List some common designations of binary 1 and 0.
10. Describe the decimal number system.
11. How is binary 10110 represented in the decimal system?
12. How is binary 1000110 represented in the decimal system?
13. Define the terms *bit* and *byte*.
14. Explain the addition of binary numbers.
15. What is the sum of binary 10110 and binary 10100?
16. What is binary 101101 minus binary 101?
17. What is the product of binary 110 times binary 10?
18. What are three common binary code systems?
19. Describe the binary-coded decimal system.
20. Describe the octal notation system.
21. Describe the hexadecimal number system.
22. What is the decimal equivalent of the octal notation number 101 001 000 101?
23. Define the term *logic gate*.
24. What is a truth table?
25. What is the major function of an exclusive or gate?
26. Explain the positive logic concept.
27. What is a digital waveform of a logic gate?

27. What is a digital waveform of a logic gate?
28. Explain the process of photolithography.
29. Explain the manufacturing process used to produce integrated circuits.
30. Define the term *microprocessor.*
31. What are the four categories of ICs?
32. What is a logic family?
33. Describe the TTL logic family.
34. Describe the advantages of CMOS logic family devices.
35. What is the DIP standard?
36. How are the connections of an IC identified?
37. What are adder circuits?
38. What is the function of a digital clock?
39. Describe the operation of a flip-flop circuit.
40. What is the function performed by latch or flip-flop circuits?
41. Describe the function of a central processing unit.
42. What are the three basic subsystems of a CPU?
43. What are the two major categories of memory circuits?
44. Describe the operation of a read-only semiconductor memory.
45. Give an example of how ROMs are used.
46. What is the function of the arithmetic logic unit?
47. Explain the process used to transmit digital data.
48. Explain the functions of a multiplexer and a demultiplexer.
49. Describe an EAROM and give an example of its use.
50. Describe a nonvolatile memory.
51. Which ARINC standard is used for transmission of digital data?
52. What is the purpose of the parity bit in the ARINC 429 code?
53. What is the ARINC 629 code?
54. What is a CSDB?
55. What is an ASCB?
56. What is the purpose of the source-destination indicator of the ARINC 429 code?
57. Explain the concept of logic troubleshooting techniques.
58. What are logic diagrams?
59. Describe the operation of a typical built-in test equipment system.
60. What are two common instruments used to test logic levels?

Electric Measuring Instruments 8

The fundamental units of electrical measurement are the ampere, volt, ohm, and watt. To measure electrical values in terms of these units, certain instruments are required.

The common electric measuring instruments are the **ammeter, voltmeter, ohmmeter,** and **wattmeter.** The unit measured by each of these instruments is clearly indicated by its name. There are many electric measuring instruments in addition to those mentioned above, but for the purposes of this chapter, the major discussion will deal with these.

Digital and analog are the two general types of meters currently in use in modern aircraft and related test equipment. Analog meters utilize an infinitely variable scale upon which a specific value is indicated. A typical analog meter might contain a scale ranging from 0 to 100. The actual indicator position has an infinite number of possibilities within this range. Digital meters and instruments contain only a finite number of possible indications. A typical digital meter might have a scale from 0 to 100 with only 100 possible readings.

The digital system is usually more accurate, even though it allows for fewer possible indications. This is because the digital system gives indications in precise figures that are easily read without error. Digital meters and instruments have also proved to be more durable and reliable than analog systems. Although there are many commonalities between digital and analog meters, digital meter systems will be discussed in a separate section of this chapter.

METER MOVEMENTS

The basic principle of many electric instruments is that of the **galvanometer.** This is a device that reacts to minute electromagnetic influences caused within itself by the flow of a small amount of current. A simple galvanometer is shown in Figure 8–1. It consists of a magnetized needle suspended within a coil of wire. When a current is passed through the wire, a magnetic field is produced, and the magnetized needle attempts to align itself with this field. Practical galvanometers cannot be constructed as simply as the one described above, but they all operate because of the reaction between magnetic and electromagnetic forces.

Any device that is designed to indicate a flow of current, particularly a very small current, and which operates on the principle of two interacting magnetic fields can be called a galvanometer. A permanent magnet pivoted so that it can turn in response to the influence of a current-carrying coil, or a current-carrying coil placed in a magnetic field and pivoted so that it can turn in response to the field produced by a current flow, can be used as a galvanometer. In any event, the rotating part must be balanced by a spring that will tend to hold it in the zero position when there is no current flow.

The most common types of electric measuring instruments employ a moving coil and a permanent magnet. This arrangement is known as the **d'Arsonval** or **Weston** movement and is illustrated in Figure 8–2. The coil, consisting of fine wire, is pivoted and mounted so that it can rotate in the magnetic field of the permanent magnet's poles. When a current flows in the coil, a magnetic field is produced. The north pole of this field is repelled by the north pole of the permanent magnet and attracted by its south pole. As shown in Figure 8–2, this will cause the coil to rotate to the right. The magnetic force causing the rotation is proportional to the current flowing in the coil and is balanced against a coil spring. The result is that the distance of rotation will increase as the current flow in the coil increases. The needle attached to the coil moves along a scale and indicates the amount of current flowing in the coil.

It is quite apparent that the d'Arsonval movement, used alone, is not suitable for the measurement of alternating current. Such current would produce rapid reversals of polarity

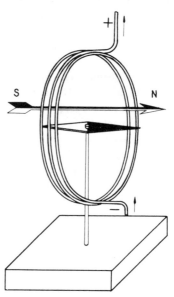

FIGURE 8–1 A simple galvanometer.

159

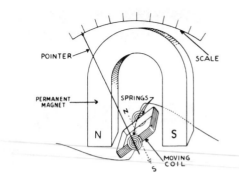

FIGURE 8–2 D'Arsonval or Weston meter movement.

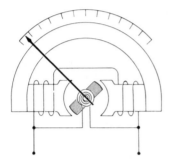

FIGURE 8–3 Dynamometer movement.

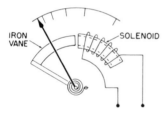

FIGURE 8–4 Iron-vane movement.

in the moving coil that would cause the needle only to vibrate. Under these conditions, no indication could be obtained.

A movement similar to the d'Arsonval movement, but suitable for ac measurements, employs an electromagnet in place of the permanent magnet. This is called a **dynamometer** movement (see Figure 8–3). The moving coil can be connected in either series or parallel with the electromagnet circuit. When a movement of this kind is used, the indicating needle will always move in the same direction, regardless of the direction of the current through the instrument. This is because the polarity of both the moving coil and the electromagnet changes when the current direction changes; hence the direction of torque (twisting force) remains the same. The movement will therefore operate with alternating current.

Another type of movement used with alternating current is illustrated in Figure 8–4. This is called an **iron-vane** mechanism, and, as the name implies, it employs an iron vane through which electromagnetic forces act to move the indicating needle. The iron vane is attached to a pivoted shaft and is free to move into the coil whenever the coil is energized.

Also mounted on the shaft is the indicating needle; hence the vane and the needle move together in response to a current flow in the coil. The movement of the vane and needle assembly is balanced by a coil spring that holds the needle in the zero position when no current is flowing.

Construction Features

Electric meters must be constructed with the utmost care and precision. This is so because some meter movements must respond to currents as small as a few millionths of an ampere. Some of the moving parts are extremely accurately machined and finished. For this reason such instruments must be handled with great care to prevent shock or vibration damage, which would result in a loss of accuracy.

Because of the sensitivity required of electric meter movements, it is necessary that the pivot shaft bearings be as nearly frictionless as possible. This is accomplished by using jewel bearings similar to those used in fine watches. Figure 8–5 shows three different types of jewel bearings. For electric measuring instruments, the V-jewel bearing is typically used because it produces the least friction.

Although the moving elements of instruments are designed to be of the lowest possible weight, the extremely small contact area between the pivot and jewel results in the large stresses for which the bearing must be designed. For example, a moving element weighing 300 milligrams (mg) [0.00066 lb], resting on the area of a circle 0.0002 in. [0.0005 cm] in diameter, produces a force of about 10 tons/in.^{2} [1406 kg/cm^2]. From this it can be seen that if an instrument is dropped or jarred, the bearing stresses can easily be increased to the level where permanent damage is done.

Some instruments are designed to withstand rather severe shocks, and these are supplied with spring-back jewel bearings such as that shown in Figure 8–5. This construction permits the pivot shaft to move axially when it is subjected to shock, with the result that the stresses are greatly reduced.

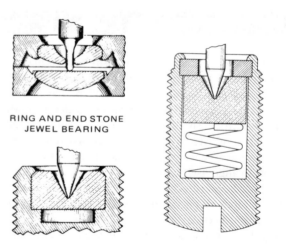

FIGURE 8–5 Jewel bearings.

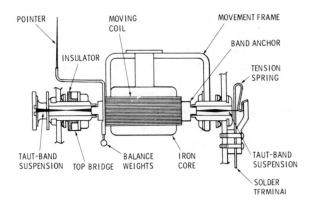

FIGURE 8–6 Taut-band movement.

Taut-Band Movement

A rather ingenious development in instrument movements has largely eliminated the friction problems and the need for pivoted bearings. In this instrument movement, the moving coil is suspended on a taut platinum-iridium band held by spring tension in the instrument frame. This type of unit is called a **taut-band movement.** Figure 8–6 shows the construction of an instrument utilizing the taut-band suspension for the moving coil.

It can be seen that the taut-band instrument is not nearly as sensitive to shock as the type having jewel bearings, because shocks are taken up by the elasticity of the taut band. The band is not subject to corrosion because of the material from which it is made. Since there are no parts rubbing against one another as in a bearing, the friction is eliminated, and the movement can respond to extremely small magnetic influences. For this reason the movement can be designed for very high sensitivity.

Design for Uniform Scale

Since the magnetic force acting upon a magnetic substance is inversely proportional to the distance between the magnet and the substance acted upon, instruments using the magnetic principle do not have a uniform scale unless special construc-

tion features are incorporated. In the Weston meter movement, a uniform scale is obtained by placing a cylindrical iron core inside the moving coil (see Figure 8–7). This arrangement results in a uniform magnetic field in the air space between the core and the poles. The coil rotates in the cylindrical space a distance proportional to the amount of current flowing in the coil windings.

It is apparent from a study of a typical instrument diagram that there is a limit to the range through which the indicating needle can act. In a conventional meter movement, such as that in Figure 8–7, this range is approximately 100°. Some meters require a greater range and must be specially constructed.

Sensitivity

As previously stated, some meters must be constructed with a high degree of sensitivity. This **sensitivity** is determined by the amount of current required to produce a full-scale deflection of the indicating needle. Very sensitive movements may require as little as 0.00005 A to produce a full-scale deflection. This value is commonly called 20,000 Ω/V, because it requires 20,000 Ω to limit the current to 0.00005 A when an emf of 1 V is applied. Movements having a sensitivity of 1000 Ω/V are commonly used by electricians when the power consumed by the instrument is of no consequence. In electronic work, where very small currents and voltages must be measured, instruments of very high sensitivity are required. Electronic measuring instruments, such as the volt-ohm meter (VOM), the multimeter, or the digital multimeter (DMM), are normally used for the measurement of resistance, currents, and voltages in electronic circuits. These instruments are designed to isolate the measuring circuit from the circuit being measured; hence very little loading is applied to the circuit being measured.

To understand the importance of sensitivity in an instrument for testing certain values where current flow is very small, it is well to consider a specific example. In the circuit in Figure 8–8, a 100-V battery is connected across two resistors in series. Each resistor has a value of 100,000 Ω, making the total resistance of the circuit 200,000 Ω. Since the two resistors are equal in value, it is obvious that the voltage across each will be 50 V. If we wish to test this voltage by means of a voltmeter that has a 1000-Ω/V sensitivity, we will discover that a large error is introduced into the reading.

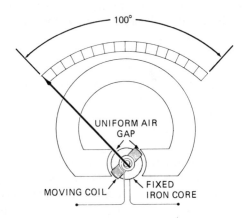

FIGURE 8–7 Iron core used to provide a uniform field.

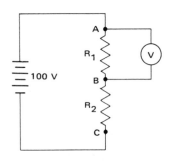

FIGURE 8–8 Demonstration of a need for high sensitivity in a voltmeter.

Assume that the voltmeter has a range of 100 V and that it is connected across R_1, between points A and B. Since the voltmeter has a sensitivity of 1000 Ω/V, its total resistance will be 100,000 Ω. When this is connected in parallel with R_1, the resistance of the parallel combination becomes 50,000 Ω, and the total resistance of the circuit is now 150,000 instead of 200,000 Ω. With the resistance between A and B 50,000 Ω and the resistance between B and C 100,000 Ω, the voltage drop will be 33.3 V between A and B and 66.7 V between B and C. It is apparent, then, that the voltmeter used would not be satisfactory for this test.

If we connect a voltmeter with 20,000-Ω/V sensitivity across R_1, we will obtain a much more accurate indication of the operating voltage. The voltmeter has an internal resistance of 2,000,000 Ω, and this resistance, combined in parallel with R_1, will produce a resistance of 95,238 Ω. This resistance in series with the 100,000 Ω of R_2 will produce a voltage drop of approximately 48.7 V across R_1 and 51.3 V across R_2. The reading of the voltmeter is then 48.7 V, which is probably as accurate as necessary for normal purposes.

THE AMMETER

Most of the electric measuring instruments in use today require very little current to produce a needle movement or digital indication. A resistance is used to restrict the current applied to the meter's sensitive circuitry when a relatively high voltage or current is being measured. If calibrated resistors are used for the restriction, the accuracy of the meter remains high, and the meter's indication range is extended.

When a resistance is connected in parallel with the terminals of a meter, it is called a **shunt resistance.** A shunt resistance, also called an **instrument shunt,** can be defined as a particular type of resistor designed to be connected in parallel with a meter to extend the current range beyond some particular value for which the instrument is already competent. In general, the word *shunt* means "connected in parallel."

A simple ammeter with low capacity can be constructed by using relatively large wire in the moving coil. If the wire is large enough to carry the full amperage of the circuit in which the meter is used, it is not necessary to incorporate a shunt resistance. The shunt resistance is employed in most ammeters because it makes it possible to use the same meter movement for a wide range of current measurement.

A typical ammeter circuit with a shunt resistance is shown in Figure 8–9. If we assume that a current of 0.01 A causes a full-scale deflection of the indicating needle and that the resistance of the movement is 5 Ω, we can calculate the voltage required to produce a full-scale deflection. By applying Ohm's law, we find this to be 0.05 V. This instrument can be made to measure almost any current value by using a shunt resistance of the correct value. Suppose that it is necessary to use the ammeter where the current range is from 0 to 30 A. Since 30 A must flow through the parallel combination of the meter and the shunt resistance and only 0.01 A can flow through the meter, then 30 − 0.01 A, or 29.99 A, must flow through the shunt resistance. We know that 0.05 V across the meter provides a current of 0.01 A; hence we must find a resistance that will cause a voltage drop of 0.05 V when a current of 29.99 A is flowing through it. By Ohm's law,

$$R = \frac{0.05}{29.99} = 0.00167 \ \Omega$$

If we wish to use the same meter movement for a range of 500 A, the value of the shunt resistance can be determined as for a 30-A range. Since 0.01 A will flow through the instrument at full-scale deflection, 499.99 A must flow through the shunt. The resistance of the shunt must be such that 499.99 A will cause a voltage drop of 0.05 V. This value is obtained by dividing 0.05 by 499.99. The required resistance is found to be approximately 0.0001 Ω. It is very difficult to construct a resistance of exactly 0.0001 Ω, and temperature changes also cause some variations; hence practical ammeters for high amperage employ a movement less sensitive than that described above.

Greater accuracy can be obtained with a sensitive movement by incorporating a series resistance into it. For example, if a resistance of 995 Ω is connected in series with the movement, then the value of the shunt resistance can be increased to approximately 0.02 Ω. This will provide much greater accuracy because the higher resistance of the shunt reduces the error factor. Typical external ammeter shunts are shown in Figure 8–10.

Milliammeters, which are used to measure current values in thousandths of amperes, do not necessarily require shunt resistance. If the instrument has a sensitivity of 100 Ω/V, it has a range of 0 to 10 milliamperes (mA) without a series or shunt resistance. A shunt resistance can be incorporated to increase the range to any desired value. To measure current in units smaller than the milliampere, a **microammeter** is used. One microampere (μA) is one-millionth of an ampere. An ammeter with a sensitivity of 20,000 Ω/V has a full-scale deflection of the indicating needle with a current of 50 μA through the movement. Such a meter can be used to measure current in a range of 0 to 50 μA.

An ammeter designed to measure a wide range of values may contain internal shunts. Each internal shunt is typically controlled through the meter's range switch. When the meter is measuring a high-current circuit, the range switch is placed in the appropriate position, and the correct shunt resistor is added in parallel to the ammeter circuit. If a different current range is desired, the switch is repositioned, and a different shunt is connected in parallel with the meter movement.

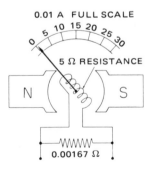

0.01 A FULL SCALE

5 Ω RESISTANCE

N   S

0.00167 Ω

FIGURE 8–9 Ammeter circuit.

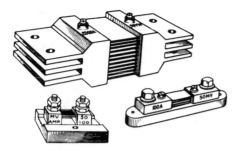

FIGURE 8–10 External ammeter shunts.

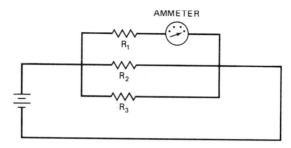

FIGURE 8–12 Ammeter connected to measure the current through R_1.

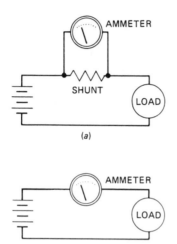

FIGURE 8–11 Ammeter connected in a circuit (a) with an external shunt and (b) with an internal shunt.

Those ammeters used to measure only a narrow range of current flows typically employ external shunts as illustrated in Figure 8–10.

The proper method for connecting an ammeter in a circuit is shown in Figure 8–11a. Note that the ammeter and shunt are in parallel with each other and in series with the load. An ammeter of the proper type uses a negligible amount of power for its operation; hence it will not interfere with the operation of the load. The instrument must not be connected in parallel with the source of power. The ammeter and its shunt are designed to offer as little resistance as possible in a circuit; hence if they are connected in parallel with the power source, they will act as a direct short circuit. This will not only prevent the operation of the circuit but also in most cases cause irreparable damage to the instrument and to the power source.

An ammeter requiring no shunt, or containing an internal shunt, is illustrated in Figure 8–11b. In this circuit the ammeter is still connected in series with the load, and no external shunt is required.

If the current in a portion of a circuit is to be measured, the ammeter must be placed in series with that portion of the circuit. As illustrated in Figure 8–12, the ammeter will measure only the current that flows through the meter. The current that bypasses the meter through a different portion of the circuit will not be measured. In this illustration, the meter measures only the current through R_1, not the current of the entire circuit.

An ammeter of the correct capacity and type can be used to determine how much current a particular load in an aircraft will draw. If it is desired to find out how much current flows in a starter circuit, it is merely necessary to disconnect one of the power cables to the starter motor and connect the ammeter between the starter cable and the terminal on the starter motor. Care must be taken to see that the ammeter is connected with the correct polarity. In a system with a negative ground, the positive (+) terminal of the ammeter is connected to the power cable from the starter relay, and the negative (−) terminal of the ammeter is connected to the power terminal of the starter motor.

The ammeter used to check the current in a light aircraft's starter system should have a range up to approximately 500 A because the initial current flow is very high. After the ammeter is properly connected with wire as large as the starter cable, the starter switch can be closed. It will be noted that there is a very high surge of current at first, and this rapidly falls off to a much lower value.

When connecting an ammeter, or any other meter, in an electric circuit, it is always essential to choose a meter with a range high enough to ensure that the meter is not overloaded. If the maximum current flow of the circuit is unknown, always install a meter of very high indicating range. If a very low indication is measured by this test, a lower-range meter can be installed to more accurately determine the circuit's current flow. When a meter containing internal shunts is used, this matter becomes very simple. Always install the meter with the range selector set to the highest value. Then, if necessary, slowly move the selector switch to a lower value in order to achieve a needle indication at approximately mid-scale.

THE VOLTMETER

A **voltmeter** of the moving-coil type actually measures the current flow through the instrument; but since the current flow is proportional to the voltage, the instrument dial may be marked in volts. The meter movement is adapted to the measurement of voltage by the use of series resistance.

Figure 8–13 shows a schematic diagram of a voltmeter circuit. Assuming that the meter movement has a full-scale

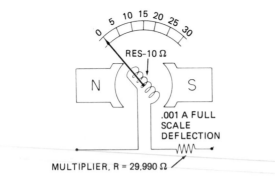

FIGURE 8–13 Voltmeter circuit.

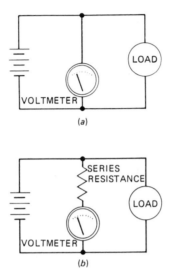

(a)

(b)

FIGURE 8–14 Voltmeter connected in a circuit (a) with an internal multiplier and (b) with an external multiplier.

deflection at a current of 0.001 A and an internal resistance of 10 Ω, it is easily determined that 0.01 V is the maximum that can be applied to the instrument without the addition of a series resistance. If we wish to give the instrument a range of 0 to 30 V, we use Ohm's law to find the required series resistance. Since the current through the instrument must be 0.001 A for a 30-V reading, we proceed as follows:

$$R = \frac{30}{0.0001} = 30,000 \ \Omega$$

Since the internal resistance of the movement is 10 Ω, this value must be subtracted from the total required resistance. The series resistance required is then 29,990 Ω.

Voltmeters usually have the necessary series resistance built into the instrument itself. The range of such an instrument can be increased by the use of additional series resistances called **multipliers.** Resistances of this type are used with test instruments when the instrument must be capable of measuring a wide range of voltages. For example, to double the range of a voltmeter, it is necessary merely to add a series resistance equal to the total resistance of the instrument.

The proper method of connecting a voltmeter with an internal multiplier is shown in Figure 8–14a. The meter is con-

nected in parallel with the power source being measured. Figure 8–14b shows a voltmeter and an external multiplier connected in a circuit. The meter is in series with the multiplier, and the multiplier and voltmeter are in parallel with the power source being measured.

As stated in the discussion of Ohm's law, parallel resistances have equal voltage drops. Since a voltmeter is a resistance, it must be placed in parallel with any item whose voltage is to be measured. When measuring the voltage in a portion of a circuit, the voltmeter must be placed in parallel with the portion to be measured. Figure 8–15 illustrates the connections necessary to measure the voltage available to R_2. The voltmeter in parallel with R_2 will measure the voltage drop across R_2 and not the voltage of the entire circuit.

If the resistance of a voltmeter is too low, it disturbs the conditions of the circuit, and an accurate reading cannot be obtained. This is particularly true in circuits having a low current flow, such as electronic circuits. In such circuits it is necessary to use very sensitive (high-resistance) instruments.

Since a voltmeter has sufficient resistance so that it can be connected in parallel with the power supply, there is no damage to the instrument if it is connected in series with the load. The only effect is to prevent the operation of the circuit.

Voltmeters are used in airplanes so that the pilot, or another member of the crew, can be kept informed concerning the operation of the electric system. They are not usually installed in small airplanes with single generator systems, but where it is necessary to operate two or more generators in parallel, the voltmeter is essential to aid in balancing the output of the generators.

The voltmeter is a valuable instrument for troubleshooting and checking electric and electronic circuits. In every case the technician must be sure that the voltmeter being used is of the correct range and that it is the proper type for the current in the circuit, whether alternating or direct current.

If a particular electric unit is not functioning, the first step is to determine whether electric power is being delivered to the unit. This is quickly accomplished by testing with the voltmeter. In a system with negative ($-$) ground, the positive probe or alligator clip connected to the voltmeter is touched to the terminal of the unit being checked, and the negative probe is touched to the metal of the airplane. If power is reaching the test point, the voltmeter should read system voltage.

Circuits can be tested while "hot," that is, with power on, or they may be tested by connecting the voltmeter with the power off and then turning the power on and observing the

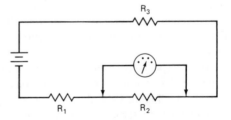

FIGURE 8–15 Voltmeter connected to measure the voltage applied to R_2.

response of the voltmeter. In testing hot circuits, care must be taken to avoid causing short circuits by allowing some metal object to bridge between the terminal being tested and the metal of the aircraft. The test leads of the voltmeter should be well insulated, and the test probes or alligator clips should be insulated except at the points where electrical contact is to be made.

On large aircraft where 208-V 400-Hz circuits are to be tested, great care must be taken to avoid contacting hot parts of circuits with the bare hands or any other part of the body. Severe shocks will occur at any time that a circuit is completed through the body. Working procedures established by the company operating large aircraft are usually such that the danger of shock is minimized.

For testing the normal operating circuits of an aircraft's electrical system, the sensitivity of the test voltmeter need not be high. A voltmeter with a sensitivity of 1000 Ω/V will usually suffice.

A voltmeter of high sensitivity must be used when testing delicate circuits, that is, circuits that operate on very low voltage or very low current. In cases where a short-term or changing voltage is to be measured, the speed of the test meter is also an important consideration. An accurate reading can only be obtained if the voltmeter can respond quickly enough to measure the voltage before it changes.

THE OHMMETER

The **ohmmeter,** as its name implies, is an instrument for measuring resistance. The type most commonly used by aircraft mechanics and electronics technicians employs a moving-coil galvanometer similar to those described for ammeters and voltmeters. To make the movement capable of measuring resistance directly, it is necessary merely to provide a source of electric power and a suitable resistance. Remember, the test instrument responds only to current through the meter's coil; but since current is a function of a circuit resistance, a galvanometer can be used as an ohmmeter.

Figure 8–16 is a schematic diagram of a simple ohmmeter circuit. The principle of operation follows Ohm's law, and a 3-V battery provides the power necessary for operation. The meter movement has a sensitivity of 1000 Ω/V and an internal resistance of 10 Ω. The total series resistance throughout the circuit must be 3000 Ω to provide a full-scale deflection of the indicating needle when the emf of the power source is

3 V. This is attained by placing a 2500-Ω fixed resistance and a 490-Ω variable resistance in series with the battery and the meter movement. The variable resistance makes it possible to compensate for the lowering of the battery voltage over a period of time.

When the test probes are in contact with each other, the indicating needle moves to the full-scale position. This point is marked zero because it indicates that there is zero resistance between the test probes. If a resistance of 3000 Ω is placed between the test probes, the needle travels halfway across the scale. This point on the scale is marked 3000 Ω. If the test probes are separated, that is, placed to measure the resistance of the air between them, the indicating needle remains at the extreme left side of the scale. This point is marked infinity (∞) because the resistance of the air is so great that no measurable current passes through the circuit; hence, for practical purposes, the resistance is infinite.

The range of resistance readings on the scale of the ohmmeter described above is from zero to infinity, but the practical range is from approximately 100 to 30,000 Ω. The scale divisions for the very high resistances are so close together that the probability of error increases tremendously as the value of the reading becomes higher. The basic range of the ohmmeter can be changed by the use of resistances as multipliers. For higher resistances, it is necessary to use a higher voltage or a more sensitive movement. In either case, the current-limiting resistance must be increased.

An ohmmeter designed for the measurement of very low resistances must be connected so that the resistance to be measured acts as a shunt resistance across the test probes. A circuit for this type of ohmmeter is shown in Figure 8–17. The meter movement is connected in series with a battery, a switch, a fixed resistor, and a variable resistor. If we assume that the meter has an internal resistance of 5 Ω and that 1 mA will produce a full-scale deflection, then a total circuit resistance of 4500 Ω will provide for a full-scale deflection when a 4.5-V battery is used to power the instrument. To test a resistance, the switch is closed and the indicating needle moves to the extreme right of the scale; that is, it indicates infinite resistance. If we place a 5-Ω resistor between the test probes, the needle will take a position at half scale. This is because

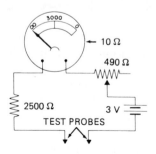

FIGURE 8–16 A simple ohmmeter circuit.

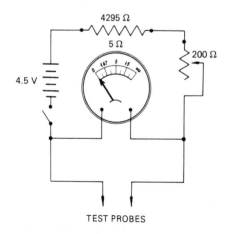

FIGURE 8–17 Ohmmeter for testing low resistance.

0.5 mA is now passing through the meter and 0.5 mA through the resistor. If a 15-Ω resistor is placed between the probes, the indicating needle will take a position three-fourths the distance from the zero end of the scale. This is because 0.75 mA will be flowing through the meter and 0.25 mA through the resistor.

To measure the resistance of any electric item such as a resistor, coil, or length of wire, it is necessary merely to connect the test probes of the ohmmeter to the terminals of the item being tested. Before the test is made, the ohmmeter should be adjusted so the scale reading is zero when the probes are connected together and infinity when the probes are separated. This is done by adjusting the meter's resistance to compensate for any variance in battery voltage. The meter's resistance is changed by adjusting a variable internal multiplier. Over time, even if a battery is not used, its voltage decreases. To obtain accurate indications, one must always "zero" an analog ohmmeter prior to use. Most digital ohmmeters do not require this zeroing adjustment. Typically, an ohmmeter contains several internal multipliers, which can be connected to the circuit by using the meter's range selector. The range selector switch can be adjusted to allow one instrument to accurately measure several different resistances. The range of the ohmmeter used should be such that the indicating needle will move to a point in the center two-thirds of the scale when the unit being tested is connected to the test probes. With some instruments, the manufacturer may suggest a different portion of the scale as providing the most accurate results.

An ohmmeter will not be damaged by measuring a resistance while the meter is set on the wrong scale; however, the meter indication may be less accurate or difficult to read.

In testing units connected in a circuit, all electric power in the circuit must be turned off. It is good practice to disconnect or remove power sources from a circuit or unit when testing portions of the circuit with an ohmmeter. Electric power in a circuit being tested with an ohmmeter will usually cause damage to the instrument and will certainly prevent a correct reading of resistance values. It is often necessary to isolate one side of a unit from a circuit to prevent other parts of the circuit from affecting the reading. If a unit is tested while it is still connected to the circuit, an inaccurate reading is likely.

The ohmmeter is a most useful instrument for testing or checking electric circuits and appliances. It not only measures resistance but also is an excellent **continuity tester.** Testing continuity is the process whereby the meter is used to determine if the circuit has a complete *(continuous)* current path. Continuity tests are often performed on items such as lightbulbs or relays to determine if each item has a continuous path or an open circuit.

To use an ohmmeter as a continuity tester, it is necessary merely to contact the terminals of the circuit being tested with the test probes of the ohmmeter. This is illustrated in Figure 8–18, which shows how a section of circuit wiring can be checked. Usually, the wiring for a number of circuits is included in a bundle called a **harness,** and it is not possible to inspect the individual wires visually. The particular circuit to be tested can be identified at each end by means of coding on the wire and in the circuit diagram provided in the aircraft maintenance manual. After the wire to be tested is identified,

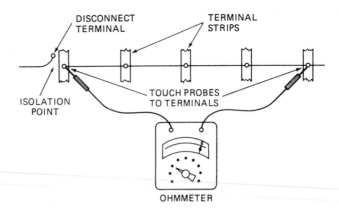

FIGURE 8–18 Use of an ohmmeter as a continuity tester.

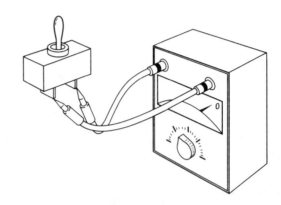

FIGURE 8–19 Measuring the continuity of a switch.

the probes of the ohmmeter are touched to the two terminals of the wire simultaneously. If the wire is in good condition, the ohmmeter will read near the zero resistance point.

Although the procedure described above is valid, it often becomes difficult to perform. For example, if the wire to be checked is routed from the wing tip to the instrument panel, the ohmmeter test leads will be too short to connect to both ends of the suspect wire. A voltage check, with the system power on, is typically the easiest way to troubleshoot an open wire. Troubleshooting techniques are discussed in Chapter 13. An ohmmeter is best used to test the continuity of a component such as the switch illustrated in Figure 8–19. The ohmmeter should indicate near zero resistance (less than 1 Ω) when the switch is turned on and infinite resistance when the switch is turned off.

AC MEASURING INSTRUMENTS

The principal types of ac instrument movements are classified as follows: **iron-vane meters, hot-wire meters, thermocouple meters, dynamometers,** and **meters using rectifiers.** The iron-vane movement and the dynamometer were discussed earlier in this chapter. These movements are satisfactory for alternating current of relatively low frequencies within a range of 15 to 1000 Hz. For frequencies of over 150 Hz, the dynamometer-type movements must have corrections applied.

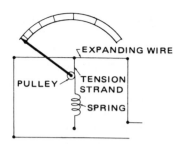

FIGURE 8–20 A hot-wire meter.

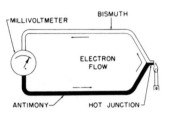

FIGURE 8–21 Thermocouple principle.

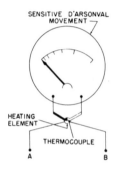

FIGURE 8–22 Thermocouple meter.

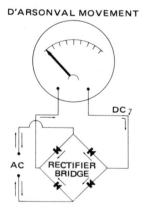

FIGURE 8–23 Rectifier-type meter.

The **hot-wire meter** was one of the first movements used for high-frequency (HF) ac measurements that was not adversely affected by the frequency of the current. The principle of this meter is illustrated in Figure 8–20. The actuating element is a wire that expands as its temperature increases. This thermal-sensitive wire usually consists of platinum silver or platinum iridium. The ends of this wire are connected

to the source of power. As the current flows through the wire, it heats and expands, and this expansion allows the tension spring to move the tension strand, which is wrapped around a small pulley on the pointer shaft. As the tension strand moves, the pointer shaft is rotated, and the pointer moves across the face of the dial. The hot-wire movement has several undesirable characteristics and therefore is in limited use on modern equipment.

A satisfactory movement for the measurement of HF ac values is the **thermocouple meter.** A thermocouple circuit consists of two junctions of two dissimilar metals in one circuit. When one junction is heated, an emf develops, and a current flows in the circuit. This action is illustrated in Figure 8–21. The emf developed by a thermocouple depends on the nature of the metals and the difference in temperature between the heated junction (hot junction) and the cold junction.

The **thermoelectric effect** (production of an emf by the application of heat) of a thermocouple is caused by the atomic structure of the metals employed.

If conductors of antimony and bismuth are connected at two points, as in Figure 8–21, there is a momentary flow of electrons from the antimony to the bismuth at both junctions. Thereafter, however, if the junctions have the same temperature, there is no flow of current through the circuit, because the difference of potential at one junction is equal to the difference of potential at the other junction. But when heat is applied at one of the junctions, the potential difference will increase at that point because the heat increases the tendency of the electrons to flow from the antimony to the bismuth. Since the emf at the hot junction is greater than the emf at the cold junction, a current will continue to flow for as long as there is a difference in temperature between the two junctions. If a sensitive meter is inserted in the circuit at the cold junction, the current flow in the circuit can be measured.

A schematic diagram of a thermocouple-meter circuit is shown in Figure 8–22. The terminals of the source of current are connected to the instrument at A and B. Current then flows through the heating element and causes a rise in temperature at the thermocouple. The emf produced by the thermocouple causes a current to flow through the sensitive meter movement, and the indicating needle gives a reading proportional to the current produced by the thermocouple.

One of the most common instruments for use with relatively low-frequency alternating currents is a dc meter movement connected in a circuit with a full-wave rectifier, as shown in Figure 8–23. An instrument such as this is often called a **rectifier instrument.** A full-wave rectifier is used to change the alternating current to direct current. When such a meter is used for both ac and dc measurements, two scales are usually provided. This is because the movement of the needle, when measuring ac values, is proportional to the average value of the current rather than the effective value.

When a rectifier-type meter is used with HF alternating current, the error increases in proportion to the frequency. This error is caused by the capacitive effect of the rectifier elements. At extremely high frequencies, the rectifier will pass a substantial amount of current in both directions; hence the current will flow through the instrument in both directions.

FIGURE 8–24 An inductive pickup meter. *(Amprobe Instrument)*

The result is that the current in one direction is reduced by the current in the opposite direction.

Inductive Pickups

Many ac meters use the principle of electromagnetic induction to measure the current and frequency of a circuit. These meters contain a test probe called an **inductive pickup** or a **current transformer** that wraps around one wire of the circuit to be tested. A meter employing an inductive pickup is shown in Figure 8–24. An inductive pickup receives its current from the electromagnetic wave that forms around any wire carrying a current. If this magnetic field is constantly changing, as in an ac circuit, a voltage/current will be induced into the inductive pickup. The pickup then sends a signal to the measuring instrument, which measures the circuit's current flow. Inductive pickups are commonly used for ac ammeters and wattmeters. They can also be used on circuits operated by a pulsating direct current.

Hall-Effect Probes

A **Hall-effect probe** is a type of inductive pickup that can be used to measure either ac or dc current. The Hall-effect probe resembles a current transformer in that the probe surrounds the wire of the circuit being tested. The Hall-effect probe produces a voltage proportional to the current being measured. The probe's voltage is then sent to the test instrument. A typical Hall-effect probe may produce 1 mV for every amp of ac or dc current it detects.

The Wattmeter

Wattmeters are not frequently used by the aircraft maintenance technician, but a short discussion of the principles of

such meters will aid in the understanding of the measurement of electric power.

The unit for the measurement of electric power is the watt. One watt is the power expended when a current of 1 A is flowing under the pressure of 1 V. In an electric circuit, the power in watts is equal to the product of the voltage and the amperage. This is true in dc circuits and in ac circuits when the voltage and current are in phase.

Since electric power involves both amperage and voltage, the wattmeter must be capable of multiplying these values. A schematic diagram of a wattmeter connected in a circuit is shown in Figure 8–25. The construction of the instrument is similar to that of a dynamometer movement, but the circuits for the magnetic field and the moving coil are separate. One of the windings must provide a field proportional to the current in the circuit, and the other must produce a field proportional to the voltage. Since the current winding must be heavy enough to carry the current of the circuit, the current coil is stationary. The moving coil carries the voltage because this winding must have a high resistance and is relatively light in weight.

The current circuit is connected in series with the load circuit, and the voltage circuit is connected in parallel with the load circuit. A current-limiting resistor is connected in series with the voltage coil. When the circuits are connected in this manner, the strength of the stationary field is proportional to the load current, and the strength of the moving-coil field is proportional to the voltage across the load. The indicating needle moves a distance proportional to the product of the voltage and the amperage; hence the scale can be marked directly in watts.

Frequency Counters

When testing circuits, it is often important to know the frequency of the alternating current. **Frequency counters** are instruments used to "count" the electrical pulses of a given voltage. Most frequency counters can measure not only the frequency of an alternating current but also the frequency of a pulsating direct current, a square waveform, or a triangle waveform. Frequency counters are connected in parallel to the circuit being measured. This allows the instrument to monitor the circuit's voltage.

A typical frequency counter will change the incoming waveform into a standard square wave with the same

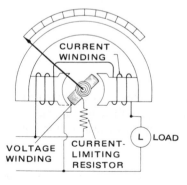

FIGURE 8–25 Wattmeter circuit.

frequency as the input signal. The frequency of the standard wave is then compared with a frequency of known value that is generated by the frequency counter. This comparison allows the instrument to determine the value of the measured frequency.

Some type of time gate circuit must be used in a frequency counter. Since the waveform being measured is constantly changing, the test instrument must measure the input for a given length of time (gate time). Frequency counters therefore measure an average frequency. A typical gate time is 1 s.

THE MULTIMETER

In practice, the functions of a voltmeter, ohmmeter, and ammeter (or milliammeter) are usually combined in an instrument called a **multimeter** or **volt-ohm-milliammeter** (VOM). This combination instrument makes it possible to take a wide range of electrical measurements with just one basic instrument. One type of multimeter is shown in Figure 8–26. This instrument will measure ac and dc voltage up to 1000 V, resistance from zero to infinity, and dc current from zero to 10 A. One meter movement is used to provide all the indications by means of the various scales on the dial. For

FIGURE 8–26 A typical analog multimeter. *(Simpson Electric Co.)*

each type of indication, the instrument is adjusted by means of a rotary switch and by plugging the test leads into the proper jacks.

The circuit diagram for the multimeter in Figure 8–26 is shown in Figure 8–27. Note that the rotary switch contacts are represented by the solid, bold horizontal lines. In the circuit diagram the switch is set in the 500–1000-V range. The resistance-measuring circuits are powered by a 1.5-V battery and a 9-V battery. An adjustment is provided through a variable resistor (R20) so that the meter can be set for zero before making resistance tests.

When using a multimeter, the technician must follow the instructions provided in the instrument manual. In testing an unknown voltage or amperage, it is important that the meter range be set above the highest level likely to be encountered. It is good practice to set the highest range available to start and then adjust downward until the reading falls in the upper one-third of the scale. This practice will avoid damage to the instrument from overload and provide for maximum accuracy. The meter movement is provided with automatic protection against overload. When voltages above the 30-V range are tested, the technician must exercise care in order to avoid electrical shock. This is particularly important in the 300- to-1000-V range.

DIGITAL METERS

Digital meters have become commonplace in modern aircraft. They are lighter, more reliable, and generally less expensive than analog meters. Digital meters, used as benchtop or portable multimeters, are also commonly employed by aircraft technicians. A **digital multimeter** (DMM) is shown in Figure 8–28. In many cases the DMM is considered more accurate than the analog meter because of its display of precise numerical values. Digital displays require less interpretation by the operator; therefore, there is less chance for error. Since the display of a digital meter is a solid-state device, there is no sensitive meter movement that can be damaged during rough handling.

Digital meters use one or more integrated circuit (IC) chips to process input data, initiate displays, and perform any required calculations. Light-emitting diodes or liquid crystals are used for displays on most digital meters. When measuring voltage, the DMM sends the input signal (from the test leads) through an analog-to-digital converter. The digital signal is then compared with an internal reference voltage by means of the meter's microprocessor. The microprocessor interprets the signal and sends output data to the numerical display. When current is measured, a resistor is placed in series with the load inside the meter. The voltage drop over the resistor is measured, and the microprocessor converts the information into the proper signals for the numerical display. When measuring resistance, the meter supplies a reference voltage to the resistance being measured. A current flow results. This current is measured by the meter to determine the load resistance.

The resolution of a digital meter will determine how small

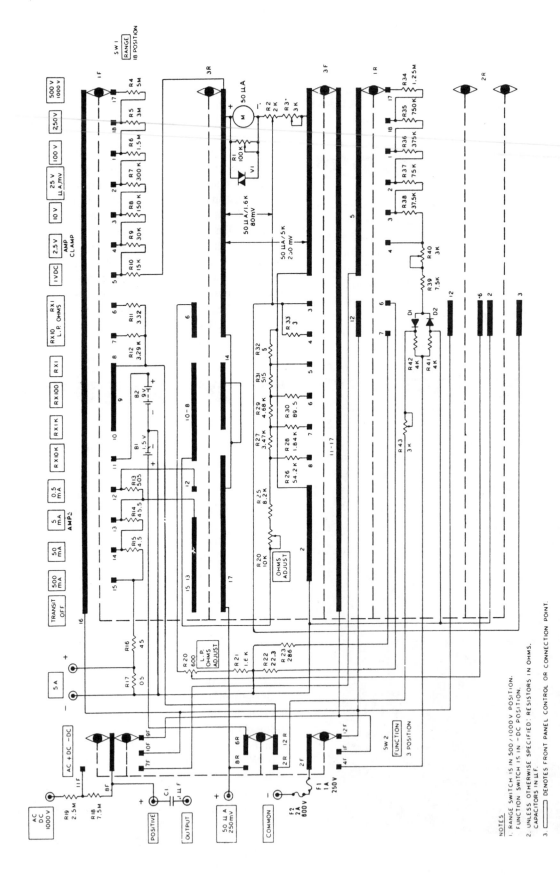

FIGURE 8–27 Electrical schematic for a Simpson 260 multimeter.   (Simpson Electric Co.)

a measurement the meter can make. In other words, a meter with high resolution may be able to measure down to 1 mV (0.001 V); a meter with poor resolution may only be able to measure down to 0.1 V. One limiting factor of a meter's resolution is the number of *digits* on the meter display. Typically, handheld DMMs have between 3 and $4\frac{1}{2}$ counts. A 3-digit meter can display up to three full digits (0 to 9). A $4\frac{1}{2}$-digit unit can display four full digits and one 1/2 digit (1 or blank). The 1/2 digit is located at the far left of the display. For example, a 3-digit display could indicate up to 999 V; a $3\frac{1}{2}$-digit display could indicate up to 1999 V.

Many digital meters incorporate a diode test function. During a diode test the meter supplies a voltage to the test probes. If the diode is operating correctly, the meter will indicate the voltage drop across the diode when the red probe is connected to the anode and the black probe is connected to the cathode of the diode. The voltage drop should be approximately 0.3 V or 0.7 V, depending on the type of diode. When the meter probes are reversed, the meter should indicate an open circuit.

Safety Precautions

Whether you are using an analog meter or a digital meter, always follow the manufacturer's safety instructions for the test equipment and the equipment being tested. Always check your meter test leads to ensure that their insulation is in good condition. Be extremely cautious when testing high-voltage or high-current circuits, and connect and disconnect test leads with the circuit's power turned off. Always disconnect the *hot* (red) test lead first, and connect the hot lead last. Never test any circuit while standing in water or wearing wet shoes and clothes. Large metallic jewelry can easily create a short circuit. Be careful with metal watchbands, bracelets, rings, and necklaces; if connected between + and − voltages, they can quickly heat up and cause severe burns. Don't assume anything; always double-check to make sure the voltage is off. Never work alone.

THE OSCILLOSCOPE

The **oscilloscope** is one of the most important measuring instruments used in analyzing complex electronic circuits. The oscilloscope is a sophisticated voltmeter with a two-dimensional graph display that can be used to measure the voltage (amplitude) and frequency (time) of an electrical signal. This allows the operator to view voltage or frequency changes over time; therefore, either a constant or a changing signal can be measured. With a multitrace oscilloscope, it is possible to compare two separate signals. An oscilloscope is typically used to analyze signals that change too fast to be monitored by a common multimeter. Measurement of a rapidly changing signal is often necessary when troubleshooting radio or digital circuitry. A typical oscilloscope is shown in Figure 8–29.

Most electrical signals can be easily connected to an oscilloscope with either probes or cables. Nonelectrical phenomena such as sound or temperature can be measured by using specialized probes called **transducers.** A transducer is a calibrated device that measures one form of energy and converts it into a voltage. A microphone is a type of transducer that converts sound waves into electrical signals. Examples of

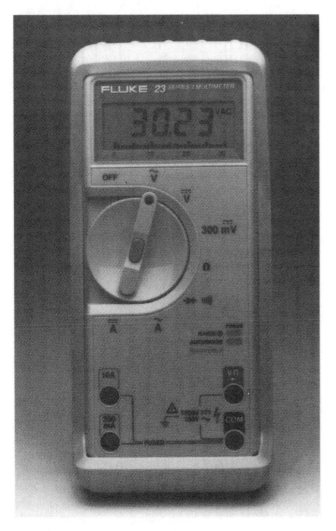

FIGURE 8–28 A typical handheld digital multimeter. *(Reproduced with permission, John Fluke Mfg. Co.)*

FIGURE 8–29 A typical oscilloscope. *(B & K Precision/ Dynascan Corporation)*

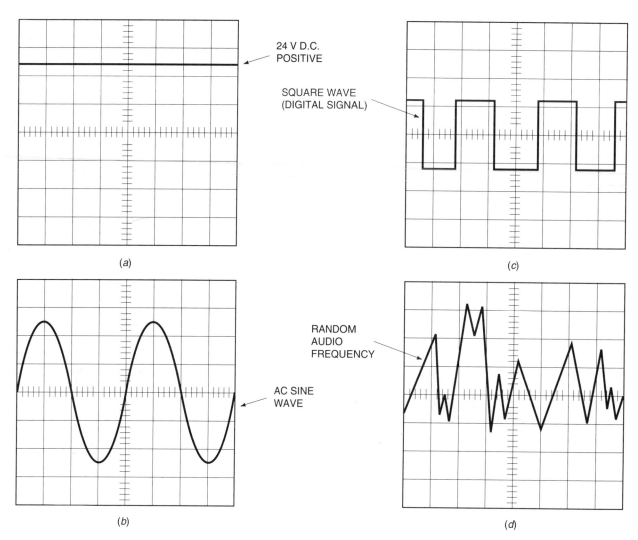

FIGURE 8–30 Various waveforms shown on an oscilloscope screen. (*a*) Positive 25 V dc; (*b*) ac sine wave; (*c*) square wave for a digital signal; (*d*) audio frequency sound wave.

various waveforms on an oscilloscope screen are shown in Figure 8–30.

The display system for an oscilloscope is composed of circuitry that drives a cathode-ray tube (CRT). The basic functional blocks of an oscilloscope are the vertical section, the trigger section, the horizontal section, and the display section, as shown in Figure 8–31.

Each part of the oscilloscope is named for its function. The vertical section controls the movement of the electron beam in the CRT along the vertical axis of the display. The horizontal section controls the movement of the electron beam along the horizontal axis of the display. The trigger section is a timing circuit that determines when the CRT will activate and display on the graph. The CRT is the display section; it graphs the results of the electrical input.

Cathode-Ray Tube (CRT)

The waveform produced by the CRT is projected by an electron beam against a phosphor screen. The movement of the beam on the screen is produced by two sets of deflection

plates inside the CRT; one set of plates controls the horizontal movement of the electron beam, and the other set controls the vertical movement. When a voltage is applied to one or more of the plates, the beam is attracted toward the positive plates and repelled away from the negative plates. The resulting movement of the electron beam across the phosphor screen leaves a tracing of the path on the screen for a short time.

When the beam is held stationary, a dot appears on the screen. If the beam is rapidly moved horizontally across the screen, a **sweep** line appears. If a signal or voltage to be measured is applied to the input of the oscilloscope, then it will indirectly be applied to the vertical deflection plates of the CRT. The beam of the CRT will move up or down on the screen as the signal voltage goes positive (increases) or negative (decreases) in value. As the signal goes positive in voltage, the beam moves upward on the screen, and as the signal goes negative, the beam moves downward. By adjusting the **sweep time,** or rate at which the beam moves horizontally across the screen, it is possible to choose the amount of signal to be viewed and determine the frequency at which the signal

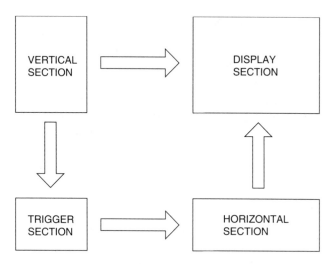

FIGURE 8–31 Diagram of the basic functional blocks of an oscilloscope.

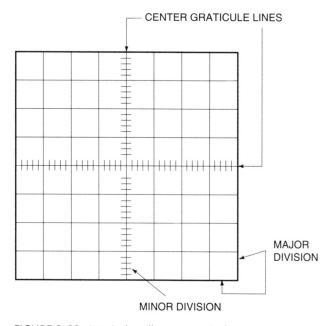

FIGURE 8–32 A typical oscilloscope graticule.

changes. Any signal that varies in voltage will appear as some type of waveform on the screen, depending on how the signal varies.

The **graticule** is the grid of lines, with the major divisions spaced at 1-cm intervals. The graticule can be painted on the face of the CRT or on a faceplate that is then installed on the CRT display. The graticule is used as a reference for voltage and frequency measurements (Figure 8–32).

The Vertical Section

The CRT represents an electrical signal as a waveform. The vertical displacement of the electrical signal is called the *y axis* and is the voltage level of the waveform. The waveform's change over time is represented horizontally on the CRT and is called the *x axis.* The vertical control section for the oscilloscope supplies the information for the *y*-axis dis-

play. The input signals are converted into deflection voltages that are used by the display section to deflect the electron beam. The vertical section also provides internal signals for the trigger section of the oscilloscope.

Triggering is accomplished by synchronizing the sawtooth generator that produces the horizontal sweep with the signal waveform that is being measured. If the sawtooth generator does not start and stop at the same point in time on the signal waveform, then an out-of-synchronization condition will occur. An out-of-sync condition will generate an instable display on the CRT. The instable display will appear as multiple waveforms on top of each other or as a moving waveform. In order to synchronize the sawtooth generator with the input signal, a **trigger pulse** is applied to the sawtooth generator. The trigger pulse can come from the vertical section or from an external source.

The Horizontal Section

The horizontal amplifier controls the horizontal sweep rate and allows the operator to choose the amount of input signal to be displayed. The horizontal section contains a sweep generator that produces a sawtooth waveform, or ramp, that is used to control the scope's sweep rates. The sweep generator is calibrated in time and often called the **time base.**

The Trigger Section

The oscilloscope draws a waveform on the CRT using the vertical and horizontal sections, which provide reference signals. The trigger section determines when the oscilloscope should start to "draw" the waveform. The waveform shown on the CRT screen is constantly being "redrawn." If the oscilloscope did not start to redraw the waveform at the same point each time, the display would not make much sense.

The trigger section ensures a stable display by recognizing a particular voltage point on the input waveform. Once this point is recognized, the sweep generator is activated, and a new waveform is drawn in the correct spot on the CRT.

Specialized instruments are essential in the troubleshooting and testing of electronic circuitry. However, if they are not used properly, great damage can be done to them and/or to the circuitry. Operators of these instruments must make sure they know how the instruments are to be connected, and they must make sure that the voltage to be tested is not above the range of the instrument. They must also be aware of the effects the test instrument has on the circuit itself. It has been pointed out that a common voltmeter will often produce effects that make the readings erroneous. The same can be true for oscilloscopes or other instruments.

Many electronic test instruments have been developed for special purposes. Operators of any instrument must thoroughly understand all operating procedures before connecting the instrument to the circuit. They should always read and be sure they understand the manufacturer's instructions before using any test instrument.

REVIEW QUESTIONS

1. Describe the differences between digital and analog meters.
2. Give three general classifications of electric measuring instruments.
3. Explain the operation of a galvanometer.
4. Describe the d'Arsonval or Weston meter movement.
5. Compare the Weston movement with the dynamometer movement.
6. For what type of current measurement is the iron-vane movement most suitable?
7. Describe the type of pivot bearing most commonly used in electric instrument movements.
8. What is a taut-band movement?
9. Explain a meter sensitivity of 20,000 Ω/V.
10. What is the reason for using a very sensitive instrument to test voltages in a circuit?
11. When are meter shunts required?
12. Explain the difference between a meter circuit connected as an ammeter and one connected as a voltmeter.
13. What important precaution must be observed when connecting an ammeter in a circuit?
14. How must a voltmeter be connected with respect to a load in order to determine the voltage applied to the load?
15. A meter movement has a full-scale deflection at 1 mA and an internal resistance of 15 Ω. In this case what value should the multiplier resistor have if the movement is to be used in a voltmeter that has a range of 0 to 10 V?
16. If you wish to use the movement described in the foregoing question for an ammeter with a range of 0 to 100 A, what should be the value of the shunt resistance?
17. What are the design characteristics of an ohmmeter that is used to measure low resistance?
18. Describe a Hall-effect probe.
19. What is a common instrument used to measure low-frequency alternating current?
20. Explain the operation of a thermocouple.
21. What are frequency counter instruments used for?
22. What is the function of a wattmeter?
23. Why is the wattmeter capable of indicating power consumption?
24. What are inductive pickup-type meters?
25. Describe a multimeter.
26. List and explain the important operations to be performed before connecting a multimeter for a test.
27. What is an oscilloscope?
28. What is the purpose of the trigger circuit in an oscilloscope?

Electric Motors 9

An electric motor is a device that changes electric power to mechanical energy. Electric motors can be classified in many ways; the number of different types of motors is so great, however, that it would be impossible to describe them with simple classifications. There are a few basic features that are common to all dc motors, and these will help to indicate the type of motor to be used for a specific purpose. DC motors are described in part by the type of internal winding they have. There are **series-wound, shunt-wound,** and **compound-wound** motors, named according to the relationship between the field coil connections and the armature winding. Motors of all types are usually rated according to horsepower. Usually, the data plate will also show the voltage and amperage. Additional information on dc motors includes rpm, type of duty, and some other points descriptive of the motor design.

AC motors are classified according to horsepower, phase, operating frequency, and type of construction. Usually, the power factor is also stated. In any event, all the characteristics of a motor must be considered in making a selection for a particular duty.

Electric motors are used in aircraft and spacecraft for many purposes. Among the many units and systems requiring electric motors are engine starters, cowl flaps, intercooler or heat-exchanger shutter or control valves, heat-control valves, landing gear, flaps, trim tab, flight controls, fuel pumps, hydraulic pumps, vacuum pumps, controllable propellers, gyrostabilizing units, navigation devices, and tracking devices. It is not intended that this text cover the details of all electric motors, but a thorough exposition of motor theory and functions will be given. This should enable the student to work out an understanding of any installation encountered.

MOTOR THEORY

Magnetic Attraction and Repulsion

Electric motors utilize the principles of magnetism and electromagnetic induction. The repulsion of like magnetic poles and the attraction of unlike poles work together to produce the torque that causes a motor to rotate. In the chapter of this text describing generator theory, it was shown that if a conductor is moved across a magnetic field, a voltage will be induced in the conductor. In that case, the movement of the conductor in the field causes a current to flow. Conversely, if a current from an external source is passed through a conductor while it is in a magnetic field; the conductor will tend to move across the field; hence the flow of current causes a movement of the conductor (see Figure 9–1).

The direction in which a current-carrying conductor in a magnetic field tends to move may be determined by the use of the **right-hand motor rule.** This rule is applied as follows: *Extend the thumb, index finger, and middle finger of the right hand so that they are at right angles to one another, as shown in Figure 9–2. Turn the hand so that the index finger points in the direction of the magnetic flux and the middle finger points in the direction of the current flowing in the conductor. The thumb will then point in the direction of the conductor movement.*

The movement of a conductor in a magnetic field, as described above, is caused by the flux that the current produces around the current-carrying conductor. This flux reacts with the flux of the magnet, thus imposing on the conductor a force that causes it to move. A study of the fundamental principles of electricity will show that a coil of wire will have magnetic polarity when a current is passed through it, and if a soft-iron

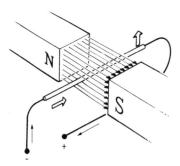

FIGURE 9–1 Current-carrying conductor in a magnetic field.

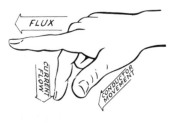

FIGURE 9–2 Right-hand motor rule.

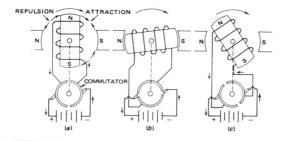

FIGURE 9–3 Principle of the dc motor.

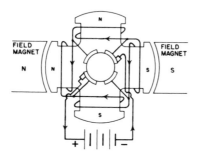

FIGURE 9–4 DC motor with a four-pole armature.

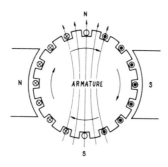

FIGURE 9–5 Cross-section diagram of an armature.

core is placed in the coil, the result is an electromagnet. If this electromagnet is placed between the poles of a field magnet and is free to rotate, the flux of the electromagnet will react with the flux of the field magnet and produce torque, which will cause the electromagnet to turn (see Figure 9–3a). The north pole of the electromagnet is attracted by the south pole of the field magnet and repelled by the north pole. The electromagnet, called the armature, will continue to rotate until it is lined up with the field. At this point it would normally stop because the conditions of repulsion and attraction would be satisfied. In an electric motor, however, the armature polarity is reversed at this point through the action of the **commutator,** a switching device that reverses the connections to the input current (see Figure 9–3b). It will be noted that the flux reversal takes place just before the armature becomes aligned with the field, thus causing the armature to continue to rotate as it attempts to line up with the new conditions (Figure 9–3c). The flux reversal in the armature takes place each time the armature becomes nearly aligned; hence it continues to rotate for as long as electric energy is applied.

A simple motor of the type described above does not de-

liver a smooth flow of power, because the torque is high when the armature is at right angles to the field poles, and there is no torque at the moment the armature is in line with the field poles. In order to have the motor deliver smooth power, the armature is provided with additional coils so that there will always be a high torque. Figure 9–4 shows a motor with four armature poles. With this arrangement the torque on one set of poles will increase as the torque on the other set decreases, and the motor will deliver a reasonably smooth power output. The addition of more coils will provide still smoother power. This is normally accomplished by winding the coils around the armature through oppositely positioned slots (see Figure 9–5). If the motor had four field poles, the sides of the armature coils would be spaced a distance of one-fourth the circumference of the armature.

The action of a drum-wound armature in a magnetic field is illustrated in Figure 9–5. This diagram represents a cross section of an armature with the conductors in the armature slots shown as small circles. The cross in a small circle indicates current flowing away from the observer, and the dot indicates current flowing toward the observer. By applying the right-hand motor rule, the direction of the torque on the armature can be determined. For example, the current on the left of the armature is flowing into the page, and the field flux is from left to right. The right-hand rule then indicates that the conductor would move up through the field. On the opposite side of the armature, the conductor would move down through the field, and the armature would turn in a clockwise direction.

Another method for finding the direction of torque is to determine the direction of the armature flux. This can be done by using the left-hand rule for electromagnets. If the left hand is held with the fingers pointing in the direction of the current around the armature, the thumb will be pointing upward. This, of course, indicates that the north pole is at the top of the armature. Since the north field pole is at the left, the top of the armature will be repelled and will move to the right. Since the south field pole is at the right, it will attract the north pole of the armature and add to the force that moves the top of the armature to the right. In a similar manner the bottom of the armature is moved to the left.

The action of the commutator continually switches the input current to new sections of the armature winding so that the top of the armature is always a north pole; hence the armature continues to rotate in an effort to align itself with the field poles.

Counter EMF and Net EMF

It has been pointed out in previous discussions that a conductor moving across a magnetic field will have an emf induced within itself. Since the conductors in the armature of a motor are cutting across a magnetic field as the armature rotates, an emf is produced in the conductors, and this emf opposes the current being applied to the armature from the outside source. This induced voltage is called **counter emf,** and it acts to reduce the amount of current flowing in the armature. The **net emf** is the difference between the applied emf and the counter emf.

FIGURE 9–6 Essential parts of a dc motor.

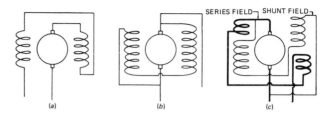

FIGURE 9–7 Schematic diagrams of different types of dc motors.

Counter emf plays a large part in the design of a motor. Motors must be designed to operate efficiently on the net emf, which is only a fraction of the applied emf; hence the resistance of the armature coils must be relatively low. Before a motor gains speed, the current through the armature is determined by the applied emf and the armature resistance. Since the armature resistance is low, the current is very high. As the speed of the motor increases, the counter emf builds up and opposes the applied emf, thus reducing the current flow through the armature. This explains the facts that there is a large surge of current when a motor is first started and that the current then rapidly falls off to a fraction of its initial value.

With some electric motor installations, the starting current is so high that it would overheat and damage the wiring or the armature, and so resistance must be inserted into the circuit until the motor has gained speed. The resistance may be automatically cut out as the speed of the motor increases, or it may be controlled manually.

An armature winding, a field, a commutator, and brushes are essential parts of a dc motor (see Figure 9–6). The field may be produced by a permanent magnet or by an electromagnet. The electromagnetic field is almost always employed because such a field lends itself to a wide range of operation. The armature receives current through the commutator and brushes and becomes an electromagnet. The magnetic forces of the armature and field interact to produce a rotational force.

Types and Characteristics of DC Motors

DC motors are series-wound, shunt-wound, or compound-wound, depending on the arrangement of the field windings with respect to the armature circuit (see Figure 9–7).

In a **series motor** the field coils are connected in series with the armature, as shown in Figure 9–7a. Since all the current used by the motor must flow through both the field and the armature, it is apparent that the flux of both the armature and the field will be strong. The greatest flow of current through the motor will take place when the motor is being started; hence the starting torque will be high. A motor of this type is very useful in installations in which the load is contin-

ually applied to the motor and in which the load is heavy when the motor starts. In aircraft, series motors are used to operate engine starters, landing gear, cowl flaps, and similar equipment. In each case the motor must start with a fairly heavy load; the high starting torque of the series motor is particularly well suited to this condition.

If a series motor is not connected mechanically to a load, the speed of the motor will continue to increase for as long as the counter emf is substantially below the applied emf. The speed may increase far above the normal operating speed of the motor, and this may result in the armature flying apart because of the centrifugal force developed by the rapid rotation. A series motor should always be connected mechanically to a load to prevent it from "running away."

The reason for the increase in speed when a series motor is not driving a load can be understood if the behavior of the field in such a motor is considered. As the speed of the motor increases, the counter emf increases. As the counter emf increases, however, the field current decreases. Remember that the field is in series with the armature and that since the counter emf causes the armature current to decrease, it must necessarily cause a decrease in the field current. This weakens the field so that the counter emf cannot build up sufficiently to oppose the applied voltage. A current continues to flow through both the armature and the field, and the resulting torque increases the armature speed still further.

In a **shunt motor** the field coils are connected in parallel with the armature (see Figure 9–7b). The shunt field must have sufficient resistance to limit the field current to that required for normal operation because the counter emf of the armature will not act to reduce the field current. Since the voltage applied to the field at operating speed will be practically the same as the voltage applied to the motor as a whole, regardless of counter emf, the resistance of the field must be many times the resistance of the armature. This is usually accomplished by winding the field coils with many turns of fine wire. The result of this arrangement is that the motor will have a low starting torque because of a weak field. The reason for the weak field is that the armature, owing to its low resistance, draws most of the current when the motor is first starting.

As the armature of a shunt motor gains in speed, the armature current will decrease because of counter emf, and the field current will increase. This will cause a corresponding increase in torque until the counter emf is almost equal to the applied emf, at which time the motor is operating at its normal speed. This speed is almost constant for all reasonable loads.

When a load is applied to a shunt motor, there is a slight reduction in speed. This causes the counter emf to decrease and the net emf across the armature to increase.

Since the resistance of the armature is low, a slight rise in net emf will cause a comparatively large increase in armature current, which in turn increases the torque. This prevents a further decrease in speed and actually holds the speed to a point only slightly less than the no-load speed. The current flow increases to a level sufficient to hold the speed against the increased load. Because of the ability of the shunt motor to maintain an almost constant speed under a variety of loads, it is often called a **constant-speed motor.**

Shunt motors are used when the load is small at the start and increases as the motor speed increases. Typical loads of this type are electric fans, centrifugal pumps, and rotary inverters.

When a motor has both a series field and a shunt field (Figure 9–7c), it is called a **compound motor.** This type of motor combines the features of series and shunt motors; that is, it has a strong starting torque like the series motor but will not overspeed when the load is light. This is because the shunt winding maintains a field that allows the counter emf to increase sufficiently to balance the applied emf. When the load on a compound motor is increased, the speed of the motor will decrease more than it does in a shunt motor, but the compound motor provides speed that is sufficiently constant for many practical applications.

Compound motors are used to operate machines subject to a wide variety of loads. In aircraft they are used to drive hydraulic pumps, which may operate from a no-load condition to a maximum-load condition. Neither a shunt motor nor a series motor would satisfactorily fulfill these requirements.

MOTOR CONSTRUCTION

Characteristics of Aircraft Electric Motors

The power-to-weight ratio of aircraft electric motors must be high; that is, a small motor must deliver a maximum amount of power for a minimum of weight. A commercial motor may weigh as much as 100 lb/hp [60.8 kg/kW], but for aircraft purposes there are motors that weigh less than 5 lb/hp [3 kg/kW]. Reduced weight is attained by operating the motors at high speeds and high frequencies and with relatively high currents. This necessitates the use of heat-resistant insulation and enamels in the armature and field windings and perhaps ram air or cooling fans to help dissipate the motor's heat.

Some fractional-horsepower motors used in aircraft rotate at over 40,000 rpm [4138 rad/s] with no load and at about 20,000 rpm [2069 rad/s] with a normal load. Since horsepower means the rate of doing work, it is apparent that a motor turning at 20,000 rpm develops twice the power of a similar motor turning at 10,000 rpm [1035 rad/s]. To reduce the effect of centrifugal force on the armature of the motor rotating at a very high speed, the armature diameter is made to be relatively small compared with its length.

Continuous- and Intermittent-Duty Motors

Many electric motors used in aircraft are not required to operate continuously. Because the heat developed in a short time is not sufficient to cause any damage, a motor in this type of service is designed to deliver more power for its weight than a motor used for continuous service. If such a motor were used continuously, it would overheat and burn the insulation and thus become useless. Motors designed for short periods of operation are called *intermittent-duty motors,* and those that operate continuously are called *continuous-duty motors.* The type of duty for which a motor is designed is sometimes stated on the nameplate, and if it is not on the nameplate, the type of duty can be found in the manufacturer's specifications.

Reversible DC Motors

On light aircraft, motors used for the operation of landing gear, flaps, cowl flaps, and certain other types of apparatus must be designed to operate in either direction and are therefore called **reversible motors.** Reversible 28-V dc motors are also found on transport-category aircraft; they are used for controlling various fuel and hydraulic valve assemblies.

The voltage polarity applied to the field and armature windings of any motor will determine that motor's direction of rotation (clockwise or counterclockwise). To reverse the rotation of a dc motor containing an electromagnetic field, the polarity of the voltage applied to the field or the armature must be reversed. This will reverse the magnetic field of one of the two coils and change an attractive force into a repulsive force (or vice versa), hence reversing the motor's rotation. Reversing a motor by this method would require a complex external circuit such as that illustrated in Figure 9–8. A

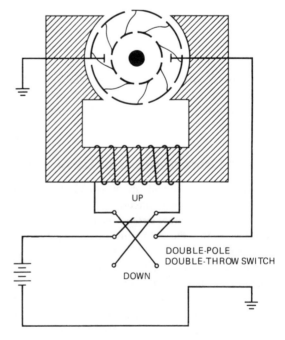

FIGURE 9–8 Reversing a dc motor with external switching.

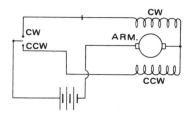

FIGURE 9–9 Circuit schematic for a reversible motor with two field windings.

simpler method is normally employed that provides a double field winding known as a **split field.** A schematic diagram of the circuit for a split-field motor is shown in Figure 9–9. Note that a separate circuit is provided for each field winding. This makes it possible to change the direction of the motor at will by placing the switch in the desired position. The motor is reversed by changing field polarity in relation to the armature polarity when the different field windings are energized.

Reversible dc motors are controlled directly by single-pole double-throw (SPDT) switches or indirectly by relays controlled by similar switches. The use of relays is dictated by the amount of current that the motor draws while in operation. Any motor requiring more than 20 to 30 A will operate more satisfactorily with a relay-controlled circuit.

The separate field coils of a reversible motor are usually wound either in opposite directions on the same pole or on alternate poles. Since the field coils are in series with the armature, they must be wound with wire of a size large enough to carry the entire motor current. Remember that the entire load current passes through both the field and the armature.

The brushes in a reversible motor are usually held in box-type holders in line with the center of the motor shaft. With this arrangement the brushes are perpendicular to a plane tangent to the commutator at the point of brush contact, and the brushes will wear evenly regardless of the direction of motor rotation. On small motors the field and brush housing is sometimes made in one piece. The brush holders are inserted through openings at the end of the housing and are insulated from the housing by composition bushings. Each brush assembly consists of the brush, a helical spring, a flexible connector inside the spring, and a metal contact. When a brush is installed in the motor, it is held in place by a screw plug (see Figure 9–10).

On some light-duty dc motors the field coil is replaced by a permanent magnet. To reverse the rotation of this type of motor, one need only reverse the polarity of the applied voltage. This will reverse the magnetic field of the armature (not the field); therefore, the motor will reverse its direction of rotation. Permanent-magnet reversible motors are commonly used to power light aircraft flap systems.

Brakes and Clutches

Many motor-driven devices used in aircraft must be designed so that the operated mechanism will stop at a precise point. For example, when landing gear is being retracted or extended, it must stop instantly when the operation is complete. If the driving motor is connected directly to the operated

mechanism, a great amount of strain will be imposed upon the motor when the mechanism is forced to stop. This strain is due to the momentum of the armature and other moving parts. In installations requiring an instantaneous stop, a brake and clutch mechanism is employed to prevent damage when the operating mechanism is stopped.

One type of brake mechanism for actuator motors is illustrated in Figure 9–11. This brake consists of a drum mounted on the armature shaft and internal brake shoes controlled by a magnetizing coil. The coil is placed inside the brake shoes, and when the motor current is turned off, the coil is de-energized and the brake shoes are forced against the drum by spring pressure. Conversely, when the power is turned on, the coil pulls the brake shoes away from the drum.

A disk-type brake, commonly used in actuator motors, consists of a rotating disk mounted on the armature shaft and a cork-lined braking surface on the stationary structure of the motor. A magnetizing coil is used to release the brake when the motor is energized, and a spring engages the brake when the current to the motor is turned off. A small amount of end play is allowed in the armature assembly mounting to provide clearance when the brake is released. When the brake coil is energized, the entire armature assembly moves slightly in a direction such that the brake disk will move away from the braking surface. When the current is turned off, a spring

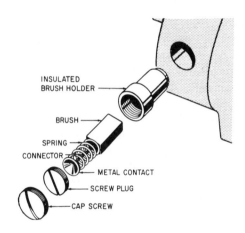

FIGURE 9–10 Brush and holder assembly for a small motor.

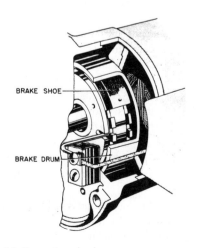

FIGURE 9–11 Drum-type brake.

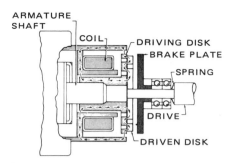

FIGURE 9–13 Brake and clutch assembly.

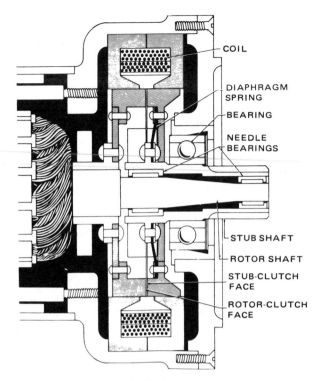

FIGURE 9–12 A magnetic clutch.

moves the assembly in the opposite direction, and the friction produced between the brake disk and the cork-lined braking surface causes the armature to stop very quickly.

Clutches of several types have been designed for the purpose of disengaging the motor from the load when the power is cut off. All such clutches are engaged by magnetic attraction when the power is turned on and disengaged by spring action. A typical magnetic clutch is shown in Figure 9–12. Two clutch faces are located within the clutch coil. One of the faces is mounted solidly on the armature shaft, and the other is connected through a diaphragm spring to the drive mechanism. When the clutch coil is energized, the two faces are magnetized with opposite polarities; hence, they are drawn together firmly. The friction thus produced causes the driven mechanism to turn with the motor. When the power is cut off, the diaphragm spring separates the faces, thus disengaging the motor.

Some actuator motors are provided with a combination brake and clutch. A magnetizing coil is located in the end of the motor housing, as shown in Figure 9–13. This coil, when energized, magnetizes a driving disk attached to the armature shaft. A driven disk is keyed to the output shaft, and when power is turned on, this disk moves against spring pressure until it engages with the driving disk. When the current is cut off, the driven disk is pulled away from the driving disk by the spring and is pressed against the brake plate at the opposite face, thus causing the driven mechanism to stop immediately.

Motors subject to sudden heavy loads are usually equipped with an overload release clutch. A clutch of this type is called a **slip clutch,** and its function is to disconnect the motor from the driven mechanism when the load is great enough to cause damage. This clutch consists of two groups of disks, alternately arranged, with one group splined to the motor drive and the other group splined to the motor-driven mechanism. These disks are pressed together by one or more springs designed to create sufficient pressure to cause the disks to rotate as one unit when the load is normal. When the load is excessive, the disks slip, thus preventing damage due to excessive torque.

Limit Switches and Protective Devices

Because of the limited distance of travel permitted in the driven mechanism, reversible actuator motors are usually limited in their amount of rotation in each direction. It is essential, therefore, that the motor circuits be provided with switches that will cut off the power when the driven mechanism has reached the limit of its travel. Switches of this type are called **limit switches** and are actuated by cams or levers linked or geared to the driven mechanism. The adjustment of these switches is critical because severe damage may result if the motor continues to run after the limit of operation is reached. Stripped gears and broken shafts are often the result of improperly adjusted limit switches. If the driven mechanism is strong enough to withstand the torque imposed by the motor, the fuse or circuit breaker in the motor circuit will usually cut off the current to the motor.

Adjustment of each of the limit switches is accomplished by running the motor to the limit of travel and then adjusting the switch-actuating mechanism so that it has just opened the switch. The switches should be adjusted to open slightly before the extreme limit is reached.

Some actuator motors are provided with a thermal circuit breaker, or **thermal protector,** which is designed to protect a motor from overload and excessive heat. This device is mounted on the motor frame, and when heat reaches a predetermined limit, the circuit breaker will open and cut off the current to the motor. After the motor has cooled sufficiently, the circuit breaker will automatically close, thus permitting normal operation.

Figure 9–14 is a schematic diagram of a reversible-motor circuit with a thermal protector and a coil for operating the clutch and brake. A circuit of the type shown would be used for operating cowl flaps, oil cooler shutters, air valves, and a variety of other devices. Both of the limit switches shown in Figure 9–14 are normally closed. Since they open only when the motor has reached the limit of travel in one direction or

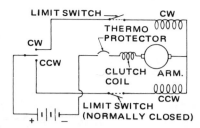

FIGURE 9–14 Schematic diagram of a reversible-motor circuit.

the other, it is readily apparent that there will never be a time when both switches are open. Notice that the thermal circuit breaker and the clutch coil are both in the ground (negative) side of the circuit and will therefore be in operation for either direction of travel.

DC Motor Construction

An exploded view of a typical dc actuator motor is shown in Figure 9–15. The principal sections of the motor assembly are the armature, the field coils and field frame, the brake as-

sembly, and the thermal-protector assembly. The armature is a standard drum type wound on a laminated soft-iron core. Also mounted on the armature shaft are the commutator at one end and the brake-lining disk at the other.

The field for the motor is provided by two poles formed to fit around the armature with a clearance of about 0.01 in. [0.025 cm]. The field coils are double-wound to provide for the reversal of field polarity necessary to reverse the motor rotation. Thermal protectors are connected in the circuit for each field (see Figure 9–16).

The brake assembly consists of a coil, a brake armature, and a brake lining mounted on the lining disk on the motor armature. The brake armature is a disk held in place by the motor studs, which pass through slots on the outer periphery of the armature. When the motor is not energized, the brake armature is held against the brake lining of the motor armature by a coil spring. This prevents the motor from turning. When the motor is energized, the magnetic brake coil draws the brake armature away from the brake lining, and the motor is free to turn.

Both ac and dc actuator motors have been manufactured in very small sizes for use in aircraft and missiles. Figure 9–17

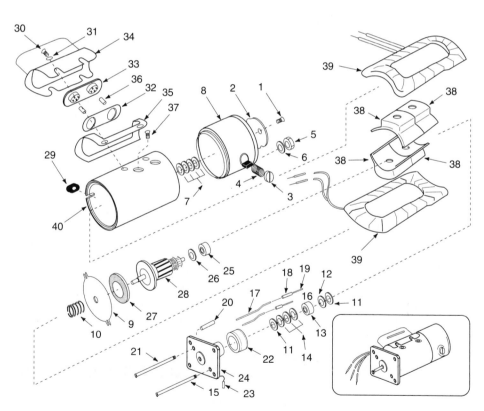

FIGURE 9–15 Exploded view of a dc actuator motor. (1) Screw, (2) name plate, (3) brush retainer, (4) brush assembly, (5) nut, (6) washer, (7) shim washers, (8) motor cap, (9) brake armature, (10) brake armature spring, (11) spacer washer, (12) shim washer, (13) ball bearing, (14) shim washers, (15) motor assembly stud, (16) insulating sleeve, (17) wire, (18) insulating sleeving, (19) brush connector, (20) insulating sleeving, (21) motor assembly stud, (22) brake-coil assembly, (23) base-registering pin, (24) motor-base assembly, (25) ball bearing, (26) shim washer, (27) brake lining, (28) armature assembly, (29) motor lead grommet, (30) thermal-protector-case screw, (31) washer, (32) thermal-protector retainer, (33) thermal protector, (34) thermal-protector case, (35) thermal-protector gasket, (36) insulating sleeving, (37) field-pole screw, (38) field pole, (39) field winding, (40) motor housing.

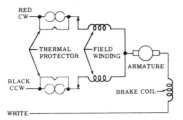

FIGURE 9–16 Thermal protectors wired in a motor circuit.

FIGURE 9–17 An electrically driven linear actuator. (*AiResearch Division, Garrett Corporation*)

is a photograph of an actuator assembly that can be operated by either a dc or an ac motor. Even though the actuator and motor assembly is very small, it can exert tremendous force.

Starter Motor

A typical direct-cranking starter motor for small aircraft engines is shown in Figure 9–18. The armature winding is of heavy copper wire capable of withstanding very high amperage. The windings are insulated with a special heat-resistant enamel, and after they are placed in the armature, the entire assembly is double-impregnated with a special insulating varnish. The leads from the armature coils are crimped in place in the commutator bars and then soldered with high-melting-point solder. An armature constructed in this manner will withstand the severe loads imposed for brief intervals while the engine is being started.

The field-frame assembly is of cast-steel construction, with the four field poles held in place by countersunk screws threaded into the pole pieces. The pole pieces are closely fitted to the inside contour of the field frame to provide the best possible magnetic circuit, because the field frame carries the magnetic flux from one field to the others. In other words, the field frame acts as a conductor for the magnetic lines of force; hence it is a part of the magnetic circuit from the field poles. Since a motor of this type is series-wound, the field windings must be of heavy copper wire of a size sufficient to carry the high starting current.

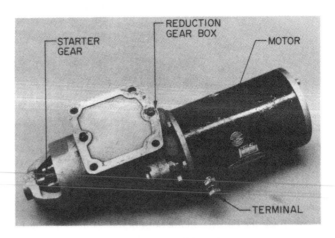

FIGURE 9–18 Typical starter motor for light aircraft.

An exploded view of a starter motor and drive is shown in Figure 9–19. This complete assembly consists of six major components. These are the **commutator end head assembly,** the **armature,** the **frame-and-field assembly,** the **gear housing,** the **Bendix-drive assembly,** and the **pinion housing assembly.**

The gear cut on the drive end of the armature shaft extends through the gear housing, where it is supported by a roller bearing. The gear mates with the teeth of the reduction gear that drives the Bendix shaft. The shaft is keyed to the reduction gear, and the Bendix drive is held in position on the shaft by a roll pin. The shaft is supported in the gear housing by a closed-end roller bearing and in the pinion housing by a graphitized bronze bearing.

When the armature turns the reduction gear, the Bendix-drive pinion meshes with the flywheel ring gear on the engine. This is done because of the inertia of the Bendix reduction gear; that is, when the armature begins to turn, the reduction gear is still at rest. This creates a relative motion between the reduction gear and the armature. Since the gear is mounted on a "threaded" shaft, the gear moves along the threads as the armature rotates and the starter engages into the engine flywheel gear.

A detent pin engages in a notch in the screw threads, which prevents demeshing if the engine fails to start and the starting circuit is de-energized. When the engine starts and reaches a predetermined speed, centrifugal force moves the detent pin out of the notch in the screw shaft and allows the pinion to demesh from the flywheel gear.

AC MOTORS

Theory of Operation

The basic principles of magnetism and electromagnetic induction are the same for ac and dc motors, but the application of the principles is different because of the rapid reversals of direction and changes in magnitude characteristic of alternating current. Certain characteristics of ac motors make most types more efficient than dc motors; hence such motors are used commercially whenever possible. During recent years

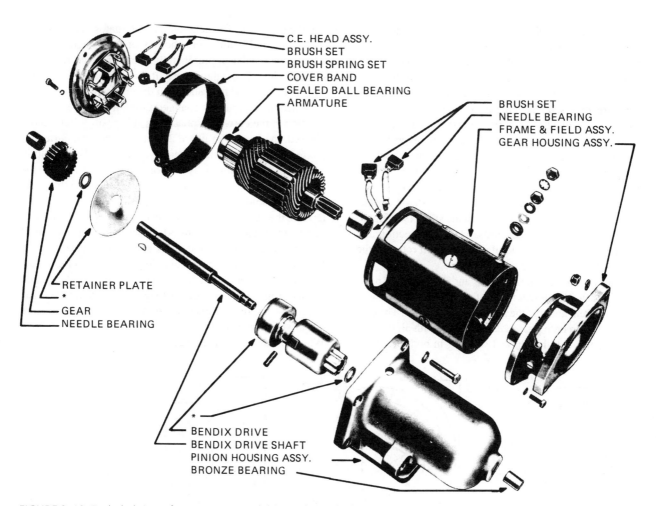

FIGURE 9–19 Exploded view of starter motor and drive. *(Prestolite)*

Labels in figure:
C.E. HEAD ASSY.
BRUSH SET
BRUSH SPRING SET
COVER BAND
SEALED BALL BEARING
ARMATURE

BRUSH SET
NEEDLE BEARING
FRAME & FIELD ASSY.
GEAR HOUSING ASSY.

RETAINER PLATE
*
GEAR
NEEDLE BEARING

*
BENDIX DRIVE
BENDIX DRIVE SHAFT
PINION HOUSING ASSY.
BRONZE BEARING

ac power systems have been developed for large aircraft with the result that a much larger amount of electric power is available on aircraft than would be available with dc systems of the same weight. Thus one of the main advantages of the ac power system is that it provides more power for less weight.

There are three principal types of ac motors. These are the **universal** motor, the **induction** motor, and the **synchronous** motor. There are many variations of these types, including combinations of features to meet different requirements. Among such motors are repulsion motors, split-phase motors, capacitor motors, and synchronous motors that utilize induction principles for starting torque.

A **universal motor** is identical with a dc motor and can be operated on either alternating or direct current. Since the direction of current flow in the field and the armature changes simultaneously when alternating current is applied to a universal motor, the torque continues in the same direction at all times. For this reason the motor will turn steadily in one direction regardless of the type of current applied. Typical universal motors are those used in vacuum cleaners, small electric appliances, and electric drill motors. Universal motors are not used in aircraft electric systems, because the alternating current has a frequency of 400 Hz, and at this frequency very substantial energy losses occur in a universal motor.

The **induction motor** has a wide variety of applications because of its operating characteristics. It does not require special starting devices or excitation from an auxiliary source and will handle a wide range of loads. It is adaptable to almost all loads when an exact and constant speed is not required.

The essential parts of an induction motor are the **rotor** and the **stator.** The stator is in the form of a shell with longitudinal slots on the inner surface (see Figure 9–20). The stator

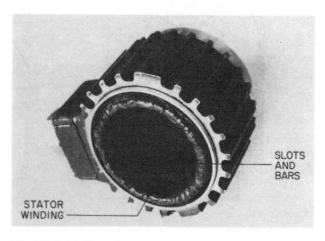

SLOTS
AND
BARS

STATOR
WINDING

FIGURE 9–20 Stator for an ac motor.

windings are placed in these slots in a manner similar to the placement of the winding of a dc armature.

If a source of direct current is connected to two terminals of a stator winding, it will be found that sections of the interior surface of the stator have a definite polarity. If the dc connections are reversed, the polarity of the stator will also reverse. When an alternating current is applied to the connections of the stator, the polarity of the stator will reverse twice each cycle. Ordinarily, the stator of an ac motor has two or three separate windings, depending on whether the motor is designed for two-phase or three-phase current. When multiphase currents are applied to the windings of a stator, a rotating magnetic field is established within the stator (see Figure 9–21). As the current in each phase changes direction and magnitude, the combined field of the stator will rotate at the frequency of the alternating current.

If we study carefully the diagrams and the graph for position in Figure 9–21, we will find that phase 1 is positive with maximum current and that the stator field is vertical. The current is negative in both phase 2 and phase 3, with all the current flowing through the phase-1 winding in both the stator

and the generator. The generator, or alternator, is represented by the inverted Y coils at the bottom of each diagram. In position 1 we see that approximately one-half the current flows through phase 2 and the other half through phase 3. This results in the vertical field shown in the diagram.

When the current has changed through an angle of 30° to position 2, the current in phase 1 is still positive but has decreased, the current in phase 2 is at zero, and the current in phase 3 has increased in the negative direction. This results in a field produced entirely by the poles of phases 1 and 3 in the stator, and the position of the field is 30° clockwise from vertical. If we study the diagrams for positions 3 and 4 and determine the current flow through each phase winding, we will find that the stator field turns 30° farther for each position. If the current values are plotted for a complete cycle, we will find that the field rotates through 360° for each cycle.

The rotor in an induction motor consists of a laminated iron core in which are placed longitudinal conductors (see Figure 9–22). In a **squirrel-cage** rotor these conductors are usually copper bars connected together at the ends by rings. When this assembly is placed in the rotating field produced

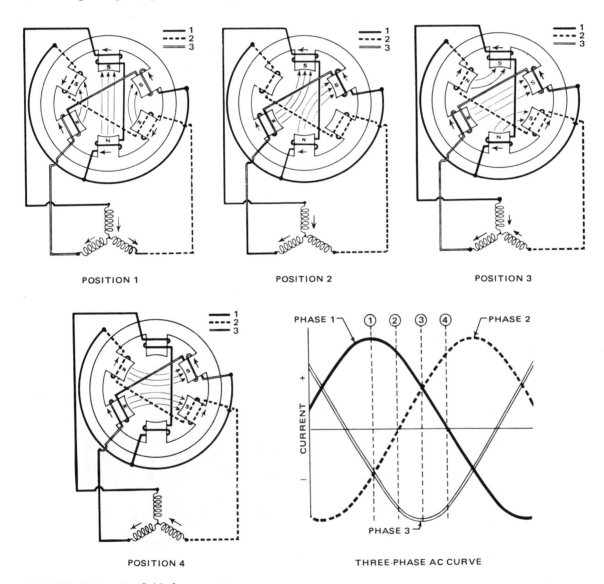

FIGURE 9–21 Rotating field of an ac motor.

FIGURE 9–22 Squirrel-cage rotor.

by the stator, a current is induced in the conductors. Since the conductors are short-circuited, there is a flow of current from those on one side of the rotor, through the rings at the ends of the rotor, to the conductors on the other side. This current produces a magnetic field that is at an angle to the field of the stator. If the rotor field came into line with the stator field, there would be no torque; hence the rotor field must always be a few degrees behind the stator field. The percentage of difference in the speeds of the stator and rotor fields is called the **slip.** It must be emphasized that this slip is absolutely necessary. The only field provided initially by the input of current to the motor is the field produced by the stator. The rotor has no electrical connection with the external power, and the only way it can produce a field is by having current induced within itself as the flux of the rotating stator field cuts across it. The interaction of the rotor field with the stator field then produces the torque that causes the rotor to turn.

When the motor is mechanically connected to a load, the load tends to slow the rotation of the rotor. This causes the slip to increase, and the rotor conductors cut a greater number of lines of force per time interval, thus increasing the rotor current and the rotor field. This stronger field produces an increased torque, which enables the motor to carry the increased load.

There is another effect that must be considered when a load is applied to an induction motor. This is a lowering of the power factor caused by the inductive reactance of the rotor. When the rotor is turning at almost synchronous speed, that is, the speed of the stator field, the frequency of the rotor current and the inductive reactance of the rotor are low. As the load is applied to the motor, the slip increases, and there is a corresponding increase in the frequency of the rotor current. This increases the inductive reactance of the rotor, and the power factor of the motor consequently decreases. It will be remembered that the power factor is equal to the cosine of the phase angle between the voltage and the current and that inductive reactance increases this phase angle. In order to maintain system efficiency, motors must be designed to keep the phase angle to a minimum.

When the load on an induction motor becomes so great that the torque of the rotor cannot carry it, the motor will stop. This is called the **pull-out point.**

Improvement of Starting Qualities

An induction motor will start satisfactorily under no load without any special starting devices. However, when such a motor is connected directly to a substantial load that must be moved when the motor starts, it is usually necessary to add resistance in the rotor circuits. There are several methods for accomplishing this, but the explanation of one method is sufficient for this text.

From the study of alternating current, it will be remembered that the power factor for alternating current flowing in a purely resistive circuit is 100 percent. On the other hand, alternating current flowing in a purely inductive circuit would have a power factor of 0 percent if such a circuit were possible. Therefore, the addition of resistance to an inductive circuit will have the effect of improving the power factor. To add the necessary resistance to a rotor circuit for starting purposes, two squirrel-cage windings are used. One of these windings is of copper and has low resistance, and the other is of German silver and has high resistance. When the starting current is applied to the motor, the high-resistance winding produces the starting torque because of its high power factor. As the rotor gains speed, the effect of the high-resistance winding decreases, and the effect of the low-resistance winding increases. When the motor is operating at normal speed, it has the advantage of a low-resistance rotor winding.

Split-Phase Motors. Single-phase induction motors have no torque when the rotor is at rest; hence it is necessary to incorporate into them devices for providing a starting torque. This can be accomplished by providing the motor with two separate windings and using an inductor or a capacitor to change the phase of the voltages applied to the different windings. This is known as **phase splitting.** A motor having devices for this purpose is called a **split-phase motor.** Figure 9–23 shows a motor circuit in which a capacitor is used to cause the current in one winding to lead the current in the other winding. This, in effect, causes the motor to act as a two-phase motor during starting.

Split-phase motors of the capacitor type are used extensively in industry for low-power applications such as drill presses, grinders, small lathes, and small saws. In large aircraft the split-phase motor is used as an actuator for various types of comparatively small loads, such as small blower fans.

As shown in Figure 9–23, a capacitor is often used in order to provide starting torque. When the motor has attained a certain rpm, a centrifugal switch opens and cuts out the capacitor circuit. Motors that employ a capacitor for starting or for continuous operation are often called **capacitor motors.**

Repulsion Motors

A **repulsion motor** utilizes the repulsion of like poles to produce the torque for operation. The rotor is wound like an armature and employs a commutator and brushes. The brushes are short-circuited across the commutator at an angle that causes the induced current in the windings to produce a polarity in the rotor that will be in opposition to the polarity of the stator. That is, a north pole produced in the rotor will be near a north pole in the stator. The rotor is therefore caused to rotate because of the repulsion between the like poles. As the rotor turns, the brushes on the commutator remain in the same position, and so the polarity of the rotor remains in the

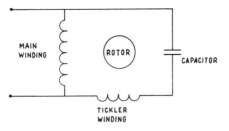

FIGURE 9–23 Circuit for a capacitor motor.

same position, even though the rotor turns. The repulsion principle is used in some motors to provide starting torque, after which the motor operates as an induction motor.

Synchronous Motors

Synchronous motors, as the name implies, rotate at a speed that is synchronized with the applied alternating current. These motors have some features in common with induction motors and a construction similar to that of alternators. The stator consists of a laminated soft-iron shell with coils wound through slots on the inner surface. A three-phase synchronous motor has three separate windings in the stator and produces a rotating field like the stator of an induction motor. The rotor may be a permanent magnet in a very small synchronous motor, but in larger motors the rotor is an electromagnet excited by an external source of direct current.

The theory of operation of a synchronous motor is very simple. If a magnet is free to turn and is placed in a rotating field, it will align itself with the field and rotate at the same speed. If no load is placed on such a motor, the center of the rotor poles will be exactly in line with the center of the stator field poles. In practice this does not occur because of friction. Friction and load cause the center of the rotor poles to lag behind the center of the field poles formed by the stator. The angle between the rotor field and stator field is called the **lag,** and it increases as the load on the motor is increased. If the load becomes so great that it overcomes the magnetic reaction, the pull-out point is reached and the motor will stop. At this time the incoming current will increase to a short-circuit value, and the torque will become negligible.

When operating within its load limits, a synchronous motor will rotate at the same speed as the alternator supplying the current, provided that the alternator has the same number of poles as the motor. Since the speed of a synchronous motor depends entirely on the frequency of the current supply, such motors are useful when constant speeds and frequencies are desired. One of the common uses of synchronous motors is to change the frequency of alternating current. Since the motor will turn at a precisely constant speed, it can be used to drive an alternator through a differential gear system to provide an exact frequency of any desired value.

Synchronous motors are commonly used on airplanes in the electric tachometer. A three-phase alternator is connected to a drive on the engine, and the alternator output is connected to a synchronous motor in the tachometer indicator. (Alternators will be discussed in Chapter 11.) The frequency of the current is directly proportional to the engine speed;

hence the synchronous motor in the indicator will rotate at a speed proportional to engine speed. The indicating needle is coupled to the synchronous motor through a permanent magnet and drag cup. The distance that the needle moves along the rpm scale is proportional to the speed of the motor.

A synchronous motor differs from an alternator in that it has a high-resistance squirrel-cage winding placed in the rotor to give a good starting torque. This winding causes the motor to start as an induction motor and run as a synchronous motor. When the motor has reached synchronous speed, it is turning with the magnetic field, and the conductors of the squirrel-cage winding are not cutting lines of force. If the rotor tends to hunt or oscillate, however, the squirrel-cage winding will have an induced current, which tends to dampen the oscillations and prevent hunting.

Motor Losses

The efficiency of electric motors of any type is largely determined by the losses of power resulting from friction, resistance, eddy currents, and hysteresis. The power used to overcome the friction of bearings is called the **friction loss.** This loss may also include the loss due to wind friction, which is sometimes called **windage loss** and is comparatively high when a motor is equipped with a fan to provide cooling by forced ventilation. The power used to overcome the resistance of the windings is called **resistance loss,** or **copper loss.** Copper losses are dissipated in the form of heat.

The currents induced in the armature core and the field poles are called **eddy currents** and are responsible for considerable loss in the form of heat. These losses are reduced by constructing the armature and field cores of laminated soft iron, with the laminations insulated from one another.

Hysteresis losses occur when a material is magnetized first in one direction and then in the other in rapid succession. The effect of hysteresis is to cause the change in strength of the magnetic flux to lag behind the magnetizing force and is presumably due to the friction between the molecules of the material as they are shifted in direction by the magnetizing force. Hysteresis losses are noticeable because of their heating effect. Any condition that produces heat in a motor causes a loss of power, or energy, because heat is one of the principal forms of energy and requires power to produce it.

The construction of electric motors with laminated armatures and field-pole cores helps to solve cooling problems because much of the heat encountered during operation is the result of the losses described above. This type of construction is particularly important for high-speed actuator motors. Actuator motors must have a high power-to-weight ratio, and to attain this it is necessary to operate them at relatively high speeds. For this reason all losses must be reduced to a minimum.

Single-Phase Reversible AC Motors

Single-phase reversible ac motors are found on transport-category aircraft and are used to drive valve assemblies and other relatively small actuators. If large amounts of mechan-

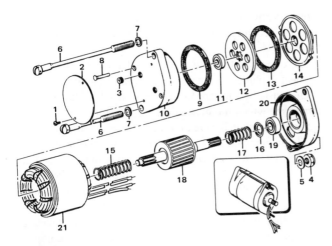

FIGURE 9–24 Exploded view of a single-phase reversible ac motor. (1) Identification-plate screw, (2) identification plate, (3) adjusting screw, (4) main assembly nut, (5) washer, (6) main assembly screw, (7) washer, (8) aligning pin, (9) brake lining, (10) end bell, (11) ball bearing, (12) break disk, (13) brake lining, (14) brake armature, (15) brake spring, (16) spring retainer washer, (17) compression spring, (18) rotor assembly, (19) ball bearing, (20) motor base, (21) stator assembly.

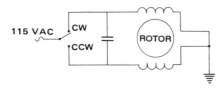

FIGURE 9–25 Circuit for a split-phase reversible ac motor.

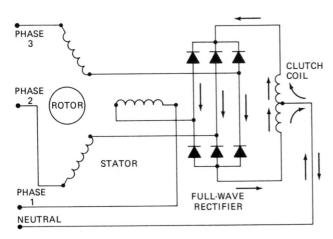

FIGURE 9–26 Circuit for a three-phase ac motor with a full-wave rectifier to provide direct current for the clutch coil.

ical energy are needed, three-phase motors are used. Single-phase reversible ac motors typically contain two stator windings. The motor's direction is determined by the current flow through the clockwise or counterclockwise winding. This can be regulated with a double-pole double-throw (DPDT) switch in the external circuit. The construction of a single-phase reversible ac motor consists principally of a squirrel-cage rotor, a double-wound stator, and a brake assembly (see

Figure 9–24). The core of the rotor is constructed of laminated soft iron. Slots are provided in the surface of the core for the copper bars that form the squirrel cage. At each end the copper bars are soldered or welded to copper rings.

The double-wound stator provides the split field that is necessary to establish torque for starting under load. The stator leads are brought to the outside of the motor, where they are connected to a capacitor, as shown in Figure 9–25. Note that when the control switch is placed in the clockwise position, current flow will be through the clockwise coil directly and through the capacitor to the counterclockwise coil. This causes the current in the counterclockwise coil to lead the current in the clockwise coil, thus creating a clockwise-rotating field. Conversely, when the switch is placed in the counterclockwise position, the rotating field is in a counterclockwise direction, and the motor turns accordingly.

Three-Phase AC Motors

Three-phase ac motors for aircraft are quite similar to three-phase induction motors of the commercial or industrial type. The principal difference is that the aircraft motor operates at a frequency of 400 Hz, thus making it possible to employ a motor of lighter weight for the same power output. Three-phase ac motors are used on large transport-category aircraft to drive hydraulic pumps, large blower fans, fuel boost and transfer pumps, and other systems requiring large amounts of mechanical energy.

The **three-phase induction motor** consists essentially of a three-phase Y-wound stator and a conventional squirrel-cage rotor. The three-phase stator produces a rotating field as explained previously, and this field induces a current in the rotor. The rotor current creates a field that opposes the stator field, with the result that the rotor attempts to turn at a speed that will keep it ahead of the stator field.

One type of three-phase actuator motor for aircraft is internally wired as shown in Figure 9–26. It will be noted that the motor has a Y-wound stator; however, the neutral connections from each phase winding are individually connected to three separate legs of a full-wave rectifier. The rectifier output is directed through a clutch coil that is split to accommodate the alternating current in the neutral line.

The effect of this type of internal circuit is to allow for a high surge of current when the motor is started and a comparatively low current as soon as the motor gains full operating speed. Since the rotor windings are in series with the clutch coil, the clutch coil will receive the benefit of the high starting current to engage the clutch. When the clutch is engaged, it requires a relatively low current to hold it in that position. This low current is the result of the inductive reactance developed by the stator windings and the clutch coil.

Since the variety of electric motors is extensive and many look exactly alike, even though they have different characteristics and specifications, the technician must be certain that a replacement motor in any system is identified by the correct part number. A motor installed in a system where the electrical characteristics do not match those of the motor will very likely be damaged and may cause damage to other elements of the circuit.

INSPECTION AND MAINTENANCE OF MOTORS

General Inspection Procedures

In Chapter 10, which discusses generators, instructions are given for the inspection and maintenance of generators. Many of the instructions apply to motors as well because of the similarities between the two.

Preflight inspections of motors are usually operational. The switches for the various motor-driven units are turned on, and if operation is satisfactory, no further inspection is made. It is obvious that landing-gear actuators cannot be tested in this manner, but if the latest pilot's report is satisfactory, only a visual inspection need be made.

Depending on the amount of operation to which a motor has been subjected, inspections should be carried out on it at intervals set forth in the manufacturer's operation manual. This type of inspection will probably include a check of the mounting, electrical connections, wiring, brushes, brush springs, and commutator. For ac motors it is not usually necessary to consider brushes and the commutator, because there are none, except in universal motors.

The construction of many small dc actuator motors makes it difficult to inspect the commutator without removal and disassembly. But because these motors are usually of the intermittent type, the wear on the commutator is negligible. A periodic inspection of the brushes and replacement, if necessary, will assure satisfactory operation until the time for overhaul. New brushes should be seated as outlined in Chapter 10, covering generator maintenance. If the construction of a motor makes it impossible to seat the brushes after they have been installed in the motor, they may be ground to the correct curvature with No. 000 sandpaper wrapped around a piece of wood or metal that has been turned to the diameter of the commutator. Usually, brushes for small motors can be ordered specifically for a particular model; the brush face will be already ground to the correct curvature. The seating of the brushes can be checked by removing the brushes from the motor after a few minutes of operation and examining the area that has been polished by the commutator.

Removal of the cap or cover band will give access to the brushes and holders. If a brush is held on a pivoted arm, it can be removed by lifting the arm and removing the brush screw. For a brush in a box-type holder, merely lift the brush spring and slide the brush out of the holder.

Removal and Installation

Because of the many different types of electric motors, specific instructions cannot be given here for their removal and installation. For any particular motor on an aircraft, the technician should consult the maintenance or overhaul manual supplied by the manufacturer.

When the removal of a motor is necessary, the technician must give due consideration to the driven mechanism. In many cases a gear-train assembly must be removed with the

motor. In any event, a brief visual inspection will usually enable the technician to determine the procedure to be followed.

Care must be taken to make sure that electric wiring and connector plugs are not damaged when a motor is disconnected. It is best to tape or otherwise insulate disconnected terminals, which might accidentally become short-circuited if the battery switch were to be inadvertently turned on.

If the removal of a motor leaves an opening through which dirt or other foreign matter can gain access to vital parts of a mechanism, the opening should be covered with a cloth or a plate. This is particularly important when removing a starter motor from an engine. If a nut, a bolt, or some other object should fall inside the engine, great damage may be caused, and the engine may require a complete overhaul.

The important points to be considered in the installation of a motor are as follows:

1. See that the mounting area is clean and properly prepared. Install the correct type of gasket if a gasket is required.
2. Be very careful not to cause damage when moving the motor into place. A nick or scratch in the mounting could develop into a crack and eventually cause failure.
3. Tighten screws or hold-down bolts evenly and with the correct torque. Make sure that nuts, bolts, or studs are properly safetied.
4. See that electrical connections are clean; then tighten and safety them as required.

Disassembly and Testing

The disassembly, inspection, overhaul, and assembly of an aircraft electric motor should be performed in accordance with the manufacturer's instructions. When performing maintenance, the following general rules apply:

1. Use the proper tools for each operation.
2. Mark and lay out parts in an order that will aid in assembly.
3. Do not use excessive force in any operation. If parts are stuck, determine the cause. If necessary, use a soft mallet to disengage parts. Sometimes parts are joined by means of metal pins or keys, which may be overlooked by the technician; be sure that such devices are removed before attempting to separate parts that have been joined in this manner.
4. When bearings are pressed on a shaft, or when they are stuck because of corrosion, use a bearing puller for removal.
5. Use an arbor press for the removal and installation of bearings and bushings that are press-fit. The use of this device is recommended, but if an arbor press is not available, a fiber tube that fits the inner or outer race of the bearing may be used.
6. Keep all parts of an assembly clean. The workbench should be free from dirt and grease. When greasy parts are removed, they should be cleaned in a nonrust solvent.

*CAUTION: Never use carbon tetrachloride for cleaning parts. The fumes from this solvent are very poisonous.

The testing of the parts of an electric motor is carried out in the same manner as tests for generator parts, as discussed in Chapter 10. A growler is used to test armatures for shorted or open coils. An ohmmeter or continuity tester is used to test for a ground between the armature windings and the core. Field coils can be tested with an ohmmeter or a continuity tester for open circuits, short circuits, and grounds.

After a motor has been assembled, it should be given an operational test before it is installed in an airplane. First, the armature should be turned by hand to see that it rotates freely; there must be no roughness or unusual noise when this is done. The motor should then be operated with a low load for about 10 min to seat the brushes. The value of the voltage applied should be according to overhaul specifications. During the time that the motor is being tested, it should be observed closely for excessive heating and vibration. Directions for testing specific motors are usually included with the manufacturer's maintenance and overhaul instructions. Motors should be disassembled only when these instructions are available.

REVIEW QUESTIONS

1. Define *electric motor*.
2. Describe series-wound, shunt-wound, and compound-wound motors.
3. What is the principal characteristic of a series-wound motor?
4. For what type of load would a series-wound motor be most suitable?
5. What is the principal characteristic of a shunt-wound motor?
6. Explain why a typical dc motor rotates when connected to a proper power source.
7. What determines the direction of rotation of a dc motor?
8. Why does the current being drawn by a shunt-wound motor decrease as the motor rpm increases?
9. What may happen to a series motor if it is connected to power without having a load?
10. What is the resistance requirement for the field winding of a shunt-wound motor?
11. What are the three principal types of ac motors?
12. Where are three-phase ac motors typically used on the aircraft?
13. Why does the rotor of an induction motor react with the field of the stator?
14. Describe a squirrel-cage rotor.
15. What is the pull-out point of an ac motor?
16. Explain how the starting torque of an induction motor may be improved.
17. What is motor slip?
18. What is the principal characteristic of a synchronous motor?
19. Name some of the internal motor losses that occur in the operation of an ac motor.
20. How is reduced weight attained in the design of motors for use in aircraft?
21. Why are heavy windings used in the armature of a dc starter motor?
22. Describe the field winding of a typical reversible motor.
23. Explain the operation of a magnetic drum brake assembly.
24. Why is it necessary to disengage an actuator motor from its drive when it is turned off?
25. Explain the adjustment of the limit switches in an actuator circuit.
26. List some of the typical precautions that must be observed in the installation of electric motors.
27. List general rules for the disassembly of an electric motor.

10 Generators and Related Control Circuits

Ever since the first aircraft to use any kind of electric equipment was launched, the electrical loads on airplanes and other flying devices have increased. Today modern jet airliners are equipped with scores of different electrical systems, each requiring a substantial amount of electric energy.

Generators were the first means of supplying electric power for aircraft. Currently, generators or generator derivatives called alternators are found in a wide variety of sizes and output capacities. A typical alternator used on a large commercial aircraft can produce over 6000 W of electric power. On multiengine aircraft, one generator is driven by each engine to allow for redundancy in the event of a generator failure.

An electric generator can be defined as a machine that changes mechanical energy into electric energy. On aircraft the mechanical energy is usually provided by the aircraft's engines. Light aircraft use 14- or 28-V dc generators. Large aircraft typically employ generators that produce an alternating current of 208 or 117 V at 400 Hz. Compared with a 28-V dc system, a higher-voltage ac system will develop several times as much power for the same weight; hence it is a great advantage to use ac systems where heavy electrical loads are imposed.

GENERATOR THEORY

Electricity is produced in a generator by electromagnetic induction. As explained in Chapter 1, it is a fundamental principle that when there is a relative movement between a magnetic field and a conductor held perpendicular to the line of flux, an emf is produced in the conductor. If the ends of the conductor are connected together, the emf will cause a current to flow, as shown in Figure 10–1. The direction of current flow is determined by the direction of the magnetic flux and the direction in which the conductor is moved through the flux.

A simple way to determine the direction of current flow is to use the **left-hand rule for generators.** *Extend the thumb, index finger, and middle finger so they are at right angles to one another, as illustrated in Figure 10–2. Turn the hand so the thumb points in the direction of movement of the conductor and the index finger points in the direction of the magnetic flux. Then the middle finger will be pointing in the direction of the current flow.* Remember, current flow is from negative to

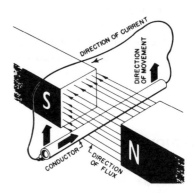

FIGURE 10–1 Generator action.

FIGURE 10–2 Left-hand rule for generators.

positive. Flux direction is considered to be from north to south.

Simple AC Generator

A simple ac generator can be constructed by placing a single loop of wire between the poles of a permanent magnet and arranging it so that it can be rotated as shown in Figure 10–3. The current is taken from the wire loop by means of brushes, which make continuous contact with the collector rings (slip rings). One collector ring is connected to each end of the wire loop. In Figure 10–3, the sides of the loop are designated *AB* and *CD*. As the loop rotates in the direction indicated by the arrow, side *AB* will be moving up through the magnetic field. If we apply the left-hand rule for generators, we find that a voltage is induced that will cause current to flow from *A* to *B* in one side of the loop and from *C* to *D* in the other side of the loop. This is because *AB* is moving **up** through the field and *CD* is moving **down** through the field.

The voltage induced in the two sides of the loop add together and cause the current to flow in the direction *ABCD*, through the external circuit, and then back to the loop. As the

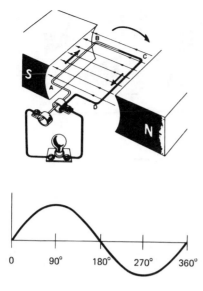

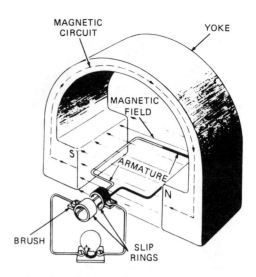

FIGURE 10–4 Essential parts of an ac generator.

FIGURE 10–3 A simple ac generator and associated voltage wave.

loop continues to rotate toward a vertical position, the sides will be cutting fewer lines of flux, and when it reaches the vertical position, the sides of the loop will not be cutting any lines of flux but will be moving parallel to them. At this position, no voltage is induced in the loop because a conductor must cut across flux lines in order to induce a voltage. By rotating the loop through the vertical position and back to the horizontal, a voltage will be induced again, but it will be in the opposite direction in the loop because side *AB* will now be moving **down** through the field and side *CD* will be moving **up** through the magnetic field. Soon the loop is once again in the vertical position, and no flux lines are being cut. When the loop is exactly perpendicular to the magnetic flux lines, no voltage is being produced. The current flow then repeats its cycle as long as the loop is rotated inside the magnetic field. The voltage waveform produced by this type of generator is called the sine wave.

By examining the sine wave of Figure 10–3, it can be seen that the voltage is at zero when the loop is in a vertical position, and then it climbs to a maximum value when the loop is in the horizontal position. This is indicated on the sine curve from 0 to 90°. As the loop continues to turn, we find that the voltage is maximum at 90°, zero at 180°, maximum at 270°, and zero again at 360°.

Essential Parts of a Sample AC Generator

The essential parts of a simple ac generator are shown in Figure 10–4. These are a **magnetic field,** which may be produced by a permanent magnet or by electromagnet field coils; a rotating loop or coil called the **armature** or **rotor; slip rings;** and **brushes** by which the current is taken from the armature. The poles of the magnet are called **field poles.** In most generators, these poles are wound with coils of wire called **field coils.** The path of the magnetic flux is called the magnetic circuit and includes the yoke connecting the field poles as well as the armature.

Value of Induced Voltage

The voltage induced in a conductor moving across a magnetic field depends on two principal factors: the strength of the field (the number of lines of force per unit area) and the speed with which the conductor moves across the lines of force. In other words, the voltage depends on the number of lines of force cut per second. For example, if a conductor cuts lines of force at the rate of 100,000,000 lines per second, an emf of 1 V will be established between the ends of the conductor.

Simple DC Generator

DC generators are needed for many aircraft electrical systems, for battery charging, and for various other applications. For this reason an ac generator will not meet all power requirements unless a means of rectifying the alternating current is provided. Figure 10–5a shows an elementary type of dc generator that is quite similar to the simple ac generator explained previously.

A pulsating direct current can be obtained from the illustrated generator by using a commutator in place of the slip rings found on the ac generator. A **commutator** is a switching device that reverses the external connections to the armature at the same time that the current reverses in the armature. The commutator in Figure 10–5a is a split ring that turns with the armature. One end of the rotating loop is connected to one half of the ring, and the other end of the loop is connected to the opposite half of the ring. The two sections of the commutator are insulated from each other. Two brushes are placed in a position relative to the commutator so that as the commutator turns, the brushes pass from one segment of the commutator to the other at the same time that the current is reversing; there is then practically no emf between the two segments. This system of changing the alternating current of the armature to direct current in the external circuit is called **commutation.**

Referring to Figure 10–5a, observe that the side of the loop moving up through the field will always be connected to

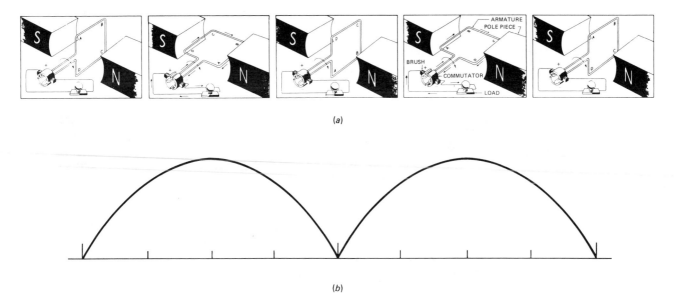

(a)

(b)

FIGURE 10–5 A simple dc generator and associated voltage wave.

the positive brush and that the side of the loop moving down through the field will always be connected to the negative brush.

As illustrated in Figure 10–5b, the current from the generator will then be traveling in one direction in the external circuit, but it will pulsate; that is, it will vary in intensity from zero to maximum and back to zero through each half turn of the armature. A current of this type is called a pulsating direct current and is not suitable for many uses.

Elimination of DC Ripple

Since the pulsating direct current of the simple dc generator is not satisfactory for all purposes, it is necessary to construct a generator that will produce an almost constant voltage. This is accomplished by increasing the number of coils in the ar-

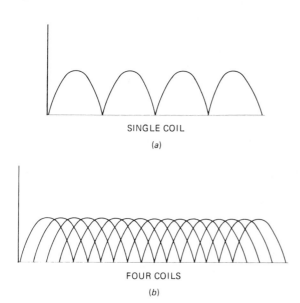

SINGLE COIL

(a)

FOUR COILS

(b)

FIGURE 10–6 Comparing voltages from a single-coil armature and a multiple-coil armature.

mature and/or the number of field coils. Figure 10–6a illustrates the nature of the voltage from a single-coil generator, and Figure 10–6b shows the curve for a generator with four armature coils. Notice the great difference in the nature of the voltage.

Armature coils are wound on a laminated soft-iron core. The iron core concentrates the field flux and greatly increases the voltage generated. The laminations reduce the effects of eddy currents induced in the core.

The current from any dc generator will have a slight pulsation known as **commutator ripple,** but this ripple does not interfere in ordinary electric circuits for purposes such as lighting and operating electric motors. For radio circuits, the commutator ripple must be eliminated because it causes a hum in the radio output. A capacitor of correct capacitance shunted across the dc power leads of a radio receiver will greatly reduce the amount of ripple. For a more effective filter, an inductance or choke coil is connected in series with the dc line along with the capacitor in parallel with the dc line. Remember that an inductance coil opposes any change in the current flow. A capacitor, a choke coil, or a combination of the two connected in a circuit to reduce ripple is called a **ripple filter.**

Residual Magnetism

An electromagnet will produce a much greater field strength per given size and weight than a permanent magnet. For this reason field coils are used to provide the *magnetism required* for the generation of current. In dc generators the field coils are usually energized by current from the generator. Fortunately, any substance that has been magnetized will retain a certain amount of magnetism. Materials such as soft iron give up most of their magnetism very quickly when removed from the magnetizing influence. However, they do retain a small amount, which is known as **residual magnetism.**

It is this residual magnetism that makes it possible to start a generator without exciting the field from an outside source of magnetism. The residual magnetism in the field poles causes a weak voltage to be generated when a generator begins to rotate. This small voltage is then used to power the generator's electromagnetic field, thus increasing the field strength.

The increase in field strength causes a corresponding increase in generator output voltage, and a mutual increase in field strength and voltage continues until the voltage reaches the proper value for the generator. If the residual magnetism should be lost because of excessive heat or shock, it can be restored to the field by passing a direct current through the field windings in the correct direction. This procedure is called **flashing the field** and is discussed in a later section of this chapter.

Characteristics of DC Generators

DC generators are classified as **shunt-wound, series-wound,** or **compound-wound,** according to the manner of connecting the field coils with respect to the armature.

The internal connections for a **shunt-wound generator** are shown in Figure 10–7, where it can be seen that the field coils are connected in parallel with the armature. In this type of generator, it is necessary to have a resistance or some other means of regulation in the field circuit to prevent the development of excessive voltage. If such a generator were to run without a load, the entire output would go through the field coils, thus producing a very strong field. This field would, of course, increase the voltage of the generator, and the field strength would also increase. The result would be a continued increase of both the field strength and the voltage until the generator burned out.

Shunt generators without field-current regulation are satisfactory only for operation at a constant speed and with a constant load. In practice, it is doubtful that such a generator could be used except experimentally.

A **series-wound generator** contains a field winding that is in series with respect to the armature winding. A diagram of a series-wound generator is shown in Figure 10–8. In this

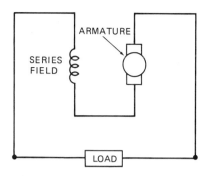

FIGURE 10–8 Diagram of a simple series-wound generator.

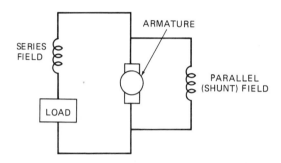

FIGURE 10–9 Diagram of a simple compound-wound generator.

type of generator the resistance of the load controls the field current. If the load resistance decreases because more electrical load is applied, the field current increases and the generator's output voltage increases. If the electrical load decreases (increasing resistance), the current through the load and the generator's field decreases and the generator's output voltage decreases. From these relationships it can be seen that an unregulated series-wound generator will not maintain a constant voltage output. Series-wound generators that are not regulated can be used in situations where a constant rpm and a constant load are applied to the generator, but they are not suitable for aircraft applications.

A **compound-wound generator** combines the features of the series- and shunt-wound generators. As illustrated in Figure 10–9, this generator contains a field winding in series and one in parallel with respect to the armature. In this type of generator, when the load increases (decreasing resistance), the series field current increases and the parallel field current decreases. The output voltage remains constant. If the load decreases (increasing resistance), the series field current decreases and the parallel field current increases. Once again, output remains constant.

In theory, the series-wound, shunt-wound, and compound-wound generators each have certain advantages and disadvantages. On modern aircraft, however, all generators contain some means of controlling the output voltage and current. The shunt-wound generator used in conjunction with a voltage regulator is the most common type of dc generator system for aircraft. The voltage regulator adjusts the current to the shunt field in order to maintain the necessary output under a variety of rpm and load conditions. Voltage regulator circuits will be discussed later in this chapter.

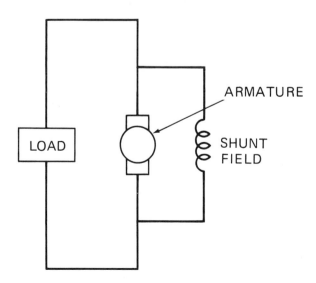

FIGURE 10–7 Diagram of a simple shunt-wound generator.

Generator Theory   **193**

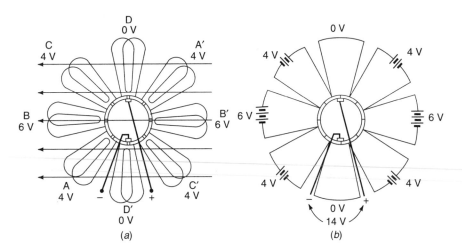

FIGURE 10–10 Armature circuit with battery analogy.

Analysis of an Armature Circuit

In a previous paragraph it was explained that a practical generator has many coils of wire in the armature. These coils are connected to the commutator segments in such a manner that they are in series with one another. Figure 10–10a shows the connections for a typical commutator in a two-pole generator. Assume that the armature has eight coils of two turns each wound around the armature through oppositely positioned slots. If the magnetic flux is horizontal, no voltage will be induced in the vertical coils because the coil sides will be moving parallel to the lines of force and will not be cutting any of them. The coils in positions B and B' will be cutting across a maximum number of flux lines and will therefore have a maximum emf induced in them.

For the purpose of illustration, we shall assume that this emf is 6 V. The coils at positions A, A', C, and C' will then have an induced emf of approximately 4 V each. The result is that three voltage-producing coils are connected in series in each half of the armature.

Figure 10–10b shows a battery analogy of the armature circuit. In each of the two series circuits in the armature, there are two 4-V coils and one 6-V coil. The total emf from each series is 14 V, and since the two circuits are connected in parallel, the amperage will be twice that of one series circuit.

The armature-winding arrangement illustrated in Figure 10–10 is known as **progressive lap winding.** There are several different types of windings used for generators and motors, but the one shown here is adequate for the purpose of this discussion.

Armature Reaction

Since an armature is wound with coils of wire, a magnetic field is set up in the armature whenever a current flows in the coils, as in Figure 10–11a. This field is at right angles to the generator field shown in Figure 10–11b and is called **cross magnetization** of the armature. The effect of the armature field is to distort the generator field and shift the neutral plane, as illustrated in Figure 10–11c. Remember, the neutral plane is the position where the armature windings are moving parallel to the magnetic flux lines. This effect is known as **armature reaction** and is proportional to the current flowing in the armature coils.

The brushes of a generator must be set in the neutral plane; that is, they must contact segments of the commutator that are connected to armature coils having no induced emf. If the brushes were contacting commutator segments outside the neutral plane, they would short-circuit "live" coils and cause arcing and loss of power. Armature reaction causes the neutral plane to shift in the direction of rotation, and if the brushes are in the neutral plane at no load, that is, when no ar-

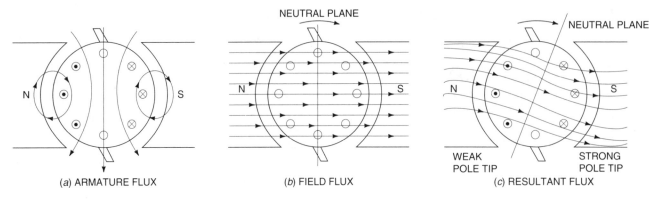

FIGURE 10–11 Armature reaction.

mature current is flowing, they will not be in the neutral plane when armature current is flowing. For this reason it is desirable to incorporate a corrective system into the generator design.

There are two principal methods by which the effect of armature reaction is overcome. The first method is to shift the position of the brushes so that they are in the neutral plane when the generator is producing its normal load current. In the other method, special field poles, called **interpoles,** are installed in the generator to counteract the effect of armature reaction.

The brush-setting method is satisfactory in installations in which the generator operates under a fairly constant load. If the load varies to a marked degree, the neutral plane will shift proportionally, and the brushes will not be in the correct position at all times. The brush-setting method is the most common means of correcting for armature reaction in small generators (those producing approximately 1000 W or less). Larger generators require the use of interpoles.

Interpoles. The use of interpoles is the most satisfactory method for maintaining a constant neutral plane in a generator. The windings of the interpoles are in series with the load; hence the interpole effect is proportional to the load. The polarity of the interpoles is such that their effect is opposite to that of the armature field; that is, each interpole is of the same polarity as the next field pole in the direction of rotation. With this polarity, the interpole may be said to pull the generator field into the correct position. A typical interpole system is shown in Figure 10–12.

In many generators a **compensating winding** is used to help overcome armature reaction. This winding consists of conductors embedded in the field-pole faces with one coil surrounding sections of two field poles of opposite polarity (see Figure 10–13). The compensating winding is in series with the interpole windings; hence it works with the interpoles and increases their effectiveness. The sparkless commutation obtained by the use of interpoles and a compensating winding increases the life of the brushes and commutator, reduces radio interference, and greatly improves the efficiency of the generator.

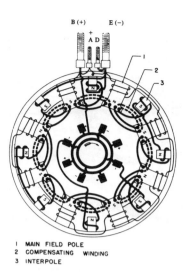

1 MAIN FIELD POLE
2 COMPENSATING WINDING
3 INTERPOLE

FIGURE 10–13 Generator with interpoles and compensating winding.

DC GENERATOR CONSTRUCTION

Aircraft dc generators have for the most part been replaced by dc alternators on modern aircraft. However, there are still several dc generators currently in operation on older aircraft. Two of these generators are illustrated in Figure 10–14. One of these is a high-output generator used on large aircraft with heavy electrical loads; the other is a dc generator typical of those found on smaller light aircraft.

Armature Assembly

The armature assembly (Figure 10–15) consists of a laminated soft-iron core mounted on a steel shaft, the commutator at one end of the assembly, and armature coils wound through the slots of the armature core. The core is made of many soft-iron laminations coated with an insulating varnish and then stacked together. The purpose of the laminations is to eliminate or reduce the eddy currents that would be induced in a solid core. The effect of these currents was explained in Chapter 9. The laminations for the armature core are stacked together in such a manner that the slots are lined up so that the armature coils can be placed in them. Before the coil windings are installed, insulating paper or fabric is placed in the slots to protect the windings from wear and abrasion.

Insulated copper wire of a size large enough to carry the maximum armature currents is wound in coils through the slots of the armature.

Each end of the copper wire is then connected to a segment of the commutator. If the generator contains two brushes, the armature wire ends connect 180° apart; if four brushes are used, the winding ends connect to the commutator segments 90° apart.

After an armature is wound, the coils are held in place by means of nonmetallic wedges placed in the slots. On some models, bands of steel are placed around the armature to prevent the windings from being thrown out by centrifugal force when the armature is driven at high speeds.

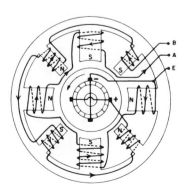

FIGURE 10–12 Generator circuit with interpoles.

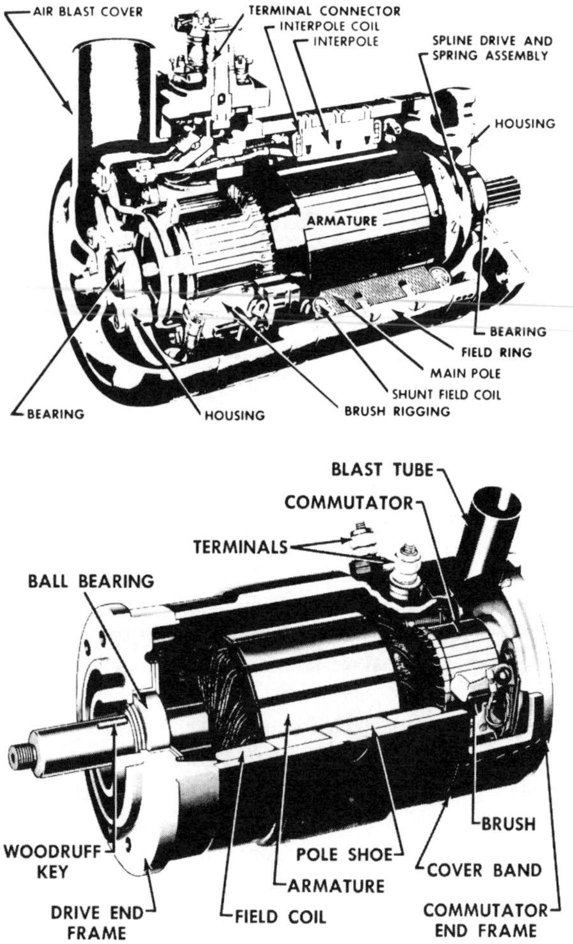

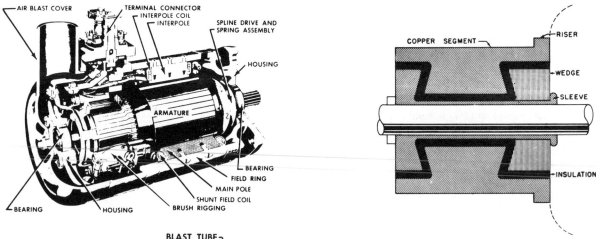

FIGURE 10–14 Two typical dc generators. (a) High-output; (b) low-output.

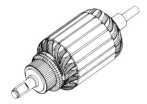

FIGURE 10–15 A typical armature assembly.

The commutator consists of a number of copper segments insulated from the armature structure, and from each other, with mica. The segments are constructed to be held in place by wedges located between the shaft and the segments. A cross section of a typical commutator is shown in Figure 10–16. Each commutator segment has a riser to which is soldered a lead from an armature coil. The surface of the commutator is cut and ground to a very smooth cylindrical surface. The mica insulation between the segments is undercut approximately 0.020 in. [0.051 cm] to make certain that it does not interfere with the contact of the brushes with the commutator.

FIGURE 10–16 Cross section of a commutator.

Field-Frame Assembly

The heavy iron or steel housing that supports the field poles is called the **field frame, field ring,** or **field housing.** It not only supports the field poles but also forms a part of the magnetic circuit of the field. The pole shoes are held in place by large countersunk screws that pass through the housing and into the shoes.

Small generators usually have two to four poles mounted in the field-frame assembly, and large generators can have as many as eight main poles and eight interpoles. The pole pieces are rectangular and in most instances are laminated. The main shunt field windings consist of many turns of comparatively small insulated copper wire. Series windings, such as those on the interpoles, consist of a few turns of insulated copper wire large enough to carry the entire load current without overheating. A typical field-frame assembly is shown in Figure 10–17.

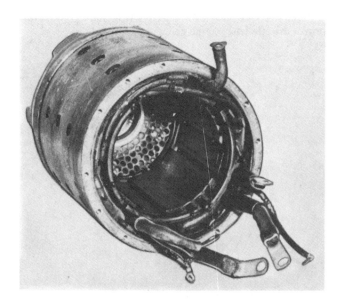

FIGURE 10–17 Field-frame assembly.

FIGURE 10–18 Brush holder assembly.

Brush Rigging

The brush rigging assembly (Figure 10–18) is located at the commutator end of the generator. The brushes are small blocks of a carbon and graphite compound soft enough to give minimum commutator wear but sufficiently hard to provide long service. Special brushes have been designed for generators operated at extremely high altitudes. These are needed because arcing increases at high altitudes and will cause the rapid deterioration of ordinary brushes.

As brushes wear, they slide in their metal holders and are held firmly against the commutator by means of springs. The tension of these springs should be sufficient to provide a brush pressure of approximately 6 psi [41 kPa] of contact surface. A flexible lead is connected from the brush to the brush frame to ensure a good electrical connection. Brushes of similar polarity are connected together electrically with a metal strip or wire.

End Frames

The generator **end frames** support the armature bearings and are mounted at each end of the field frame. The frame at the commutator end of the generator also supports the brush rigging assembly. The frame at the drive end is flanged to provide a mounting structure. On some generators the end frames are attached to the field-frame assembly by means of long bolts extending entirely through the field frame. On others the end frames are attached by machine screws into the ends of the field frame.

Generator bearings are usually of the ball type, prelubricated and sealed by the manufacturer. Prelubricated bearings do not require any service except at overhaul or in case of damage. The bearings fit snugly into the recesses in the end frames and are held in place by retainers attached to the end frames with screws.

Cooling Features

Since a generator operating at full capacity develops a large amount of heat, it is necessary to provide cooling. This is accomplished by means of passages leading through the generator housing between the field coils. In high-output generators there are also cooling air passages through the armature. Cooling air is forced through the passages either by a fan mounted on the generator shaft, or by pressure from a ram air duct leading into an air scoop mounted on the end of the generator, or by bleed air from the compressor of a turbine engine. Openings are provided in the end frame opposite the fan or air fittings to allow the heated air to pass out of the generator housing.

STARTER-GENERATORS

A **starter-generator** is a combination of a generator and a starter in one housing. A typical starter-generator is shown in Figure 10–19. Starter-generators are typically employed on small turboprop and turbine-powered aircraft such as the Beechcraft King Air. Most starter-generators contain at least two sets of field windings and only one armature winding. While in the start mode, the starter-generator employs a low-resistance series field. At this time a high current flows through both the field and armature windings, producing the high torque required for engine starting.

While in the generator mode, the starter-generator is capable of supplying current to the aircraft's electrical system. A typical starter-generator can supply a direct current of up to 300 A at 28.5 V while in the generator mode. To generate electric power, the shunt winding of the starter-generator is energized, and the series field is de-energized. The shunt winding is a relatively high-resistance coil that produces the magnetic field to induce voltage into the armature. The voltage produced in the armature sends current to the aircraft bus, where it is distributed to the various loads of the aircraft.

FIGURE 10–19 A typical starter-generator. *(Lear Siegler, Inc., Power Equipment Division)*

It should be noted that several types of starter-generators are currently in use. Some employ two separate field windings as stated above; others use only one (shunt) field winding. If only one field winding is used, special circuitry is needed in the generator control unit (GCU) to increase starting torque to an appropriate level. The technician should become familiar with any specific starter-generator before beginning maintenance procedures.

One advantage of the starter-generator is that only one drive gear mechanism is used for both the start and the generator modes. Therefore, the starter motor drive gear need not be engaged to or disengaged from the engine drive gear. Also, the starter-generator reduces both size and weight as compared to a conventional system that employs two units: a starter and a generator. The main disadvantage of starter-generators is that they are unable to maintain full output at a low rpm. Most starter-generators therefore must be used on turbine-powered aircraft that consistently maintain a relatively high engine rpm.

Starter-Generator Components

Starter-generators are designed to provide torque for engine starting and generate dc electric power for the aircraft's electrical systems. The starter-generator shown in Figure 10–19 contains a self-excited four-pole generator. Four interpoles and a compensating winding are used to help overcome armature reaction. An integral fan is used to draw air through the unit during rotation. The cooling air is required to maintain temperature limits during high power generation. A clutch damper may be used on some units to connect the armature to the starter-generator's drive shaft. This clutch provides friction damping of any torsional loads that may be applied to the armature during operation. Changes in torsional loads occur whenever the aircraft's electric equipment is turned on or off. If the armature is connected directly to the engine, without a clutch, the torsional loads may overstress

the drive shaft and cause generator failure. Some starter-generators employ a drive shaft **shear section,** which is used to protect the engine's gearbox in the event the generator mechanically fails and cannot rotate. In this situation the shear section breaks (shears) and disconnects the generator from the drive gear.

GENERATOR CONTROL

Ways to Monitor Generator Output

There are two ways to monitor the output of a generator. The voltage produced by the generator can be indicated by a voltmeter, or the current flow from the generator can be displayed by an ammeter. One or both of these instruments are usually mounted on the aircraft's flight deck. If only one instrument is used, the ammeter is preferred. A voltmeter should never be used without an ammeter.

The ammeter can be placed in the generator output lead, as shown in Figure 10–20a, or in the battery positive lead, as shown in Figure 10–20b. Ammeters located in the generator output lead measure only the current leaving the generator; therefore, they are calibrated in a positive scale only. A 30-A generator would require an ammeter scaled from 0 to 30 A. Any reading on this type of ammeter indicates that current is leaving the generator and flowing to the bus.

Ammeters located in the battery positive lead must be calibrated from a negative value to a positive value. A typical ammeter of this type reads from −60 A to 0 to +60 A. This is necessary because the current can flow through the ammeter either to or from the battery. If the battery is discharging, the ammeter will indicate a negative value. If the battery is charging, the ammeter will indicate a positive value. When this type of ammeter is used, the indications must be positive when the charging system is working properly.

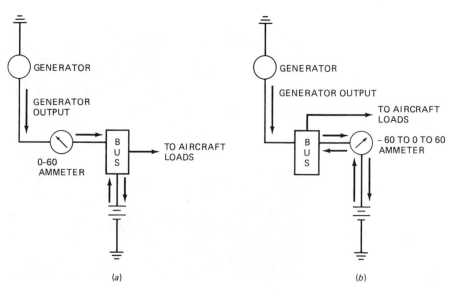

FIGURE 10–20 Ammeter placement in a generator system. (a) Located in the generator output lead; (b) located in the battery lead.

A voltmeter is often necessary to correctly monitor a generator's output on multiengine aircraft. The voltage of an operating generator must be slightly higher than battery voltage. This is necessary to ensure that the battery receives a charging current from the generator. Generators produce nearly 14 V for systems using 12-V batteries and 28 V for systems using 24-V batteries. With multigenerator systems it often becomes necessary to monitor the output voltage of each generator in order to determine which, if any, generator has failed. Some multiengine aircraft contain only one voltmeter, which can measure either bus voltage, right generator voltage, or left generator voltage by means of a control switch. This saves weight and space on the instrument panel.

Before making any determination as to the condition of a generator or generator system, be sure to run the aircraft and monitor all related instruments. At that time consider the two different types of ammeters, specifically what they measure. This procedure will help to ensure a proper system diagnosis.

Principles of Voltage Regulation

In the section of this chapter describing generator theory, it was explained that the voltage produced by electromagnetic induction depends on the number of lines of force being cut per second by a conductor. **In a generator, the voltage produced depends on three factors: (1) the speed at which the armature rotates, (2) the number of conductors in series in the armature, and (3) the strength of the magnetic field.** In order to maintain a constant voltage from a generator under all conditions of speed and load, one of the foregoing conditions must be varied in accordance with operational requirements.

It is obvious that the speed of a generator cannot be varied according to load requirements if the generator is directly driven by the engine. Also, it is impossible to change the number of turns of wire in the armature during operation. Therefore, the only practical way to regulate the generator voltage is to control the **strength of the field.** This is easily accomplished, because the strength of the field is determined by the current flowing through the field coils, and this current can be controlled by a variable resistor in the field circuit outside the generator.

The simplest type of voltage regulation is accomplished as shown in Figure 10–21. In this arrangement a rheostat (variable resistor) is placed in series with the shunt field circuit. If the voltage rises above the desired value, the operator can reduce the field current with the rheostat, thus weakening the field and lowering the generator voltage. An increase in voltage is obtained by reducing the field circuit resistance with the rheostat. All methods of voltage regulation in aircraft electrical systems employ the principle of a variable or intermittent field resistance. Modern voltage regulators have been developed to such a high degree of efficiency that the emf of a generator will vary only a small fraction of a volt throughout extreme ranges of load and speed.

Voltage regulators or controls for modern aircraft are usually of the solid-state type; that is, they employ transistors and diodes as controlling elements. Because there are still

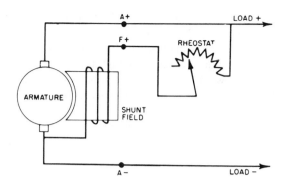

FIGURE 10–21 A simple voltage regulator (a variable resistor).

many older airplanes in use that employ vibrator-type and variable-resistance voltage regulators, we shall examine these in the following sections.

Vibrator-Type Voltage Regulator

A generator system using a vibrator-type voltage regulator is shown in Figure 10–22. A resistance that is intermittently cut in and out of the field circuit by means of vibrating contact points is placed in series with the field circuit. The contact points are controlled by a voltage coil connected in parallel with the generator output. When the generator voltage rises to the desired value, the voltage coil produces a magnetic field strong enough to open the contact points. When the points are open, the field current must pass through the resistance. This causes a substantial reduction in field current, with the result that the magnetic field in the generator is weakened. The generator voltage then drops immediately, causing the voltage coil electromagnet to lose strength so that a spring can close to the contact points. This allows the generator voltage to rise, and the cycle is then repeated. The contact points open and close many times a second, but the actual time that they are open depends on the load being carried by the generator and the generator (engine) rpm. As the generator load is increased, the time that the contact points remain closed increases, and the time that they are open decreases. Adjustment of the generator voltage is made by increasing or decreasing the tension of the spring that controls the contact points.

Often temperature greatly affects the generator output; therefore, this adjustment should be made only under specific conditions set up by the manufacturer.

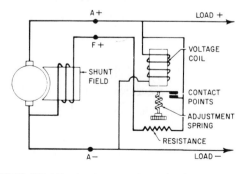

FIGURE 10–22 Vibrator-type regulator circuit.

Because the contact points do not burn or pit appreciably, vibrator-type voltage regulators are satisfactory for generators that require a low field current. In a system in which the generator field requires a current as high as 8 A, the vibrating contact points would soon burn and probably fuse together. For this reason a different type of regulator is required for heavy-duty generator systems.

If the regulating resistance becomes disconnected or burned out, the generator voltage will fluctuate, and excessive arcing will occur at the contact points. When inspecting a vibrator-type voltage regulator, make sure that the connections to the resistance are secure and that the resistance is in good condition.

Carbon-Pile Voltage Regulator

The carbon-pile voltage regulator derives its name from the fact that the regulating element (variable resistance) consists of a stack, or **pile,** of carbon disks (see Figure 10–23). Usually, the carbon pile has alternate hard carbon and soft carbon (graphite) disks contained in a ceramic tube with a carbon or metal contact plug at each end. At one end of the pile, a number of radially arranged leaf springs exert pressure against the contact plug, thus keeping the disks pressed firmly together. For as long as the disks are compressed, the resistance of the pile is very low. If the pressure on the carbon pile is reduced, the resistance increases. By placing an electromagnet in a position where it will release the spring pressure on the disks as the voltage rises above a predetermined value, a stable and efficient voltage regulator is obtained.

The carbon-pile voltage regulator is connected in a generator system the same way any other regulator is connected, that is, with a resistance in the field circuit and an electromagnet to control the resistance. The carbon pile is in series with the generator field, and the voltage coil is shunted across the generator output. A small manually operated rheostat is connected in series with the voltage coil to provide for a limited amount of adjustment, which is necessary when two or more generators are connected in parallel to the same electrical system.

Equalizing Circuit

When two or more generators are connected in parallel to a power system, the generators should share the load equally. If the voltage of one generator is slightly higher than that of the other generators in parallel, that generator will assume the greater part of the load. For this reason an equalizing circuit must be provided that will cause the load to be distributed evenly among the generators. An equalizing circuit includes an equalizing coil wound with the voltage coil in each of the voltage regulators, an equalizing bus to which all equalizing circuits are connected, and a low-resistance shunt in the ground lead of each generator (see Figure 10–24). The equalizing coil will either strengthen or weaken the effect of the voltage coil, depending on the direction of current flow through the equalizing circuit. The low-resistance shunt in the ground lead of each generator causes a difference in potential between the negative terminals of the generators that is proportional to the difference in load current. The shunt is of such a value that there will be a potential difference of 0.5 V across it at maximum generator load.

Assume that generator 1 in Figure 10–24 is delivering 200 A (full load) and that generator 2 is delivering 100 A (half load). Under these conditions there will be a potential difference of 0.5 V across the shunt of generator 1 and 0.25 V across the shunt of generator 2. This will make a net potential difference of 0.25 V between the negative terminals of the generators. Since the equalizing circuit is connected between these points, a current will flow through the circuit. The current flowing through the equalizing coil of voltage regulator 1 will be in a direction to strengthen the effect of the voltage coil. This will cause more resistance to be placed in the field circuit of generator 1, thus weakening the field strength and causing the voltage to be reduced. The drop in voltage will result in the generator taking less load. The current flowing through the equalizing coil of voltage regulator 2 will be in a direction to oppose the effect of the voltage coil, thus causing a decrease in the resistance in the field circuit of generator 2. The generator voltage will increase because of increased current in the field windings, and the generator will take more of the load. To summarize, the effect of an equalizing circuit is to lower the voltage of a generator that is taking too much of the load and to increase the voltage of the generator that is not taking its share of the load.

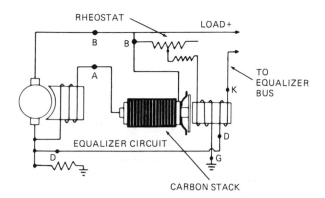

FIGURE 10–23 Carbon-pile voltage regulator circuit.

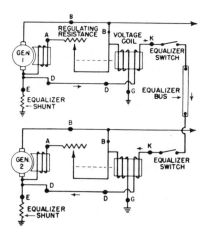

FIGURE 10–24 Equalizing circuit.

Equalizing circuits can correct for only small differences in generator voltage; hence the generators should be adjusted to be as nearly equal in voltage as possible. If the generator voltages are adjusted so that there is a difference of less than 0.5 V between any of them, the equalizing circuit will maintain a satisfactory load balance. A periodic inspection of the ammeters should be made during flight to see that the generator loads are remaining properly balanced.

Reverse-Current Cutout Relay

In every system in which the generator is used to charge batteries as well as to supply operating power, an automatic means must be provided for disconnecting the generator from the battery when the generator voltage is lower than the battery voltage. If this is not done, the battery will discharge through the generator and may burn out the armature. Numerous devices have been manufactured for the purpose of automatically disconnecting the generator, the simplest being the **reverse-current cutout relay.** Figure 10–25 is a schematic diagram illustrating the operation of such a relay.

A **voltage coil** and a **current coil** are wound on the same soft-iron core. The voltage coil has many turns of fine wire and is connected in parallel with the generator output; that is, one end of the voltage winding is connected to the positive side of the generator output, and the other end of the winding is connected to ground, which is the negative side of the generator output. This is clearly shown in the diagram. The current coil consists of a few turns of large wire connected in series with the generator output; hence it must carry the entire load current of the generator. A pair of heavy contact points is placed where it will be controlled by the magnetic field of the soft-iron core. When the generator is not operating, these contact points are held in an open position by a spring.

When the generator voltage reaches a value slightly above that of the battery in the system, the voltage coil in the relay magnetizes the soft-iron core sufficiently to overcome the spring tension. The magnetic field closes the contact points and thus connects the generator to the electrical system of the airplane. As long as the generator voltage remains higher than the battery voltage, the current flow through the current coil will be in a direction that aids the voltage coil in keeping the points closed. This means that the field of the current coil

will be in the same direction as the magnetic field of the voltage coil and that the two will strengthen each other.

When an airplane engine is slowed down or stopped, the generator voltage will decrease and fall below that of the battery. In this case the battery voltage will cause current to start flowing toward the generator through the relay current coil. When this happens, the current flow will be in a direction that creates a field opposing the field of the voltage winding. This results in a weakening of the total field of the relay, and the contact points are opened by the spring, thus disconnecting the generator from the battery. The contact points may not open in normal operation until the reverse current has reached a value of 5 to 10 A.

Generally speaking, the tension of the spring controlling the contact points should be adjusted so that the points will close at approximately 13.5 V in a 12-V system and at 26.6 to 27 V in a 24-V system.

Current Limiter

In some generator systems a device is installed that will reduce the generator voltage whenever the maximum safe load is exceeded. This device is called a **current limiter** and is designed to protect the generator from loads that will cause it to overheat and eventually burn the insulation and windings.

The current limiter operates on a principle similar to that of the vibrator-type voltage regulator. Instead of having a voltage coil to regulate the resistance in the field circuit of the generator, the current limiter has a current coil connected in series with the generator load circuit (see Figure 10–26).

When the load current becomes excessive, the current coil magnetizes the iron core sufficiently to open the contact points and add a resistance to the generator field circuit. This causes the generator voltage to decrease, with a corresponding decrease in generator current. Since the magnetism produced by the current-limiter coil is proportional to the current flowing through it, the decrease in generator load current also weakens the magnetic field of the current coil and thus permits the contact points to close. This removes the resistance from the generator field circuit and allows the voltage to rise again. If an excessive load remains connected to the generator, the contacts of the current limiter will continue to vibrate, thus holding the current output at or below the minimum safe

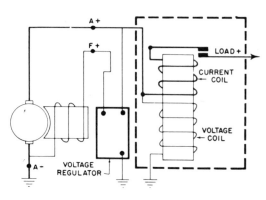

FIGURE 10–25 Reverse-current cutout relay circuit.

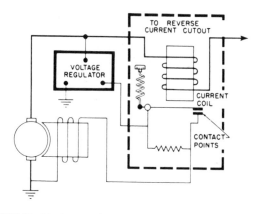

FIGURE 10–26 Current-limiter circuit.

limit. The contact points are usually set to open when the current flow is 10 percent above the rated capacity of the generator.

The current limiter described above should not be confused with the fuse-type current limiter. The fuse-type limiter is merely a high-capacity fuse that permits a short period of overload in a circuit before the fuse link melts and breaks the circuit.

Two-Unit Control Panel

Generator systems for light aircraft often have the generator control units mounted on a single panel. When the voltage regulator and reverse-current cutout relay are mounted on a single panel, the combination is called a two-unit control panel or box (see Figure 10–27). In the voltage regulator on this panel, an extra coil, called an **accelerator winding,** is wound with the voltage coil. This coil is connected in series with the field-regulating resistance and is wound in a direction opposite to that of the voltage winding. Its purpose is to reduce the magnetism of the core when the contact points open; this reduction causes the points to close more quickly than they would without the neutralizing effect of a reverse coil. The result of this arrangement is that the contact points vibrate more rapidly and produce a steadier voltage from the generator.

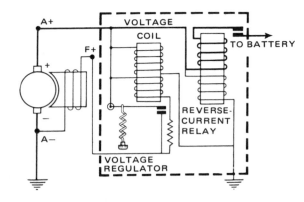

FIGURE 10–27 Two-unit generator control.

Three-Unit Control Panel

A three-unit control panel consists of a voltage regulator, a current limiter, and a reverse-current cutout relay mounted on a single panel (see Figure 10–28). This combination will provide for both voltage regulation and protection from excessive loads. A photograph of such a panel with the cover off is shown in Figure 10–29.

The three-unit control panel has proved very successful for the control of 12- and 24-V generator systems. Because of its dependability and low cost, it was used almost exclusively in light-aircraft generator systems before the development of transistor regulators.

The wiring diagram in Figure 10–28 is only one of several possible arrangements. In some systems the voltage regulator is placed in the ground side of the generator field circuit, but the results are the same in either case.

Starter-Generator Control Systems

The control circuits of a starter-generator are relatively complex; they must control current for both starting and generating operations. These circuits are contained in a device that is often referred to as a **generator control unit** (GCU). The GCU discussed here and some of the information presented were produced by Lear Siegler, Power Equipment Division; this GCU is used in conjunction with the starter-generator discussed earlier. The GCU is pictured in Figure 10–30; it contains a voltage regulator, the various control circuits for the start and generator modes, and protection circuits used during abnormal operating conditions. The GCU electronic components are contained on three printed circuit boards. Each board is mounted on a forged aluminum base, which acts as a heat sink for those components which must dissipate heat. The entire unit is enclosed by an aluminum cover that allows for all external connections to be exposed.

GCU Functions

The start mode of a starter-generator is controlled through a circuit independent of the GCU. During starting, battery or auxiliary power unit (APU) power is sent to the starter-generator via a starter contactor. The starter contactor is energized

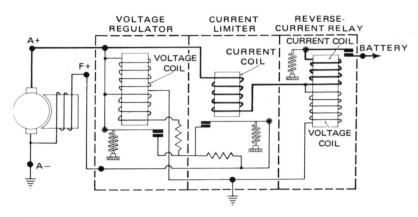

FIGURE 10–28 Three-unit generator control.

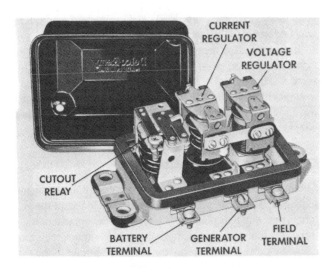

FIGURE 10–29 A three-unit control panel.

FIGURE 10–30 A typical generator control unit for a starter-generator. *(Lear Siegler, Inc., Power Equipment Division)*

by the engine start switch on the flight deck. During the generator mode, the GCU controls the generator output, generator and system protection, and self-test functions. If a fault is detected in the generator system, the GCU will illuminate the appropriate annunciator fail light. The generator control unit is capable of performing the following 10 functions.

1. *Voltage Regulation*   The voltage regulator section of the GCU maintains a constant generator voltage under various loads, temperatures, and rotational speeds. The current of the generator field circuit is controlled through the field transistor. This transistor varies the field current pulse time to vary generator output.

2. *Generator Line Contactor Control*   The generator line contactor control provides a means of connecting the generator output to the aircraft's dc load bus. This circuit operates with a time delay to ensure that generator voltage is nearly equal to bus voltage immediately following initial engine starting. Several inhibiting signals are also employed to ensure proper contactor positioning (open or closed) when failure conditions are sensed.

3. *Overvoltage Protection*   The overvoltage protection circuit prevents damage to aircraft equipment in the event an excessive generator output occurs. If the generator output voltage exceeds the preset limits, an integrator starts to function. This integrator is used as an inverse time delay, so that a slight overvoltage condition is allowed for a much longer time than a severe one before a trip occurs. In this way, unusually large but momentary voltage transients will not cause a nuisance trip of the field relay. If a severe overvoltage is sensed, the generator is de-energized and the line contactor is opened. A completely separate circuit is used to open the generator line contactor as soon as the voltage exceeds 40 V dc. This feature not only provides redundant protection for utilization equipment but also allows a faster response of the line contactor after a failure. Unlike the overvoltage with inverse time delay, this function is not latched, so that manual reset is not required after a temporary overvoltage condition.

4. *Overload and Undervoltage Protection*   The overload and undervoltage protective functions cooperate to de-energize the system in the event of an overload condition. An overload condition is sensed by the GCU as either a generator overcurrent condition as indicated by an excessive generator interpole voltage or an undervoltage condition. When the GCU senses either condition, an internal time delay is initiated. If the overload condition continues for a period of approximately 10 s, the GCU trips the field relay, de-energizing the generator and opening the line contactor.

5. *Reverse-Current and Differential Voltage Protection*   The reverse-current protection function senses generator interpole voltage to determine whether the generator is acting as a load on the system rather than a power source. If because of a failure or during a normal engine shutdown current begins to flow into the generator, this is sensed and the line contactor is opened. An inverse time delay is used to quickly open the contactor under severe conditions, while more time is allowed during normal shutdowns. This prevents needless cycling of the contactor during a transient condition. The circuit is not latched, and so no reset is required to reclose the contactor after reverse current is sensed. The contactor is held open owing to differential voltage sensing once reverse current has been detected. The differential voltage function also operates on generator buildup to keep the generator line contactor from closing until the generator output voltage is within 0.5 V dc of the bus voltage.

6. *Reverse-Polarity Protection*   The reverse-polarity protection function protects the utilization equipment from reverse-polarity buildup of the generator. This protection trips the field relay to de-energize the generator.

7. *Anticycle Protection*   The anticycle protection feature prevents more than one reset attempt of the generator field relay for each activation of the generator control switch. Because the generator output voltage is used for GCU control power, and this voltage disappears after a trip, the system

would repetitively build up in voltage and trip again if a fault existed in the system.

8. *Latching Field Relay Control* A magnetic latching field relay is used to de-energize the generator after a fault condition has been sensed. The field relay is used to de-energize the generator by opening the generator shunt field excitation path and open the line contactor by opening its power input. The field relay is tripped by a protection function such as overvoltage, overload, undervoltage, reverse polarity, or open ground wire sensing; it may also be tripped by an external switch applying a ground signal to the GCU.

9. *Flash and Start Relay Control* The field-flashing relay and the associated circuitry ensure that the generator output can be built up from the residual voltage without help from any other power source. The residual voltage bootstraps the generator upward to a point where the field relay is reset and then to a higher voltage to energize the field-flashing relay. Once the field-flashing relay is energized, the field flash path is broken, but the normal voltage regulator circuitry is able to operate at this voltage level, and so the generator continues to build up to the normal operating voltage.

10. *Overvoltage and Overload Protection Self-Test* Provisions are made within the GCU to enable it to periodically exercise the overvoltage, overload, and undervoltage protection circuits. A passive failure of the circuitry would not otherwise be discovered until that function was required to operate. If an external test switch applies generator output voltage to the GCU, the protection will be biased to a point where it will operate, even though normal voltage appears on the generator output. If a trip of the channel results within a few seconds after the voltage has been applied, the circuit is working correctly.

GENERATOR INSPECTION, SERVICE, AND REPAIR

Generator Load Balancing

When it is desired to balance the load among the generators in any system, the technician should always follow the procedure set forth by the manufacturer of the aircraft. This procedure will be found in the manufacturer's service or maintenance manual.

The balance procedure is usually begun by checking all generators with a precision voltmeter. This is done after the generators and engines are warmed up to normal operating temperature. Under these conditions all generators are adjusted to exactly the same output voltage (28 to 29 V for a 24-V system). A substantial load is then turned on, and the ammeters are examined. All generator loads should be within ± 10 percent of one another. If the generator loads are not within these limits, the generator with the greatest error should be adjusted first. A small movement of the paralleling potentiometer on the voltage regulator should produce an instant change in the load current for the generator being adjusted. If the load on one generator is reduced, the other

generators will pick up load. All ammeters should be watched while the adjustments are being made.

If the aircraft generators cannot be properly balanced, the system should be repaired before the aircraft is returned to service. The process of balancing generators is often referred to as **paralleling the generators.** If both generators are producing equal voltage, they will carry equal current loads if connected in parallel.

Generator Troubleshooting

The first step in troubleshooting a generator circuit on an airplane is to determine what type of system is in use. If the system is on a small airplane, it is likely that the control unit is a three-element type, that is, that it contains a voltage regulator, a reverse-current cutout relay, and a current limiter. If the system contains a 24-V generator or starter-generator, it is likely that a more complex system of control is employed. Always refer to the current manufacturer's data and electrical wiring diagrams prior to any troubleshooting.

The two most likely indications of generator system failure are (1) no or low voltage and (2) battery discharge. If the aircraft's ammeter indicates a battery discharge, the generator may not be producing the proper voltage. The voltage at the generator's output terminal can then be measured, and if approximately 2 to 6 V is present with the system operating, the generator is operating from residual magnetism only. This means that there is no current through the generator field coils and some component of the field circuit is defective. Likely suspects include wiring connections, the generator master switch, the voltage regulator, and the generator's field. If zero volts is measured between the generator's output terminal and ground with the system operating, the generator has lost its residual magnetism.

If it becomes necessary to determine if the voltage regulator or the generator is defective, simply bypass the voltage regulator circuit. This is done by connecting a voltage directly to the generator's field circuit and monitoring generator output. If current is fed into the field of a rotating generator, that generator should produce a normal or above-normal output voltage. Since there are two different methods of wiring a generator's field, one must first determine which method is being used. With one method, the field positive is connected to the voltage regulator, as illustrated in Figure 10–31a; with the other method, the field negative is connected to the voltage regulator, as shown in Figure 10–31b. To bypass the voltage regulator, the correct voltage signal (positive or negative) must be connected to the generator F terminal with the regulator disconnected and the generator armature rotating. If the voltage measured at the generator output terminal to ground is less than system voltage, the generator is defective. If the generator output voltage is at or above the normal value, the generator is good and the voltage regulator most likely is defective. There may be some defective conditions that this test will not detect; however, for the most part, bypassing the regulator is valid. Many factors must be considered when one is troubleshooting a generator system; always become familiar with the system before troubleshooting begins.

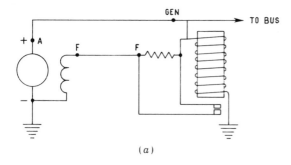

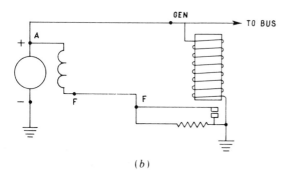

FIGURE 10–31 Different arrangements for field control circuits. (*a*) Controlling field positive voltage; (*b*) controlling field negative voltage.

Starter-Generator Maintenance

All starter-generator maintenance and inspection procedures should be performed in accordance with current manufacturer's data. Most manufacturers require periodic inspections of starter-generators to ensure proper system operation. These inspections can be performed as often as every 200 flight hours. The interval between starter-generator overhauls should not exceed 1000 h unless specified by the aircraft manufacturer. The typical brush life for starter-generators ranges from 500 to 1000 h. Both the brushes and the commutators must be inspected to ensure that they do not wear beyond operational limits.

The following information is intended as a general reference of typical troubleshooting inspection techniques. The specific details of the starter and generator circuits will vary significantly between installations. However, in general, all starter-generators require that current be sent to both the armature and the field for starting and that current be sent only to the field for generating purposes. The units that control the current to and from the starter-generator can be bypassed to determine proper system operation. This procedure should be performed only after achieving a complete understanding of the system. This will ensure that no components will be damaged by the bypass test.

If the generator is determined to be defective, one or more of the following procedures can be employed. The starter-generator can be removed from the engine by loosening the attachment bolts or the quick-attach-detach (QAD) adapter. A QAD adapter is mounted to many generators and starter-generators to allow for easy removal and installation of the unit. This eliminates long maintenance times for generator

replacement. Typically, only one bolt or latching mechanism must be loosened to remove a unit that uses a QAD adapter. Always be careful to properly support the starter-generator during removal and installation. The unit's drive shaft *shear section* can be damaged if the starter-generator is allowed to hang, unsupported, by the drive shaft spline.

Cleaning the starter-generator should be accomplished by removing the brush inspection cover and using compressed air, at about 40 psi, to blow away any carbon and copper dust. This dust collects around the electrical windings and brush assembly as a result of normal brush and commutator wear.

Stoddard solvent should be used to clean the exterior of the unit. Once the unit is cleaned, a thorough visual inspection should be performed. Items that require special consideration are the brush assembly, the commutator, the cooling system, and the drive shaft.

The brushes should be checked for cracks, chips, frayed leads, and general integrity. Many brushes incorporate a wear groove in order to facilitate inspection. The amount of **wear groove** visible indicates the service life remaining on that particular brush. Figure 10–32 shows a typical wear groove reference. The brush holders and springs should also be inspected for damage or excess wear. On many units the starter-generator brushes can be replaced without brush seating if **"instant-filming"** brushes are installed. Instant-filming brushes contain a lubricating additive that improves brush conductivity and wear characteristics. These brushes can be replaced in the field without commutator resurfacing if the commutator is not excessively worn.

The commutator of the starter-generator should be inspected for burned spots, pitting, or excessive wear. The commutator mica should be undercut to 0.020 in. [0.051 cm] on most models. The cooling fan, any related vent holes, and air ducts should be inspected and cleaned thoroughly. Any loose, cracked, or bent fans should be replaced.

The drive mechanism should be inspected for excess wear or damage. Most starter-generators incorporate a damper assembly to absorb excessive shock loads. This damping assembly is subject to extreme load changes during normal operation; therefore, it should be inspected carefully to ensure proper system operation. If any defects are found, the starter-generator should be sent to an appropriate overhaul facility. The drive shaft or spline should be inspected for excess wear or damage, and any defective mounting gasket or O-ring should be replaced prior to installation of the starter-

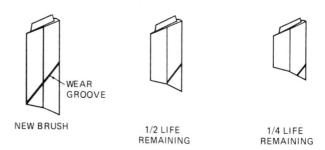

FIGURE 10–32 Typical generator brush wear groove. (*Lear Siegler, Inc., Power Equipment Division*)

generator. After installation of the unit, a thorough operational check should be performed to ensure that the system functions properly.

Flashing the Field

If a generator fails to show any voltage whatsoever when it is operating at the proper rpm, this condition is often due either to the loss or to the reversal of the polarity of the residual magnetism in the field. This can be corrected by **flashing the field.**

To flash the field of any generator, a voltage must be momentarily applied with the correct polarity to the field coil. Prior to flashing the field, always disconnect the voltage regulator from the generator. This will prevent possible damage to the regulator circuit. It is also important to observe polarity, connecting positive voltage to the field positive lead and negative voltage to the field negative lead. The internal wiring of the generator must be known to ensure proper polarity during the flashing operation.

Inspection

Generators in service should be given a periodic inspection of external connections, wiring, brushes, commutator, mounting, and performance. These inspections should be carried out according to the manufacturer's instructions; however, a good general rule is that generator inspection should be performed at least every 100 flight hours. The following inspections are considered essential:

1. Inspect the generator terminal connections to see that they are clean and tight.

2. Inspect the flange mounting for cracks or looseness of mounting bolts. See that there are no oil leaks around the mounting.

3. Remove the cap or band that covers the brushes and commutator. Blow out any accumulation of carbon dust with dry compressed air. Inspect the brushes for amount of wear, and see that they slide freely in the holders. If a brush is binding in the holder, it should be removed and cleaned with a clean cloth moistened with unleaded gasoline or a good petroleum solvent. If brushes are worn in excess of the tolerances specified by the manufacturer, they must be replaced. Inspect the tension of the brush springs by lifting them with the fingers. A weak spring should be adjusted or replaced. If a spring scale of the proper range is available, it may be used to measure the brush-spring tension; the spring tension should measure within the manufacturer's specifications.

4. Inspect the commutator for cleanliness, wear, and pitting. A commutator in good condition should be smooth and of a light chocolate color. If there is a slight amount of roughness on the commutator, it can be removed with No. 000 sandpaper or a special abrasive stick manufactured for cleaning commutators and seating brushes. If an abrasive stick is used, the proper application is to hold the end of the stick against the commutator while the generator is running. This is done until the commutator is smooth, clean, and bright. After smoothing, all sand and dust particles should be blown

out with compressed air. Never clean a commutator or seat brushes with a metallic abrasive paper. The metal particles may become lodged between the commutator segments and short-circuit the armature. Dirt may be removed from the commutator with a cloth moistened with a petroleum solvent. Oil on the brushes and commutator indicates a faulty oil seal in the engine. If this condition exists, the generator should be removed, disassembled, and thoroughly cleaned. If the end frame of the generator has an oil drain vent, it should be inspected to see that it is open. Before the generator is reinstalled, the engine oil seal at the generator drive must be replaced.

5. Inspect the area inside the commutator end of the generator case for lead particles. If particles of lead are visible, it is likely that the armature has been overheated. This may have been caused by overloading of the generator for a sustained period, by short-circuited coils in the armature, by short-circuited segments of the commutator, or by the sticking of the reverse-current-relay points. The generator should be removed and a new or rebuilt armature installed. Before the generator is reinstalled, the cause of armature failure should be determined and corrected.

6. Always inspect the mounting of the generator to the aircraft. The structure must be free of cracks and all bolts correctly torqued. The mounting bracket bolt holes should be carefully inspected. If any elongated holes are found, the structure must be repaired or replaced. Inspect the generator belt for wear and/or cracks. The correct belt tension is also very important. Typically, an installed belt should deflect about $1/2$ in. when a moderate pressure is applied by hand.

Disassembly

The disassembly procedure for generators cannot be discussed in detail in this text, inasmuch as it varies among different makes and models of generators. However, if it becomes necessary to disassemble a generator, the technician should refer to the instructions furnished by the manufacturer of that particular model. If these instructions are not available, the technician can proceed as follows with reasonably good results:

1. Remove the strap or cap that covers the brushes and commutator.

2. Remove the brushes, and disconnect the flexible leads from the brush holders. Mark the brushes for their proper position in the brush rigging.

3. Disconnect the field and terminal leads, and mark the connections so that they can be reconnected correctly.

4. Remove the screws or bolts that attach the end frames to the field frame. Some generators have a nut and washer that holds the armature shaft in the bearing; in this case the nut and washer must be removed before the end frames are taken off.

5. When both end frames are free, remove them and take out the armature. *Note:* Further disassembly can be accomplished as required, but for inspection and cleaning purposes, the removal of brushes, end frames, and armature is usually sufficient.

6. After disassembly blow the brush dust from the field assembly, using dry compressed air.

7. Use a cloth moistened with a good petroleum solvent to clean the field-frame assembly, commutator, and brush rigging. Do not immerse any of these parts in gasoline or some other cleaning solvent.

Repair of the Commutator

If the commutator is slightly rough or pitted, it can be smoothed with No. 000 sandpaper. After smoothing, blow out all sand and dust particles with dry compressed air. If the commutator is very rough or badly pitted, the armature should be placed on a metal turning lathe and a light cut taken across the surface of the commutator. This is most easily accomplished with equipment especially designed for the purpose. The cut on the commutator should be only deep enough to remove the irregularities on the surface. This cut will also correct any eccentricity that has developed as a result of uneven wear.

After the commutator has been turned on a lathe, it is necessary to undercut the mica insulation between the segments to a depth of approximately 0.020 in. [0.051 cm]. To assure a clean cut to the required depth, use a cutting tool slightly wider than the thickness of the mica. After undercutting the mica, smooth any burrs or sharp edges on the commutator segments with No. 000 sandpaper.

Testing

For testing armatures, a device called a **growler** is used. This device consists of many turns of wire wound around a laminated core with two heavy pole shoes extended upward to form a V into which an armature can be placed. Actually, the growler is nothing more than a large, specially designed electromagnet. Figure 10–33 shows an armature being tested on a growler. The power supply for the growler is standard 110-V alternating current. The current causes a noticeable hum when an armature is placed between the pole shoes; hence the name *growler.*

When placed on a growler, an armature forms the secondary of a transformer. The winding of the growler is the primary. The rapidly moving field produced by the winding of the growler induces an alternating current in the windings

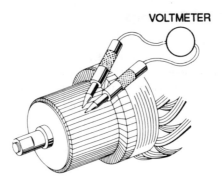

FIGURE 10–34 Testing for an open coil in an armature.

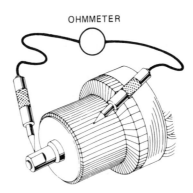

FIGURE 10–35 Testing an armature for a short to ground.

of the armature. By connecting a test meter between segments of the commutator and the armature on the growler (see Figure 10–34), it can be determined whether an open circuit exists in any of the coils. The meter will indicate a given voltage when connected across segments of a good coil. To test for a short circuit in the windings, a thin strip of steel is placed on the armature segments, and the armature is slowly rotated between the poles of the growler. If there are no shorts, a weak magnetic attraction will be noticed. One or more shorted coils will cause a strong vibration of the metal strip at certain points on the armature surface.

To test for a ground between the windings and the core of the armature, an ohmmeter can be used. Connect one test probe of the meter to the armature shaft and one probe to the commutator segments as illustrated in Figure 10–35. If the commutator is shorted to the armature shaft, zero resistance will be indicated. If the armature is operational, infinite resistance will be indicated on the ohmmeter.

To test a field coil for continuity, the probes of an ohmmeter are connected to the terminals of the coil. A shunt field coil should show low resistance, approximately 2 to 30 Ω, depending on the type of generator in which the coil is used. A series field coil should show practically no resistance, because it carries the entire load to the generator and the internal resistance of the generator must be as low as possible.

The field coils should also be tested for shorts to the field frame. An ohmmeter should indicate an infinite resistance between any field connection and the field frame. If zero resistance is indicated, the field is shorted to the case and should be replaced.

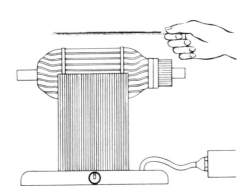

FIGURE 10–33 Armature testing using a growler.

Service of Bearings

As stated previously, modern aircraft generators are usually equipped with prelubricated sealed bearings. During normal service inspections, it is not necessary to lubricate or otherwise service bearings of this type. If a bearing seizes or becomes rough, it should be replaced with a new one. Bearings can be checked by rotating the armature of the assembled generator by hand. The armature should turn freely and smoothly. If any roughness is noted, the bearings should be replaced.

Seating of New Brushes

If new brushes are to be installed in a generator, they must be seated so that the face of each brush will have maximum surface contact with the commutator. The brushes should be installed after the generator is assembled and then seated as follows: Place a strip of No. 000 sandpaper around the commutator with the sand surface against the brush face, and turn the armature in the normal direction of rotation. This causes the sandpaper to grind the face of each brush on a contour with the commutator. When the face of each brush is ground sufficiently to make maximum contact with the sandpaper of the commutator, remove the sandpaper and blow out all sand and brush particles with dry compressed air.

Another method of seating brushes recommended by some manufacturers is as follows: Mount the generator on a test stand so that it can be rotated at normal operating speeds. Install the brushes in their proper positions, and run the generator at approximately 1500 rpm [157 radians per second (rad/s)]. Fold a strip of No. 000 sandpaper over the end of a rigid piece of insulating material with the sand surface outside. Hold the sand surface against the commutator while the generator is rotating. Fine sand particles will be carried across the face of each brush, and the brush faces will be shaped into the contour of the commutator. After the brushes are seated, blow out all sand and dust with dry compressed air.

A third method of seating brushes, which has proved very satisfactory, is to use an abrasive stick specially designed for the purpose. This abrasive stick should be used in the same manner as the sandpaper described in the above paragraph. As the abrasive stick is held against the commutator, small particles of abrasive material are carried under the brushes, and the brush faces are ground to the contour of the commutator.

Installation

There are two basic types of driving mechanisms used on aircraft generators, direct-gear and belt-driven. To install a gear-driven generator, remove the mounting-pad cover and install the proper gasket over the studs. Fit the generator spline or gear into place, being careful not to damage it. Tighten the nuts on the hold-down studs, applying torque as recommended for the size of the nuts used. Connect the generator cables to the proper terminals, and see that all connections are clean and tight. If the generator employs an air duct for cool-ing, make sure that it is connected properly to avoid the possibility of its coming loose during operation.

To install a belt-driven generator, mount the unit in the appropriate location and install all hardware. Then place the drive belt around the aircraft and generator drive pulleys. Once the belt is in the proper location, position the generator to tighten the drive belt, and secure all hardware and safety equipment as necessary.

After generator installation has been completed, the entire charging system should be tested. Both a static and a dynamic check should be performed to ensure that the generator functions properly under various load and rpm conditions. If all voltage and amperage readings are within specifications, the aircraft can be returned to service.

REVIEW QUESTIONS

1. Explain the electrical principle by which electricity is produced in a generator.
2. How may the direction of current flow in an armature be determined?
3. Name the essential parts of a dc generator.
4. What determines the voltage value in a generator?
5. How may commutator ripple be reduced?
6. How is residual magnetism utilized in a generator?
7. Give two classifications for dc generators with respect to internal circuit connections.
8. Explain armature reaction.
9. How may armature reaction be reduced?
10. Describe the functions of a starter-generator.
11. What types of aircraft typically employ starter-generators?
12. Describe the current flow path through a starter-generator during the generator mode.
13. Describe the armature assembly for a typical aircraft generator.
14. What method is used to reduce the shock load on the shaft of an aircraft generator?
15. Compare the shunt field winding in an aircraft generator with the series winding.
16. Give the approximate brush pressure required for aircraft generators.
17. What type of bearing is commonly used in aircraft generators?
18. What service is required for prelubricated sealed bearings?
19. Describe means for cooling aircraft generators.
20. For voltage regulation, which of the following is varied: rpm, field strength, or number of armature windings?
21. Describe the action of a vibrator-type voltage regulator.
22. What is the function of the carbon pile in a carbon-pile voltage regulator?
23. Describe the operation of a carbon-pile voltage regulator.
24. Describe the operation of an equalizing circuit.

25. Why is a reverse-current cutout relay required in a generator system?

26. Explain the operation of a reverse-current cutout relay.

27. What voltage is required of a generator in a 24-V system?

28. Explain the operation of a current limiter used with a vibrator-type voltage regulator.

29. What is the purpose of an overvoltage relay in a generator system?

30. What unit is used to aid in the installation of many starter-generators?

31. How often are inspections required for starter-generators?

32. Give the procedure for balancing generator load in an airplane.

33. What is meant by *flashing the field*?

34. If a generator produces only 2 or 3 V, what is likely to be the trouble?

35. What trouble may exist if there is no voltage from a generator?

36. List typical generator inspection procedures.

37. If you wish to overhaul a generator, what information should you have available?

38. How would you smooth the commutator of a generator?

39. What is the proper method for seating new generator brushes?

40. What is used to test the armature of a dc generator?

41. How would you test an armature for grounded windings?

42. Describe the use of an ohmmeter to test a field for grounded windings.

43. Describe the steps for the installation of a typical gear-driven aircraft generator.

11 Alternators, Inverters, and Related Controls

There are two major types of alternators currently used on aircraft, the **dc alternator** and the **ac alternator.** DC alternators are most often found on light aircraft where the electric load is relatively small. AC alternators are found on large commercial airliners and many military aircraft. Since these aircraft require large amounts of electric power, the use of ac systems creates a valuable weight savings. Through the use of transformers, the transmission of ac electric power can be accomplished more efficiently and therefore with lighter equipment. With the transmission of electric power at relatively high voltages and low current, the power loss is kept to a minimum. As discussed in Chapter 6, transformers are used to step up or step down ac voltage.

In large aircraft, ac power is used directly to perform the majority of power functions for the operation of control systems and electric motors for a variety of purposes. On light aircraft, most electric devices operate on 14- or 28-V dc power. If a small amount of alternating current is desirable for specific applications, an inverter is used to convert the dc voltage into an ac voltage. The ac voltage is then used to power only those particular items requiring alternating current for proper operation.

AC GENERATION

Principles of AC Generation

The principle of **electromagnetic induction** has previously been explained as it relates to both dc and ac generators. To repeat briefly, when a conductor is cut by magnetic lines of force, a voltage will be induced in the conductor, and the direction of the induced voltage will depend on the direction of the magnetic flux and the direction of movement across the flux.

Consider the simple generator (alternator) illustrated in Figure 11–1. A bar magnet is mounted to rotate between the faces of a soft-iron yoke on which is wound a coil of insulated wire. As the magnet rotates, a field will build first in one direction and then in the other. As this occurs, an alternating voltage will appear across the terminals of the coil. The wave shape of the ac voltage will roughly approximate a sine wave.

Aircraft Alternators

Almost all alternators for aircraft power systems are constructed with a **rotating field** and a **stationary armature.** Since a steady voltage must be provided for the aircraft's electrical system, the field strength of the alternator must be varied according to load requirements. For this purpose a **regulator** is employed that can furnish a variable direct current to the rotor (field) winding of the alternator, and a voltage-regulator system is used to change this current as required to maintain a constant alternator output voltage. This variable regulator current must be supplied by a dc source.

Aircraft alternators and generators have many similarities; both units change mechanical energy into electric energy. The major differences between a dc alternator and a dc generator are the various design features. Since a generator has a rotating armature, all the output current must be supplied through the commutator and brush assembly. An alternator, having a stationary armature, can supply its output current through direct connections to the aircraft bus. This system of directly contacting the alternator output to the bus eliminates the problems caused by poor connections between a rotating commutator and stationary brushes. At high power levels, rotating contacts are too inefficient to be practical; therefore, alternators, as opposed to generators, are preferred on most aircraft.

Principles of Aircraft Alternators

The aircraft alternator is a **three-phase** unit rather than the single-phase type shown in Figure 11–1. This means that the **stator** (stationary armature) has three separate windings, effectively 120° apart. The field rotates and is called the **rotor.** The schematic illustration in Figure 11–2 will serve to

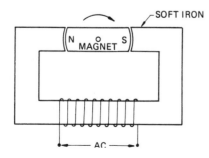

FIGURE 11–1 Simple ac generator.

210

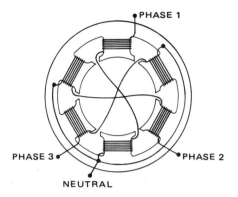

FIGURE 11–2 A Y-connected stator winding.

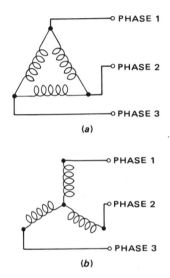

FIGURE 11–3 Diagrams of (a) delta- and (b) Y-connected stators.

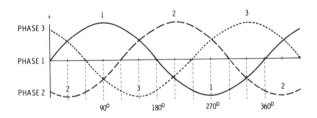

FIGURE 11–4 Output of a three-phase alternator.

indicate how the stator windings are arranged, although the windings in an actual stator will appear different. Also, it will be found that some stators will be wound in the Y configuration, and others will be wound in the delta configuration. Schematic diagrams of these arrangements are shown in Figure 11–3.

The output of a three-phase alternator is shown in Figure 11–4. Note that there are three separate voltages 120° apart; that is, each voltage attains a maximum value in the same direction at points 120° apart. As the rotor of the alternator turns, each phase goes through a complete cycle in 360° of

rotation; that is, each voltage reaches maximum in one direction, passes through zero, reaches maximum in the opposite direction, and then returns to the starting point in 360°.

Alternator System for Light Aircraft

The alternator system for light aircraft is similar to a power system with a dc generator; however, the actual output of the alternator is alternating current. To use this current in a light-aircraft power system, it is necessary to convert it into direct current. This is accomplished by means of a **three-phase, full-wave rectifier.** A rectifier for three-phase alternating current consists of six silicon diodes if the rectifier is designed for full-wave rectification. A schematic diagram of a delta-wound stator with a three-phase, full-wave rectifier is shown in Figure 11–5. The arrowheads, which represent the diodes, point in a direction opposite the actual electron flow. Under the conventional system (current flow from positive to negative), the arrowheads would point in the direction of flow. In the diagrams in Figure 11–6, it can be seen how the current produced in each phase of the stator is rectified.

The three separate voltages produced by each phase of the armature overlap, as seen in Figure 11–6. Once the current is rectified, the voltage curves remain overlapped; however, since the stator is wired in parallel, only the strongest voltage reaches the alternator output terminals. As illustrated in Figure 11–6, the effective voltage is an average of the voltage values above the intersection of the individual voltage curves. The effective voltage is equal to the rated output voltage of the alternator. This value averages near 14 V for a 12-V battery system and 28 V for a 24-V battery system. The dc ripple voltage values actually range from approximately 13.8 to 14.2 V or 23.8 to 24.2 V. However, the dc ripple voltage changes value so quickly and so little that for all practical purposes, the voltage of the aircraft electrical system is considered to be the effective voltage of the alternator.

A typical electric power circuit is shown in Figure 11–7. Since the rectifier is mounted in the end frame of the alternator, the alternator output terminals are marked for direct current.

Light-Plane Alternator

A typical alternator for light aircraft is shown in Figure 11–8. Units similar to this are manufactured by such companies as the Ford Motor Company, Prestolite, the Chrysler

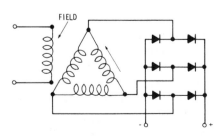

FIGURE 11–5 A schematic of a three-phase, full-wave rectifier in an alternator stator circuit.

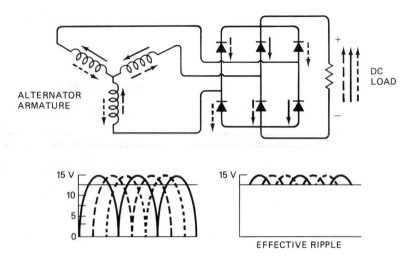

FIGURE 11–6 Rectification of a three-phase current.

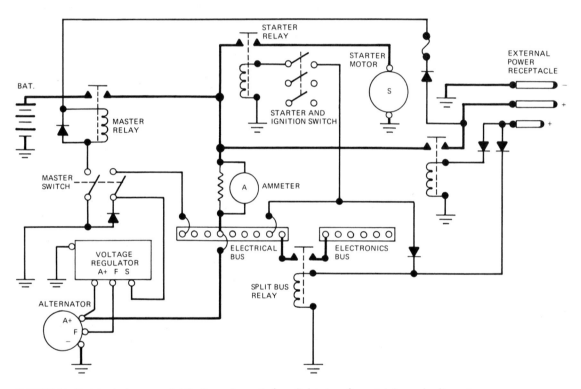

FIGURE 11–7 A typical power distribution schematic for a light aircraft containing a dc alternator.

Corporation, and the Delco-Remy Division of General Motors Corporation. The particular type of alternator to be used in an aircraft system can be determined from the aircraft manufacturer's parts catalog or from the catalog prepared by the manufacturer of the alternator.

The alternator is a comparatively simple device and is designed to give many hours of trouble-free service. The principal components are the three-phase stator (armature windings), the rotor (field windings), and the rectifier assembly. The rotating field winding provides the electromagnetic field, which is used to excite the stator windings. A brush set and slip-ring assembly is used to transfer current to the rotating field. Since the field coil requires a relatively low

amperage (approximately 4 A maximum) to power the electromagnet, the brushes are smaller and longer-lasting than those found on dc generators. The brush assembly of a dc generator often carries well over 50 A. The stationary armature receives an induced voltage, which is connected to the rectifier assembly. The rectifier consists of six diodes connected to form the three-phase, full-wave rectifier.

A typical alternator for light aircraft has a rotor with 8 or 12 poles alternately spaced with north and south polarity. This provides the rotating field within the stator. The strength of the rotating field is controlled by the amount of current flowing in the rotor winding. This current is governed by the voltage regulator. The output of the stator is applied to a

FIGURE 11–8 A typical alternator for a light airplane.

FIGURE 11–9 A gear-driven alternator with drive gear removed.

full-wave rectifier consisting of six diodes mounted within the alternator housing. The output of the alternator is, therefore, direct current as it is supplied to the aircraft electric power system.

Alternators for light aircraft may be driven by a belt and pulleys, or they may be gear-driven and flange-mounted on the engine. In the latter case the engine manufacturer must provide the correct mounting and gear drive for the alternator.

Figure 11–9 shows a gear-driven alternator. The internal construction of the alternators is the same and consists of the components shown in Figure 11–10. The drive end head contains a prelubricated bearing, an oil seal, a collar and shaft seal, and a blast tube connection for ventilation.

The rotor is mounted on a shaft with a ventilating fan on the drive end. The slip rings, slip-ring end bearing inner race, and spacer are on the other end of the shaft. The rotor windings and winding leads are treated with high-temperature epoxy cement to provide vibration and temperature resistance. High-temperature solder is used to secure the winding leads to the slip rings.

The stator of the alternator has a special electric lead that is connected to the center of the three-phase windings. This lead may be used to activate low-voltage warning systems or relays. The center connection of the three-phase stator is not always needed for external circuitry; in this case the connec-

tion is insulated and secured to the stator core. The entire stator assembly is then coated with a heat-resistant epoxy varnish.

Six diodes rectify the ac voltage produced in the armature. There are three positive and three negative diodes, each mounted into an assembly plate. The positive and negative diode plates are insulated from each other, and the positive plate is insulated from the alternator case. In many alternators the diodes are not individual units; all six diodes must be replaced as one assembly. Each diode is connected to the alternator stator by means of a high-temperature solder or a solderless crimp-type terminal.

The brush end housing provides the mounting for the rectifiers and rectifier assembly plates, the output and auxiliary terminal studs, and the brush and holder assembly. The brush end housing also contains the roller bearing, the outer race assembly, and a grease seal.

The brush and holder assembly contains two brushes, two brush springs, a brush holder, and insulators. Each brush is connected to a separate terminal stud and is insulated from ground. The brush and holder assembly can be easily removed for brush inspection or replacement purposes. In

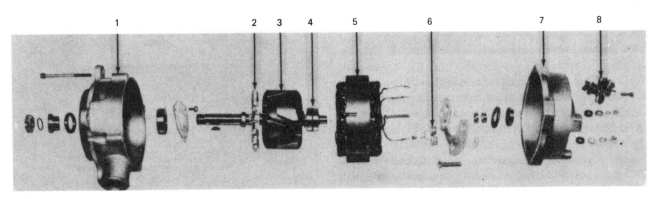

FIGURE 11–10 Components of an alternator. (1) Drive end housing, (2) cooling fan, (3) rotor (field coil), (4) slip rings, (5) stator (armature assembly), (6) rectifiers, (7) brush end housing, (8) brush assembly.

some cases the brush assembly can be removed only after the alternator has been disassembled. These brushes are not inspected during a routine aircraft inspection.

Maintenance of Alternators

Maintenance of alternators follows the principles of good mechanical and electrical practice and should be accomplished according to the instructions given in the maintenance manual for each particular unit requiring service. In general, the disassembly procedure is similar to that for other generators. Care must be taken to assure that the parts are marked and identified in such a manner that they can be reassembled correctly.

The rotor winding can be tested with an ohmmeter or continuity tester. The reading is taken with the test probes of the instrument applied to the slip rings. The resistance of the rotor winding should be relatively low and within the limits specified by the manufacturer. Grounding of the rotor winding can be tested by connecting one test probe of an ohmmeter to the rotor shaft and the other to one of the slip rings. The reading should indicate infinite resistance. If current flow is indicated, the rotor must be replaced.

The stator windings can be tested by checking between the stator leads with the ohmmeter. The reading in each case should be within specifications. Normally, the reading will show low resistance. If the resistance is above or below the limits specified by the manufacturer, the stator must be replaced. To test for grounded windings in the stator, the ohmmeter is connected between one stator lead and the stator frame. The ohmmeter should show infinite resistance.

To test for open windings in the stator, one test probe of the ohmmeter is connected to the auxiliary terminal or to the stator winding center connection. The other probe is connected to each of the three stator leads, one at a time. The ohmmeter should show continuity in each case, and the resistance should be in the range specified by the manufacturer.

A visual inspection of an armature or field winding may also indicate defects. If the coil shows signs of overheating or contains loose windings, a shorted or faulty circuit is likely. Be especially careful when performing the ohmmeter test; the defect may be intermittent and difficult to locate. Any stator or rotor containing loose windings must be replaced. If these coils have not yet failed, they will soon.

VOLTAGE REGULATORS

Unlike dc generators, dc alternators require only two means of control, a voltage regulator and a current limiter. The reverse-current cutout relay is not needed, because the alternator's rectifier resists any current flow into the armature. The current limiter for most dc alternators is a simple circuit breaker. The **voltage regulator** may be a vibrating type, as discussed with regard to generators, or a transistorized unit. In either case, the voltage regulator controls alternator output by varying the alternator's input. Specifically, the voltage

regulator increases the field circuit's resistance to decrease the alternator's output. Conversely, a decrease of the field circuit's resistance will increase the alternator's output.

Transistorized Voltage Regulators

One type of **transistorized voltage regulator** contains a field relay that supplies current to the transistor; the transistor controls the current to the field. A transistor contains no moving parts; therefore, there are no contact points to fail and/or change resistance. As the contact points of a vibrating-type regulator become pitted, the accuracy of the regulator decreases and the unit eventually fails; therefore, transistorized regulators are generally considered more accurate and reliable.

Some voltage regulators contain a field relay to "turn on" the regulator and use a transistor to regulate the alternator output voltage. One example of this type of regulator consists of a field relay, a transistor, a voltage regulator winding, a diode, and resistors. A voltage regulator of this type is shown in Figure 11–11.

The field relay is similar to relays previously discussed. Essentially, the relay is an electromagnetically controlled switch used to connect the alternator output to one terminal of the alternator field. The other side of the field circuit is controlled by the transistor and the voltage regulator coil.

The field control circuit of the transistorized voltage regulator is shown in a simplified form in Figure 11–12. In this circuit the field relay is controlled by the alternator master switch. The alternator master switch is often one-half of a dual master switch. The other half of the switch controls a solenoid, which connects the aircraft battery to the bus. When the alternator master switch is closed, the field relay closes and connects the transistor base connection to the aircraft ground (negative connection). In a typical system this will allow about 4.5 A to flow through the field winding.

In the discussion of transistors in Chapter 6, it was explained that a *pnp* transistor becomes a good conductor

FIGURE 11–11 A semi-solid-state voltage regulator.

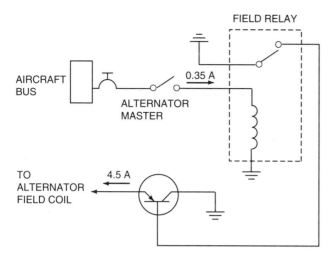

FIGURE 11–12 Simplified schematic of a transistorized voltage regulator.

through the emitter-collector section when the base circuit has negative bias. In the case of the regulator under discussion, the base circuit carries approximately 0.35 A when the voltage regulator contact points are closed.

When the alternator voltage attains the value for which the regulator is adjusted, the regulator contact points are opened by the magnetic force of the voltage relay, thus cutting off the base current to the transistor. The emitter-collector section then becomes nonconductive, and the field current is blocked. The alternator voltage then drops, and the regulator points close again to provide bias for the transistor-base circuit. Field current can then flow through the transistor, and the voltage rises to the regulated value. This cycle continues, with the regulator contact points vibrating rapidly (about 2000 times per second) to maintain the alternator voltage at the required value.

In this type of voltage regulator, the transistor carries the field current (4.5 A), while the contact points control a much lower current (0.35 A). By applying a relatively low current through the contact points, the reliability of the voltage regulator is significantly increased.

A more complete schematic diagram of the voltage regulator and alternator circuits is shown in Figure 11–13. The diagram of the alternator shows the stator, in which the alternating current is generated; the field coil and slip rings, through which current flows to the field coil; and the six-diode rectifier. When the alternator master switch is closed, the battery and alternator are connected to the field-relay coil to produce a magnetic field that closes the relay contacts. When this happens, current flows from ground, through the emitter-collector circuit of the control transistor and the F_1 terminal of the regulator, and to the F_1 terminal of the alternator. After passing through the alternator field, the current enters the F_2 terminal of the regulator and passes through the closed field-relay contact points and out the *BAT* battery terminal of the regulator. From this point it flows to the positive (+) terminal of the alternator, through the rectifier network, and to ground.

When the field winding of the alternator is energized, a dc voltage will be delivered to the system by the alternator, provided, of course, that the alternator is running. As alternator voltage increases, current flow through the two windings of the voltage regulator coil will increase. When the voltage reaches the value for which the regulator is adjusted, the contact points in the regulator section open. This lowers the emitter-base current in the transistor, and the transistor lowers the current to the alternator field.

With the contact points open, the alternator field winding receives less current, and the alternator voltage immediately decreases. With a lower alternator output, the spring at the end of the contact arm closes the points, and the cycle repeats as previously explained. The high rate of vibration of the contact points provides a steady voltage, for all practical purposes. Since the voltage regulator contact points are held closed by a spring when the voltage is below the desired value, and the points are opened as a result of alternator voltage reaching this value, an increase of spring tension will cause the alternator voltage to increase. Voltage adjustments are therefore made by turning the screw that controls the spring tension.

It is important to note that the voltage regulator contact points carry only about 0.35 A when the alternator field current is over 4 A. On regulators without transistor control, full generator field current must pass through the regulator contact points. Since vibrating contact points are burned by higher amperages, the use of a transistor makes it possible to increase the life of the contact points because of the lower current through the points.

In the diagram in Figure 11–13, observe that the **accelerator winding** on the voltage regulator is connected to the regulator's F_2 terminal (a positive voltage point) and through a resistor and the contact points to ground. This winding will therefore carry less current when the contact points are open. The effect of this arrangement is to reduce the magnetic pull on the contacts as soon as the contact points open, thus making the spring more effective in closing them again. When the contact points are closed, the magnetic effect of the accelerator winding is added to the total magnetic force again, and the points reopen very quickly. The effect of the accelerator winding thus causes the contact points to vibrate (open and close) much more rapidly than they would with the shunt coil only. This is the reason for the term *accelerator winding*.

The diode in the regulator is connected directly across the field winding. If the voltage contacts opened without a diode in this circuit, the sudden interruption of field current and the resulting high voltage induced in the field winding would damage or destroy the power transistor. The diode is connected in such a manner to offer a high resistance to any applied voltage and a low resistance to any voltage of reversed polarity. The high voltage induced in the field relay, as the contact points open, is a reverse-polarity voltage; thus the diode will short any current produced in this manner. The shorted current is therefore unable to damage the transistor.

In the use of transistors, it is important to note that high temperatures can cause improper functioning of and permanent damage to the transistors. For this reason the transistors

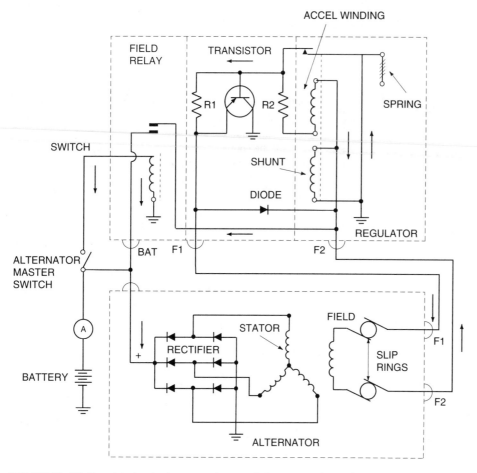

FIGURE 11–13 Transistorized voltage regulator and alternator schematic.

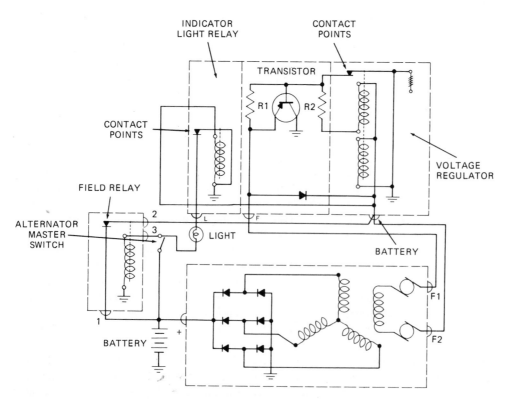

FIGURE 11–14 Transistorized voltage regulator with an indicator light circuit.

used with voltage regulators or in any other circuits must be kept at safe operating temperatures. The transistors used with voltage regulators usually have heavy metal bases, which act as heat sinks to carry the heat away from the active elements. In any installation the maximum safe operating temperature for the transistor must be known, and provision must be made to assure that this temperature is not exceeded.

Some aircraft utilize an indicator light to show when the alternator is not charging the battery. One example of this type of system employs a three-terminal voltage regulator, an external field relay, and an indicator light. The circuit for such a system is shown in Figure 11–14. An examination of this circuit shows that the alternator master switch connects the battery to the field-relay winding and causes the contact points to close. The battery is also connected to the indicator light. The indicator light circuit is completed to ground through the contact points of the indicator light relay. These points are closed when the relay is not energized; hence the light will be on. Since the field relay closes when the alternator master switch is turned on, the battery is connected to the field winding of the alternator. The alternator output voltage then opens the indicator light relay, thus turning off the indicator light.

Solid-State Voltage Regulators

A circuit diagram to illustrate the operation of a completely solid-state voltage regulator is shown in Figure 11–15. This regulator has no moving parts and is generally considered more reliable. In the description that follows, each item will be explained in terms of actual current flow from negative to positive. For example, when the battery is furnishing current to the circuit, current flows from the negative terminal (ground) through the circuit into the positive terminal of the battery. When the battery is being charged, current is flowing into the negative terminal of the battery and out the positive terminal.

In the circuit in Figure 11–15, when the alternator master switch is closed, the battery and alternator are connected to the relay and through the relay to the positive terminal (A) of the regulator. There is then a complete circuit from ground through the resistor R_2, the base of the **power transistor** TR_1, the diode D_1, and the resistor R_1 and back to the battery and the positive terminal of the alternator. If the output of the alternator is below the voltage for which the regulator is set, transistor TR_1 will have forward bias, and current will flow from the F_1 terminal of the alternator to the F terminal of the regulator and through the emitter-collector circuit of TR_1. The circuit is completed through D_1, R_1, and the relay to the alternator. This current flow excites the field of the alternator, and the output of the alternator quickly rises to the desired level. In the circuit in Figure 11–15, it can be seen that there is a circuit from ground through R_6, R_5, the **zener diode,** R_3, and R_1 to A. There is also a circuit from the zener diode through the emitter-base circuit of the **control transistor** TR_2 and through R_1 to terminal A of the regulator. The zener diode blocks the flow of current from R_5 until the voltage between ground and A reaches approximately 14.5 V. At this point the zener diode begins to conduct and applies a forward bias through the emitter-base circuit of TR_2, the control transistor. TR_2 then becomes conductive, and current flows through the emitter-collector section from ground. This current flow is from ground through R_2, TR_2, and R_1 and out A. The effect of this is to short-circuit the emitter-base circuit of TR_1, and this causes TR_1 to stop conducting field current for the alternator. The alternator voltage immediately drops, and the zener diode stops conducting, thus removing the forward bias from TR_2, which also stops conducting. This returns the forward bias to TR_1, which starts conducting field current again, and the cycle repeats. This cycle repeats about 2000 times per second, thus producing a reasonably steady voltage of approximately 14.5 V from the alternator.

Transistorized Voltage Regulator Operating Theory

The two key points to understand with respect to the operation of the transistorized voltage regulator are the zener diode operation and the control of the power transistor by the control transistor. The zener diode can be compared to a relief valve that opens at a given pressure in a hydraulic system. When the zener diode conducts current, it causes the control transistor to shut off the power transistor. The reason that the control transistor can stop the flow of current through the emitter-base circuit of the power transistor is that there is a difference in the voltage drops across the emitter-base circuits of the two transistors when the control transistor's emitter-base circuit is conducting. The diode D_1 causes approximately a 1-V drop in potential across the emitter-base circuit of the power transistor when the circuit is conducting. When the emitter-collector circuit of the control transistor begins to conduct, there is no appreciable voltage drop across the control transistor; hence a 1-V reverse bias becomes effective across the emitter-base circuit of the power transistor. This, of course, stops the emitter-base current in the power transistor.

Adjustment of alternator voltage output is accomplished through the variable resistor R_5. A change in the resistance of this resistor will change the voltage level across the zener

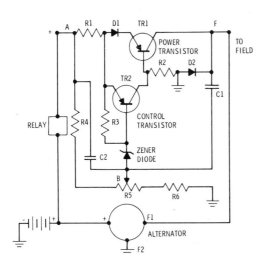

FIGURE 11–15 A schematic of a dc alternator and transistorized voltage regulator.

diode, thus raising or lowering the level of alternator output voltage required to cause the zener diode to conduct.

The resistor R_1 and capacitor C_1 act to reduce the time required for the field voltage to change between maximum and minimum values. This prevents overheating of the transistors. The capacitor C_2 reduces the voltage variations that appear across the resistors R_4 and R_5, thus making the regulator more accurate. Resistor R_3 prevents leakage current from the emitter to the collector in the control transistor. Resistor R_4 is a special temperature-sensitive type that acts to increase the alternator voltage slightly at a lower temperature. This aids in maintaining adequate charge current for low-temperature operation. Diode D_2 aids in controlling field current flow as the power transistor rapidly turns the field current on and off.

There are many modern voltage regulators that also monitor system voltage for any improper values. Figure 11–16 shows a Cessna **alternator control unit** (ACU). This ACU not only regulates alternator output voltage but also turns off the charging system if an overvoltage condition exists. The ACU also controls a low-voltage light, which illuminates in the event of a charging system failure.

The ACU normally maintains system voltage between 28.4 and 28.9 V. This is accomplished by changing the field circuit resistance through the ACU. If system voltage rises above 28.9 V, the ACU automatically opens the alternator field circuit and shuts off the alternator. This in turn establishes a low-voltage condition (battery voltage only) and causes the low-voltage warning light to illuminate. If system voltage should drop below 25.2 V, the low-voltage warning light will illuminate.

The low-voltage warning circuit can be tested by turning on a heavy load, such as pitot heat and landing light, and momentarily turning off the alternator master switch. While the system is operating on battery voltage only, the low-voltage sensor should illuminate. The load must be applied for this test to ensure that the battery voltage is below 25.2 V.

FIGURE 11–16 An alternator control unit.

Troubleshooting a DC Alternator System

Typical procedures for troubleshooting a light-aircraft alternator system can be found within the electrical section of the aircraft maintenance manual. It must be remembered that there are variations with different systems; hence it is important to consult the manufacturer's instructions for a particular system. A general rule to remember in all cases is that alternator voltage is controlled by varying the amount of field excitation current. The voltage regulator can therefore be bypassed by supplying field current directly to the alternator. If the engine is run while the regulator is bypassed, the alternator should produce a relatively high output voltage and

---

BE AWARE

DIFFERENCES BETWEEN AIRCRAFT AND AUTO ALTERNATORS USING A FORD BELT-DRIVEN 12 V OR 24 V ALTERNATOR FOR A COMPARISON

Aircraft alternators include features not found on automotive alternators.

1. Although alternators are birotational, aircraft engines turn opposite of automotive. This means cooling fans must be canted in the opposite direction. Also, pulleys and belt sizes vary due to coming-in speed.

2. The through bolts are of a higher tensile strength utilizing an antirotational device in the form of a lock tab. The rectifier assembly has a heavy-duty diode with higher voltage and amperage capacity. Also, one excites at 90 PIV and the other at 150 PIV. Radio suppression is designed for 108 frequencies and up, which is the VHF, and 108 and down, which is FM band.

3. The brushes have a higher graphite content and they utilize a tin plate on the brush leads to prevent corrosion.

4. The stator is of the Delta wind rather than the "Y" wind, and it does not utilize the stator terminal. The aircraft unit also carries "H" insulation, which is capable of 200° Centigrade temperatures. It is also rated at 60 amperes instead of 55.

5. The rotor has a shorter shaft and a smaller thread size. Because of the opposite rotation, it is wound in the opposite direction. It also utilizes "H" insulation and Havel varnish.

6. The front and rear housings are the same as automotive. With this brief description, I hope I have enlightened you on the differences between aircraft and automotive alternators. Using automotive units in an aircraft creates a potential safety hazard, as well as a short alternator life and unreliability.

If you suspect an automotive unit on an aircraft, check with your nearest FAA-approved aircraft accessory shop or your local FAA General Aviation District Office.

NOTE: The above article was submitted by an FAA certificated Aircraft Accessory Shop.

FIGURE 11–17 Portion of FAA Alert #63. Differences between aircraft and automotive alternators.

output current. If no voltage is produced during this test, the alternator or related circuitry is defective. All radios and sensitive electronic equipment should be turned off during this test to ensure their protection from an overvoltage condition.

Whenever troubleshooting an alternator circuit, remember that alternator output voltage must be slightly higher than battery voltage. If the alternator is turned on and the bus voltage fails to increase approximately 2 to 4 V above battery voltage, the charging system or related circuitry is defective. This voltage measurement is typically performed between the aircraft's main bus and ground.

Be Aware. Although there are several similarities between some aircraft dc alternators and automotive alternators, they are different. Automotive alternators and voltage regulators must not be substituted for aircraft components. A portion of the FAA Alert #63 is shown in Figure 11–17. This illustrates several differences between an aircraft alternator and an automotive alternator. When installing replacement parts on any aircraft, always use FAA-approved units.

AC GENERATORS

AC generators, often called alternators, are used as the principal source of electric power in almost all transport-category aircraft. The ac system supplies almost all the electric power required for the aircraft. Where dc is needed, rectifiers are used. For emergency situations ac generators driven by **auxiliary power units** (APUs) or **ram air turbines** (RATs) are often used.

AC power systems produce more power per weight of equipment than dc systems; however, all ac generators require a constant-speed drive to maintain a constant ac frequency. A **constant-speed drive** (CSD) is a type of automatic transmission that maintains a constant output rpm with a variable input rpm. Since heavy aircraft use large amounts of electric power, the employment of a constant-speed drive and an ac generator is practical. On light aircraft, where a relatively small amount of electric power is used, an ac generator requiring a constant-speed drive is simply too heavy. Modern light aircraft use dc alternators to produce electric power. A dc alternator produces an ac voltage and uses internal rectifiers to produce a dc output.

The stationary part of the alternator circuit is called the **stator,** and the rotating part is called the **rotor.** The stator is actually a stationary armature, and the rotor is a rotating field, which may be produced by either a permanent magnet or an electromagnet. As the rotor turns, the magnetic flux cuts across the stator poles and induces a voltage in the stator winding. The induced voltage will reverse polarity every half revolution of the rotor because the flux will reverse in direction as the opposite poles of the rotor pass the stator poles. One complete revolution of the rotor in a two-pole alternator will produce 1 cycle of alternating current; that is, one complete ac sine wave will be produced for each complete revolution of the rotor.

The number of cycles of alternating current per second is called the **frequency.** Since a two-pole alternator produces 1 cycle per revolution (cpr), it is apparent that an alternator produces 1 cycle of alternating current from each pair of poles in the rotor. If we wish to determine the frequency of any given alternator, we proceed as follows: Divide the number of poles by 2, and multiply the result by the speed in rpm to obtain the number of cycles per minute. To find the cycles per second, divide the cycles per minute by 60.

Let us assume that we wish to determine the frequency of an alternator having four poles and turning at 1800 rpm. Dividing 4 by 2 gives 2 cpr, and 2 cpr multiplied by 1800 rpm equals 3600 cycles per minute. Dividing 3600 by 60 (60 s/min), we obtain 60 Hz (cycles per second).

Typically, alternators use an electromagnet for the field, which receives a direct current to excite the armature. An alternator with a four-pole electromagnetic rotor is illustrated in Figure 11–19. A permanent-magnet rotor is not satisfactory, because the strength of the field flux always remains the same, and voltage regulation cannot be accomplished. For this reason an electromagnetic field is always used for aircraft applications.

Alternators are classified according to voltage, amperage, phase, power output (watts or kilovolt-amperes), and power factor. The phase classification of an alternator is the number of separate voltages that it will produce. Usually, alternators are single-phase or three-phase, depending on the number of separate sets of windings in the stator. Three-phase alternators are typical for most aircraft applications. Three-phase alternators are constructed with three separate armature windings spaced so that their voltages are 120° apart.

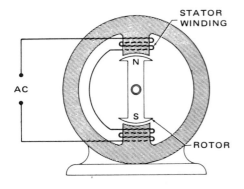

FIGURE 11–18 A simple two-pole ac alternator.

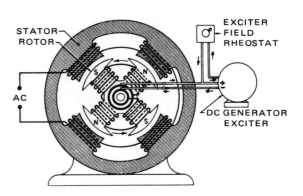

FIGURE 11–19 An ac alternator with a dc-excited field.

High-Output Brushless Alternators (Generators)

High-output brushless alternators were developed for the purpose of eliminating some of the problems of alternators that employ slip rings and brushes to carry exciter current to the rotating field. High-output alternators are often referred to as ac generators. When ac systems are discussed, the two terms, *alternator* and *generator,* are often considered equivalent. For the purpose of this text, the terms *ac generator* and *ac alternator* will be used interchangeably during our discussion.

Among the advantages of a brushless alternator are the following:

1. Lower maintenance cost, since there is no brush or slip-ring wear.
2. High stability and consistency of output, because variations of resistance and conductivity at the brushes and slip rings are eliminated.
3. Better performance at high altitudes, because arcing at the brushes is eliminated.

The theory behind the brushless alternator is to use electromagnetic induction to transfer current from the stationary components of the generator to the rotating components. Unlike the alternator in Figure 11–19, the brushless system induces current into the rotor using magnetic flux lines. This principle eliminates the need for the rotating contacts of slip rings and brushes.

Typically, brushless alternators use a three-phase, Y-connected armature. The voltage across any single phase is 120 V, whereas the voltage across any two of the main output terminals is 208 V. This is illustrated in Figure 11–20. One terminal of each separate stator winding is connected to ground, and the other terminal of the winding is the main output terminal. For aircraft circuits requiring 115/120 V, single-phase power, the circuit is connected between one main phase and

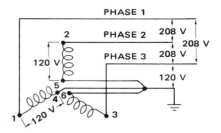

FIGURE 11–20 A Y-connected stator winding for an ac alternator.

ground. For three-phase power circuits such as those for motors, all three main phases are connected to the motor.

Modern brushless alternators are called **permanent magnet generators** (PMGs). The PMG gets its name from the permanent magnet within the generator, which initiates the production of electric power. As seen in Figure 11–21, there are actually three separate generators within one case: (1) the **permanent magnet generator,** (2) the **exciter generator,** and (3) **the main generator.** Each of these three units is an essential part of the modern brushless alternator.

The permanent magnet, which is connected to the rotor, is used to induce an alternating current into the stationary PMG three-phase armature winding (Figure 11–21). The generator control unit (GCU) rectifies a 1200-Hz ac armature current and sends a dc voltage to the exciter field winding. The exciter field induces an alternating current into the exciter armature. The exciter armature is connected to the rotating rectifier, which changes the alternating current to direct current and sends a current to the main generator field. The main field induces an ac voltage into the main generator armature.

The main generator armature is a three-phase winding that produces 120 volts across a single phase and 208 V across two phases. This armature is connected to the output terminals of the generator and hence supplies the electric power for the aircraft systems. As seen in Figure 11–21, the GCU

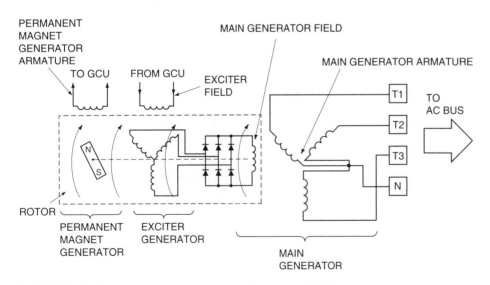

FIGURE 11–21 Diagram of a permanent magnet generator (PMG).

monitors the main generator output and in turn regulates the exciter field current as needed. If more generator output is required, the GCU will increase the exciter field current; this will, in turn, increase the exciter armature output and the main field current. A stronger main field will increase the main armature's output. If less generator output is needed, the GCU will weaken the exciter field current, and the generator output will decrease.

Constant-Speed Drive System

In an ac power system it is usually necessary to maintain a fairly constant speed in the ac generator. This is because the frequency of the ac generator is determined by the speed with which it is driven. It is especially important to maintain constant generator speed in installations in which the generators operate in parallel. In this case it is absolutely essential that generator speed be kept constant within extremely close limits.

In order to provide constant-speed generator operation in modern ac electrical systems, it is common practice to use a **constant-speed drive** (CSD), such as that manufactured by the Sundstrand Aviation Electric Power Division of the Sundstrand Corporation.

CSD units are manufactured in many designs to fit a variety of applications. The principle of operation for all CSDs is essentially the same.

The complete CSD system consists of an axial gear differential (AGD), whose output speed relative to input speed is controlled by a flyweight-type governor that controls a variable-delivery hydraulic pump. The pump supplies hydraulic pressure to a hydraulic motor, which varies the ratio of input rpm to output rpm for the AGD in order to maintain a constant output rpm to drive the generator and maintain an ac frequency of 400 Hz.

A typical ac generator and constant-speed drive assembly is shown in Figure 11–22. In this view the CSD is on the left end of the assembly, and the generator is at the right end. The generator is cooled by an oil spray delivered by the CSD section. Most CSDs are equipped with a **quick attach/detach** (QAD) adapter. This unit allows the technician to remove and replace a generator and CSD assembly in a matter of minutes. The QAD ring is mounted to the CSD by means of several bolts through the mounting flange. To remove the generator from the aircraft, one must only release the QAD using one fastener. This mounting technique allows for quick line repair of an aircraft that has a defective generator CSD.

A normal operating temperature for the cooling oil is approximately 200°F. In order to maintain the correct cooling capacity, the oil level should be monitored periodically. A sight glass, as shown in Figure 11–22, is employed on most constant-speed drives to allow the technician a quick check of the oil level. In the case of an in-flight oil loss or an overtemperature condition, a warning indicator will light up on the flightdeck. In this situation the CSD should be disengaged immediately and inspected upon landing.

Most constant-speed drive units are equipped with an electrically activated generator-drive disconnect mechanism. This disconnect couples the CSD input shaft to the CSD input spline. The CSD disconnect is operated manually from the aircraft's flightdeck or automatically by a generator control unit. The disconnect is activated in the event of certain generator system failures.

Integrated Drive Generator

The **integrated drive generator** (IDG) is a state-of-the-art means of producing ac electric power. As illustrated in Figure 11–23, the IDG contains both the generator and the CSD in one unit. This concept helps to reduce both the weight and the

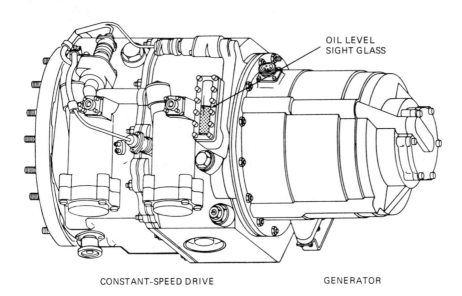

CONSTANT-SPEED DRIVE   GENERATOR

FIGURE 11–22 A typical constant-speed drive and generator assembly. *(Sundstrand Corporation)*

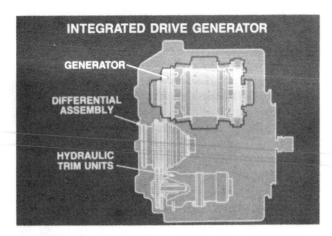

FIGURE 11–23 An integrated drive generator. *(Sundstrand Corporation)*

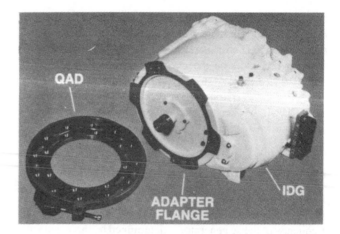

FIGURE 11–24 An integrated drive generator (IDG), an adapter flange, and a quick attach/detach (QAD) ring. *(Sundstrand Corporation)*

size of the traditional two-unit system. The CSD, containing hydraulic trim units and a differential assembly, converts the variable engine rpm to an alternator input speed of 12,000 rpm. Older alternators typically rotate at 8000 to 9000 rpm. The 30 percent increase in alternator speed, along with improved cooling features, has allowed for a reduction in alternator size without a decrease in output power.

The IDG used on many Boeing 757 aircraft is produced by the Sundstrand Corporation. This unit is capable of producing 90 kilovolt-amperes (kVA) continuously, 112.5 kVA for a 5-min overload, and 150 kVA for a 5-s overload. The output voltage is 120 V ac at 400 Hz. This IDG is shown in Figure 11–24.

Three electrical subassemblies make up the generator (alternator) portion of the IDG: the permanent magnet generator (PMG), the exciter generator and rectifier assembly, and the main generator. This generate (alternator) operates in the same way as the PMG described previously.

Generator Cooling

Owing to their compact size, most ac generators require some means of cooling during operation. Older and/or less powerful generators are typically cooled by means of ram air forced through the units. Newer systems use oil as the cooling agent. The oil is sent from the CSD through the generator and then through an air/oil heat exchanger. The air cools the oil, which is once again cycled through the CSD and generator. The use of oil cooling allows for a higher-speed rotor within the generator section. A higher-speed rotor means a lighter, more compact generator.

Generator Control Units

Aircraft electric power control systems include functions such as voltage regulation, current limiting, protection for out-of-tolerance voltage and frequency conditions, and crew

FIGURE 11–25 A typical generator control unit (GCU). *(Sundstrand Corporation)*

alerting. The major component used to perform these functions is called the **generator control unit** (GCU). A typical GCU is shown in Figure 11–25. The GCU regulates generator output by sensing the aircraft system's voltage and comparing it with a reference signal. The voltage regulator then sends an adjusted current flow to the exciter field of the main generator. This, in turn, controls the main generator's output voltage.

Protection circuitry monitors various electrical system parameters including overvoltage and overcurrent conditions, frequency, phase sequence, and current differentials. If a fault occurs, the protection circuitry then operates corresponding electric relays in order to isolate defective components. In the case of a generator system failure, the GCU senses partial loss of electric power and automatically sends the appropriate signal to the **bus power control unit** (BPCU). In this event the BPCU will automatically isolate any defective generator and reconnect the load bus to another power source. Bus power control units will be discussed in greater detail in Chapter 12.

INVERTERS

An **inverter** is a device for converting direct current into alternating current at the frequency and voltage required for particular purposes. Certain systems and equipment in aircraft electrical or electronic systems require 26-V 400-Hz ac power, and others require 115-V 400-Hz power. To provide this power, it is often necessary to employ an inverter.

Inverters are typically used on large aircraft for emergency situations only. In this case the aircraft employs engine-driven generators (alternators) to supply needed ac power during normal operating conditions. If all ac generators should fail, the inverter would then be used to convert battery dc power into ac power available for essential ac loads.

Many light aircraft employ static inverters during normal operating conditions. These aircraft require a relatively small amount of alternating current and therefore utilize engine-driven dc generators or alternators as their main electric power source. Aircraft using inverters to get the alternating current they need include the Beechcraft King Air, most Cessna 421 and 310 aircraft, and many small business jets, such as the Lear Jet 23. These aircraft use alternating current to power a variety of components, including engine instruments, heated windshields, and lighting circuits. In some cases these components are feasible only if operated by ac power; therefore, the inverter is essential. There are two basic types of inverters, rotary and static. Modern aircraft employ static inverters because of their reliability, efficiency, and weight savings over rotary inverters.

Rotary Inverters

For many years inverters were simply special types of motor-generators; that is, a constant-speed motor was employed to drive an alternator that was designed to produce the particular type of power required.

A typical **rotary inverter** is shown in Figure 11–26. The rotary section of this unit consists of a dc motor driving an ac generator. The rotors of the motor and the alternator are dynamically balanced and are mounted on the same shaft. Fans are also mounted on the shaft to provide for air cooling.

A four-pole, compound, compensating field winding and a wave-wound armature are utilized in the motor. A damper winding in the salient poles of the alternator aids in maintaining output waveshape under single-phase operating conditions.

This particular inverter utilizes an input voltage of 26 to 29 V dc. The output is 115 V, single-phase; 115 V, three-

FIGURE 11–26 A rotary inverter. *(Bendix Corporation, Electric and Fluid Power Division)*

phase; and 200 V, three-phase. Frequency is 400 Hz for all phases.

Maintenance of rotary inverters is similar to that for motors and generators. Maintenance practices are set forth in the manufacturer's maintenance or service manual.

The control box on top of the inverter should not be disassembled in the field. If it appears to be defective, the entire inverter should be sent to a repair shop that is equipped to perform the electrical and electronic work and tests that may be required.

Static Inverters

A static, or solid-state, inverter serves the same functions as other inverters. However, it has no moving parts and is therefore less subject to maintenance problems than the rotary inverter.

The internal circuitry of a **static inverter** contains standard electric and electronic components, such as crystal diodes, transistors, capacitors, and transformers. By means of an oscillator circuit, the inverter develops the 400-Hz frequency for which it is designed. This current is passed through a transformer and filtered to produce the proper waveshape and voltage. The unit shown in Figure 11–27 utilizes an input voltage of 18 to 30 V dc and produces an output of 115-V single-phase alternating current with a frequency of 400 Hz. The unit weighs 18.5 lb [8.4 kg].

Static inverters are easily removed for testing. If they require repair, they should be sent to an approved facility that is equipped to perform the work required.

FIGURE 11–27 A static inverter. *(Bendix Corporation, Electric and Fluid Power Division)*

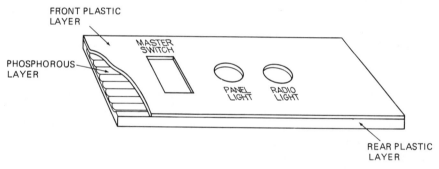

FIGURE 11–28 An electroluminescent panel.

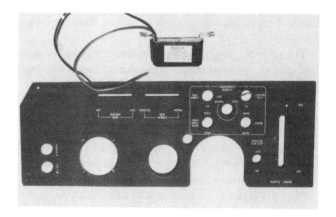

FIGURE 11–29 An electroluminescent panel and associated static inverter.

Because of the miniaturization of electronic components, static inverters have become relatively small and lightweight. This has made it possible for light single-engine aircraft to employ an ac electrical system. An electroluminescent (EL) panel is a high-efficiency lighting system powered by alternating current. The panel is constructed of phosphorous material laminated between two clear plastic layers, as shown in Figure 11–28. The phosphorous material glows when connected to an ac voltage. The front plastic layer is painted black except where stencils of appropriate letters or numbers are placed. The letters and numbers therefore remain clear in order to transmit the light from the glowing phosphorous material. Figure 11–29 shows a typical static inverter and EL panel.

VARIABLE-SPEED CONSTANT-FREQUENCY POWER SYSTEMS

In an effort to simplify and improve the production of ac power for aircraft and to get away from the need for hydromechanical constant-speed drives, a number of systems have been devised for producing 400-Hz three-phase electric power through electronic circuitry. This has been made possible by the great advances in solid-state technology developed in recent years.

Variable-speed constant-frequency systems are typically referred to as **VSCF** systems. Basically, the systems employ a generator driven at a variable speed, thus producing a variable-frequency output. The rotational speed of the generator is a direct function of the engine rpm. No constant-speed drive (CSD) mechanism is needed for VSCF systems. The elimination of the mechanical CSD improves reliability of the systems and offers more flexibility on installation of the generator. The generator's variable-frequency output current is converted to a constant-frequency 400-Hz alternating current by means of solid-state circuitry. This makes the electric power suitable for aircraft use. Several military aircraft currently use VSCF systems as primary and secondary ac power sources. VSCF systems are also found on the Boeing 737-300, -400, and -500 commercial airliners.

Utilizing state-of-the-art electronic components, the VSCF systems improve reliability over the mechanical-hydraulic units of the constant-speed drive. One VSCF system produced by Sundstrand Corporation contains only two moving parts, the oil pump and the generator rotor. There are no other parts to wear or require periodic overhaul. A major breakthrough in component design for the VSCF systems was the development of the 600-A power transistor shown in Figure 11–30. This transistor has made possible VSCF systems capable of 110-kVA output.

The VSCF systems also offer greater flexibility than the typical CSD and generator configuration. The generator must still be mounted to the engine drive mechanism; however, the control units of the VSCF system can be mounted virtually anywhere on the aircraft. Elimination of the CSD therefore allows for a more compact engine nacelle. A typical integrated unit is shown in Figure 11–31. The major subassemblies are shown in Figure 11–32; the high-power inverter assembly is shown in Figure 11–33; and Figure 11–34 shows the bottom view of the inverter and the modular power pole. As seen in this figure, the 600-A power transistors are mounted in one stack and, in this case, are capable of a 60-kVA output.

Figure 11–35 is a block diagram showing the principal elements of a VSCF system found on the Boeing 737 aircraft. The brushless ac generator is similar to those described previously; however, since it is driven directly by the engine, its rotational speed and output frequency will vary as engine speed varies. The variable three-phase power is fed to the full-wave rectifier within the VSCF converter, where it is changed into direct current and filtered. This direct current is fed to the inverter circuitry, where it is formed into square-wave outputs that are separated and summed to produce three-phase 400-Hz alternating current. The functions of the VSCF converter are similar to those of a typical static inverter. The generator converter control unit (GCCU) provides VSCF control and protection through the use of a voltage regulator and built-in test circuitry.

The B-737 VSCF components are cooled by means of spray oil for the generator, spray oil and forced air for the converter assembly, and convection cooling for the GCCU. This system produces 115/200 V three-phase 400 Hz ac and is capable of a continuous 60-kVA output, or 80 kVA for 5 s. The input speed range is 4630 to 8600 rpm. The power factor limits range from .75 lag to .95 lead. The generator, converter, and GCCU together weigh only 145 lb. (65.8 kg).

The maintainability of the VSCF system is enhanced through the use of the aircraft's fault annunciator system and built-in test equipment located directly on the convertor assembly. The built-in test (BIT) feature is designed to operate at two levels. Flight-line technicians use the first test level by activating a switch on the unit. Adjacent to the switch are two lights labeled "VSCF fault detected" and "Aircraft open

FIGURE 11–30 A 600-A power transistor. *(Westinghouse Electric Corporation)*

FIGURE 11–31 An integrated VSCF system. *(Westinghouse Electric Corporation)*

phase trip." This test will inform the technician if the fault lies in the VSCF components or the aircraft wiring. The second test level is used by technicians repairing the VSCF system once it has been removed from the aircraft.

Currently, VSCF systems are in limited use; however, if the projected reliability and operating costs are accurate, VSCF systems will be the next-generation electric power supply systems for modern aircraft.

FIGURE 11–32 The major subassemblies of a VSCF system.   *(Westinghouse Electric Corporation)*

FIGURE 11–33 The power inverter assembly of a VSCF system. *(Westinghouse Electric Corporation)*

FIGURE 11–34 The main power pole assembly of a VSCF inverter.   *(Westinghouse Electric Corporation)*

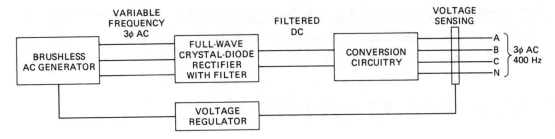

FIGURE 11–35 A block diagram of a variable-speed constant-frequency power system.

REVIEW QUESTIONS

1. Describe the operation of a typical aircraft dc alternator.

2. What are the means by which the rotor of an alternator is excited?

3. How is the production of a three-phase alternating current accomplished in an aircraft alternator?

4. In a light-aircraft electrical system, how is the direct current needed for the system obtained when the principal power source is an alternator?

5. What is a three-phase armature?

6. Compare a delta winding with a Y winding.

7. In a three-phase system, how many degrees separate the phases?

8. What is the composition of a three-phase, full-wave rectifier?

9. Describe a three-phase alternator for light aircraft.

10. By what means are alternators for light aircraft driven?

11. Describe the tests that may be made on a small alternator in order to determine that it will operate satisfactorily.

12. What is the advantage of using a transistorized voltage regulator?

13. What are the two key points that describe the operation of a transistorized voltage regulator?

14. How is voltage adjusted when a transistorized voltage regulator is used?

15. Why is it not necessary to include a reverse-current cutout relay in a system using an alternator?

16. What is the function of the zener diode in a transistorized voltage regulator?

17. What unit is used in a transistorized voltage regulator to regulate alternator voltage?

18. Describe the function of the alternator master switch.

19. Describe the differences between aircraft and automotive alternators.

20. Describe the theory of operation for a transport category aircraft's brushless alternator.

21. Explain how field excitation is accomplished in a brushless generator.

22. What design feature makes the brushless alternator self-starting?

23. What factor must be controlled when ac generators are operated in parallel?

24. What is an integrated drive generator?

25. How are large ac alternators typically cooled?

26. What is the meaning of *kVA*?

27. What is the purpose of the solid-state GCU?

28. Describe the voltage-regulator function of a solid-state GCU.

29. How is it possible to keep an alternator at a constant speed when the engine by which it is driven changes rpm?

30. Give a brief explanation of the operation of a CSD.

31. By what means does a CSD sense oil loss or an overtemperature condition?

32. How can a technician determine if an IDG has the correct oil level?

33. Explain the basic principles of a variable-speed constant-frequency electric power system.

34. What is the advantage of a VSCF system?

35. Explain the purpose of an inverter.

36. What is the advantage of a static inverter?

37. Describe the principle of a static inverter.

12 Power Distribution Systems

Modern aircraft require a consistent and reliable supply of electric power. There are four common sources of electric power used during normal aircraft operations. These sources are dc alternators, dc generators, ac alternators (generators), and the aircraft's storage battery. As discussed in Chapter 11, the aircraft's battery is typically used for emergency operations and any intermittent system overloads. DC alternators are typically used on piston engine aircraft. DC starter-generators are used on medium-sized turbine-powered aircraft. AC alternators are used on transport-category aircraft and some military aircraft. Some form of electrical distribution system must be employed on every aircraft containing an electrical system. A simple power distribution system consists of a basic copper conductor, called a **bus bar** or **bus.** This type of system is found on most single-engine aircraft. The bus is a conductor designed to carry the entire electrical load and distribute that load to the individual power users. Each electric power user is connected to the bus through a fuse or circuit breaker.

On almost all aircraft the bus bar is connected to the positive output terminal of the generator and/or battery. The negative voltage is distributed through the metal structure of the aircraft. The metal airframe (negative side of the voltage) is often referred to as the **ground;** hence this type of distribution is often called a **negative-ground system.** In all negative-ground aircraft, the positive voltage is distributed to any given piece of electrical equipment through an insulated wire, and the negative voltage is connected through the airframe. Since only one wire (and the ground) is needed to operate electrical equipment, this is known as a **single-wire system.** Single-wire systems are possible only where the airframe is constructed from a conductive material, such as aluminum. On composite aircraft, some type of ground (negative) conductor is required. In some cases two wires (one positive, one negative) are used; in other cases a ground plane is added to the structure of the aircraft. Power distribution on composite aircraft will be discussed later in this chapter.

Larger, more complex aircraft typically contain several bus bars. Each bus has the specific task of distributing electric power to a given group of electrical loads. Bus bars are often categorized as ac and dc, left and right, and essential and nonessential distribution buses. On multiengine aircraft each engine-driven alternator typically employs its own distribution bus. These generator buses are then connected to their re-

spective loads via distribution buses and associated bus ties.

As described earlier, alternators or generators are used on nearly every aircraft to produce electric power. Since both units operate similarly, the terms *alternator* and *generator* are used interchangeably throughout the aircraft industry. Although there are obvious differences between alternators and generators, in this chapter the reader should consider the terms synonymous. The following pages present the FAA recommendations concerning power distribution systems, examine the various types of systems, and present their related control circuits.

REQUIREMENTS FOR POWER DISTRIBUTION SYSTEMS

General Requirements

The general requirements for power distribution systems on normal, utility, and acrobatic aircraft are set forth by **FAR Part 23. FAR Part 25** establishes the requirements for transport-category aircraft. The specific design details of any aircraft are agreed upon by the manufacturer and the FAA prior to aircraft certification. The federal aviation regulations (FARs) set forth only basic guidelines upon which an aircraft's certification is based.

The electric power system is one of the most critical systems found on modern aircraft. A complete electrical system failure would be catastrophic. The following guidelines related to electric power systems are designed to prevent such a failure.

Electric power sources must function properly when connected in combination or independently, except that alternators may depend on a battery for initial excitation or for stabilization. No failure or malfunction of any electric power source may impair the ability of any remaining source to supply load circuits essential for safe operation of the aircraft, except that an alternator that depends on a battery for initial excitation or for stabilization may be stopped by failure of that battery.

Each electric power source control must allow for independent operation of each source. However, controls associated with alternators that depend on a battery for initial excitation do not need to provide independent operation between the alternator and its battery. A design of this type makes it possible to disconnect an alternator in a multi-generator (parallel) system without affecting the operation of other alternators or generators in the system. Of course, with one alternator or generator disconnected, there will be a subsequent load increase on the active generator(s).

There must be at least one generator in an electrical system if the system supplies power to circuits that are essential for safe operation of the aircraft. Each generator must be able to deliver its continuous rated power. If the design of the generator and its associated circuit is such that a reverse current could flow from the battery to the generator, a reverse-current cutout relay must be provided in the circuit to disconnect the generator from the other generators and the battery when enough reverse current exists to damage the system. Alternator systems do not require a reverse-current cutout relay because the diode rectifiers in the alternators prevent a reverse current flow.

Generator voltage control equipment must be able to regulate the generator output within rated limits on a continuous and dependable basis. Each generator or alternator must have an overvoltage control designed and installed to prevent damage to the electrical system, or equipment supplied by the system, that could result if the generator were to develop an overvoltage condition. There must be a means to warn the flight crew immediately in case any generator in the system should fail.

There must be a means to indicate to appropriate flight crew members the electrical quantities in the system essential for safe operation of the aircraft. Generally, one or more ammeters are required. For dc systems, an ammeter that can be switched into each generator output lead can be used, and if there is only one generator, the ammeter may be in the battery positive lead. Often on twin-engine aircraft one ammeter is used with the capability of being switched to monitor either generator output current or battery current. As previously discussed, any ammeter that measures battery current must be capable of both positive and negative current readings, because the battery can either supply or receive current. Those ammeters used to measure generator output need only indicate positive current values; the current flows from the generator to the aircraft bus.

If provisions are made for connecting an external power source to the aircraft and that external power can be electrically connected to equipment other than that used for engine starting, means must be provided to ensure that no external power source having a reverse polarity or a reverse phase sequence can supply power to the aircraft's electric power system. This is usually accomplished by the use of a plug with different-sized prongs. This makes it impossible to insert the plug in the receptacle incorrectly.

On many aircraft a diode is used to prevent any reversed-polarity current from entering the external power receptacle. The diode is placed in series with the circuit that controls the external power contactor (solenoid). If the external power system supplies current of incorrect polarity, the diode will be reverse-biased, and the contactor will not close. This occurs because a reverse-biased diode acts as an open switch; an open in the contactor circuit prevents the contactor from clos-

ing. Therefore, any reversed-polarity external power supplied to the receptacle will never be connected to the power distribution system.

The power sources and the electrical system must be able to supply the following power loads in probable operating combinations and for probable durations:

1. Loads connected to the system with the system functioning normally
2. Essential loads, after failure of any one prime mover, power converter, or energy storage device (battery)
3. Essential loads after the failure of any one engine on two-engine aircraft
4. Essential loads after the failure of any two engines on aircraft with three or more engines
5. Essential loads for which an alternate source of power is required, after any failure or malfunction in any one power supply system, distribution system, or other utilization system

Further requirements for electrical systems in transport-category aircraft specify that the generating capacity for the system and the number and kinds of power sources must be determined by a **load analysis.** The generating system includes electric power sources, main power buses, transmission cables, and associated control, regulation, and protective devices. The system must be designed so that power sources function properly when independent and when connected in combination with other sources. No failure or malfunction of any power source can create a hazard or impair the ability of remaining sources to supply essential loads. The design of the system must be such that the system voltage and frequency, as applicable, at the terminals of all essential load equipment can be maintained within the limits for which the equipment is designed, during any probable operating condition. System transients (variations in voltage and frequency) due to switching, fault clearing, or other causes must not make essential loads inoperative and must not cause a smoke or fire hazard.

There must be means accessible in flight to appropriate crew members for the individual and collective disconnection of the electric power sources from the system. The system must include instruments such as voltmeters and ammeters to indicate to appropriate crew members that the generating system is providing the electrical quantities essential for the safe operation of the system.

It must be shown by analysis, tests, or both that the aircraft can be operated safely in **VFR (visual flight rules)** conditions for a period of not less than 5 min with the normal electric power sources, excluding the battery, inoperative; with critical-type fuel, from the standpoint of flameout and restart capability; and with the airplane initially at the maximum certificated altitude. Most commercial aircraft will far exceed this 5-min minimum. For example, a Boeing 727 with a fully charged battery can operate all essential electrical systems for approximately 30 min without supplemental power from any generator.

Need for Protective Devices

Short circuits in electrical systems constitute a serious fire hazard and also may cause the destruction of electric wiring and damage to units of electric equipment. For these reasons adequate protective devices and systems must be provided. Such devices include fuses, circuit breakers, and cutout relays.

In the generating system, the protective devices must be of a type that will de-energize faulty power sources and power transmission equipment and disconnect them from their associated buses with sufficient rapidity to provide protection against hazardous overvoltage and other malfunctions.

All resettable circuit protective devices should be so designed that when an overload or circuit fault exists, they will open the circuit irrespective of the position of the operating control. This means, of course, that a circuit protective device must not be of a type that can be overridden manually. This type of circuit breaker is said to be **trip-free.** Protective devices must be clearly identified and accessible for resetting in flight if they are in an essential circuit. Resetting can be done only after the fault is corrected.

When fuses are used in an aircraft electrical system, spare fuses must be provided for use in flight in a quantity equal to at least 50 percent of the number of fuses of each rating required for complete circuit protection. If only one fuse of a particular rating is used in an aircraft system, then one spare should be carried for that rating.

Protective devices are not required in the main circuits of starter motors or in circuits where no hazard is presented by their omission. Each circuit for **essential loads** must have individual circuit protection; however, individual protection for each circuit in an essential-load system is not required.

All fuses, circuit breakers, switches, and other electric controls in an airplane must be clearly identified so that the pilot or some other member of the crew can quickly and easily perform in flight any necessary service to a unit. A **master switch** must be provided that will make it possible to disconnect all power sources from the distribution system. By means of relays, the actual disconnection should be made as near to the power source as possible.

On many aircraft more than one master switch can be employed to allow for the isolation of certain electric equipment. For example, an avionics master switch controls the electric power to all avionics equipment. This is typically accomplished through a relay that controls power to an avionics bus bar. Other master switches are commonly used to operate individual generators and galley power (if applicable).

Electrical Load

The **electrical load** of an aircraft is determined by the load requirements of the electric units or systems that can be operated simultaneously. It is essential that the electrical load of any aircraft be known by the owner or operator, or at least by the person responsible for maintenance of the aircraft. No electric equipment can be added to an aircraft's electrical system until or unless the total load is computed, and it is found that the electric power source for the aircraft has sufficient

capacity to operate the additional equipment.

To determine the electrical load of an aircraft, an **electrical-load analysis** is made. One way to do this is by adding all the possible loads that can be operating at any one time. (Electrical-load analysis is discussed in detail in the next section.) Loads may be **continuous** or **intermittent,** depending on the nature of the operation. Examples of continuous loads are navigation lights, the rotating beacon, the radio receiver, radio navigation equipment, electric instruments, electric fuel pumps, electric vacuum pumps, and the air-conditioning system. These are units and systems that can be operated continuously during flight.

Intermittent loads are those that are operated for 2 min or less and are then turned off. Examples of intermittent loads are landing gear, flaps, emergency hydraulic pumps, trim motors, and landing lights. These units and circuits for other electrically operated devices are normally operated for a very short period of time.

In computing the electrical load for an aircraft, all circuits that can or may be operated at any one time must be considered. The total **probable continuous load** is the basis for selecting the capacity of the power source. It is recommended that the probable continuous load be not more than 80 percent of the generator capacity on aircraft where special placards or monitoring devices are not installed. This permits the generator or alternator to supply the load and also keep the battery charged.

Aircraft that employ an ammeter to monitor the charging system status can operate continuous electrical loads up to 100 percent of the generator capacity. In this case some means of indicating an alternator failure is also required. This purpose is typically served by a generator or alternator failure light.

During periods when a heavy intermittent load such as landing gear is operated, an overload will probably exist, and the overload will be met for a short time by the battery and generator together. The operator of the aircraft should understand that prolonged operation under overload conditions will cause the battery to discharge to the extent that it cannot provide emergency electric power.

On twin-engine aircraft where two generators are used to supply the electric power, the capacity of the two generators operating together is used when power requirements are computed. The probable continuous load is not excessive if the two generators can supply the power. When the total continuous load is greater than the capability of one generator to supply, it is necessary to provide for load reduction if one of the generators or one engine fails. The load should be reduced as soon as possible to a level that can be supplied by the operating generator.

The load condition during operation can be determined by observing the ammeter and voltmeter. When the ammeter is connected between the battery and the battery bus so that it will indicate CHARGE or DISCHARGE, it will be known that the system is not overloaded as long as the ammeter shows a charge condition. In this case a voltmeter connected to the main power bus will show that the system is operating at the rated system voltage. If there is an overload, the ammeter will show a discharge, and the voltmeter will give a low reading, the value of which is determined by the amount of the overload.

When the ammeter is connected in the generator output lead and the system is not current-limited, an overload will be indicated when the ammeter reading is above the 100 percent mark. The ammeter should be **"red-lined"** so the pilot can determine easily when an overload exists. Most modern generator or alternator circuits contain an automatic means of controlling any overload condition; therefore, generator overloads typically do not exist.

When the ammeter is connected in the generator output lead and no output current is being produced, the ammeter will indicate zero current flow. On single-engine aircraft this indicates that the battery is supplying all the electric power. On multiengine systems, if only one alternator fails, the battery and the other generator(s) will supply the needed electric power. If this condition overloads the operating generator(s), the pilot may then shut off some nonessential equipment and reduce the load to a suitable level.

The principal concern of the aviation maintenance technician with respect to electrical load in an aircraft is a situation where it is desired to add electric equipment. If the addition of such equipment has been tested and approved by the FAA for a particular installation, instructions will be available from the manufacturer of the equipment or the aircraft setting forth all requirements for the installation. These instructions should be followed carefully.

Electrical-Load Analysis

Prior to installing any electric equipment in an aircraft, the technician must perform an **electrical-load analysis.** This is done to ensure that the aircraft's electric power system will not be overloaded by the addition of the new equipment. The goal is to compare the sum of all continuous electrical loads with the generator's (alternator's) maximum output. If the total continuous load is less than the rated generator output, more equipment may be added; however, the maximum generator output may never be exceeded by a continuous load.

There are basically two means to determine an aircraft's electrical load: via measurement or via summation of the individual loads. To measure electrical loads, an accurate ammeter must be placed in the generator output lead. The airplane ammeter is typically incapable of giving readings to an accuracy of 1 A, which is desirable for this test. Start the aircraft engine, and allow the battery to regain full charge. Next, turn on all the aircraft's continuous electric equipment, and monitor the ammeter. The ammeter will measure the total electrical load. This value can then be compared with the generator's rated output.

To find the total electrical load through summation, each individual electric current load must be known. The aircraft's service manual may provide this information, or it can be obtained from the data plate of each individual unit. In either case be sure to sum all continuous electrical loads. Then compare the generator's maximum rated output with the sum of the electrical loads. The total continuous load must always be equal to or less than the generator's maximum rated output. If "extra" capacity is available from the generator, more electric equipment can be added without a need for restricting the operation of certain electrical loads. If restriction(s) are necessary, a placard must be located on the aircraft's instrument panel to alert the pilot to each restriction. When summing electrical loads, be sure to use actual current draws, not circuit breaker (or fuse) ratings. A circuit breaker must be capable of sustaining a higher value than the actual load; therefore, an inaccurate sum will be obtained when using circuit breaker values.

If it becomes necessary to add equipment that may exceed the generator's maximum output, placards can be used to inform the pilot of the appropriate load configurations. That is, a placard will give the pilot the necessary load data to ensure that the electric power limits are not exceeded. For example, a placard may read, "Do not operate the air conditioner and windshield heat simultaneously." This placard would be placed near the windshield heat and air conditioner control switches.

A Simple Electrical System

A simple electrical system for a light aircraft consists of a battery circuit, an alternator circuit with associated controls, an engine starter circuit, a bus bar with circuit breakers, control switches, an ammeter, lighting circuits, and radio circuits. A schematic diagram of the basic power distribution system is shown in Figure 12–1. The high-current-carrying cables in this system are connected from the battery to the main battery relay, from the battery relay to the starter relay, and from the starter relay to the starter. The ground leads for the starter and the battery are also of heavy cable.

The main alternator power cables are also considerably larger than the normal circuit wiring; however, they are usually smaller than the cables required to carry full battery current. This is because the battery is used for starting the engine, and the starting current is very large. During operation of the aircraft, the battery is connected to the system but is not supplying power. Instead, it is taking power from the alternator in order to maintain a charge. All the normal load currents are supplied by the alternator during flight. The distribution bus receives power from the alternator and/or battery during different operating modes. The bus then distributes the electric current through the individual circuit breakers to their respective loads. As shown in the schematic (see Figure 12–1), the circuit breakers are connected directly to the distribution bus. This is done to prevent any accidental short to ground of an unprotected wire. It is always desirable to protect as much wiring as possible. Any wires that are not protected by a fuse or circuit breaker must be as short as practical and protected by insulated covers or "boots" at all terminal connections.

Although the schematic diagram in Figure 12–1 shows an entire aircraft electrical system, this is not typical. Most manufacturers prefer to divide aircraft schematics into individual systems. This becomes a necessity when dealing with large, complex aircraft. If an entire electrical system were represented in one schematic, the diagram would be extremely cluttered and too difficult to read. The following paragraphs will discuss schematic diagrams of several power

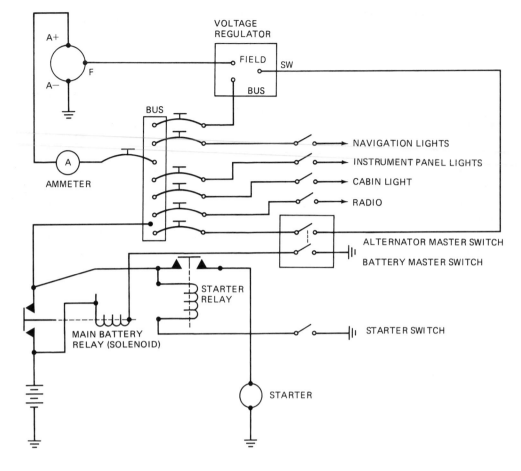

FIGURE 12–1 Basic power distribution system.

FOR TRAINING PURPOSES ONLY

distribution systems. Individual circuits, as they may be represented in a typical aircraft maintenance manual, will be presented in Chapter 13.

MAIN POWER DISTRIBUTION SYSTEMS

Single-Engine Aircraft

The Piper Tomahawk aircraft alternator and battery power systems are shown in Figure 12–2. This schematic is typical of a single-engine power distribution system. In the Piper Tomahawk, and in most other light aircraft, the master solenoid coil is switched on the negative side of the circuit. The master switch contains two independent poles and throws. The battery master, on the left half of the switch, connects the ground (negative voltage) to the master solenoid. The solenoid's negative lead is switched to ensure proper system operation in case of an electrical short to ground. That is, if wire number P2A should short to ground, the master solenoid will remain closed. If the solenoid is closed, battery power is connected to the starter solenoid and the distribution bus, thus creating no immediate danger. The alternator master switch,

on the right side of the combination master switch, connects the voltage regulator to the bus, turning on the alternator. In many aircraft the alternator side of the master switch can be operated only if the battery master is also turned on. This is done to ensure that the battery is connected to the bus prior to the alternator.

There are two notes listed on the bottom left side of this diagram. Always refer to any notes or effective serial numbers prior to using a schematic for maintenance purposes.

Twin-Engine Aircraft

The simplified power distribution schematic of the Cessna 421 twin-engine aircraft is shown in Figure 12–3. This system employs a diode in series with the wire connecting the main and emergency power distribution buses. This diode will allow current flow from the main bus to the emergency bus, but not in the reverse direction. This is done to isolate the main bus in the event that it should short to ground. In that configuration the emergency bus could still receive battery power without being affected by the short circuit.

This schematic also contains a diode in parallel with the battery relay coil. If a diode is placed in parallel with an electromagnetic coil, it is used to "clip voltage spikes." As explained in Chapter 6, when a current starts to flow in a coil, or

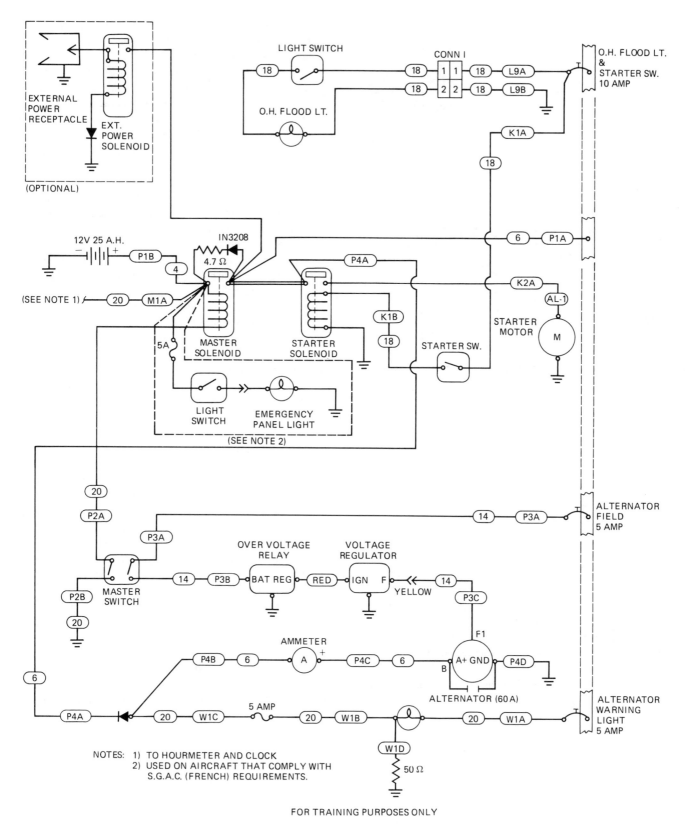

NOTES: 1) TO HOURMETER AND CLOCK
       2) USED ON AIRCRAFT THAT COMPLY WITH
          S.G.A.C. (FRENCH) REQUIREMENTS.

FOR TRAINING PURPOSES ONLY

FIGURE 12–2 Typical single-engine aircraft power distribution systems.   *(Piper Aircraft Corporation)*

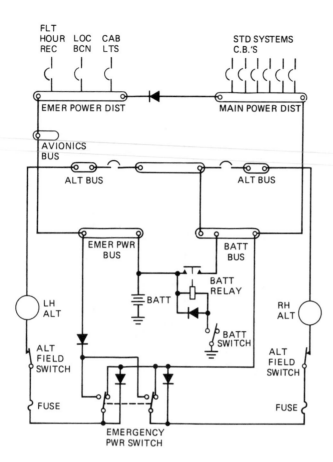

FIGURE 12–3 Typical light twin-engine aircraft power distribution system. *(Cessna Aircraft Corporation)*

when the current flow is stopped, the inductance of the coil creates a voltage opposing the applied voltage. Thus whenever the switch is opened or closed within the relay or solenoid coil circuit, a **voltage spike** or **transient voltage** is produced. This reverse-polarity voltage spike will damage sensitive electronic equipment if it is allowed to enter the electrical system. The diode in parallel with the relay's coil will short together any reverse-polarity voltage spike; however, the applied voltage will be unaffected. A bidirectional zener diode can also be used for this purpose. The zener diode conducts and short-circuits the relatively high-value transient voltage. The lower system voltage is unaffected. Remember, the zener diode is a voltage-sensitive device.

In general, diodes of all types are becoming more popular in aircraft power distribution systems. If a diode is placed in parallel with a coil, it is used to prevent damage from induced voltage spikes. If a diode is placed in series, it is used to create a one-way current path between units.

The power distribution system for a light twin-engine Piper airplane is shown in Figure 12–4. Since the airplane is equipped with all the avionic equipment necessary for electronic navigation and optimum flight performance, it is necessary for the alternators to have a comparatively high capacity.

The electrical system shown in Figure 12–4 includes a 24-V, 17-Ah battery enclosed in a sealed stainless-steel battery box. Two 24-V, 70-A alternators driven by the engines

supply all the normal power requirements of the aircraft and its equipment. The battery supplies power for starting the engines and for emergency peak loads.

The alternators are paralleled by using one voltage regulator to control the field current for both alternators. The circuit diagram in Figure 12–4 shows how this is accomplished.

An overvoltage relay in the system serves as a safety valve in case either one or both of the alternators should produce dangerously excessive voltage. This condition would exist in the case of failure of the voltage regulator. In the event that the main voltage regulator fails and the overvoltage relay disconnects the alternator fields from the system, an auxiliary voltage regulator is available. Failure of the alternators can be detected by a discharge indication for the battery and a zero output on both alternator test positions.

The output of each alternator is checked by pressing a PRESS-TO-TEST switch and observing the ammeter in the overhead switch panel. The test switches are shown as LEFT ALTERNATOR SWITCH and RIGHT ALTERNATOR SWITCH in the circuit diagram in Figure 12–4.

Electrical switches for the various systems, including the MASTER SWITCH, are located on the aircraft instrument panel. The circuit breakers, located below the switches, automatically open their respective circuits in case of an overload. The circuit breakers can be reset merely by pressing the RESET button. If a circuit breaker continues to disconnect, the trouble should be located and repaired before another attempt is made to operate the circuit.

The power distribution system for a gas-turbine-powered airplane with two engines, the Beechcraft Super King Air 200, is shown in Figure 12–5. This schematic diagram is presented to show the complexity of a modern aircraft electrical system and the many functions that require electric power.

The two-engine-driven starter-generators receive power from the main battery bus for starting purposes. During the generator mode, the starter-generator output current is directed to the right and left generator buses, respectively. The two generator buses are connected to the **feed buses** (numbers 1 to 4) through diodes and circuit breakers. This arrangement allows for both generators to power the four feed buses during normal operations or remain isolated during accidental short circuits to ground.

The two generator buses are connected to the isolation bus through **isolation limiters.** The isolation limiters, which are often referred to as **current limiters,** are simply high-amperage fuses. The isolation limiters can carry 325 A before opening. These isolation limiters open during overload conditions. For example, if an overload exists (a direct connection to ground) on the right generator bus, the right-side isolation limiter will open and disconnect the battery (via the battery and isolation buses) from the right generator bus. At the same time, the right generator will be disconnected from the right generator bus by the right generator control unit. The diodes placed between the right generator bus and the four feed buses will be reverse-biased in this event and therefore will isolate the feed buses and prevent current flow from the feed buses to the right generator bus. The right generator bus is therefore completely isolated, and the rest of the electrical system operates normally. Under these conditions the

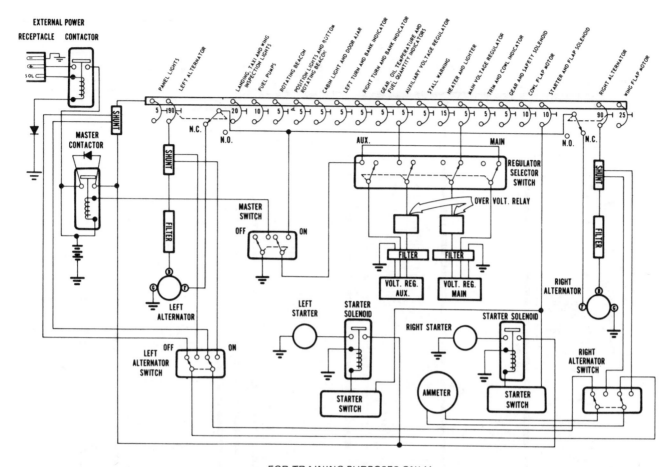

FOR TRAINING PURPOSES ONLY

FIGURE 12–4 Typical light twin-engine aircraft power distribution system. *(Piper Aircraft Corporation)*

right generator cannot supply power to the system, and all nonessential loads must be eliminated.

The King Air 200 electrical system is typical of those found on moderately sized corporate aircraft. Isolation of overloaded circuits and distribution buses becomes essential for safe aircraft operation. Each of the aircraft's electrical loads may be powered by a minimum of two different means (the right and left generators). All essential electrical loads may be powered by one of three different means, the right or left generator or the aircraft battery.

POWER DISTRIBUTION ON COMPOSITE AIRCRAFT

Composite aircraft present an interesting challenge when it comes to electric power distribution, control of static electricity, and lightning strikes. Several experimental composite aircraft are currently being flown throughout the world. Most of these aircraft are light single-engine planes with limited electrical systems. On transport-category aircraft, some components are made from composite materials, but the fuselage and wing structure are made of aluminum. The Beechcraft Starship is currently the only production composite aircraft.

This plane is a twin-engine, turboprop corporate aircraft. The power distribution system of this composite aircraft will be addressed here.

The entire fuselage and wing assemblies of the Beechcraft Starship are made from composite materials that have too high of a resistance to easily carry current. To counteract this high-resistance effect, a **ground plane** is integrated into the composite airframe. The ground plane is made of an aluminum mesh material. This material is similar to an aluminum window screen. The aluminum mesh is bonded into the composite material during the manufacturing process. The ground plane is located toward the inside of the aircraft structure for ease of bonding to electric equipment. The mesh runs throughout the airframe, including structural parts, bulkheads, floorboards, instrument panels, and electric equipment shelves. Virtually any portion of the aircraft that has electric equipment has a ground plane integrated into the composite material.

Two methods are used to connect electric equipment to the ground plane, **direct electrical bonding** and **indirect electrical bonding.** The direct method is used where electric equipment is mounted adjacent to the ground plane. To properly ground a component when the direct method is used, one must first remove a thin layer of composite material, paint, or any resistive coating to expose the wire mesh. The wire mesh

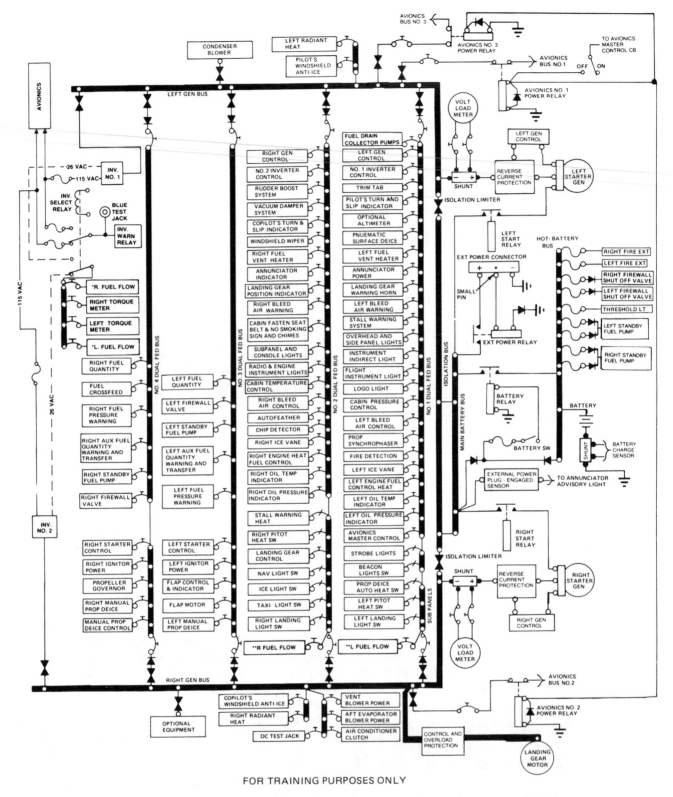

FOR TRAINING PURPOSES ONLY

FIGURE 12–5 Power distribution system for a twin-engine turboprop aircraft. *(Beech Aircraft Corporation)*

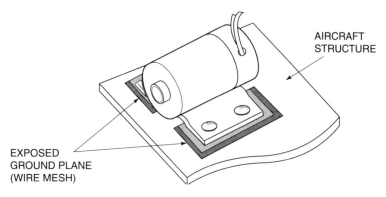

FIGURE 12–6 A typical direct electrical ground connection on a composite aircraft.

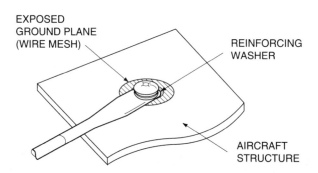

FIGURE 12–7 Connecting a bonding jumper to a composite aircraft.

is then coated with an anticorrosive agent, and the electric component is mounted directly to the ground plane. As seen in Figure 12–6, it is very important to remove as little material as possible during this process and still provide a sufficient area for a proper ground connection. The exposed area is refinished with a protective coating after component installation.

The indirect method is used in areas of the aircraft that are not adjacent to the ground plane. The indirect method uses a flexible metal strap called a **bonding jumper** to connect the ground plane to the electric component. The bonding jumper is attached to the ground plane in a manner similar to the one described above. The jumper is then attached to the component requiring an electrical ground (see Figure 12–7).

Lightning protection for a composite aircraft requires the installation of aluminum wire, which is interwoven in the outer ply of the aircraft skin. If lightning strikes the aircraft, the current is distributed over a large area through the aluminum wire. Since lightning typically enters the airframe at one extremity and leaves at another, the aluminum wire covers the entire structure of the aircraft. All sections of this lightning diversion wire must be connected by a low-resistance attachment. The lightning protection wire must not be used for electrical grounding. Only the aircraft's ground plane is designed to carry the current of electric equipment.

The distribution of the positive voltage for composite aircraft is virtually identical with such distribution on metal aircraft. Positive-voltage distribution on the Starship employs an automated control system powering five separate distribution buses. During normal operation, the buses are fed by the right and left 28-V, 300-A generators. The battery is being charged. In the event of a generator failure, the automated system isolates the generator and performs load shedding as needed.

LARGE-AIRCRAFT ELECTRICAL SYSTEMS

Large-aircraft electrical systems have many similarities to those systems found on small aircraft. On large aircraft there is typically one battery and two or more ac generators (alternators), which supply power to several distribution buses. The ac generators are connected to the ac buses. The dc battery is connected directly to the battery or emergency bus. The ac power produced by the generators is converted to direct current where needed for special applications. Essential lighting, flight control systems, and communication and navigation radios are high-priority electrical systems. Galley power, nonessential lighting, and various other comfort systems are considered low-priority electrical systems. These nonessential systems are usually turned off during a partial generator system failure. In the case of a complete generator failure, the battery will supply power for all essential electric equipment. A fully charged battery will normally supply approximately 20 to 30 min of emergency power.

Power Distribution Systems

Modern large aircraft use both ac and dc electric power. The output of a typical generator is **three-phase 115 V ac;** this is converted by **transformer-rectifier units** (TRUs) where dc power is needed. A TRU incorporates a step-down transformer and a full-wave rectifier. Its output is 28 V dc. Most large aircraft contain two or more static inverters, which are used for emergency situations (generator failure). Each inverter is capable of converting direct current, supplied by the battery, into ac power, which is distributed by the essential ac bus. The static inverters supply a relatively small amount of ac power; however, their output is adequate to power all essential ac equipment.

There are two basic configurations that are used to distribute electric power, the **split-bus system** and the **parallel system.** The split-bus system is used on most twin-engine commercial aircraft, such as the Boeing 737, 757, and 767; the McDonnell Douglas MD-80; and the Airbus Industrie A-300 and A-310. In a split-bus system the engine-driven generators can never be connected to the same distribution bus simultaneously. Under normal conditions each generator supplies power only to its associated loads.

In a parallel electrical system, the entire electrical load is equally shared by all the working generators. Parallel ac power distribution systems are typically found on commercial aircraft containing three or more engines, such as the Boeing 727 and early 747s, the Lockheed L-1011, and the McDonnell Douglas DC-10 and MD-11. A modified parallel system is used on some modern four-engine aircraft, such as the Boeing 747-400. All four generators are not necessarily paralleled in this system. The right-side generators and the left- side generators can be connected, or they can be isolated from each other by means of a split system breaker.

The Split-Bus System

The **split-bus electrical system** contains two completely isolated power-generating systems. Each system, the left and the right, contains its own ac generator, transformer-rectifiers, and distribution buses. The right and the left generators power their respective loads independently of other system operations. In the event of a generator failure, the operating generator is connected in such a manner as to feed all the essential electrical loads, or the APU (auxiliary power unit) generator may be employed to carry the electrical load of the inoperative generator.

Figure 12–8 shows a simplified schematic of a typical split-bus system. This schematic shows the **external power contactor** (EPC) closed and a ground power supply connected to the aircraft. The **bus tie breakers** (BTBs), 1 and 2, are closed, connecting external power to both transfer buses and their respective electrical loads. In this configuration the **generator breakers** (GBs) are open, thus disconnecting the generators from the electrical system.

It should be noted that the various contactors, such as the BTBs and GBs, are controlling three-phase current. The contactors are therefore actually made up of a set of three contacts, one contact to open or close the "hot" wire for each phase. In some cases the contactors may have four contacts, one for each phase and one for the neutral wire. The buses also consist of three distinct units, one for each phase. The power distribution diagrams presented here are simplified diagrams and do not show the actual wiring for each phase of the ac power generated.

In the case where the APU would be used to supply electric power to the entire aircraft, the EPC would open and the APU generator breaker would close. This would distribute the electric power from the APU generator to both transfer buses.

If both engine-driven generators are operating, the current flow is from each generator to its respective transfer bus, as illustrated in Figure 12–9. At this time BTBs 1 and 2 are open, GBs 1 and 2 are closed, and the transfer relay is in its normal position. It can be seen from this distribution diagram that the two generators operate completely independently of each other.

If one engine-driven generator fails, the opposite generator is connected to both transfer buses in order to power the entire electrical system. Under this configuration, nonessential loads are automatically removed from the system in order to avoid a generator overload. The schematic in Figure 12–10 shows a failure in generator 1. The current from generator 2 is divided to transfer buses 1 and 2 via the correct positioning of the transfer relays. In this case transfer relay 1 is automatically activated to its abnormal position, which connects generator bus 2 with transfer bus 1. At this same time GB 1 opens to disconnect the failed generator from the electrical system. This entire process is controlled automatically within several microseconds, and the flight continues uninterrupted.

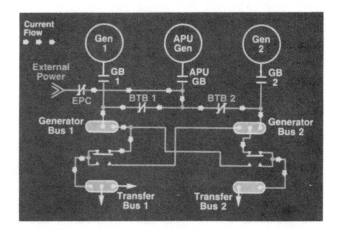

FIGURE 12–8 A typical split-bus system for a large aircraft. *(Sundstrand Corporation)*

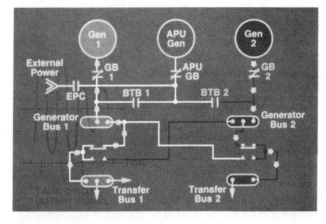

FIGURE 12–9 Split-bus system with both generators operating. *(Sundstrand Corporation)*

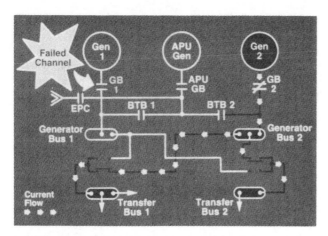

FIGURE 12–10 Split-bus system with generator 1 failed. *(Sundstrand Corporation)*

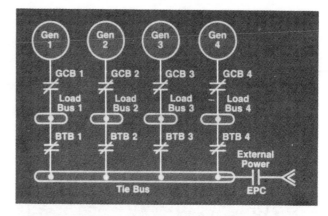

FIGURE 12–11 A four-generator, parallel power distribution system. *(Sundstrand Corporation)*

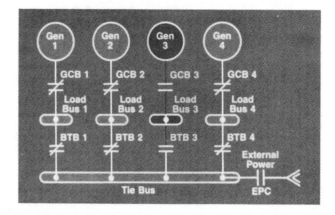

FIGURE 12–12 Parallel power distribution system with bus 3 isolated. *(Sundstrand Corporation)*

On certain aircraft, if a primary generator fails, the flight crew may elect to employ the APU generator. In this situation the APU must be started and its generator connected by closing the APU generator breaker (GB). This automatically closes BTB 1 and repositions transfer relay 1 to its normal position. Thus the APU generator is connected to transfer bus 1, and once again two independent generators are operating to supply all the aircraft electrical power.

The major advantage of a split-bus system is that the generators operate independently; that is, generator output frequencies and phase relationships need not be so closely regulated. Parallel systems require strict operating limits. Split-bus systems are, in effect, more tolerant of frequency variance.

Parallel Electrical Systems

In a **parallel electric power distribution system,** all ac generators are connected to one distribution bus. This type of system maintains equal load sharing for three or more ac generators. Since the generators are connected in parallel to a common bus, all generator voltages, frequencies, and their phase sequence must be within very strict limits to ensure proper system operation.

The simplified block diagram in Figure 12–11 represents a typical **four-generator parallel system.** This diagram shows a normal operating configuration; all four generator circuit breakers (GCBs) and bus tie breakers (BTBs) are closed. All four generators are synchronized and connected in parallel by the **tie bus.** The tie bus is often referred to as a **synchronizing bus;** its purpose is to connect the output of all operating generators. The **load buses** are used to distribute the generator current to the various electrical loads.

If one generator fails, its receptive GCB opens. This isolates that generator from its load bus; however, that load bus still continues to receive power while connected to the tie bus. In case of a load bus overload, the bus is isolated by the opening of its GCB and BTB. This mode of failure is illustrated in Figure 12–12. In this diagram, load bus 3 has been isolated. The likely suspicion is that load bus 3 has been

shorted to ground and the bus power control unit has automatically tripped the appropriate contactors. This manner of isolation takes place whenever one or more load buses are faulty.

If two or more generators fail, their respective generator circuit breakers open, and they are isolated. The remaining generator(s) supply the power to the entire system. In this situation nonessential electrical loads are automatically disconnected from the system. This prevents an accidental overload of the operating generator(s).

Split Parallel System

A **split parallel electric power distribution system** is illustrated in Figure 12–13. This system allows for flexibility in load distribution and yet maintains isolation between systems when needed. When closed, the split system breaker connects all generators together, thus paralleling the system. When open, the split system breaker isolates the right- and left-hand systems, thus creating a more flexible parallel system.

A split parallel system is used on the Boeing 747-400 aircraft. As seen in Figure 12–14, this system employs four engine-driven generators and two auxiliary power unit (APU)

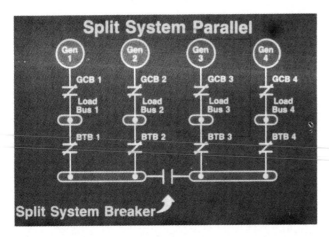

FIGURE 12–13 A split parallel power distribution system. *(Sundstrand Corporation)*

generators, and it can accept two separate external power sources (EXT 1 and EXT 2). The B-747-400 uses an automated power distribution control system that features a no-break power transfer. The no-break power transfer will be discussed later in this chapter. As seen in the schematic of the system, the four integrated drive generators (IDGs) are connected to their respective ac buses through generator control breakers (GCBs). The ac buses are paralleled through the bus tie breakers (BTBs) and the split system breaker (SSB). When the SSB is open, the right system operates independ-

ently of the left. With this system any generator can supply power to any load bus, and any combination of the IDGs can operate in parallel.

The external power or APU power can be connected to the ac buses through their associated contactors. The ac **ground-handling** (GH) buses are powered by closing the ground-handling relay (GHR) to either the APU or EXT power. The dc GH buses receive power from the transformer-rectifier (TR) located directly above them on the diagram (Figure 12–14). The ground-handling buses are used to power lighting and miscellaneous equipment for cargo loading, aircraft fueling, and cleaning. The GH buses are not powered during normal flight.

The **ground service** (GS) buses are controlled from the flight attendants' station located at the number 2 left door of the aircraft. The control switch energizes the ground service relay (GSR), which connects the GS buses to whichever is currently on line, the APU or EXT power. The ground service buses are used to light the interior of the aircraft and power the main battery charger and other miscellaneous systems required for maintenance, cleaning, and initial start-up of the aircraft.

DC Electrical Systems

Large commercial aircraft employ both an ac power distribution system and a dc system. The dc system incorporates redundancy and isolation capabilities for safety. Loading of the dc buses is so arranged that a complete loss of power to one

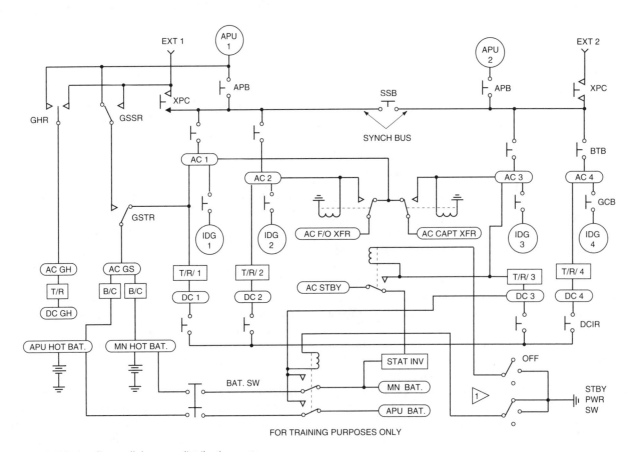

FOR TRAINING PURPOSES ONLY

FIGURE 12–14 A split parallel power distribution system.

system is unlikely. During a power loss situation, the essential dc loads can be powered by any transformer-rectifier (TR) via the essential dc bus. In the event of a complete generator system failure, the essential dc power would be supplied by the aircraft's battery. An ac inverter would also be powered in an emergency situation in order to operate all essential ac loads.

A simplified diagram of a **Boeing 727** dc power distribution system is shown in Figure 12–15. As illustrated, the 1 and 2 ac load buses and the essential ac bus are connected to TR units, which furnish power to the corresponding dc buses. The battery charger TR unit receives power from the ac transfer bus. This is a typical means of in-flight battery charging. DC buses 1 and 2 are connected together by a 100-A current limiter. DC bus 1 is tied to the essential dc bus through a diode that permits current flow only from dc bus 1 to the essential dc bus. The essential dc bus supplies current to the battery bus, the battery transfer bus, and the hot battery bus. The **hot battery bus** is always connected directly to the battery.

The complete power distribution system for the Boeing 727 is illustrated in Figure 12–16. It can be seen from this diagram that the B-727 power distribution system is a parallel system containing three engine-driven ac generators. Each generator can be connected to carry the entire aircraft electrical load individually during emergency situations or synchronized with the other two generators during normal operation. In this aircraft the essential ac and dc power is normally supplied by generator 3; however, any generator may perform this function. The rotary switch used to control essential power is shown near the bottom center of the diagram. In this type of system, the most critical electrical loads would be connected to the hot battery bus. This bus is the least likely source to lose power in the event of a catastrophic electrical system failure.

Power Distribution Hierarchy

All aircraft electrical systems are designed with a **bus hierarchy.** That is, each system is designed so that the most critical components are the least likely to fail. On all aircraft the most critical components must operate from battery power. Less critical components can operate from other power sources, such as an aircraft generator. For example, look at the Boeing 727 schematic in Figure 12–16. Here the least critical ac loads are powered by ac bus 1, 2, or 3. The least critical dc loads are powered by dc bus 1 or 2. The next most critical systems are powered by the **essential** (ESS) ac and dc buses, shown on the left side of the diagram. The essential ac bus can receive power from any ac generator. The essential dc bus can be powered either by dc bus 1 or 2 or by the essential TR unit. The most critical electrical loads on the aircraft are powered by the standby buses, located at the top left of the diagram. These buses will still receive power from the battery even if all three generators fail.

On the Boeing 747-400 the least critical loads are connected to ac and dc buses 1, 2, 3, and 4 (Figure 12–14). More critical loads may be connected to the captain's and first officer's **transfer buses** (AC CAPT XFR and AC F/O XFR). The most critical ac loads are connected to the **ac standby bus** (AC STBY); dc loads are connected to the **main battery bus** (MN BAT). This hierarchy allows for safe operation of the aircraft even in the unlikely event that all engine-driven generators fail.

Control of the Power Distribution Systems

On modern aircraft employing a parallel or split-bus system, a centralized means of controlling the power distribution between individual load buses is essential. For example, if a generator fails or a bus shorts to ground, the appropriate bus ties and generator circuit breakers must be set to the correct position. Or in the event of a system overload, the control unit must reduce the electrical load to an acceptable level. This is called **load shedding.** The aircraft's galley power is usually the first nonessential load to be disconnected. Also, the control unit must automatically reconnect any essential loads to an operable bus. This power manipulation must take place within a fraction of a second to ensure an uninterrupted flight. To achieve this goal, modern aircraft employ a solid-state **bus power control unit** (BPCU).

The BPCU receives data from the generator control units (GCUs), the ground power control unit (GPCU), and the various bus ties and circuit breakers of the system. As discussed in Chapter 11, GCUs are used in conjunction with each aircraft generator to monitor and regulate generator activities. If a GCU detects a malfunction, it will inform the BPCU. The BPCU will then ensure the appropriate power distribution system configuration.

The BPCU also receives input information concerning system loads from **load controllers.** Load controllers are electric circuits that sense real system current and provide control signals for the generator's constant-speed drive rpm governor. The constant-speed drive output rpm in turn affects generator output frequency. Load controllers receive their input signals from current transformers, such as the one shown in Figure 12–17. **Current transformers** consist of three inductive pickup coils that provide current-sensing signals. The main power leads carrying the three-phase alternating current from each generator are routed through the corresponding holes in a current transformer. As the alternating current travels through the wire, the corresponding magnetic field induces a voltage into the current transformer (Figure 12–18). The electrical signals from the current transformer, in conjunction with the GCU and BPCU, are used to control protection circuitry and supply signals to load meters in the cockpit.

The BPCU is the main control computer for all generator and electric power distribution. The BPCU receives input signals from several current transformers in order to monitor the electrical system. If the BPCU detects an abnormal condition, it opens and/or closes the appropriate bus tie and/or circuit breaker. As illustrated earlier in this chapter by the schematics of parallel and split-bus systems, circuit breakers are located throughout the aircraft's electrical system and are used to isolate or connect various generators and/or distribution buses. Circuit breakers operate automatically according to GCU and BPCU signals or manually via cockpit controls.

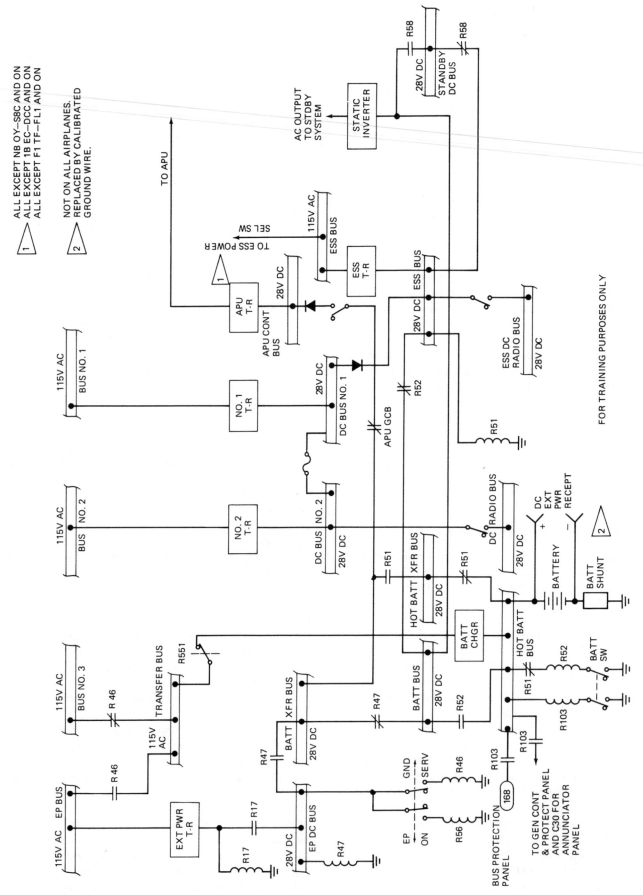

FIGURE 12–15 Boeing 727 dc power distribution system. *(Boeing Company)*

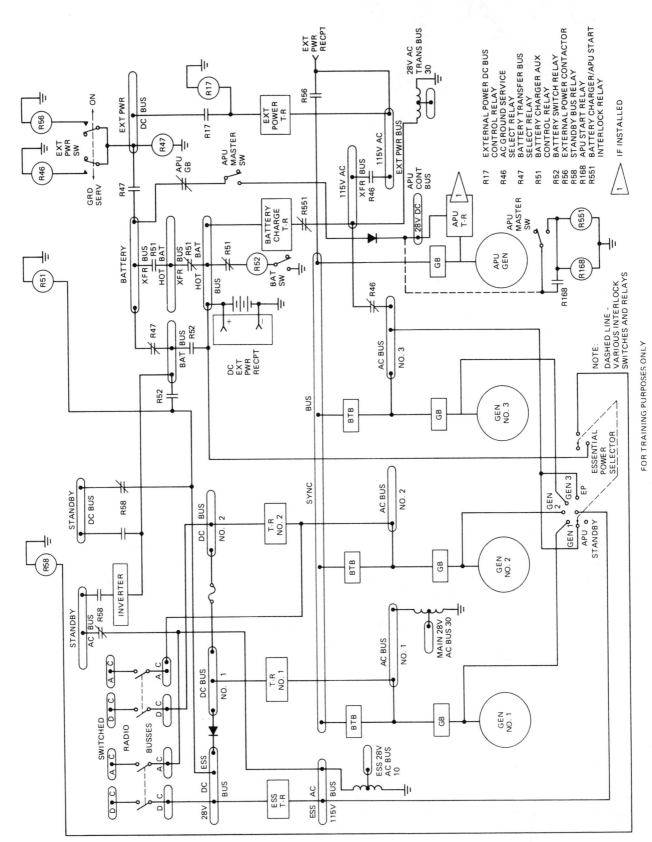

FIGURE 12–16 Boeing 727 power distribution system. (Boeing Company)

FIGURE 12–17 Current transformer. *(Sundstrand Corporation)*

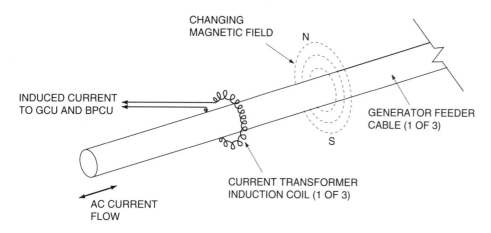

CHANGING MAGNETIC FIELD

N

INDUCED CURRENT TO GCU AND BPCU

GENERATOR FEEDER CABLE (1 OF 3)

S

CURRENT TRANSFORMER INDUCTION COIL (1 OF 3)

AC CURRENT FLOW

FIGURE 12–18 Current transformer induction coil.

Bus tie breakers are another type of unit used to connect or disconnect main electrical distribution points. A bus tie breaker (BTB) is similar to an electric solenoid in that it is used as a remote control switch. Each BTB is usually controlled by the BPCU.

On modern commercial passenger jets, the BPCU performs control, test, protection, and fault identification functions. The schematic in Figure 12–19 shows the GCUs, the BPCU and its related sensors, and the current transformer assemblies (CTAs). Figure 12–20 shows the generator breakers (GBs) and the bus tie breakers (BTBs), which are used to control system loads via signals from the BPCU.

BPCUs are basically small computers designed for a specific function. Each aircraft typically contains two BPCUs for redundancy in the event of a failure. Each BPCU constantly monitors its input and output data using a digitally coded message. If a system fault occurs, the BPCU initiates the necessary corrective action and records the fault in a nonvolatile memory. The nonvolatile memory is part of the built-in test equipment (BITE) of the BPCU. Any fault data stored by the BITE system can be recalled at a later time by a line technician. This process greatly reduces maintenance time and enhances system reliability. Built-in test equipment was discussed in Chapter 7 and will be discussed again in Chapter 13.

On some aircraft the automated power distribution system provides for a **no-break power transfer** (NBPT). A no-break power transfer means that the automated system can change the ac power source without a momentary interruption of electric power. For example, when external power is being used and the aircraft is preparing to depart the gate, the engines are started and the main generators are brought on line. During an NBPT, the generator control units monitor the power source currently on line (external power) and the power source requested by the flight crew (main generators). If the power requested is within specifications, both power supplies are paralleled for a split second, and no power interruption occurs. If the requested power is out of limits, the GCUs try to adjust the system and then connect the requested power to the buses. If the power system cannot be adjusted to the correct tolerance for paralleling, the requested power source will be rejected, or there will be a momentary power interruption.

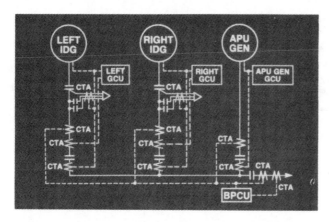

FIGURE 12–19 Schematic of a power distribution control system. *(Sundstrand Corporation)*

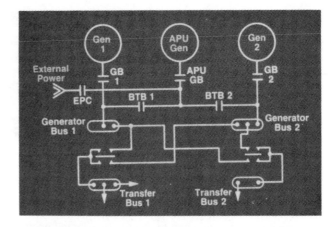

FIGURE 12–20 Power distribution control system showing generator breakers (GBs) and bus tie breakers (BTBs). *(Sundstrand Corporation)*

REVIEW QUESTIONS

1. What is the purpose of a bus bar?
2. Describe a power distribution system for a large aircraft.
3. What FAR establishes the general requirements for aircraft power distribution systems?
4. What are the various means of monitoring a power distribution system?
5. What safety precautions must be provided in an external power circuit?
6. What is the purpose of a diode in an external power contactor circuit?
7. What is the purpose of an electrical-load analysis?
8. What loads are considered for an electrical-load analysis?
9. What is the minimum time an emergency battery system must supply power?
10. Discuss the need for protective devices in aircraft electrical systems.
11. Where should protective devices be located?
12. What is the purpose of a master switch?
13. What types of electrical loads may be considered as intermittent loads?
14. What must be done before adding electric equipment to an aircraft system?
15. What is a trip-free circuit breaker?
16. What are the basic requirements for circuit protection identification?
17. Describe the function of a dual master switch.
18. When can a total continuous load equal 100 percent of alternator output?
19. When are placards required on electrical loads?
20. What is the difference between an ammeter located in an alternator output lead and one located in a battery output lead?

21. Why are electrical schematics within an aircraft's maintenance manual separated into different systems?
22. What elements of an electrical system would be found on a power distribution system schematic?
23. Describe the difference between a power distribution system on a single-engine aircraft and one on a twin-engine aircraft.
24. What is used to clip voltage spikes?
25. What is a feed bus?
26. Describe the function of a current limiter.
27. What is a ground plane?
28. Describe the two methods used to connect electric equipment to a ground plane.
29. What is a bonding jumper?
30. In the event of a partial power system failure, what happens to nonessential electrical loads?
31. What is the function of a transformer-rectifier unit?
32. Describe the basic operation of a split-bus power distribution system.
33. What types of aircraft typically employ a split-bus system?
34. Describe the operation of a parallel power distribution system.
35. What are the disadvantages of a parallel power distribution system?
36. When are parallel systems used?
37. Describe the basic dc power distribution system for a large commercial airliner.
38. What is the purpose of the hot battery bus?
39. What units are used to control the power distribution system found on large aircraft?
40. What is a no-break power transfer?

✈

13 Design and Maintenance of Aircraft Electrical Systems

Modern aircraft depend upon the proper functioning operation of their electrical systems for safe and satisfactory operation. Electrical systems are required for power-plant control, systems control, navigation, communications, flight control, lights, galley operation, and other functions. With many aircraft, flight operations cannot be conducted safely without certain **essential** electrical systems. It is therefore apparent that the proper maintenance of aircraft requires that the electrical systems be kept in the best possible condition through inspections, testing, and the exercise of approved maintenance procedures.

To attain reliability in electrical systems, it is essential that great care be exercised in the selection of components and materials and that each part be installed in such a manner that it will not be subjected to damaging conditions of any kind. For commercial and other civil aircraft, the requirements for the installation and approval of electric components and materials are established by the Federal Aviation Administration (FAA) and published in Federal Aviation Regulations (FARs). The regulations and directives of the FAA should always be observed in the maintenance of civil aircraft. For specific types of aircraft and equipment, the manufacturer's overhaul and maintenance manuals should be followed. During the design and manufacture of aircraft, the manufacturer makes certain that the requirements of the FAA are met to assure safe aircraft that can be certificated for civil use. For military aircraft, missiles, and spacecraft, specifications are established by the appropriate agency in cooperation with the manufacturers. The purpose of this chapter is to point out requirements established by all agencies for the correct installation of electrical systems and components and to describe the types of wiring and other components that make up a complete electrical system.

REQUIREMENTS FOR ELECTRICAL SYSTEMS

General Requirements

In general, requirements for aircraft electrical systems are established to assure that the systems will perform their functions reliably and effectively. The requirements for normal, utility, and acrobatic aircraft are set forth in **FAR Part 23. FAR Part 25** gives the requirements for transport-category aircraft. Various changes are made in these requirements from time to time, and it is the responsibility of the FAA, manufacturers, and maintenance personnel to ensure that required changes are incorporated in certificated aircraft.

In this section it is not possible to list all current requirements in detail; however, we shall consider the principal factors that assure safe and effective electrical systems. For the current requirements for an aircraft that a technician may be inspecting and maintaining, the appropriate manufacturer's bulletins and FAR should be consulted.

Electrical systems for all aircraft must be adequate for the intended use. Electric power sources, their transmission cables, and associated control and protective devices must be able to furnish the required power at the proper voltage to each load circuit essential for the safe operation of the aircraft. Compliance with the foregoing requirement must be substantiated by an electrical-load analysis (measurement or summation) that accounts for the electrical loads applied to the electrical system in probable combinations and for probable durations.

Electrical systems, when installed, must be free from hazards in themselves, in their methods of operation, and in their effects on other parts of the aircraft. They must be protected from fuel, oil, water, and other detrimental substances and from mechanical damage such as abrasion or physically applied force. The systems must be designed so that the risk of electric shock to the crew, passengers, and ground personnel is reduced to a minimum.

Electric equipment in a system must be so designed that in the event of a fire in the engine compartment, during which the surface of the fire wall adjacent to the fire is heated to 2000°F [1093°C] for 5 min or to a lower temperature substantiated by the applicant, the equipment that is essential to continued safe operation of the aircraft and located behind the fire wall will function satisfactorily and will not create an additional fire hazard.

Requirements for Transport Aircraft

All systems and equipment installed in **transport-category aircraft** must meet certain basic safety requirements, and these are set forth in FAR Part 25. All systems must be designed so they will perform their intended functions under foreseeable operating conditions. The electrical system and associated components, when considered separately and in relation to other systems, must be designed so that the occurrence of any failure condition that would prevent the continued safe flight and landing of the aircraft is extremely improbable. Any failure that would reduce the capability of the aircraft or the ability of the crew to cope with adverse operating conditions must be improbable.

Warning information must be provided to alert the crew to unsafe operating conditions, thus enabling them to take appropriate corrective action. Systems, controls, and associated monitoring and warning equipment must be designed to minimize crew errors that could cause additional hazards. Compliance with requirements must be shown by analysis and, where necessary, by appropriate ground, flight, or simulator tests. The analysis must consider possible modes of failure, including malfunctions and damage from external sources. It must deal with the probability of multiple failures and undetected failures and the resulting effects on the aircraft and occupants, considering the stage of flight and operating conditions. The analysis must also consider the crew warning cues, the corrective action required, and the capability of detecting faults.

Installations

The electric equipment, controls, and wiring for an aircraft must be installed so that operation of any one unit or system of units will not adversely affect the simultaneous operation of any other electric unit or system essential to the safe operation of the aircraft. For this reason individual electric circuits are wired in parallel with respect to each other. If one circuit of a parallel group is turned off or fails, it will not affect the remaining circuits. It is also important to ensure that circuits critical to flight safety are fused separately. If only one critical circuit is connected to any given fuse or circuit breaker, its failure will not adversely affect other circuits. Individual control (on/off) switches must also be employed on all circuits critical to flight. That is, the appropriate crew member must be able to turn off or on any critical circuit without adversely affecting others. In short, each critical circuit must contain its own independent switch and circuit protective device.

Cables and wires must be grouped, routed, and spaced so that damage to essential circuits will be minimized if there are faults in heavy-current-carrying cables. This means that cables that might be subject to burning in case of a short circuit should not be grouped with essential-circuit cables, because the burning of a shorted cable could also damage an essential circuit to the extent that it would not be operable.

The installations designed for an aircraft by the manufacturer are usually acceptable; however, changes are sometimes required, and these are called to the attention of the aircraft owner or operator by means of the manufacturer's bulletins or Airworthiness Directives (AD) issued by the FAA.

Typical Schematic Diagrams

The maintenance publications for any aircraft must contain information explaining the operation of electrical systems. To fully understand the operation of any electrical system, the technician must become familiar with the wiring of that system. The **schematic diagram** is an electrical road map that identifies the various wires and electric components of a particular system. Electrical schematic diagrams for light aircraft are often contained in the **maintenance manual.** The maintenance manual also describes each system's operation and maintenance practices. The electrical schematics for larger, more complex aircraft are contained in a separate **wiring diagram manual.** Electrical systems that are not produced or installed by the aircraft's manufacturer are not typically included in these data. Schematics of "add-on" electric equipment must be obtained from the manufacturers of those particular items.

The manufacturers of corporate and transport-category aircraft typically follow the **Air Transport Association** (ATA) specifications for categorizing data in the maintenance and wiring diagram manuals. Some, but not all, general aviation aircraft manuals follow ATA specs. **ATA specification 100** is a detailed number code of the various items found on a typical aircraft. This code dictates what components are covered by the various chapters of a manual. Knowledge of this standard will help technicians find a specific electrical schematic diagram or system description. Some of the various chapters that might apply to electrical systems include chapters **20 (standard practices airframe), 24 (electric power),** 31 (indicating and recording systems), 33 (lighting), 34 (navigation), 39 (electrical/electronic panels and multipurpose components), 74 (engine ignition), and 77 (engine indicating). Keep in mind that this is only a partial list; virtually any part of an aircraft could have a related electrical or electronic system. Each of the ATA chapters is divided into sections that describes in detail the various parts of an aircraft system.

Schematic diagrams usually represent the electrical configurations of one or more systems. Schematics *do not* show physical configurations of components within an electrical system. That is, schematic diagrams do not represent the location of electric components within the aircraft or with respect to other components of the system. Most civilian aircraft schematics are generally similar; however, there are several differences in diagrams drawn by different manufacturers. Some schematics indicate wire size within the wire code number. This becomes helpful when replacing defective wires. Often the individual components of an electrical system are identified on the schematic; other diagrams number the components and use an identification list. There are literally hundreds of symbols used to represent the various components of aircraft electrical systems. For the most part, these symbols are standardized; however, some variance does occur among manufacturers. The appendix of this text includes the electrical and electronic symbols from various aircraft manufacturers.

The schematic diagram of an electrical system seldom identifies the electric wiring within a component or "black box." For example, a diagram of an aircraft starter circuit may represent the starter as an empty circle. The schematic diagram of the starter motor will show the internal wiring of the motor. **Black box** is often the term used to refer to an electric component, a communication radio, or a generator

GMA/ATA CODE & REF DES	PART NO.	DESCRIPTION 1 2 3 4 5 6 7	UNITS PER ASSY	INSTL ZONE	USABLE ON CODE
46-01-CB56		CIRCUIT BREAKER SWITCH, LANDING LIGHT/SEE CH 24/	1	240	
-DS20	4596	LIGHT ASSY, LANDING LIGHT......................	1	410	
-GS4	131270-3	GROUND STUD................................	1	222	
-J61		RECEPTACLE, SUBPANEL DISCONNECT/SEE CHAPTER 91/.	1	231	
-P61		PLUG, SUBPANEL DISCONNECT/SEE CHAPTER 91/	1	231	

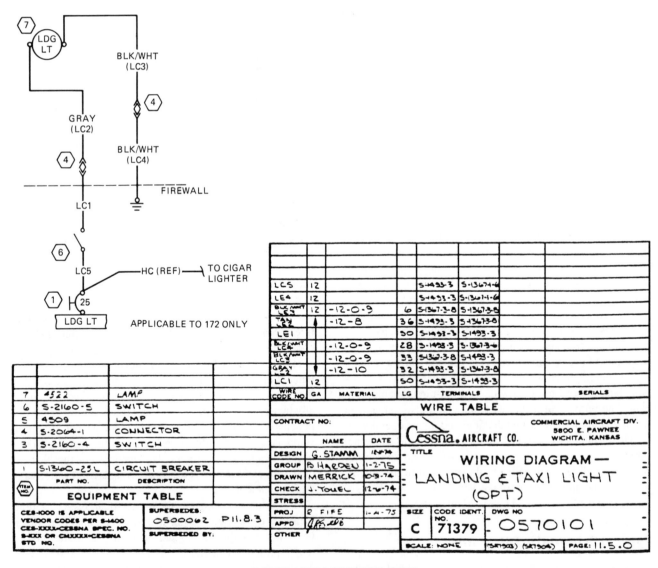

FOR TRAINING PURPOSES ONLY

FIGURE 13–1 Beechcraft landing light circuit schematic.   *(Beech Aircraft Corporation)*

FOR TRAINING PURPOSES ONLY

FIGURE 13–2 Cessna landing light circuit schematic.   *(Cessna Aircraft Corporation)*

control unit. The schematics of these components are typically available from the manufacturers of the components. The term **line replaceable unit** (LRU) means that the component is easily removed and installed on the aircraft. Black boxes are often called LRUs.

The three major general aviation aircraft producers, Beechcraft, Cessna, and Piper, each represent electrical schematics slightly differently. Figures 13–1 through 13–3 show three different schematics of a landing light system; there are several differences and several similarities. The Beechcraft schematic, Figure 13–1, includes a component description, which lists applicable part numbers, the quantity of units, and the component installation zones. An **installation zone** indicates the location of a component within the aircraft. If the zone number is referenced to a zone code chart, the component's location can be identified. This becomes extremely important when dealing with complex aircraft containing several electric components in remote locations. Beechcraft indicates a circuit's wire size in the last two digits of the **wire code.** For example, L5A18 indicates an 18-gauge wire.

The Cessna Corporation identifies schematic components in a separate equipment table, as illustrated in Figure 13–2. The Cessna equipment table references the part numbers of all major components in the system. The wire table contains information on wire size and any applicable color coding.

The Piper schematic diagram shown in Figure 13–3 does not contain a separate component or wire table. All the information concerning the circuit is contained within the schematic. Each component is labeled, and the wire size is overlaid on the individual wires.

In general, all three schematics are very similar; each contains the necessary information to convey the electrical layout of the circuit. Any design differences are illustrated in the schematic. For example, the Beechcraft controls the light by means of a circuit breaker switch; the Piper system connects the landing light to the connector plug by means of shielded cable. These design differences are typical among various manufacturers. Always refer to the schematic diagram prior to servicing any electric circuit. If any portion of a schematic cannot be interpreted, contact the appropriate technical representative for assistance.

Identification System for Locating Electric Components

Virtually all manufacturers of corporate and transport-category aircraft have an identification system that is used to locate components on an aircraft from the electrical schematics. The systems may vary among manufacturers and even among different aircraft produced by the same company; so, always refer to the introduction of the wiring diagram manual to find the specific details. In general, the electric components and wiring found on schematics are each assigned a number. That number can be used to locate the part using the appropriate manuals.

AIRCRAFT LIGHTS

All aircraft approved for flying at night must be equipped with various types of lights. Among these are **position (or navigation) lights, anticollision lights, landing lights, instrument lights, warning lights,** and **cabin lights.** In addition, other lights may be needed or required. Among these are taxi lights, ice-detection lights, cargo compartment lights, and all the special-purpose lights required in large passenger aircraft. All lighting equipment and installations must be approved by the FAA.

Position Lights

Each aircraft must have three **position lights:** two forward and one aft. The forward position lights are usually mounted on the tips of the wings because they are required to be as far outward as possible. The right position light is green, and the left is red. The forward position lights must show light through a 110° angle from directly forward to the right and the left, as shown in Figure 13–4.

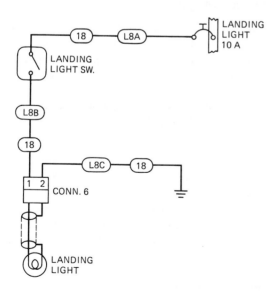

FOR TRAINING PURPOSES ONLY

FIGURE 13–3 Piper landing light circuit schematic. *(Piper Aircraft Corporation)*

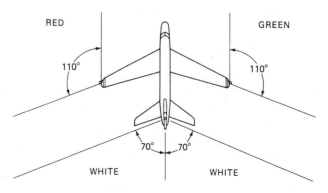

FIGURE 13–4 Arrangement of position lights.

The aft light is white and mounted as far to the rear as possible. It is common practice to mount the aft position light on the top of the vertical stabilizer (fin). The aft position light must show light through an angle of 70° on each side of the centerline of the aircraft and to the rear.

The covers or color filters used on position lights must be of a material that is heat-resistant and will not shrink, fade, or become clouded or opaque.

All position lights must be in a single circuit and must be controlled by one switch. The power source is connected through one fuse or circuit breaker. It should be noted that the term *navigation lights* is often substituted for *position lights*.

The schematic in Figure 13-5 represents a typical panel-and-position-light system. This aircraft employs transistorized dimming circuits to control the intensity of the radio and panel lights.

Anticollision Lights

An **anticollision light** is designed to make the presence of an aircraft visible to pilots and crew members of other aircraft in the vicinity, particularly in areas of high-density aviation activity, at night, and in conditions of reduced visibility. The anticollision light is of high intensity and flashes on and off not less than 40 and no more than 100 cycles/min. There are two basic types of anticollision lights, **rotating beacons** and **strobe lights.** Most modern aircraft employ strobe (flashing-type) anticollision lights, since they use no moving parts and generally produce a brighter light.

A strobe light is a glass or quartz tube filled at low pressure with xenon gas. The tube is caused to flash by applying a high voltage to two electrodes in the tube and triggering the tube with an additional circuit. The current used to fire the tube is stored in a capacitor by means of a charging circuit. This circuit converts the low voltage from the aircraft electrical system to a high voltage (300 to 500 V) to charge the storage capacitor. A trigger circuit then applies the trigger signal to the trigger terminal of the tube and causes it to fire. The duration of the flash caused by the capacitor discharge may be a little more or less than 0.001 s, but the intensity of the light is very high; thus the light can be seen for many miles. The strobe principle is the same as that of a photographer's electronic flash.

At least one anticollision light is required on all aircraft certified for night flight. Any aircraft certified after August 11, 1971, or any anticollision lights installed after that date must meet the anticollision-light requirements of AC 43.13-2A, chapter 4, paragraph 56.b.(1). These requirements stipulate that each anticollision-light system must illuminate a specific field of coverage. That is, each system must illuminate a given area around the aircraft. In order to meet this condition, most aircraft certified after August 11, 1971, employ three anticollision lights. The three lights are usually one red flashing light or rotating beacon located on top of the vertical stabilizer and two white flashing (strobe) lights on the wingtips. Since flashing-type anticollision lights produce a high-intensity light, the glass portion of any strobe lamp must be kept free from oil and grease in order to ensure proper op-

eration. Even small amounts of oil will create hot spots on the glass and form cracks in the bulb.

The schematic in Figure 13-6 illustrates the electrical system of a typical strobe-light circuit. In this system one power supply is used to light two independent flashtubes. As illustrated in Figure 13-6, shielded cable is used to connect the flashtubes to the strobe power supply. This prevents radio interference, which is created by the short pulse of current used to produce the intense flash.

Landing Lights

Landing lights for an aircraft are required to provide adequate light to illuminate the runway when the aircraft is making a landing. A parabolic reflector is utilized to concentrate the light into a beam of the desired width.

Landing lights may be attached to the stationary part of the nose gear, installed in the leading edges of the wings, or installed in the engine cowl. Some large aircraft have landing lights in the leading edges of the wings and retractable lights in the lower surfaces of the wings. The leading-edge landing lights can be turned on several miles away from the landing site, and the retractable lights are turned on shortly before landing.

Retractable landing lights are extended by means of a small but powerful motor that is able to move the lights outward and forward against the force of the airstream. Or these lights can be mounted to a portion of the landing gear. The lights then automatically extend or retract with the gear mechanism.

Taxi lights can be employed on some aircraft to improve visibility during ground operations. Taxi lights are aimed slightly higher than landing lights in order to illuminate the area directly in front of the aircraft. Both landing and taxi lights are usually of very high wattage; therefore, they are often controlled via a switching solenoid or relay.

Instrument Lights

Instrument lights are installed behind the face of the instrument panel. The lights illuminate the instruments but do not shine directly toward the pilot or copilot. All instrument lights must be shielded in this manner. Instrument lights are provided with a dimming arrangement so the intensity can be adjusted to suit the needs of the pilot. A transistorized dimming circuit is illustrated in Figure 13-5.

Warning Lights

Warning lights are provided to alert the pilot and crew to operating conditions within the aircraft systems. Red lights are used to indicate danger, amber lights to indicate caution, and green lights to indicate safe conditions. Indicator lights that are intended only for the purpose of providing information can be white.

Landing-Gear Circuits

Circuits involved in the operation of electrically powered landing gear are shown in Figures 13-7 and 13-8. Figure

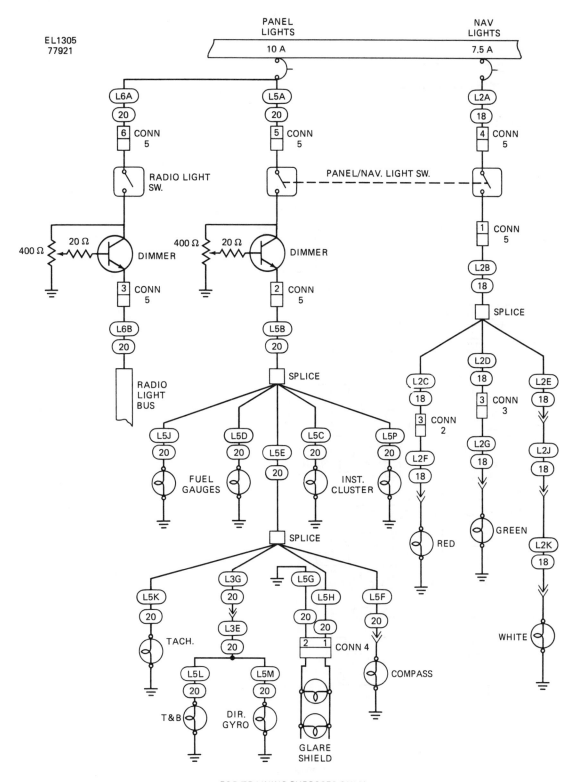

FIGURE 13–5 Panel-and-position-light circuit schematic. *(Piper Aircraft Corporation)*

13–7 shows the circuitry associated with the reversible electric motor that raises and lowers the landing gear. The control circuit is protected by a 5-A circuit breaker in the circuit-breaker panel assembly. This circuit incorporates the two **landing-gear safety switches ("squat" switches)**, which prevent the operation of the landing gear as long as the airplane is on the ground. The landing-gear safety switches are identified as S36 and S37 in the circuit. The power circuit is

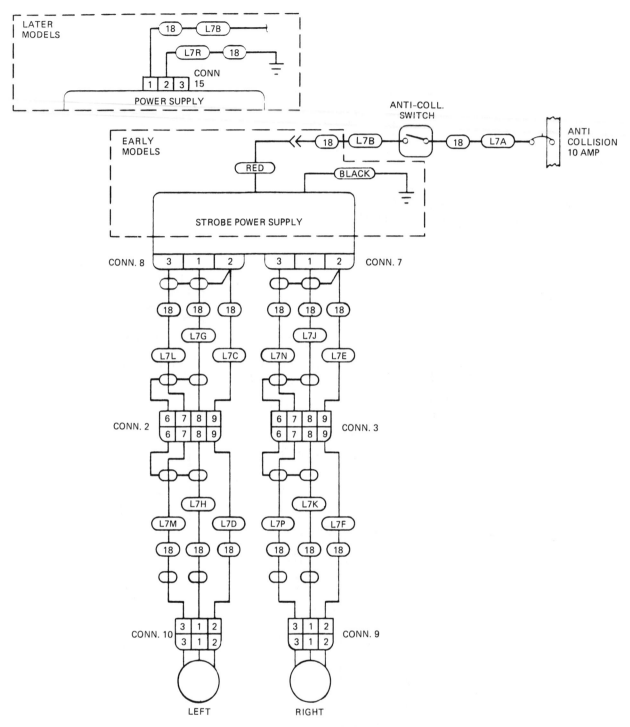

FOR TRAINING PURPOSES ONLY

FIGURE 13–6 Strobe-light circuit schematic. *(Piper Aircraft Corporation)*

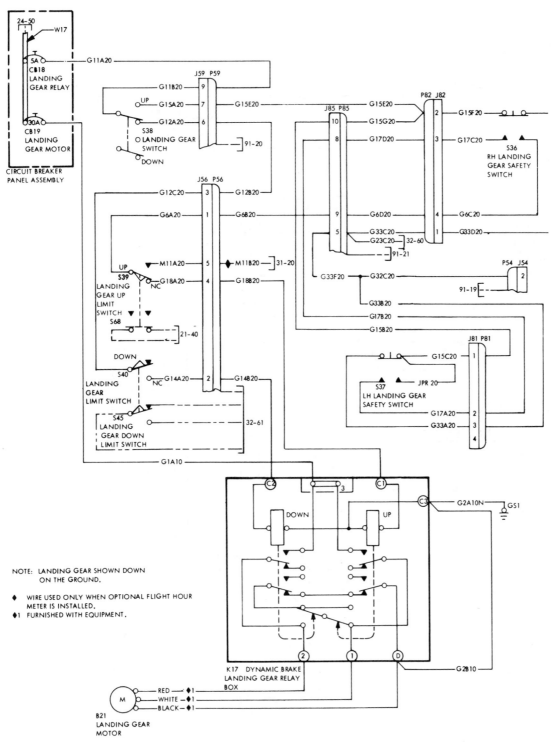

NOTE: LANDING GEAR SHOWN DOWN
ON THE GROUND.

◆ WIRE USED ONLY WHEN OPTIONAL FLIGHT HOUR
METER IS INSTALLED.
◆1 FURNISHED WITH EQUIPMENT.

FOR TRAINING PURPOSES ONLY

FIGURE 13–7 Landing-gear activating-circuit schematic. *(Beech Aircraft Corporation)*

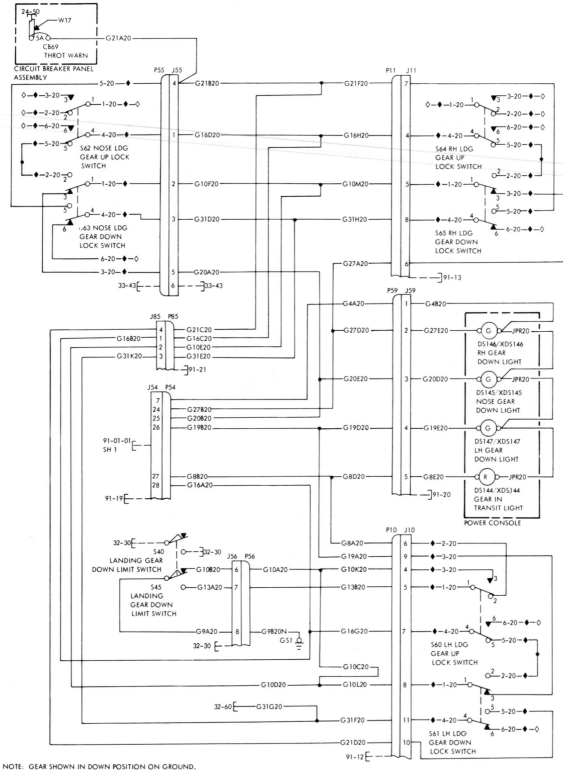

NOTE: GEAR SHOWN IN DOWN POSITION ON GROUND.
♦ STAMP REF DES OF APPLICABLE SWITCH ON HEAT
  SHRINKABLE TUBING & INSTALL ON WIRE FURNISHED
  WITH EQUIPMENT.
◊ TIE UNUSED WIRES INTO HARNESS. CUT OFF AT SPLICE
  LOCATION.

FOR TRAINING PURPOSES ONLY

FIGURE 13–8 Landing-gear position-indicating system.   *(Beech Aircraft Corporation)*

connected to a 30-A circuit breaker. This circuit supplies power to the UP and DOWN power relays, which are controlled through the control circuit. When the landing-gear switch, S38, is placed in the UP position with the airplane in flight, electric power flows from ground, through the UP relay, through the landing-gear UP limit switch (S39), through both safety switches, and to circuit breaker CB18. This causes the relay to close and direct power to the landing-gear motor. When the landing gear reaches the UP position, the UP limit switch opens and cuts off power to the UP relay, thus stopping the motor. The reverse action takes place when the landing gear is lowered.

Figure 13–8 is a circuit diagram of a landing-gear position-indicating system. This circuit operates in conjunction with the landing-gear control circuit shown in Figure 13–7. The switches shown in the circuit represent the condition when the landing gear is in the down and locked position and the aircraft is on the ground. At this time, if electric power is turned on, the three green lights will be on to indicate that all three units of the landing gear are down and locked. The red gear-in-transit light will be out.

Another landing-gear circuit is shown in Figure 13–9. The switches in the circuit are shown in the position for the gear down and the weight of the airplane resting on the landing gear. A careful study of this circuit reveals a number of

safety features. For example, it is not possible to raise the gear when the airplane is on the ground, even if the landing-gear switch is placed in the UP position. Notice that the gear-relay control coil is fed through the left-gear safety (squat) switch. When the airplane is on the ground, this switch is open; hence no current can be supplied to the relay coil. Furthermore, if the gear switch is in the UP position when the airplane is on the ground, a warning horn will sound.

When the airplane is in flight and the gear switch is placed in the UP position, the gear will rise, and when it is completely up, the UP limit switch will open and break the circuit to the control coil of the gear relay (183). When the limit switch opens, it also closes the circuit through the gear-up indicator light.

When the landing gear is up, the DOWN limit switch will be closed, thus making it possible to direct current to the DOWN side of the gear motor if the gear selector handle is placed in the DOWN position. The gear lowers, but before the gear-down indicator light can come on, three microswitches must be closed. These are the right gear-down switch, the left gear-down switch, and the nose gear-down switch. These switches are connected in series; hence no current can flow in the circuit unless all three switches are closed. In flight, if the throttle is partially closed, the warning horn will sound unless the gear is down. Note that the warning horn must obtain power through one side of the DOWN limit switch. This switch is closed except when the gear is down.

The circuit in Figure 13–9 incorporates a **press-to-test** light in order to verify correct system operation. The power during normal operation travels from ground to terminal 1 of the light, through the bulb to terminal 2, and through the UP or DOWN limit switch to the positive bus. During the test function the current travels from ground through terminal 1, through the bulb to terminal 3, and directly to the positive bus. To activate the light's test function, the lens of the bulb is depressed, which moves a switch contact inside the light socket from terminal 2 to 3. The pilot would depress this switch during flight operations if he or she suspected a landing-gear retract or extend system failure.

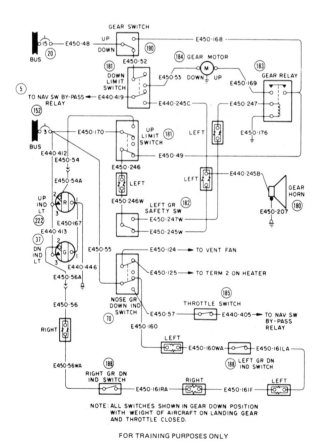

NOTE: ALL SWITCHES SHOWN IN GEAR DOWN POSITION
WITH WEIGHT OF AIRCRAFT ON LANDING GEAR
AND THROTTLE CLOSED.

FOR TRAINING PURPOSES ONLY

FIGURE 13–9 Landing-gear circuit. *(Cessna Aircraft Corporation)*

LARGE-AIRCRAFT ELECTRICAL SYSTEMS

Lighting Circuits

Like other commercial aircraft systems, lighting circuits are classified into two basic categories, **essential** and **nonessential.** In order to facilitate safety, certain flight deck and cabin lights must be powered by the essential power bus(es), or they must contain their own battery packs. These essential lights, including exit signs and escape slide lights, will remain lit even in the event of a catastrophic electrical failure. Typically, a dedicated battery pack is used to power the emergency lights. The lights can be turned on manually or set to come on in the event of primary lighting system failure.

Primary (nonessential) lighting systems incorporate several varieties of lights. Generally speaking, there are flight compartment lights, passenger compartment lights, service lights, and exterior lights. There are literally hundreds of lights that operate from either dc or ac electric power.

Flight Compartment Lights

There are four categories of flight compartment lights found on a Boeing 757. The Boeing 757 contains lighting systems typical of those found on all commercial airliners. The flight compartment floodlights consist of two sets of **fluorescent floodlights,** one mounted in the captain's glare shield and one in the first officer's shield. The captain's lights may be connected to the standby ac bus in the event of a primary ac power system failure.

There are also **incandescent floodlights** mounted in the ceiling and under the glare shield. Dimmer circuits are employed for both the fluorescent and incandescent floodlights. Each dimmer system has an override circuit that is used in the event of a dimmer failure. A schematic of the floodlights is shown in Figure 13–10. As illustrated, the position of the captain's standby relay determines the light's power source. Each light may be connected to the right ac bus or the standby ac bus. If the right ac bus is energized, the relay moves to its abnormal position, thus connecting the right ac bus power to the lights. If the right bus fails, the relay is de-energized, and the standby bus is automatically connected to the lighting circuit.

The flight compartment also contains **dome lights** mounted in the ceiling and powered by 28 V dc. The power can be supplied by either the battery bus or the ground service bus.

Panel lights are used in the flight compartment to illuminate individual instruments and panel lettering. Map and chart lights, utility lights, and threshold step lights may also be located within the flight compartment. All the lights that are in view of both pilots may be controlled by one master dim and test system. The test function is used during preflight inspection to verify that all lights are operable.

Passenger Compartment Lights

Most commercial aircraft contain several varieties of **cabin lights.** Cabin fluorescent ceiling lights with a variable intensity; sign lights, including NO SMOKING, FASTEN SEATBELT, and LAVATORY; and sidewall lights are all controlled from the flight attendants' stations. Passenger reading lights are located above each passenger seat and controlled individually by a switch adjacent to each seat. Entryway lights must also be included in all cabin compartments, along with lights in the galley and flight attendants' stations. Emergency cabin lights include exit sign lights, main aisle area lights, and emergency exit slide lights. Emergency lights typically contain their own power pack.

The fluorescent lights used in the cabin area are located in the sidewall panels or overhead panels of the aircraft. Fluorescent lights are more efficient than the incandescent lights; therefore, they are typically the system of choice. Flu-

orescent lights require a **ballast transformer** to increase system voltage. The high voltage is used to ionize the gas inside the fluorescent tube, thus producing light. A typical fluorescent lighting system schematic is shown in Figure 13–11. From this schematic it can be seen that either the ground service bus or the utility bus may supply power for the lights. A bright/dim switch activates the sidewall light control relay, which in turn directs voltage to the ballast. In the dim position, 208 V ac is sent from the sidewall light transformer to the ballast, and the lights are dim. In the bright position, an additional ac voltage is sent to the ballast. This produces a stronger ballast output, and therefore the fluorescent lights brighten.

Other Lights

Other lighting circuits common to large aircraft include service lights, cargo lights, and exterior lights. **Service lights** are located in the main and nose gear wheel wells, in electric equipment compartments, and in some engine compartments. The power for service lights is typically supplied by the ground power equipment through a ground service bus.

Cargo compartment lights are used on most commercial aircraft to aid in the handling and storage of cargo. Typically, an ac ground power source is used to provide power for cargo compartment lights. Several lights are used in each compartment in order to supply adequate illumination.

Exterior lighting circuits often include wing illumination lights, landing lights, runway turnoff lights, anticollision lights, and position lights. Figure 13–12 illustrates these lights and their respective illumination paths. Wing illumination lights are designed to illuminate the leading edges of the wings for inspection by ground service personnel or flight crew members. The leading edge of a wing is very susceptible to ice formation and is often visually inspected to ensure that ice has not accumulated.

Large aircraft often contain three or more landing lights. Figure 13–12 shows two wing-mounted lights and one nose gear landing light. On some aircraft these lights are automatically dimmed if the landing gear are retracted.

Runway turnoff lights are used to provide illumination of the area to the immediate right and left of the aircraft. These lights are used during taxiing operations to improve ground visibility.

Anticollision lights are mounted on each wingtip and on the top and bottom of the cabin. Red rotating beacons or flashing lights are typically located on the cabin. White flashing, or strobe, lights are located on the wingtips.

As noted earlier, position lights are designed to indicate an aircraft's position and direction. A white light must shine toward the rear of the aircraft, a red light must shine left, and a green light must shine right. On the aircraft shown in Figure 13–12, the wingtip contains both the white and colored position lights. The red (or green) light shines forward and to the side of the aircraft. The white light illuminates an area toward the rear of the aircraft wing.

A variety of lighting circuits are found on large aircraft. Many of these have been discussed in this chapter; however, it is impossible to cover all types and combinations of lights.

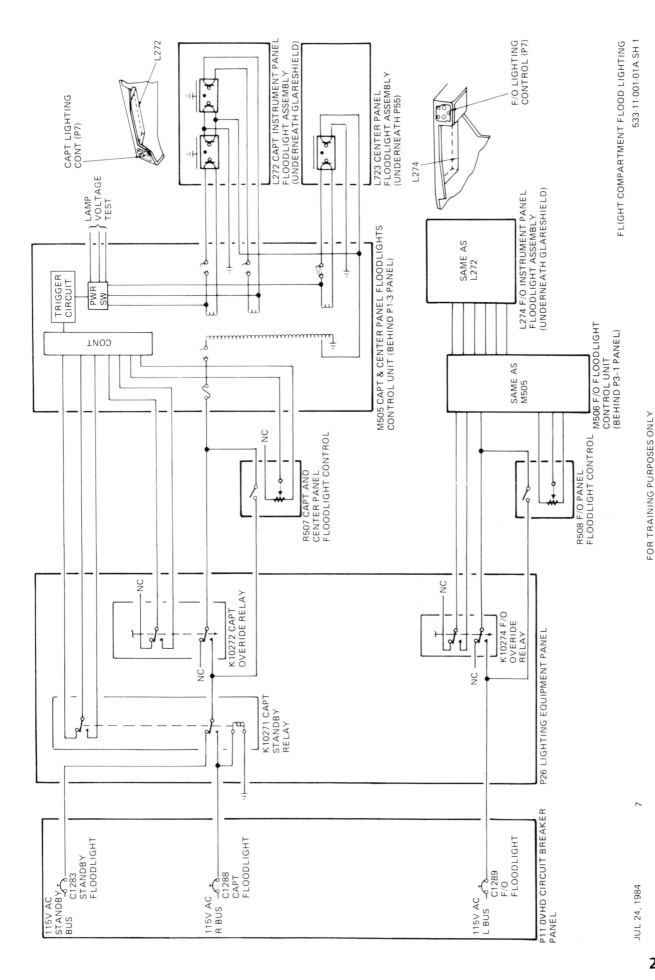

FIGURE 13–10 Floodlight circuit. *(Boeing Corporation)*

FLIGHT COMPARTMENT FLOOD LIGHTING

533:11-001-01A SH 1

FOR TRAINING PURPOSES ONLY

JUL 24, 1984

7

257

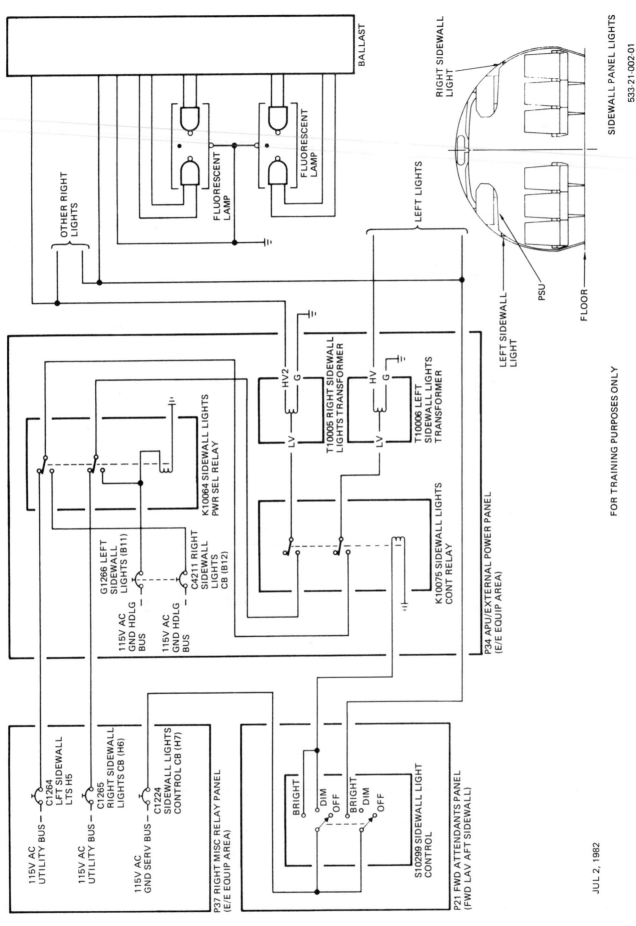

258

FOR TRAINING PURPOSES ONLY

JUL 2, 1982

FIGURE 13–11 Fluorescent lighting circuit. (Boeing Corporation)

533-21-002-01

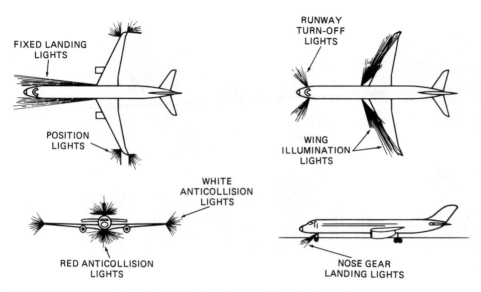

FIGURE 13–12 Typical exterior lighting locations and illumination path.

Always refer to the current manufacturer's data when servicing any lighting equipment.

Landing-Gear Control Circuits

On large aircraft, landing-gear actuators are hydraulically operated. The electronic circuitry of the system is used to provide an indication of gear position and in some cases to control the hydraulic system components. On some aircraft microswitches are tripped when the landing gear reaches its limit. These switches turn off the hydraulic pump motor and turn on the correct gear indicator in the flight compartment. These systems are very similar to those found on light aircraft.

Another means of controlling landing-gear actuator and indicator systems employs **proximity sensors.** Proximity sensors are simply inductance coils that operate in conjunction with steel targets. The inductance of a coil changes with the proximity of the steel target. As discussed earlier, the inductance of any coil is a function of the core material. If the steel target acts as the core for the proximity coil, the inductance of the coil changes as the steel target moves farther from or closer to the coil. A diagram of a proximity sensor is in Figure 13–13. The advantage of these sensors is that there are no moving switch contacts to fail; therefore, the reliability of the system is improved as compared with the performance of systems that employ limit switches.

The inductance of a proximity sensor is measured by an electronic control unit. This unit interprets the input data (some from the proximity sensors) and sends out control signals to the landing-gear actuator and indicator systems.

The Boeing 757 aircraft contains a **proximity switch electronic unit** (PSEU), which provides position sensing for landing gear, cabin doors, and thrust reversers. The system contains 70 sensors located throughout the aircraft that provide input data for the PSEU. The PSEU processes the discrete input signals and controls relays, lights, and/or other electronic components.

Built-In Test Equipment

Large aircraft often incorporate **built-in test equipment** (BITE) systems to monitor and detect faults in a variety of aircraft systems. The use of BITE systems reduces troubleshooting costs by eliminating the time required to connect carry-on test equipment, perform tests, and remove that equipment. The built-in test equipment continuously tests the various systems and stores all fault information so it can be recalled later by line technicians. Once the appropriate repair has been made, the BITE system can be used to retest the system for proper operation. Most BITE systems are capable of isolating system faults with at least a 95 percent probability of success on the first attempt.

The introduction of digital systems on aircraft has made BITE systems possible. Discrete digital signals are used as the code language for BITE systems. Built-in test equipment interprets the various combinations of digital signals to determine a system's status. If an incorrect input value is detected, the BITE system records the fault and displays the information upon request. As shown on this version of a BITE system illustrated in Figure 13–14, the fault information is displayed by light-emitting diodes on the face of the BITE unit when the appropriate button is depressed. Other BITE systems

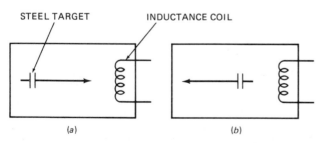

FIGURE 13–13 Proximity sensor diagram. (a) Target moving toward coil; (b) target moving away from coil.

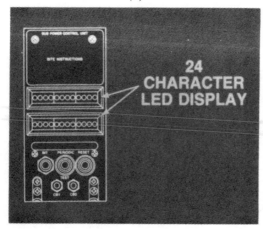

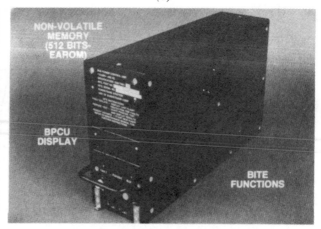

(a) *(b)*

FIGURE 13–14 Built-in test equipment. (*a*) LED display (*b*) bus power control unit with BITE display. *(Sundstrand Corporation)*

include more user-friendly displays that can be accessed from the flight deck. The proper operation and troubleshooting techniques associated with the BITE systems will be discussed later in this chapter.

Intercom and Interphone Systems

The **intercom** system is used for communication between flight crew personnel and passengers. This system typically contains a control panel and microphone at one or more flight attendants' stations and in the flight compartment. The intercom is used to inform passengers of flight details and communicate any instructions necessary to ensure a safe and comfortable flight. On most aircraft there is one central intercom amplifier, which connects to several speakers throughout the aircraft. The amplifier's volume level is automatically adjusted to compensate for varying cabin noise.

An **interphone** system provides a means of communication between flight crew members and ground service personnel. Communication during fueling, ground handling, and baggage storage is essential. On a large aircraft it is virtually impossible to communicate from the cockpit to areas outside the aircraft without some form of assistance. The interphone system contains an amplifier and several stations where a headset, containing a microphone and speaker, can be connected to the system.

The interphone system can also be used during aircraft maintenance. Figure 13–15 shows the interphone configuration for a typical MD-80. If communication is needed between maintenance personnel inside and outside the aircraft, the interphone system is typically used. The system receives power from the ground service bus; thus it can be operated without use of the aircraft's generators.

Electronic Control Units

On modern large aircraft there are several types of control units used to monitor, test, and regulate various electrical systems. These control units, commonly known as black boxes,

are miniature computers designed for a specific function. Typically, black boxes are **line replaceable units** (LRUs) designed for quick removal and installation. Employing the LRU concept has helped to reduce maintenance times and improve airline productivity. Several of these control units are found on modern commercial aircraft. The **generator control unit** (GCU) and **ground power control unit** (GPCU) have already been discussed.

Other common control systems include the **thrust management computer** (TMC), which is used to analyze engine parameters and power requests in order to control engine thrust, and the **flight management computer** (FMC), which monitors flight parameters and performs autopilot functions. The FMC regulates the movement of the control surface actuators. These actuator mechanisms provide control for most

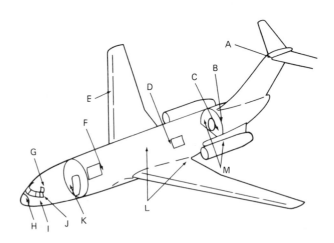

FIGURE 13–15 Typical interphone connection locations. (*A*) Vertical stabilizer; (*B*) aft accessory compartment; (*C*) aft cabin attendant station; (*D*) aft lower cargo compartment; (*E*) fueling control panel; (*F*) forward lower cargo compartment; (*G*) flight crew cabin; (*H*) forward accessory compartment; (*I*) external power control panel; (*J*) electrical/electronics compartment; (*K*) forward cabin attendant station; (*L*) main gear wheel wells; (*M*) aft fuselage (external). *(McDonnell Douglas Corp.)*

primary and secondary control surfaces, such as stabilizers, elevators, rudders, speed brakes, and spoilers.

Engine indicating and crew alerting system (EICAS) control units monitor various electrical parameters and display system status to the flight crew. The EICAS is also responsible for alerting the crew in case of an emergency situation. As illustrated in Figure 13–16, the two EICAS computers receive input data from various airframe and engine sensors. Output data are sent to warning electronic units, the standby engine indicator, and the EICAS display panels. The EICAS display panels consist of two cathode-ray tubes (CRTs). Each CRT is used to display status, caution, or warning information. On Airbus Industries aircraft a similar system is used to monitor engine and flight parameters. This system is known as **electronic centralized aircraft monitoring** (ECAM). The EICAS and the ECAM system will be discussed in greater detail in Chapter 17.

Equipment Cooling

Heat is an electronic unit's biggest enemy; therefore, most aircraft contain some means of electronic-equipment cooling. Since large aircraft contain numerous electronic LRUs, they are, for the most part, centrally located. Typically, this equipment compartment is behind and/or below the aircraft's flight deck. The use of a centralized equipment center enables cooling with a minimum of air ducts.

Cooling fans and air ducts are commonly employed to force air over the warm equipment and dispense the heat overboard. In some cases heat exchangers are used to cool the warm air and recirculate it back over the equipment. On some aircraft separate air conditioner units are used to ensure proper equipment cooling. In this case the warm air is circulated through the air conditioner, and the cool air is returned to the equipment compartment.

Most equipment-cooling systems also employ overheat and smoke detector sensors. These sensors monitor the system and provide an appropriate indication for the flight crew. The flight crew may then take the appropriate actions to ensure proper system operations.

Pressurized air can also be used to cool electronic instruments. Cooling air is forced into a plenum chamber created by an inner and outer instrument panel. As illustrated in Figure 13–17, holes in the inner panel direct air over each instrument. This process improves instrument cooling and enhances the reliability of the instruments.

Static Dischargers

During flight, aircraft produce **precipitation static** through contact with rain, dust, snow, and other particles. Precipitation static (**P-static**) can also be created by the movement of jet exhaust over the aircraft's surface. P-static may occur on either large or small aircraft; however, it is only prevalent on relatively high-speed vehicles. At high speeds the friction between the air and the aircraft's surface increases. It is this friction that produces a static electrical charge on the aircraft's surface.

The static charge itself poses little threat to flight safety; however, the discharge of the P-static back into the air creates problems. As the P-static "jumps" from the aircraft to the air, a low-frequency magnetic wave is produced. This magnetic wave creates radio interference identical with that produced by a lightning bolt discharge, except that P-static discharge creates a weaker interference signal.

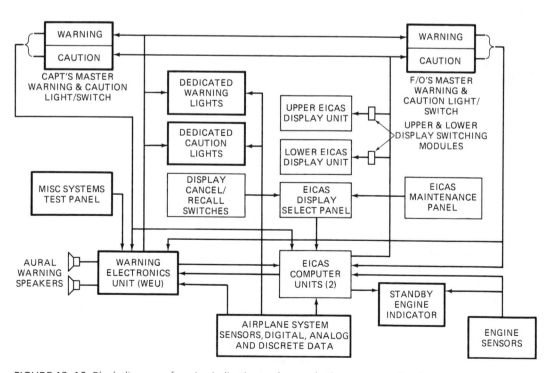

FIGURE 13–16 Block diagram of engine indicating and crew alerting system. *(Boeing Corporation)*

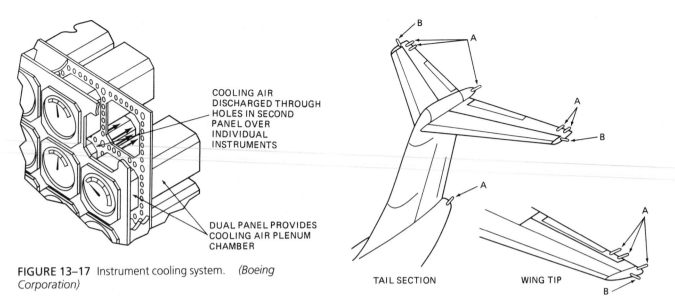

COOLING AIR
DISCHARGED THROUGH
HOLES IN SECOND
PANEL OVER
INDIVIDUAL
INSTRUMENTS

DUAL PANEL PROVIDES
COOLING AIR PLENUM
CHAMBER

FIGURE 13–17 Instrument cooling system. *(Boeing Corporation)*

TAIL SECTION   WING TIP

FIGURE 13–18 Typical location of static dischargers. *(A)* Trailing type and *(B)* tip type.

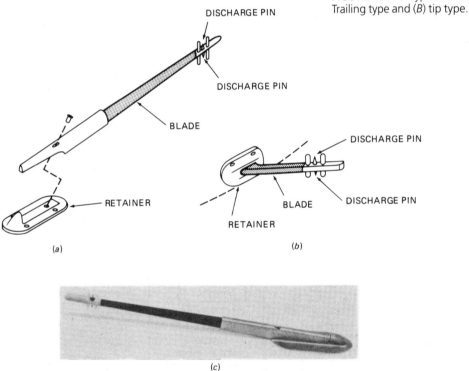

DISCHARGE PIN

DISCHARGE PIN

BLADE

RETAINER

(a)

DISCHARGE PIN

DISCHARGE PIN

BLADE

RETAINER

(b)

(c)

FIGURE 13–19 Typical installation of static discharger. *(a)* Trailing type; *(b)* tip type; *(c)* photograph of complete assembly. *(Dayton Aircraft Products)*

Radio interference can also be created as a static charge moves from one portion of the airplane to another. As discussed in Chapter 12, electrical bonding techniques are used to eliminate static discharge between aircraft components.

Recently, low-frequency navigation radios have become more popular. Since P-static discharge emits a low-frequency interference signal, there has been a renewed interest in controlling P-static. The correct use of static dischargers is vital to proper P-static control. Static dischargers reduce the threshold at which the P-static leaves the aircraft. In other words, a lesser amount of static charge must accumulate be-

fore it discharges back into the air. This lower level of discharge current produces a lower level of radio interference. If the discharge is controlled to a low enough value, the radio interference becomes negligible.

Static dischargers are located at numerous places on an aircraft, as illustrated in Figure 13–18. P-static tends to accumulate on the tips and trailing edges of wings, on control surfaces, and on horizontal and vertical stabilizers. Dischargers located in these areas have the greatest potential to reduce radio interference. As shown in Figure 13–19, dischargers are typically mounted to a retainer. The retainer is permanently mounted to the aircraft's surface. This facilitates discharger replacement in the event the discharge pins become damaged or pitted from use. Always inspect the static dischargers at the appropriate time intervals and in accordance with the manufacturer's recommendations. Figure 13–19c is a photograph of a typical static discharger.

MAINTENANCE AND TROUBLESHOOTING OF ELECTRICAL SYSTEMS

General Requirements

To ensure safe flight operations, electrical systems must be maintained in perfect working condition. The routine inspection procedures performed on all aircraft are used to detect any potential electrical system failures. During these inspections, specific electric components are inspected and tested as dictated by the manufacturer. If a malfunction or defect is found, the proper maintenance procedures are used to correct the problem.

Electrical system failure is not always detected during inspections. Often systems fail during operation and must be repaired prior to further flight. Maintenance of this type is usually more critical; that is, the aircraft downtime must be as short as possible. Unexpected maintenance causes flight delays, passenger inconvenience, and lower profit margins. Maintenance of electrical systems must be performed with both speed and accuracy. The safety of any flight often lies in the hands of the aircraft technician. Be sure to perform all maintenance procedures and electrical system inspections in accordance with the manufacturer's recommendations and to the best of your ability.

Inspection Schedules

By mandate of the FAA flight regulations, all civilian aircraft must be inspected in accordance with a schedule set forth by an approved inspection program. The **100-hour,** the **annual,** or the **periodic inspection program** can be used for light-aircraft inspections. Each of these programs is designed to instruct the aircraft technician as to which systems and components require routine maintenance and/or inspection.

Large aircraft are typically maintained according to one of the inspection programs approved by the FAA. These programs, known as **continuous airworthiness inspection programs,** include various routine service inspections and more complete maintenance procedures called *checks.* An *A-check, B-check, C-check,* and *D-check* are designed to fit the specific needs of a particular aircraft operator. *A-checks* are the simplest; routine maintenance is performed approximately every 200 hours. *D-checks* are typically complete airframe overhauls performed every 4 to 5 years. Typically under this type of system, certain portions of the aircraft are inspected at given intervals of flight time. For example, the aircraft's position lights may require an operational check during a preflight walkaround inspection; every 100 hours

the oil level of the integrated drive generators may require visual inspection, with oil added as needed. Repair or replacement of any defective flight-critical parts must be completed prior to the aircraft's return to service. During a *check* any life-limited electrical parts must be replaced or overhauled according to the manufacturer's schedule.

On large aircraft the use of built-in test equipment often facilitates an inspection. The technician can quickly and easily inspect an electrical system by examining the fault data stored by the test equipment. If a fault is stored within the system's memory, the technician can make necessary repairs during the inspection. The current trend in the aircraft industry is to employ more BITE systems wherever possible. This is being done in an effort to reduce maintenance costs and aircraft downtime.

Light aircraft are often maintained on an annual or 100-hour inspection basis. During an inspection of this type, the entire aircraft is inspected, including the electrical systems. All electrical systems, their components, and related wiring should be checked in accordance with the inspection schedule. Typically, an operational check of all electrical systems is conducted. Any defects are repaired, routine maintenance is performed, and all life-limited parts are replaced.

Life-limited parts are those that deteriorate beyond use in a given length of time. For example, emergency lighting system batteries are often considered life-limited parts; that is, they must be replaced on or before specific dates. Routine maintenance of electric components may include servicing batteries, lubricating motor bearings, and replacing generator brushes. Inspections of electrical systems include an operational check and a visual inspection. While performing a visual inspection, the technician should look for loose connectors, chafed wires, poor electrical bonding, loose bundle supports, nicked or damaged wire insulation, and any other obvious defects.

Multimeter Troubleshooting

As discussed in Chapter 8, a multimeter is a combination of three basic instruments: an ohmmeter, a voltmeter, and an ammeter. This combination instrument has made it possible for the technician to reduce his or her inventory of test instruments. Each of the three instruments contained within a multimeter performs a specific function. A typical multimeter is shown in Figure 13–20. Of the three instruments, the voltmeter is by far the most useful tool to detect an open circuit. **Open circuits** (opens) are the most common wiring defect. Open circuits are created by broken wires, defective connectors, loose terminals, and any other condition that creates a circuit disconnection (see Figure 13–21). Opens can also occur within components such as switches, fuses, circuit breakers, lamps, or motors.

Short circuits (shorts) are also common problems for aircraft electrical systems. There are two types of short circuits, a **short to ground** and a **cross short.** A **short to ground** from a positive wire creates an infinite current flow because of the extremely low resistance from the voltage positive to negative (see Figure 13–22). In Figure 13–22a, the wire broke, forming an open circuit; the conductor exposed by the break

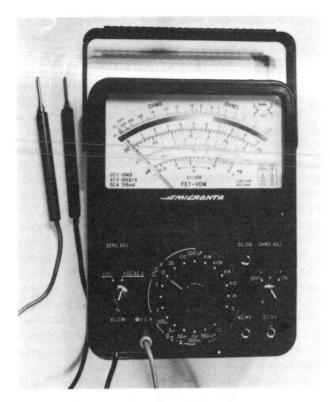

FIGURE 13–20 A typical analog multimeter.

more than one circuit operates. A cross short connects power to an "extra" circuit. Short circuits are most likely created by the friction between two wires or between a wire and the airframe. The friction wears through the insulation, and the conductor is exposed, thus creating the potential for a short circuit. An ohmmeter is usually used to troubleshoot both types of short circuits.

Voltmeter Troubleshooting

Voltmeters are always connected in a circuit in parallel with respect to that portion of the circuit to be measured. If one probe of a voltmeter is connected to a positive voltage and the other probe to a negative voltage, the meter will indicate the voltage difference between those two points. Voltage, as you may recall, is the difference in electrical pressure between two points.

If one desires to measure the voltage available to the lamp represented in Figure 13–24a, the voltmeter should be placed between points A and B. Point A is connected to the positive of the battery, and point B is connected to the negative of the battery. In this case the voltmeter would indicate 12 V.

In an aircraft most circuits are connected from the positive bus through a load to the aircraft's ground, as shown in

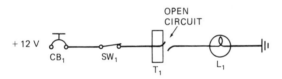

FIGURE 13–21 Diagram to illustrate an open circuit.

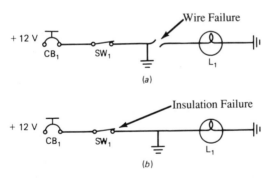

FIGURE 13–22 An illustration of a short to ground (a) created by a broken wire and (b) created by defective insulation.

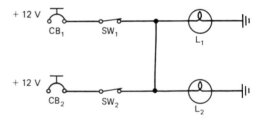

FIGURE 13–23 An illustration of a cross short. Both L_1 and L_2 illuminate when SW_1 or SW_2 is closed.

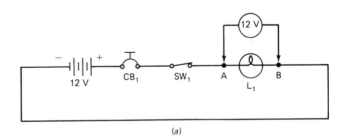

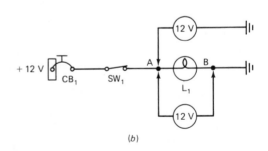

FIGURE 13–24 Measuring the voltage available to a light (a) in a two-wire system and (b) in a single-wire system.

then shorted to ground. In Figure 13–22b, the wire's insulation failed and exposed the conductor; the exposed conductor shorted to ground. The high current flow opens the circuit protector (the fuse or circuit breaker). If the circuit is not protected, the wiring will overheat and most likely melt into a disconnection. A **cross short** takes place when two or more circuits are accidentally connected together (see Figure 13–23). In this situation, when one circuit is switched on,

Figure 13–24b. Since the negative connection of the battery is connected to the aircraft's ground, a voltmeter connected between points A and B will indicate 12 V, and a voltmeter connected between point A and the aircraft's ground will indicate 12 V. The fact that the entire metal structure of the aircraft is connected to the battery negative makes the voltmeter a very versatile tool. As illustrated in Figure 13–25, a voltmeter can be connected to any convenient ground in order to find the positive voltage present in a circuit. Voltmeter V_1 indicates 12 V, V_2 indicates 12 V, and V_3 indicates 12 V. Voltmeter V_4 indicates zero volts because its probes are connected between two negative voltage points (ground to ground).

When using a voltmeter, it is important to consider voltage as consisting of two parts, a positive voltage and a negative voltage. As long as a voltmeter is connected to one positive voltage and one negative voltage, it will indicate system voltage. As illustrated in Figure 13–26, if the voltmeter is connected to two equal positive voltage values or two equal negative voltage values, it will indicate zero.

When troubleshooting a circuit with an open (disconnected) wire, the technician should place the voltmeter in various convenient places along the suspect wire. Since the positive voltage signal initiates at the aircraft bus, it is logical to first test for a positive voltage near the bus and move systematically toward the load. This concept is illustrated in

Figure 13–27. The first test is performed as represented by voltmeter V_1; the second test, voltmeter V_2; the third test, voltmeter V_3; each measures 12 V. This indicates that the circuit's positive wire is continuous (not open) from the bus through terminal 1 and the switch. Voltmeter V_4, being connected to what should be the positive side of the lamp, should read 12 V. Since V_4 indicates zero volts, the circuit must be open between the switch and the light.

When you are dealing with complex circuitry, the troubleshooting process becomes more difficult. When deciding where to connect the voltmeter, always consider the following: (1) A wire's insulation should never be removed to install a meter's test probe; therefore, take all measurements at open terminals, plug connectors, switches, fuses, or any other areas where the conductor is exposed. (2) Since an open in a wire can occur virtually anywhere, always connect the test meter to an easily accessible connection. If there is no positive voltage at that point (while you are referencing ground with the other meter probe), you can conclude that the open is between that test point and the positive bus. To further pinpoint the defective wire or connector, move the positive voltmeter probe to the next exposed terminal nearer the aircraft bus. If the voltmeter indicates zero volts, the open is between that test point and the positive bus. If the meter indicates system voltage, the open circuit is between the first and second test points, as shown in Figure 13–28. The first test was performed at the switch because it was easily accessible. From this test it was easy to determine which portion of the circuit (before or after the switch) should be tested next.

In many situations it is easiest to check voltage at the load of a circuit. In this case, if there is no voltage available to the load, the circuit is defective. If voltage is present to the input connections of the load, the load itself is defective. Be cautious: voltage consists of two parts, positive and negative, and both must be available to the load in order for it to operate.

A voltmeter may be used to determine if the negative voltage of a circuit is available to the load. As illustrated in Figure 13–29, the voltmeter should be referenced to a positive voltage source when testing for a negative signal. That is, the meter's red test probe should be connected to a point that is known to be a positive voltage. The aircraft's bus, or any other positive connection, may be used for this purpose. In this configuration, if the meter's negative probe is connected to a negative voltage, the meter will indicate system voltage. If no negative signal is present at the meter's black test probe, the meter will indicate zero volts.

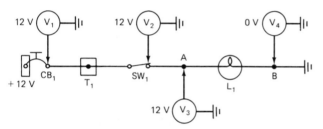

FIGURE 13–25 Voltmeters used to test voltage in a circuit.

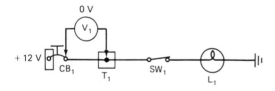

FIGURE 13–26 A voltmeter connected between two points of positive voltage.

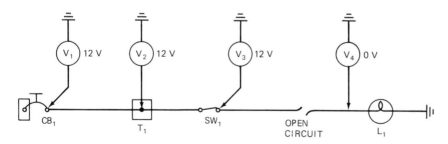

FIGURE 13–27 Placement of a voltmeter to troubleshoot an open circuit.

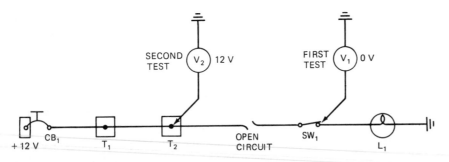

FIGURE 13–28 Voltage tests of an open circuit. Since the first test indicated 0 V, the second test is made closer to the bus.

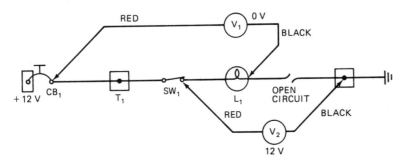

FIGURE 13–29 Using a voltmeter to test for a negative voltage. If V_1 measures 0 V, there is no negative voltage present at the meter's black probe. If V_2 measures 12 V, negative voltage is present at the black probe.

Voltmeters and Composite Aircraft

It should be noted that the new breed of composite aircraft require some special procedures when voltage is checked. As with systems on metal aircraft, both the negative and positive voltage signals must be present to all electric power users. However, on composite aircraft the negative voltage cannot be transmitted through a metal structure. Some composite aircraft use a separate wire to carry the negative (ground) voltage from a *ground bus* to each electrical load. When checking for a positive voltage on this type of system, be sure to verify that you are connected to an uninterrupted ground source, as seen in Figure 13–30. To verify an uninterrupted

ground circuit, place your voltmeter probes between a known voltage positive source and the ground wire in question. Never draw conclusions about voltage measured between two unverified points.

Some composite aircraft incorporate a *ground plane* on the inside skin of the aircraft. This ground plane is used as the negative-side voltage source. When an electric component is not working because of a lack of voltage, be sure to check the ground plane. To check for a negative voltage signal on the ground plane, use a voltmeter and reference a known positive voltage. On some aircraft special low-resistance ohmmeters are used to verify continuity of the ground plane.

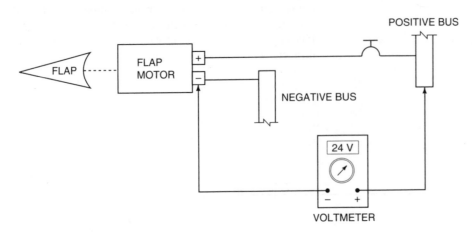

FIGURE 13–30 Testing for an uninterrupted ground circuit on a composite aircraft.

Ohmmeter Troubleshooting

Ohmmeters are best suited for two types of tests: (1) continuity checks of components removed from a circuit and (2) continuity checks of short circuits. Components such as switches, relays, lightbulbs, and transformers may all be tested with an ohmmeter. However, these components must be removed or disconnected from the circuit prior to testing.

Figure 13–31 illustrates the use of an ohmmeter for testing components. A component such as a switch, circuit breaker, or fuse must have zero resistance (when closed) to operate properly. If the ohmmeter measures infinite resistance, the component is defective.

An ohmmeter test is also valid for most power users as shown in Figure 13–32. The light tested should show relatively low resistance if it is functional. If it is defective (open), the light will show infinite resistance. In general, any power user should have a resistance equal to its rated voltage divided by its rated amperage ($R = E/I$). Any load tht has infinite resistance is defective.

Ohmmeters are often used to troubleshoot shorted circuits. For this type of troubleshooting, the circuit power must be turned off and the circuit isolated from the rest of the electrical system. In most cases this can be achieved by turning off the aircraft's battery master switch and opening the appropriate circuit breaker. Figure 13–33 shows the ohmmeter

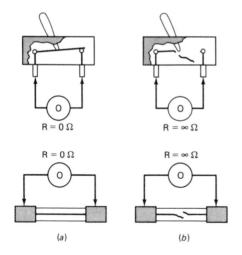

FIGURE 13–31 Testing components with an ohmmeter. (a) Zero resistance measured across operable components; (b) infinite resistance measured across defective components.

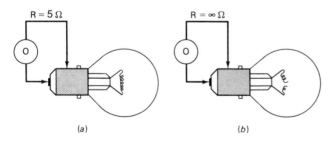

FIGURE 13–32 Testing load units with an ohmmeter. (a) A good bulb indicates low resistance; (b) a defective bulb indicates infinite resistance.

configuration for testing a wire shorted to ground. In Figure 13–33a the short to ground seems to appear at T_2; however, this is incorrect. The short is in wire segment C; but since wire segment C is connected to T_2, the meter indicates zero resistance from T_2 to ground. To pinpoint the location of the shorted circuit, isolate the various segments of wire. In Figure 13–33b, segment C is isolated from T_2. The ohmmeter now reads infinite resistance; the short no longer appears to be at T_2.

If the meter probe is moved to the end of wire segment C, it will once again measure zero resistance to ground. Figure 13–33c illustrates the final test needed to find the defective wire. In this case wire segment C is completely isolated; the switch is open, and wire segment C is removed from T_2. Since the ohmmeter indicates zero resistance, wire segment C (a positive wire) must be shorted to ground.

Ohmmeters can be used to test for open circuits; however, it is typically easier to use a voltmeter. The physical length of a meter's test leads may inhibit the use of an ohmmeter for a continuity check of open circuits. If you want to test a wire that is routed from the flight deck to the tail of the aircraft, your ohmmeter must be connected on both ends of the wire. With even a relatively small aircraft, this is impossible without extending the length of the meter's test leads. Using a voltmeter, one could simply test voltage at the tail end of the wire to ground and determine the wire's condition.

Ammeter Troubleshooting

Ammeters are typically used to test aircraft charging systems. In such cases it is often important to measure the total output amperage of an alternator or generator. Although multimeters normally incorporate an ammeter, they are not typically of high enough capacity to measure charging system current. If it becomes necessary to measure a relatively high amperage, be sure to utilize an ammeter that is capable of measuring the anticipated current. The installation of charging system ammeters is discussed in Chapter 10.

A Typical Troubleshooting Sequence

A typical sequence for troubleshooting a defective position-light circuit is as follows. First examine the schematic diagram, and operate the defective system. While operating the system, make note of exactly what operates correctly and what operates incorrectly. Examine the circuit protector of the system, and determine its condition. If the fuse or circuit breaker has opened, the circuit is most likely shorted to ground. If the fuse is continuous or the circuit breaker is closed, the defect is most likely an open. Study the circuit's schematic diagram, and determine which component or wire is a likely suspect. If the defect is a short circuit, as illustrated in Figure 13–34, an ohmmeter should be used to find the defective wire segment.

Before the ohmmeter is installed to troubleshoot the short, a portion of the circuit could be tested from the flight deck. For example, if switch 1 is turned off and the fuse (or circuit breaker) opens when the circuit is tried again, wire segments C through J are not the cause of the defect. This must be true,

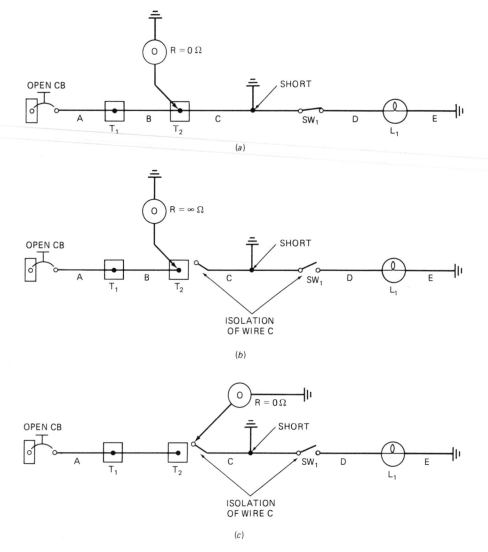

FIGURE 13–33 Using an ohmmeter to find a short to ground. (a) T_2 has zero resistance to ground when wire C is not isolated. (b) T_2 has infinite resistance to ground with wire C isolated. (c) Wire C has zero resistance to ground.

since these wire segments are disconnected from the bus when switch 1 is opened. If switch 1 is turned off and the fuse remains closed (operable), the circuit defect must be located between the switch and the position lamps (wires C through J). This must be true since the circuit protector did not open while the wire segments C through J were disconnected from the circuit.

Troubleshooting from the flight deck is an important part of the repair process. As illustrated in the last paragraph, this is done by operating the flight deck controls and drawing accurate conclusions from the results. If done properly, considerable time can be saved by studying the system's schematic and operating related electrical systems. Often this process can significantly reduce the possible defect locations, therefore improving the troubleshooting process.

If the defect in the position-light circuit is an open, as in Figure 13–35, a voltmeter should be used to locate the fault. In this case opening switch 1 would not allow the technician to draw any significant conclusions. A voltmeter must be in-stalled systematically throughout the circuit to determine which wire segment is defective.

Troubleshooting With Built-In Test Equipment

Built-in test equipment (BITE) systems found on modern commercial aircraft are designed to troubleshoot the electrical problems typically encountered during maintenance. System faults that occur during normal aircraft operations must be repaired swiftly and accurately. A typical aircraft may use several BITE units to monitor the major systems, such as electric power, environmental control systems, and flight control systems. BITE systems perform fault detection, fault isolation, and operational verification after system repair.

BITE systems provide fault detection continuously during aircraft operation. If a fault is detected, the BITE system stores the necessary defect information in a nonvolatile mem-

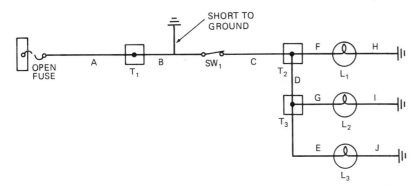

FIGURE 13–34 A typical short circuit.

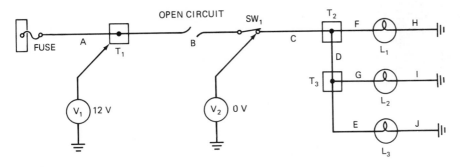

FIGURE 13–35 A typical open circuit.

ory and sends the appropriate display signal (if any) to the flight deck. If the fault requires immediate attention, the flight crew will notify the ground technician via radio transmission or upon landing. The technician must access the appropriate BITE system to perform the fault isolation test. Through correct operation, many BITE systems will display failure data and repair code information.

Several versions of built-in test equipment are in use today. Simple systems typically incorporate a **go/no go** red or green LED on the equipment **black box** or **line replaceable unit** (LRU). More complex systems use a multicharacter display and monitor more than one LRU and the associated wiring. The system in Figure 13–36 is accessed from the equipment center of the aircraft. More advanced BITE systems incorporate displays that are activated from the flight deck and have paper printouts. In addition, advanced systems may have a means to transmit data from the aircraft to the maintenance facility during flight; this means is known as ACARS (ARINC communication addressing and reporting system).

The BITE system shown in Figure 13–36 is incorporated with the bus power control unit (BPCU). This system monitors the entire electric power generation system, including the left, right, and APU generators; the constant-speed drives; and their related control units. The BIT button is depressed on this system to activate the 24-character LED fault display. Typically, a BITE system will display fault information in a coded message, as illustrated in Figure 13–37. The message is then decoded by the technician through the use of the aircraft's maintenance manual. The appropriate manual will inform the technician of any LRU to be replaced or cir-

cuit to be repaired and its location within the aircraft. The fault information on this system is displayed for 2 s, and then the display automatically advances to the next fault. This BITE system will make an appropriate indication when all fault data have been displayed, as in Figure 13–38.

After the system fault has been repaired, the BITE box

FIGURE 13–36 BITE system display. *(Sundstrand Corporation)*

Maintenance and Troubleshooting of Electrical Systems **269**

FIGURE 13–37 Typical BITE display: BPCU FAILED ERROR CODE 02. *(Sundstrand Corporation)*

FIGURE 13–38 BITE display of LAST FLT 00 END OF DATA. *(Sundstrand Corporation)*

should be reset and an operational check performed. The repaired system should be run through a complete cycle of operation. In the case of the electric power generation system, the appropriate engine and ac generator should be subjected to a variety of operating parameters. The flight deck instruments and failure indicators are monitored during the test to detect any further problems.

After the repaired system has been run, the BITE system fault display should be reactivated. This will initiate the read-

out of the nonvolatile memories, and remaining operational faults will be displayed. If the system is found to be without fault, the BITE display will respond accordingly, as in Figure 13–39.

Multipurpose Control Display Unit

A multipurpose control display unit (MCDU) is used to access a slightly more advanced BITE system. Some aircraft require that the MCDU be accessed from the equipment bay, while other aircraft require a carry-on MCDU controller, which is connected to the system on the flight deck. Many aircraft use a controller located on the instrument panel and display information on the EICAS display unit. The operation of the MCDU is similar to the operation of the previously described BITE system. The MCDU is typical of the system found on the Boeing 757 and 767 aircraft. The MCDU receives digital data in an ARINC 429 format from the thrust management, flight control, and flight management computers, along with EICAS inputs. The MCDU both monitors inflight faults and performs ground test functions. In-flight faults are directly correlated to the various flight deck effects associated with in-flight problems. A **flight deck effect** is any EICAS display or discrete annunciator used to inform the flight crew of an in-flight fault.

When the aircraft lands, the MCDU automatically records any in-flight faults (from the last flight) in a nonvolatile memory. To access this memory, the technician must first cycle the MCDU off and on again. This will result in an internal test of the MCDU. After the internal test has been completed and OK'd by the display, the technician should select the in-flight mode of operation. The unit will respond accordingly with faults listed in order of occurrence. At the end of the fault data, the unit will ask if it should display faults from previous flights. The MCDU will store faults from a maximum of 10 flights.

In the case of an MCDU located in the equipment bay, fault data appear on an LED display similar to that shown in Figure 13–37. On this type of MCDU, the top line displays the flight on which the fault occurred and the related flight deck effect; the bottom line displays the faulty LRU to be replaced. If the MCDU is accessed from the flight deck, the EICAS is used to display the message.

Central Maintenance Computer System

The latest generation of built-in test equipment is known as the **central maintenance computer system** (CMCS). This system is designed to perform in-flight and ground tests of virtually every aircraft system, each one accessed from a central location. The **control display unit** (CDU), used to access and display faults, is located in the center console of the flight deck. This type of system is found on the Boeing 747-400, a state-of-the-art four-engine transport-category aircraft. As shown in Figure 13–40, the CDU uses a CRT display. This type of display allows for a more descriptive message of faults, which are directly correlated to flight deck effects created by the same faults. A CMCS printer is incorporated to provide a written report of the fault data, and a software data loader is used to store faults on a computer disk (Figure

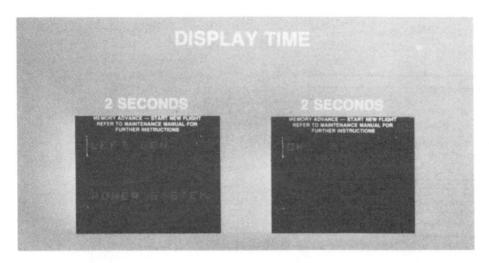

FIGURE 13–39 BITE display of a system with no recorded fault data: LEFT GEN POWER SYSTEM, OK. *(Sundstrand Corporation)*

FIGURE 13–40 A central display unit (CDU) on a Boeing 747-400 aircraft. *(Boeing Corporation)*

13–41). Aircraft equipped with ACARS are capable of transmitting fault data from the aircraft to a ground facility. ACARS will also answer all maintenance data requests from the ground facility.

There are two central maintenance computers (CMCs) located in the aircraft's equipment bay. The CMCs receive up to 50 digital ARINC 429 data inputs and various discrete inputs. Each CMC has 10 ARINC 429 outputs; one is a crosstalk bus to the other CMC. The outputs are sent to the aircraft systems through the left CMC; therefore, if only one CMC is available, it must be installed in the left slot. If the left CMC detects internal faults, output data are automatically passed from the right CMC directly through the left CMC. Figure 13–42 is a block diagram of the CMC data inputs and outputs.

During flight the CMCs receive fault data from the aircraft's electronic interface units (EIUs) and other digital and discrete systems to record in-flight failures. The EIUs

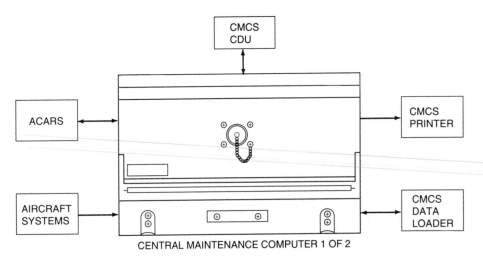

CENTRAL MAINTENANCE COMPUTER 1 OF 2

FIGURE 13–41 Central maintenance computer and data links to CDU, printer, data loader, aircraft systems, and ACARS.

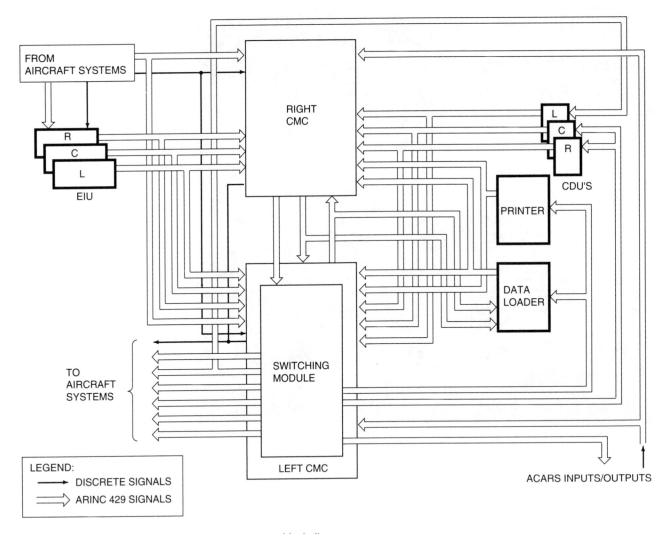

FIGURE 13–42 Central maintenance computer system block diagram.

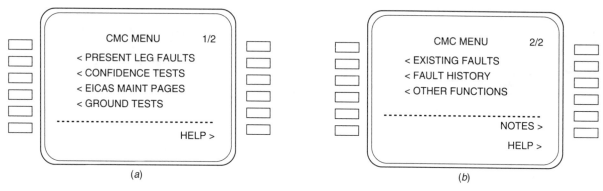

FIGURE 13–43 Central maintanance computer menu display. (a) Page 1 of 2, for line maintenance; (b) page 2 of 2, for extended maintenance troubleshooting

monitor system parameters and control the displays of the EICAS and EFIS (electronic flight instrument system). Once the aircraft is on the ground, the CMC can be interrogated for any history of in-flight faults stored in a nonvolatile memory. Up to 500 faults can be stored in the nonvolatile memory.

The **control display unit** (CDU) shown in Figure 13–43 displays two different menu pages used to initiate interrogation of the CMC. The first page is used primarily for line maintenance and operations; the second page is used for in-depth troubleshooting. A specific function is selected from the menu by pressing the button adjacent to that function. There are three basic types of faults: (1) **existing faults** (those faults active at the time of inquiry), (2) **present leg faults,** (those faults recorded during the last flight), and (3) **fault history** (those faults that were recorded during present leg or previous flights).

Depressing the button for the **ground tests** function tells the CMC to test LRUs and various systems. The **EICAS maintenance pages** function will activate the real-time display of various systems, such as electric power. This function will also allow for access of maintenance pages recorded in memory at an earlier time, called *snapshots*. The **confidence tests** function allows the technician to perform tests that are typically performed before a flight.

Centralized Fault Display System

The Airbus A-320 employs a similar central maintenance system called the **central fault display system** (CFDS). This system classifies faults into three categories: **Class 1, Class 2,** and **Class 3** failures. Class 1 failures have an operational consequence on the flight. The crew is notified by a red warning or amber caution on the electronic centralized aircraft monitoring (ECAM) system or by discrete instrument flags. The pilot must report Class 1 failures in the logbook, since they require maintenance action before the next flight. Class 2 failures are displayed to the pilot by means of the ECAM system only after landing and engine shutdown. Class 2 failures must be reported by the pilot in the log because they cannot be left uncorrected until the next scheduled maintenance. Class 2 failures are categorized by the **minimum equipment list** (MEL) to determine the number of flights allowable prior

to initiating repair. Class 3 failures are not reported to the pilot and can be left unattended until the next scheduled maintenance. Class 3 failures are displayed only during access of CFDS data.

There are several different types of built-in test equipment systems. The descriptions above give an overview of common equipment. Systems of the future promise to be even more accurate, cover more equipment, and be easier to use. BITE systems are here to stay and for good reason: they simplify complicated troubleshooting tasks. The operation of many BITE systems is relatively complex, so before using any particular system, the technician should become completely familiar with the operation of the equipment.

Electrostatic-Discharge-Sensitive Equipment

Some electronic units found on modern aircraft are extremely sensitive to stray current flows. Even a static electrical discharge from a technician to a sensitive component could damage that component. These extremely delicate components are known as **electrostatic-discharge-sensitive** (ESDS) parts. ESDS parts are identified by one or more of the symbols found in Figure 13–44. A part labeled as electrostatic-discharge-sensitive may be damaged by a static discharge of as little as 100 V. A technician walking on an aircraft's carpet, removing a coat, or simply rubbing his or her hair can accumulate a static charge well over 1000 V. A static charge of this magnitude will damage ESDS parts. Most people cannot feel an electrostatic discharge below 3000 V. A visible spark from a static discharge is typically above 12,000 V. Each of these two levels is well above the tolerance of ESDS parts. A technician may become charged and damage a component without even realizing it.

A technician and all of his or her equipment must be connected to the aircraft's ground prior to servicing any ESDS components. This will neutralize any electrostatic charge that may have accumulated. The most common way to ground a technician employs the use of a grounded wrist strap. The wrist strap, as shown in Figure 13–45, is connected around a bare wrist of the technician and connected to the aircraft's ground by means of a wire and plug. All bench technicians

CAUTION
CONTENTS SUBJECT TO DAMAGE BY STATIC ELECTRICITY
DO NOT OPEN
EXCEPT AT APPROVED STATIC-FREE WORK STATION

CAUTION
THIS ASSEMBLY CONTAINS ELECTROSTATIC SENSITIVE DEVICES

ATTENTION
STATIC-SENSITIVE DEVICES HANDLE ONLY AT STATIC SAFETY WORK STATIONS

FIGURE 13–44 Typical symbols used to identify ESDS parts.

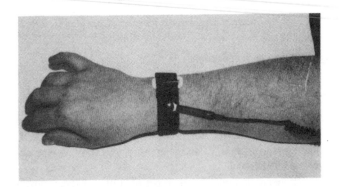

FIGURE 13–45 A typical wrist strap.

FIGURE 13–46 Storage of ESDS parts in a protective package.

and equipment used to repair ESDS units that have been removed from the aircraft are also grounded to prevent component damage.

If an ESDS component is removed from the aircraft, its connecting leads must be shorted together by means of wires, shorting clips, metal foil, or a conductive foam. Printed-circuit-board connections are also shorted together in order to keep all components at the same voltage potential. After this procedure has been completed, the unit should be placed in a special container to protect it from static electricity. A semiconductive plastic bag is often used for this purpose (see Figure 13–46). The protective containers are designed to prevent a static charge from reaching the components inside. All ESDS components should be stored and/or shipped in these protective containers.

REVIEW QUESTIONS

1. Why are dependable electrical systems essential in modern aircraft?
2. How is reliability in aircraft electrical systems attained?
3. List general requirements for aircraft electrical systems.
4. What is an essential electric circuit.
5. Give an example of a nonessential electric circuit.
6. Where are the basic requirements found for transport-category electrical systems?
7. Describe the requirements for a circuit that is critical to flight safety.
8. What precautions must be taken when wires and cables are grouped together?
9. Explain the purpose of schematic diagrams.
10. What is meant by the term *black box*?

11. List the major components of a typical position-light system.
12. Where are position lights located on an aircraft?
13. What colors for position lights are required on an aircraft?
14. What is an anticollision light?
15. What are the two common types of anticollision lights?
16. Describe the operation of a typical strobe-light system.
17. What is the purpose of the trigger signal in a strobe-light system?
18. What are the flashing requirements for an anticollision light?
19. How many anticollision lights are typically found on modern aircraft?
20. Where are anticollision lights located?
21. Where are landing lights located?
22. Describe the difference between a landing light and a taxi light.
23. Describe an instrument lighting system.
24. What is the purpose of landing-gear indicator lights?
25. How are landing-gear positions controlled?
26. Describe the operation of a press-to-test light.
27. Describe an essential lighting circuit for large aircraft.
28. What are the basic components of a fluorescent lighting system?
29. How are essential lights powered in a large aircraft?
30. What types of passenger compartment lights are found on large aircraft?
31. List the various types of exterior lights found on large aircraft.

32. What are proximity sensors?

33. What is the purpose of BITE systems?

34. Describe the functions of an interphone system.

35. Describe the functions of an intercom system.

36. What is the function of the EICAS?

37. How are electric components cooled in a large aircraft?

38. What method is used to cool panel instruments?

39. Describe the function of static dischargers.

40. Where are static dischargers located?

41. Why are static dischargers needed on high-speed aircraft?

42. What are the different types of inspection schedules for light aircraft?

43. What electrical system components are checked during an airframe inspection?

44. Describe voltmeter troubleshooting.

45. What is an open circuit?

46. What are the two types of short circuits?

47. Ammeters are typically used for troubleshooting which electrical system?

48. What is meant by the expression *troubleshooting from the flight deck*?

49. Describe a typical troubleshooting sequence.

50. What is the purpose of a centralized maintenance computer system?

51. What are the three basic types of faults presented by the control display unit?

52. What are the differences between the central maintenance computer system and the central fault display system?

53. What are electrostatic-discharge-sensitive components?

54. Describe the procedure used to protect ESDS components during aircraft maintenance.

14 Radio Theory

The transmission and reception of radio signals involve the use of electronic equipment to develop electromagnetic and electric fields that are modulated to carry the type of intelligence desired, project these fields into the atmosphere, and then intercept these fields and convert them into usable information or data. In this section the principles of radio transmission and reception are discussed.

Radio for aircraft includes communication equipment, navigation equipment, radar, and other electronic systems. In each of these areas there are many different types of electronic circuitry and devices designed to assure the safe and efficient operation of modern aircraft under all types of weather conditions and air traffic.

Radio transmitters and receivers are particularly important in the vicinity of large commercial airports. In this area the airspace can be very congested, and the pilot must be able to communicate with air traffic control. Special radios are designed for air-to-ground communication. Each airport control tower has one or more assigned frequencies. To ensure the safe operation of an aircraft, the airborne radio-communications equipment must be capable of operating at any control-tower frequency.

Most of the material in this chapter is generic and will apply to almost any type of radio receiver or transmitter. However, there are areas that apply to analog radios and do not apply to digital systems. While reading this chapter, please consider the theory to be presented in terms of analog systems. A specific section has been included to discuss the operation of digital radio technologies and how they differ from analog radios.

RADIO WAVES

Radio signals emanate from the antenna of a transmitter partly in the form of **electromagnetic waves.** Such waves are radiated from any current-carrying conductor when the current periodically changes in magnitude and direction. The radiation of an electromagnetic wave from an antenna may be compared in some respects to a sound wave sent out by the string of a banjo when it is picked. The sound wave from the banjo string is a mechanical compression and rarefaction of the air caused by the vibration of the string (see Figure 14–1).

It was explained in Chapter 1 that a magnetic field surrounds a current-carrying conductor. If the current flow in the conductor changes, the magnetic field will also change. The resulting movement of the field will cause the induction of a voltage in any conductor cut by the moving field.

During radio transmission, an electric field is generated by the antenna in addition to the electromagnetic field produced. The two fields radiate from the antenna at the speed of light, which is approximately 186,300 mi/s [300,000,000 m/s].

Since a radio wave travels at the speed of light, it can readily be understood that when a transmitter starts operation, the signal from that transmitter may be detected "instantly" hundreds or thousands of miles from the transmitter, depending

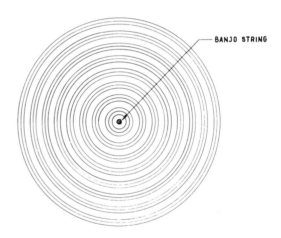

FIGURE 14–1 Sound wave emanating from a vibrating string.

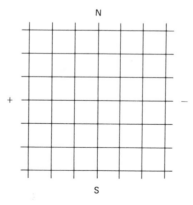

FIGURE 14–2 Electric and electromagnetic fields in a radio wave.

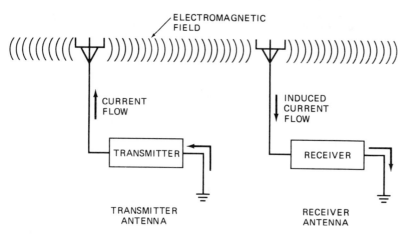

FIGURE 14-3 Radiation of an electromagnetic wave from a transmitter to a receiver.

on the power of the transmitter and the nature of the wave being transmitted.

The **electromagnetic** and **electric fields** produced by a radio transmitter antenna are at right angles to each other, as shown in Figure 14-2. The polarization of the fields with respect to vertical or horizontal positioning depends on the design and position of the antenna from which they are being emitted. The polarity of the fields reverses rapidly, with the rate of reversal established by the frequency of the wave.

The electromagnetic field radiated by an antenna accounts for the majority of the energy being transmitted. As illustrated in Figure 14-3, the electromagnetic wave travels from the transmitter's antenna to the receiver's antenna. The transmitter typically radiates an electromagnetic signal in a 360° pattern from the antenna. The electromagnetic wave travels through the air and passes the receiver's antenna. Acting as an inductor placed in a moving magnetic field, the receiver's antenna produces an induced voltage and sends current into the radio receiver circuitry. The current produced in the receiver's antenna has proportionately the same characteristics (frequency and amplitude) as the current of the transmitter's antenna. The induced-current flow is very weak and therefore must be amplified in order to produce a usable signal.

Wavelength and Frequency

The length of a radio wave depends on its **frequency.** Like an ac sine wave, the wave emanating from an antenna increases to a maximum in one direction, drops to zero, and then increases to a maximum in the opposite direction, as indicated

by the curve in Figure 14-4. The **wavelength,** indicated by the Greek letter lambda (λ), is the distance from the crest of one wave to the crest of the next. Since the wave travels at the rate of 300,000,000 m/s, the wavelength in meters is equal to 300,000,000 divided by the number of cycles per second (hertz). If a wave is produced at the rate of 1 Hz, the length of the wave will be 300,000,000 m. If 300 cycles are produced per second, the wavelength will be 1,000,000 m [328,000,000 ft]. The equation for wavelength is

$$\lambda = \frac{300,000,000}{f}$$

If the wavelength is known, the frequency may be found by the equation

$$f = \frac{300,000,000}{\lambda}$$

For example, if an aircraft's VHF communication radio is operating at 30 MHz, the wavelength is determined as follows:

$$\lambda = \frac{300,000,000}{f}$$

$$f = 30 \text{ MHz, or } 30,000,000 \text{ Hz}$$

$$\lambda = \frac{300,000,000}{30,000,000}$$

$$= 10 \text{ m}$$

The distance between crests of the 30-MHz wave is 10 m; $\lambda = 10$ m.

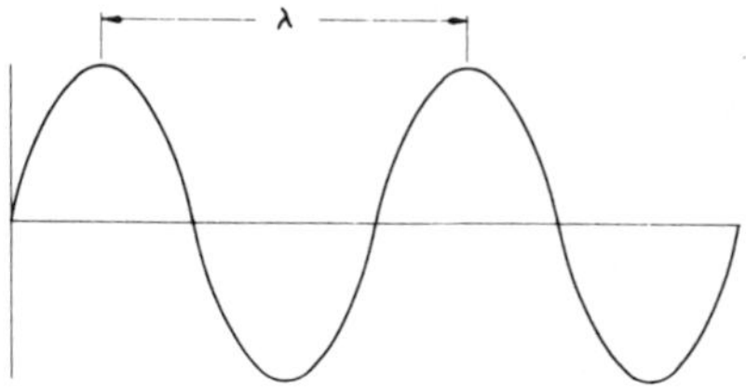

Frequency Bands

Frequencies utilized in various types of radio systems range from 3 kHz to as high as 30 gigahertz (GHz). The frequencies are divided into seven bands, and these bands are assigned to certain types of operations. The table on page 278 shows the utilization of the various bands.

Above this radio frequency spectrum lie the various light frequencies. Infrared and white light are currently being used

FIGURE 14-4 Wavelength of a sine wave.

Designation	Frequency range	Wavelength	Utilization
Very low frequency (VLF)	3–30 kHZ	100 000–10 000 m	Navigation, time signals
Low frequency (LF)	30–300 kHz	10 000–1000 m	Navigation, broadcasting, maritime mobile, fixed
Medium frequency (MF)	300–3000 kHz	1000–100 m	Broadcasting, maritime mobile
High frequency (HF)	3–30 MHz	100–10 m	Broadcasting, amateur, maritime and aeronautical mobile, citizens' band (CB)
Very high frequency (VHF)	30–300 MHz	10–1 m	FM and TV broadcasting, aeronautical navigation and communication, amateur, maritime mobile
Ultrahigh frequency (UHF)	300–3000 MHz	1 m–10 cm	TV broadcasting, radar, aeronautical and maritime mobile, navigation, radio location, space communication, meteorological
Superhigh frequency (SHF)	3–30 GHz	10 cm–1 cm	Space and satellite communication, radio location and navigation, radar

for some information transmission at frequencies between 10^9 and 10^{11} kHz. Below the radio frequency spectrum are the audible sound waves, ranging from 20 Hz to 15 kHz. Long-distance transmission of data at these frequency levels is virtually impossible, because any information transmitted would simply be heard by anyone receiving the sound wave.

The Carrier Wave

The field of electric and electromagnetic energy that carries the intelligence of a radio signal is called a **carrier wave.** The frequency of this carrier wave may be only a few hundred kilohertz or several thousand megahertz. Carrier waves are usually in the **radio frequency** (RF) range, which is in excess of 20,000 Hz. Frequencies below 20,000 Hz are in the **audio frequency** (AF) range.

In order to carry intelligence, an RF carrier wave must be modulated. This means that its form and characteristics are changed by means of some type of signal impressed upon it. Figure 14–5 shows an unmodulated carrier wave and a wave that has been modulated in amplitude by an AF signal. An RF carrier wave that has been modulated in amplitude is called an **amplitude-modulated** (**AM**) signal. If a voice signal is impressed upon a carrier, the modulation curve will follow the pattern of the voice frequencies.

Frequency modulation can be used in the VHF range and above. This type of modulation, commonly called **FM,** provides a signal that is much less affected by interference than

an AM signal. As indicated by the name, frequency modulation is accomplished by varying the frequency of the carrier wave in accordance with the audio signal desired. Figure 14–6 shows how frequency modulation affects a carrier wave.

The carrier waves emitted by a radio transmission antenna may be broken into three different categories: ground waves, sky waves, and space waves. Low-frequency waves (up to about 2 MHz) are considered ground waves. As illustrated in Figure 14–7, **ground waves** tend to be held near the earth's surface and "bend" with the curvature of the earth. Ground

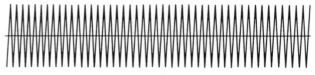

UNMODULATED RF CARRIER WAVE

WAVE FREQUENCY MODULATED

FIGURE 14–6 Frequency modulation.

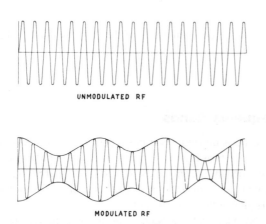

FIGURE 14–5 Carrier waves: unmodulated and modulated.

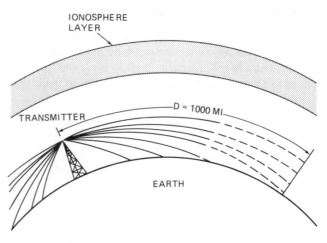

FIGURE 14–7 Ground-wave transmission pattern.

waves will travel a distance limited by the transmitter's output power, antenna design, local terrain, and current weather conditions. Typically, a relatively powerful transmitter is capable of sending ground waves a maximum of 1000 mi.

Sky waves, which are produced in frequencies from 20 to 30 MHz, tend to travel in straight lines. As illustrated in Figure 14–8, sky waves may be transmitted in a straight path or reflect off the ionosphere layer in order to reach the receiving antenna. Because of this means of travel, sky waves may produce a skip zone, where no reception is possible. Neither the line-of-sight wave nor the reflected wave can be received in the skip zone. The ionosphere density and distance from the earth determine the skip-zone range and the exact frequencies that are reflected. The ionosphere is a layer of ionized gases that surrounds the earth at an altitude of about 60 to 250 mi [96.6 to 402.6 km], varying with the time of day, the season, and the location. The density of this layer is affected mainly by the sun's solar flare activities. All these factors will determine the exact frequencies that are reflected and their angle of reflection off the ionosphere.

Space waves are found in frequencies above 30 MHz. Because of their high frequencies, space waves have a short wavelength, which allows them to travel through the ionos-

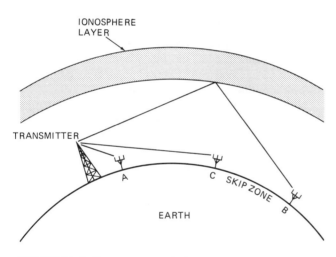

FIGURE 14–8 Sky-wave transmission pattern.

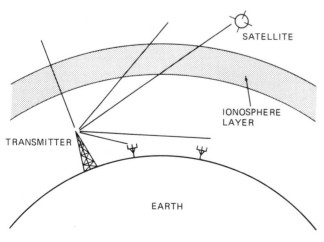

FIGURE 14–9 Space-wave transmission pattern.

phere layer. As illustrated in Figure 14–9, space waves are limited to line-of-sight reception. The properties of high-frequency space waves make satellite and space communications possible. Most aircraft communication and navigation radio frequencies are those for space waves; therefore, transmission is limited to line of sight. This means that there should be no mountains, buildings, or other objects between the transmitter and the receiver if good reception is to be expected.

Antennas

An **antenna** is a specially designed conductor that accepts energy from a transmitter and radiates it into the atmosphere. During reception, an antenna acts as a device that receives an induced current from passing electromagnetic waves. This induced current is then sent to the radio receiver circuitry. Where transmitters and receivers are built into one unit, often called a **transceiver,** the same antenna may serve for both transmitting and receiving.

The size and design of antennas vary in accordance with the frequency or frequencies of signals being handled. As frequencies increase, the length of the antenna must decrease. This is because wavelengths decrease as frequencies increase, and the length of an antenna must be matched as closely as possible to the wavelengths of the carrier waves. Typical antenna lengths are full-wave (the same length as the carrier wave), half-wave, quarter-wave, and some other fraction of the wavelength. It should be noted that antennas of these lengths will produce the strongest current flow for a given RF signal; however, shorter antennas are often used with modern amplifier systems that compensate for antenna inefficiencies. Smaller, less efficient antennas produce less drag on the aircraft, and they can be placed behind nonconductive panels to produce a **flush mounted antenna** system.

The simplest form of receiving antenna is merely a length of wire insulated from the ground and connected to the antenna coil of the receiver. The wire is cut by radio waves, and these waves induce very small voltages of many frequencies in the wire. The signal for which the receiver is tuned will pass through the receiver and to the loudspeaker in a form suitable for sound reproduction.

The correct antenna length (half-wave) for either a transmitter or a receiver is determined by using the following equation:

$$l = \frac{468}{f}$$

where l = length, in feet, and f = frequency, in megahertz. This equation gives the length for a half-wave antenna. The figure 468 in the equation is a factor derived by converting meters per second into millions of feet per second, dividing by 2, and multiplying by 0.95, the correction constant for antennas.

The equation for the length of a half-wave antenna is also expressed as

$$l = \frac{300,000,000 \times 3.28}{2 \times f}$$

Radio Waves **279**

In this equation the correction constant 0.95 is ignored. The principal objective in constructing an antenna and antenna-coupling system is to match the output impedance of the transmitter with the input impedance of the antenna system.

Remember, as discussed in Chapter 5 ("Alternating Current"), impedance is the total opposition to current flow in an ac circuit. Impedance is a function of the resistance, capacitive reactance, and inductive reactance. Since all RF waves are ac waves, their total opposition to current flow is impedance Z.

The simplest types of antennas are the **Hertz** antenna and the **Marconi,** or vertical, antenna. The Hertz antenna consists of two lengths of wire extended in opposite directions, as shown in Figure 14–10. Each length of wire is 1/4 wavelength ($\lambda/4$) long. (A study of a sine wave shows that 1/4 wavelength permits voltage or current to increase from zero to maximum in one direction.) The two lengths of wire are fed by the transmitter at the center; hence one length will become negative as the other becomes positive.

Coupling from a transmitter to an antenna is normally accomplished by means of an *LC circuit.* A typical coupling circuit for a coaxial transmission line and a dipole antenna is shown in Figure 14–11, but this is only one of the many possible arrangements.

The proper coupling of an antenna to a transmitter is essential for the maximum radiation of energy. The input impedance of the antenna must be matched as closely as possible to the internal impedance of the transmitter, and the effective length of the antenna must be adjusted to the wavelength of the signal being broadcast. The coupling circuit in the transmitter accomplishes these requirements. When a transmitter is being prepared for operation, it is necessary to determine that a signal of the maximum strength will be radiated from the antenna. The antenna output is checked by

means of a standing-wave ratio (SWR) meter. If the meter indicates poor signal radiation, the coupling circuit can be adjusted.

Since antennas are inductors, their effective length may be changed by adding an inductance coil in series or parallel with the antenna element. Inductors in parallel will decrease the antenna's total inductance. Inductors in series will increase the antenna's total inductance. The fine tuning of an antenna is typically performed by adjusting a variable inductor in the antenna coupler circuit. This is known as "peaking" the antenna.

Definition

An **amplifier** is a circuit that receives a signal of a certain amplitude and produces a signal of greater amplitude. The amplification may affect voltage or power or both, but its principal purpose is to increase the value of a weak signal so that it may be used to operate a speaker or some other electronic device.

Classification of Amplifiers

Amplifiers are classified according to function, operating level, or circuit design. The function may be to amplify power or voltage, and in this case the amplifier is described as a **power amplifier** or a **voltage amplifier.**

When an amplifier is classified according to operating level, the classification refers to the point on the characteristic curve through which the transistor operates as established by the emitter-to-base bias. A **class A** amplifier operates at a level such that the emitter-collector current flows at all times

FIGURE 14–10 The Hertz dipole antenna.

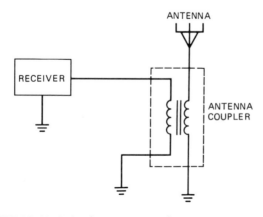

FIGURE 14–11 A simple antenna coupler.

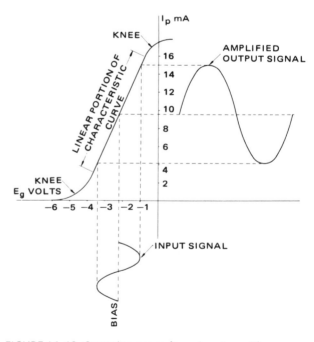

FIGURE 14–12 Operation curves for a class A amplifier.

because the voltage never reaches a sufficiently negative value to cut off the electron flow. The operation of this type of amplifier is shown by the curves in Figure 14–12. It will be seen that the transistor is biased near the center of the linear portion of the operating curve. The class A amplifier provides for a minimum of distortion of the signal; hence it is used where maximum fidelity is desired.

A **class B** amplifier is biased at approximately the cutoff point. With this arrangement only one half of the signal will be amplified, but the amplification can be carried to a much higher level than it can be by a class A amplifier because a much greater bias range is possible. Class B amplification is often used in **push-pull** amplifiers, in which two transistors are employed, one amplifying one half of the signal and the other amplifying the other half. The two amplified halves of the signal are recombined in the output circuit to produce a signal of low distortion and high power. The curves in Figure 14–13 illustrate the operating level of a class B amplifier and the curves in Figure 14–14 show how class B amplification performs in a push-pull circuit. Note that one half of the signal is amplified by transistor 1 and the other half by transistor 2. The circuit for a push-pull amplifier is shown in Figure 14–15.

In **class C** amplification the emitter-base circuit of the transistor is biased well beyond the cutoff point so that only a small portion of the positive peaks of the signal is amplified. Current flows only during approximately 120° of the cycle. The use of class C amplifiers is limited to RF circuits because only a part of the signal curve is reproduced. In an RF circuit, when the output of the class C amplifier is fed into an *LC* system, the flywheel effect of the tank supplies the missing parts of the signal curves. The E_b-I_c curves for a class C amplifier are shown in Figure 14–16.

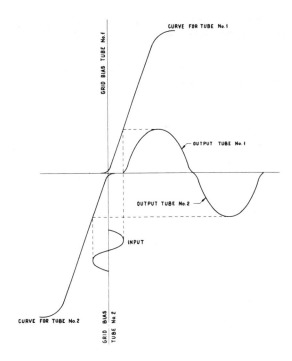

FIGURE 14–14 Class B amplification in a push-pull circuit.

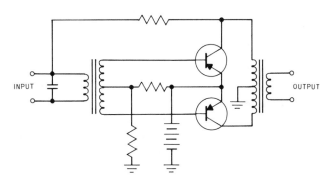

FIGURE 14–15 Circuit for a push-pull amplifier.

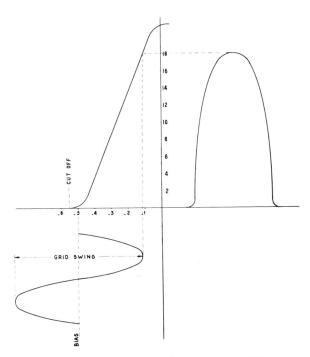

FIGURE 14–13 Operation curves for a class B amplifier.

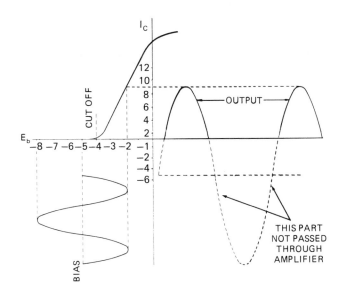

FIGURE 14–16 Operation curves for a class C amplifier.

Amplifiers **281**

FUNCTIONS OF A TRANSMITTER

A radio transmitter has several functions and one final objective. Briefly, the functions are (1) to generate an RF carrier wave, (2) to amplify the carrier wave; (3) to modulate the carrier wave with a sound-wave digital signal or some other form of information, (4) to amplify the modulated signal, (5) to couple the modulated signal to an antenna and (6) to radiate the signal into the atmosphere. All these functions except the last are performed by circuits within the transmitter system so that the final objective is accomplished. The radiation of the signal is accomplished by the antenna.

A block diagram of a typical radio transmitter is shown in Figure 14–17. This is only one of many possible arrangements.

Microphones

The purpose of a **microphone** (MIC) is to convert sound energy into electric energy. This process is completed by using the dynamic energy of the sound wave produced by the pilot. The sound wave strikes a diaphragm, and the sound energy is converted into mechanical energy. The mechanical energy is then converted into electric energy.

There are three common aircraft microphones: **carbon, dynamic,** and **electret.** Each uses a different means to create the electric signal that is sent to the radio transmitter. The **carbon MIC** contains tiny carbon granules compressed in a sealed chamber. The voice diaphragm vibrates the carbon chamber, changing the resistance of the carbon granules. A current that passes through the granules changes in amplitude as the sound wave moves the diaphragm.

The **dynamic MIC** uses the process of electromagnetic induction to produce the electric signal. The diaphragm of this MIC is connected to an inductance coil. As the pilot's voice moves the voice diaphragm, the coil moves into and out of a magnet core. This movement produces a current flow. The signal is then amplified and sent to the radio transmitter.

The **electret MIC** uses two plates, similar to a capacitor, to produce the electric current. This MIC's diaphragm moves one of the plates, and this motion changes the distance between the plates, generating a small electric current. A per-manently polarized piece of dielectric is needed between the two plates. The electret MIC is currently the most common aircraft microphone.

Use of Oscillators

An **oscillator** is a circuit designed to generate an alternating current that may be of a comparatively low frequency or of a very high frequency, depending on the design of the oscillator. Oscillators are used in radio and television transmitters to generate the RF carrier waves, in receivers to produce the intermediate frequency, and in other circuits and systems in which it is necessary to develop an alternating current with a particular frequency.

Oscillator Theory

Fundamentally, an oscillator consists of an *LC* **tank** circuit, a transistor, and a means of feedback to supply power to replace tank losses.

Examine the circuit in Figure 14–18. A battery is connected with a double-throw switch to a capacitor *C*. When the switch is thrown to position 1, the capacitor will charge to the voltage of the battery. If the switch is then placed in position 2, the battery will be disconnected from the capacitor, and the capacitor will retain its charge. Now when the switch is moved to position 3, the capacitor will be connected to the inductor *L* and will discharge through the inductance coil. When the capacitor first starts to discharge, the current through *L* will build a magnetic field that induces an opposing voltage in *L*. This slows the discharge of *C*. When *C* becomes almost discharged, the field around *L* will begin to collapse, and this will induce a voltage that keeps the current flowing. This induced voltage charges the capacitor in a direction opposite to that in which the battery originally charged it.

When the charge of the capacitor reaches a voltage equal to the induced voltage in *L*, the current flow will stop, and the capacitor will start to discharge back through *L*, with current flowing in the correct opposite direction. This action will be continuously repeated, with the energy stored first in the capacitor and then in the field of the inductance coil. An alternating current results that will degenerate to zero because of losses sustained in the circuit. If we could replace the small

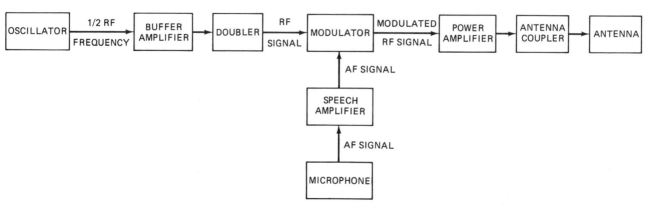

FIGURE 14–17 Block diagram of a radio transmitter.

amount of energy lost during each cycle, we could prolong the generation of alternating current indefinitely. Oscillator circuits must contain some means of applying energy to the tank circuit in order to maintain a constant frequency and power output.

The operation of an oscillator may be compared to the operation of the pendulum of a clock, as illustrated in Figure 14–19. If the pendulum is raised to the far right position and set free, it will travel in one direction to the center position, where it reaches a maximum velocity, and then swing up to the far left position. At the end of the swing, the motion of the pendulum stops and reverses direction, just as an ac current stops and reverses direction; the pendulum then swings back to the right, gaining speed to a maximum velocity and then slowing until it reaches the far right. This process then repeats. In a clock the mainspring adds power to the pendulum in order to maintain a constant frequency. The pendulum would eventually stop, owing to friction losses, if power were not added.

A **simple oscillator circuit** is illustrated in Figure 14–20. If SW_1 is closed, the transistor is forward-biased, and current flows through L_2. At this time the transistor base is connected to positive voltage through R_B, and the emitter is connected to the battery negative. The current flow through L_2 produces a magnetic field. This magnetic field induces a current into L_1, thus charging the tank circuit. The tank circuit now produces an alternating current between L_1 and C_1. This current flow applies a negative voltage to the base of the transistor for one-half of every cycle. This negative signal to the base reverse-biases the transistor emitter-base circuit and turns off the transistor. Current flow through the emitter-collector and L_2 circuit is, therefore, pulsating. The circuit is designed to allow the correct amount of pulsating current through L_2 in order to maintain a constant frequency in the tank circuit.

The exact frequency of the oscillator will be a function of the inductance of L_1, and L_2, the capacitance of C_1, and the current flow through the transistor. There are several different types of oscillators currently in use. The Armstrong, Hartley, and Colpitts oscillators are three common types. The oscillator most commonly used in transmitters is crystal-controlled in order to maintain an exact transmitting frequency. A mounted crystal serves the same function as a resonant circuit, with the thickness of the crystal determining the frequency.

A basic **crystal-controlled oscillator** circuit is shown in Figure 14–21. In this circuit the crystal takes the place of the tank circuit employed in other oscillators, and feedback is provided by means of capacitive coupling. Maximum amplitude of the RF signal is obtained when the output circuit is tuned to the frequency of the crystal.

The Buffer Amplifier

The purpose of the **buffer amplifier** is to amplify the RF signal produced by the oscillator without loading the oscillator circuit and thus producing a change in the oscillator frequency. This means, of course, that the circuits following the oscillator must not draw power from the oscillator. Because the buffer amplifier must be of a type that will not draw power, it is usually an amplifier operated as a class A type, which has no appreciable base current flow. A field-effect transistor (FET) is ideal for this application because emitter-collector current is controlled by the strength of an electric field rather than current flow through the base circuit. The RF current is coupled to the next amplifier stage through the coupling capacitor and a tank circuit. If it is desired, the tank may

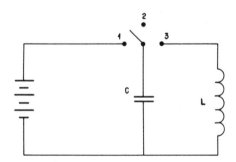

FIGURE 14–18 A simple tank circuit.

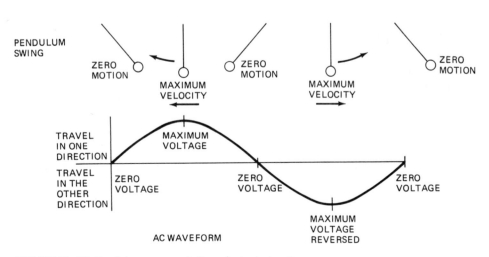

FIGURE 14–19 Pendulum representation of a tank circuit.

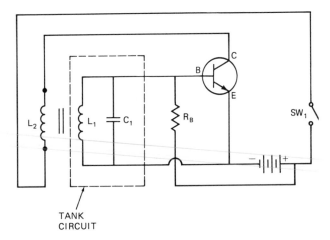

FIGURE 14–20 A simple oscillator circuit.

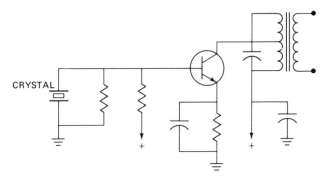

FIGURE 14–21 A crystal-controlled oscillator circuit.

be used as a frequency multiplier by having it tuned to a higher multiple of the oscillator frequency. If the tank is tuned to twice the oscillator frequency, it is called a **doubler;** if it is tuned to three times the frequency of the oscillator, it is called a **tripler.**

Frequency Multipliers

It is common practice to use a crystal in the oscillator circuit of a transmitter in order to hold the frequency at a fixed value. Since the frequency of a crystal is determined by its thickness, it will be apparent that there is a limit to the frequency that can be obtained with a crystal. This is because a crystal cannot be cut sufficiently thin for high frequencies without becoming so delicate that it cannot be handled safely.

To overcome the frequency limit, **frequency-multiplying** circuits are employed. These may be doublers or triplers, depending on the frequency of the tank circuit into which the output of the oscillator is fed.

The principal disadvantage of a frequency-multiplying circuit is that the output power is considerably less than it is from an amplifier operating straight through, that is, one in which no change in frequency takes place.

The principle of frequency multiplication may be understood by considering the action of a tank circuit. Any tank circuit has a resonant frequency that is determined by the values of its capacitance and inductance according to the equation

$$f = \frac{1}{2\pi\sqrt{LC}}$$

If the capacitor in Figure 14–22 is charged by means of the battery through the switch S in position 1 and the switch is then shifted to position 2, the electrons stored on one plate of the capacitor will start to flow through the inductance coil toward the opposite side of the capacitor. Current flow through the inductance coil will create a magnetic field in which is stored electric energy. When the current flow begins to decrease, the inductance coil tends to keep it flowing, and so the capacitor becomes charged in the opposite direction. Thus the cycle continues back and forth, with the energy being stored alternately in the electrostatic field of the capacitor and in the electromagnetic field of the inductor. Because of the resistance in the circuit, the alternating current will degenerate and disappear unless additional energy is supplied to keep it up.

In a frequency multiplier, the energy to maintain current flow is supplied by the transistor output. If the transistor is connected as a class C amplifier, the output will be in the form of widely separated pulses, as shown in Figure 14–23. In this illustration the amplifier output is shown as separate pulses with a frequency of 1000 kHz; the tank current is sustained at 2000 kHz.

The action of a frequency multiplier may be compared to the action of a child in a swing. The swing may be kept going easily by applying a short push at every second or third swing. In Figure 14–23 the short push is the amplifier pulse, and the swing in motion is the tank current.

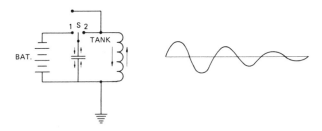

FIGURE 14–22 Operation of a tank circuit.

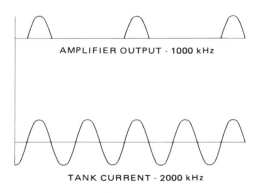

AMPLIFIER OUTPUT - 1000 kHz

TANK CURRENT - 2000 kHz

FIGURE 14–23 Voltage curves to illustrate the operation of a frequency doubler.

284 Chapter 14 Radio Theory

The Modulator

The function of the **modulator,** or modulation circuit, in a transmitter is to impress a signal on the RF carrier wave. This signal usually consists of an AF sound wave; however, the carrier may also be modulated by means of a key or digital pulse to produce code signals.

The modulation of a carrier wave is illustrated in Figure 14–24. It will be noted that the modulating wave modulates the **amplitude** of the carrier. As explained previously, this type of modulation is called **amplitude modulation** (AM). To obtain the greatest efficiency from a modulator, it is necessary that the maximum modulation be of an amplitude that will increase the unmodulated RF carrier to twice its unmodulated amplitude. Likewise, the negative peaks of modulation power should be of a value that will reduce the RF carrier to zero amplitude. When these conditions exist, the modulation is 100 percent. Figure 14–25 illustrates 100 percent modulation. If a smaller degree of modulation takes place, the full potential of the carrier is not utilized. Or if the modulating wave has too great an amplitude, overmodulation will occur, and the signal will be distorted.

The actual practice of modulation in the modulator of a transmitter is the development of the modulating signal and the application of this signal to the carrier wave. As mentioned previously, the audio signal may modulate either the amplitude or the frequency of the carrier, depending on how it is applied. In an aircraft radio, amplitude modulation is employed for all voice transmissions.

In the modulator, the audio signal is applied to the base circuit of a transistor, thus modifying the emitter-collector current. The circuit in Figure 14–26 illustrates a simple transistorized amplitude modulator. This circuit is not typical of those found in radio transmitters; however, it is appropriate to convey the basic modulation principles. In this circuit the microphone signal is sent to the emitter-base circuit of the transistor. As the microphone produces a signal of a given frequency, the transistor connects the emitter-collector circuit at that same frequency. The RF wave produced by the oscillator will travel through the transistor and change amplitude according to the frequency produced by the microphone. The RF signal sent to the emitter-collector circuit is an ac waveform; therefore, the transistor will be reverse-biased for one-half of the RF signal. The resultant transistor output is a half-wave rectified, modulated RF wave. This output signal must be sent to a tank circuit in order to reproduce an ac radio frequency signal.

The AF wave produced by the microphone is illustrated in Figure 14–27a. The modulated RF wave as it leaves the transistor is illustrated in Figure 14–27b. This signal enters a tank circuit, which produces an alternating current RF wave, as shown in Figure 14–27c. As illustrated, point A of the AF signal is the peak of the wave produced by the microphone. Points A_1 and A_2 are the peaks of the modulated RF wave. Point B of the AF wave is the lowest value produced; points B_1 and B_2 are the lowest amplitude of the modulated RF wave. From this example it is easy to see that the RF wave changes amplitude in the exact pattern as the AF wave sent to the base of the modulating transistor. Once the signal is sent through the tank circuit, amplitude modulation is complete.

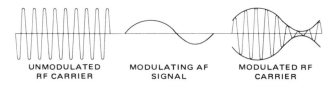

FIGURE 14–24 Modulation of an RF carrier wave.

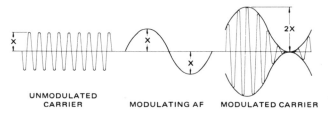

FIGURE 14–25 One hundred percent modulation.

Power Amplifiers

The function of the **power amplifier** of a transmitter is to increase the power level of the modulated signal to the point where it meets the requirements of the transmitting system. Radio transmitters employ power transistors that are designed to carry the current required. The output of the power amplifier is coupled to the antenna by means of an antenna coupler circuit.

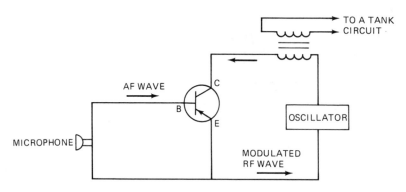

FIGURE 14–26 A simple transistorized amplitude modulator.

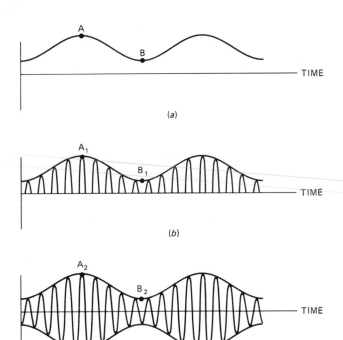

(a)

A₁

B₁

TIME

(b)

A₂

B₂

TIME

(c)

FIGURE 14–27 Waveforms of a modulator. (*a*) The AF wave; (*b*) half-wave rectified, modulated RF wave; (*c*) ac modulated RF wave.

Class C amplifiers are typically the most efficient type of power amplifier. Class C power amplifiers produce a pulsating dc output; therefore, they must be used in conjunction with a tank circuit to restore the sine wave signal. Figure 14–28*a* illustrates a class C power amplifier and related tank circuit. The signal that enters the transistor emitter-base circuit creates a forward bias only at the peak positive values of the waveform. At any lower positive or negative value, the emitter-collector circuit allows current to flow only for short intervals. This current pulse feeds the tank circuit by instantly creating a strong positive charge on the capacitor. The tank circuit then produces a sine wave signal at an amplified power level that follows the same pattern as the emitter-base signal. The output of the tank circuit is connected to an antenna coupler, which connects the sine wave to the antenna for transmission. Figure 14–28*b* shows the signal sine wave forms that enter and leave the transistor, along with the output signal of the tank circuit. Notice that the input and output waveforms are identical except for power levels.

Antenna Couplers

An **antenna coupler** is a circuit that connects the amplifier of a transmitter to its antenna. An antenna coupler may be a simple isolation transformer, or it may be much more complex, containing *LC* circuits and/or a lightning arrester circuit. The *LC* circuit is used to "tune" or "peak" the antenna to the trans-

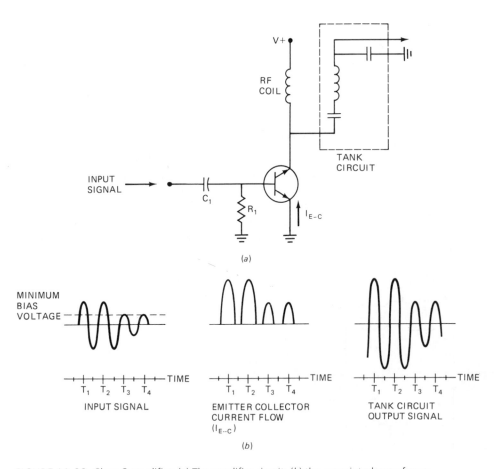

FIGURE 14–28 Class C amplifier. (*a*) The amplifier circuit; (*b*) the associated waveforms.

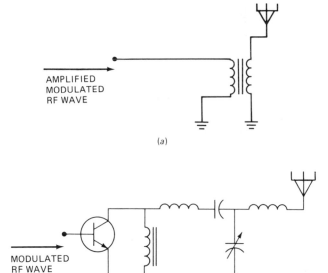

FIGURE 14–29 Antenna couplers. (a) An isolation transformer; (b) a coupler for a VHF transmitter.

mitter. Lightning arresters are used to protect radio circuits from unwanted lightning strikes on the antenna. Figure 14–29 illustrates two types of antenna couplers.

RECEIVERS

Most **radio receivers** must perform at least six functions to produce the desired results. The antenna must receive the transmitted wave and produce a current flow into the antenna coupler. The antenna's current is produced through the process of electromagnetic induction. The tuner selects and passes only one frequency and blocks the others.

One or more preamplifiers are used to amplify the weak signal received from the tuner. In some cases an amplifier is also connected directly to the antenna in order to receive extremely weak signals. Once the input signal is amplified sufficiently, the RF carrier wave is separated from the AF wave by a detector circuit. The AF wave is then amplified again and directed to the receiver output, typically a speaker. A block diagram of a simple radio receiver is shown in Figure 14–30.

Principles of Tuning

In the design and operation of electronic systems, **resonant circuits** provide the key to frequency control. When a certain frequency is to be produced, it is necessary to establish a circuit that is resonant at that frequency. Also, when a certain frequency is to be passed through a circuit and others eliminated, it is necessary to have a circuit that is resonant at the frequency to be passed. When a certain frequency is to be blocked, it is then necessary to place in the circuit a resonant tank circuit, which will block the frequency for which it is resonant. As can easily be seen, resonant circuits are most essential in radio and television receivers and transmitters.

From the study of alternating current, we know that a resonant circuit is one in which the capacitive reactance X_C is equal to the inductive reactance X_L. For any particular combination of capacitance and inductance, we know that the resonant frequency is fixed; that is, the combination can have only one resonant frequency. This frequency may be determined from the equation

$$f = \frac{1}{2\pi\sqrt{LC}}$$

As an example, let us consider the circuit in Figure 14–31. This is a series LC (inductance-capacitance) circuit containing a capacitance of 10 μF and an inductance of 250 mH. We shall determine the resonant frequency of this circuit as follows, remembering the foregoing equation:

$$f = \frac{1}{6.28\sqrt{10^{-5} \times 0.25}}$$

This may also be expressed as

$$f = \frac{1}{6.28\sqrt{2.5 \times 10^{-6}}}$$

$$= \frac{1}{6.28 \times 1.581 \times 0.001}$$

Then

$$f = \frac{1000}{6.28 \times 1.581}$$

or

$$f = 100.7 \text{ Hz}$$

This is the resonant frequency of the circuit.

In a series LC circuit, such as that just described, the

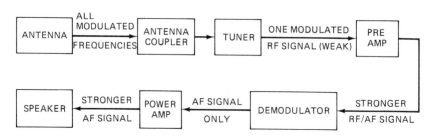

FIGURE 14–30 Block diagram of a radio receiver.

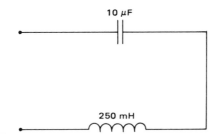

FIGURE 14–31 A series *LC* circuit.

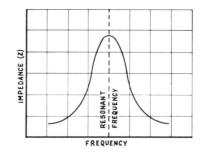

FIGURE 14–34 Relationship of current to frequency in a parallel *LC* circuit.

impedance at resonance is equal to the resistance of the circuit, inasmuch as the capacitive reactance and the inductive reactance cancel each other.

In other words, the impedance of the *LC* series circuit is at its lowest value when the frequency of 100.7 Hz (the resonant frequency) is applied. At any frequency above or below this value, the opposition to current flow (impedance) is greater. The relationship of current flow to frequency for this circuit is shown in Figure 14–32. The impedance value of the circuit is very low at the resonant frequency and increases as the frequency deviates from 100.7 Hz.

If we consider a parallel *LC* circuit, such as that shown in Figure 14–33, we find that at resonance the impedance across the parallel circuit is almost infinite, and if it were not for resistance in the circuit, the impedance actually would be infinite. Figure 14–34 shows a curve that indicates the effect of resonance in a parallel *LC* circuit. Notice that at zero frequency the impedance is very low and that it then rises as the frequency approaches the resonant value. At resonance the impedance is at a maximum; then it falls off as the frequency increases above resonance.

Filters

The characteristics of resonant circuits, as just described, make them very useful for **filtering** various frequencies in an electronic circuit. Among the types of filters used in electronic circuits are **high-pass** filters, **low-pass** filters, and **band-pass** filters. A high-pass filter tends to pass frequencies in the higher ranges and attenuate, or reduce, the current at frequencies in low ranges. A low-pass filter will pass frequencies in the lower ranges and attenuate, or reduce, the current at frequencies in the higher ranges. A band-pass filter will allow a certain band of frequencies to pass and will reduce the current at frequencies below or above the band range.

A circuit for a **high-pass filter** is shown in Figure 14–35. Notice that the capacitor is in series with the circuit and that the inductance coil is in parallel with the circuit. Since capacitive reactance decreases as frequency increases, at high-frequency levels current will appear to flow through the capacitor, and since inductive reactance increases as frequency increases, current will decrease through the inductance coil as the frequency increases. This circuit will therefore tend to pass high frequencies and eliminate or reduce low frequencies.

Figure 14–36 shows a circuit for a **low-pass filter.** In this circuit the inductance coil is in series, and the capacitor is in parallel. Low frequencies will pass easily through the inductance coil and will be blocked by the capacitor. As frequen-

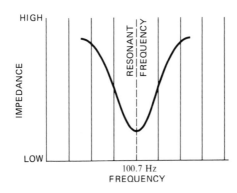

FIGURE 14–32 Relationship of current to frequency in a series *LC* circuit.

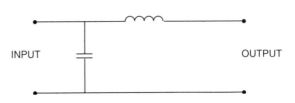

FIGURE 14–35 A high-pass filter.

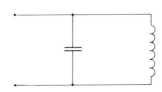

FIGURE 14–33 A parallel *LC* circuit.

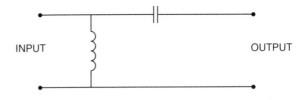

FIGURE 14–36 A low-pass filter.

cies increase, they will be blocked by the inductance coil and passed by the capacitor.

A **band-pass filter** is shown in Figure 14–37. The impedance of the series *LC* circuit is high except at or near resonant frequency. Therefore, at resonant frequency the current flow will be comparatively high. At resonant frequency the impedance across the parallel portion of the circuit will also be high, thus preventing the current from being bypassed. The bandwidth of a band-pass filter is determined by the number of circuit elements and by the resistance of the circuit: the greater the resistance, the wider the band.

A **band-reject filter** is shown in Figure 14–38. In this circuit the parallel *LC* circuit is in series with the load, and the series portion or series *LC* circuit is in parallel. With this arrangement the resonant frequencies will be bypassed and blocked from reaching the output. All other frequencies will be passed to the output of the filter.

A filter may be made a **tuning circuit** by making either the inductance or the capacitance variable. A typical tuning circuit consists of a variable capacitor used with a fixed resistor. In some cases, however, the capacitor is fixed, and the inductance is tuned by means of a slug, or movable core, within the inductor. Tuning circuits are usually designed to have fairly high selectivity; that is, they allow only a very narrow band of frequencies to pass and reject all others.

Figure 14–39 shows a simple circuit for a typical tuning unit. The radio signals cutting across the antenna induce sig-

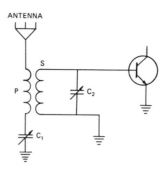

FIGURE 14–40 A tuning circuit with a series resonant circuit.

nals of various frequencies that flow through the primary winding of the antenna coil to ground. These currents produce electromagnetic waves, which induce voltages in the secondary winding of the antenna coil. Since a variable capacitor *C* is connected across the secondary coil, a maximum of current will flow only at the resonant frequency of the coil and the capacitor. Hence at resonant frequency a maximum voltage will be developed across the capacitor, and this same voltage will be applied to the emitter-base circuit of the transistor. This voltage is the input signal to the transistor, which amplifies the relatively weak signal from the tuner.

In some cases a series resonant circuit is provided in the primary system of the antenna coil, as shown in Figure 14–40. In this case maximum current will flow in the primary only at resonant frequency. This provides for increased selectivity because unwanted frequencies are largely prevented from being induced in the secondary winding of the antenna coil.

The two circuits described are in no way the only means of tuning, but they represent the basic principles used in all tuning circuits in the low and medium frequencies.

Detection

Detection of a radio signal is the process of separating the RF carrier wave from the AF intelligence wave. This is accomplished by rectifying the modulated wave to produce a dc signal and then filtering the remaining wave to remove the high-frequency carrier from the low-frequency audio wave. Detection is often referred to as **demodulation** because it is the opposite of the modulation process.

The principle of detection is illustrated in Figure 14–41. Figure 14–41*a* shows the modulated carrier wave, Figure 14–41*b* shows the wave after detection, and Figure 14–41*c* shows the audio signal after filtering. This audio signal is usually amplified before being reproduced through a speaker.

There are several types of detection circuits, ranging from simple to complex; however, they all rectify and filter a modulated signal. A simple detection circuit is illustrated in Figure 14–42. The first function performed by the detector is to rectify the modulated wave; the diode performs this function. The second step is to remove the RF signal; the capacitive filter does this by shorting together the RF input for high frequencies (remember, the RF wave is of a higher frequency than the AF wave). The input wave changes as it passes

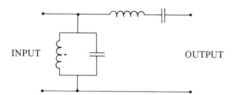

FIGURE 14–37 A band-pass filter.

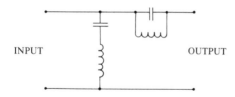

FIGURE 14–38 A band-reject filter.

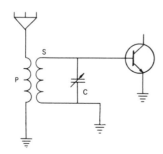

FIGURE 14–39 A simple tuning circuit.

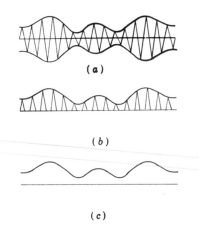

(a)

(b)

(c)

FIGURE 14–41 Principle of detection. (a) Modulated RF wave; (b) rectified and modulated RF wave; (c) AF wave.

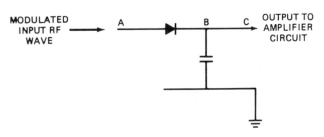

FIGURE 14–42 A simple detection circuit.

through the detector. The wave in Figure 14–41a, a modulated RF wave, represents the signal at point A. After rectification, point B, the signal may resemble Figure 14–41b. Once rectification is complete, an AF signal remains and is sent to the amplifier circuitry. The output signal, point C, is represented in Figure 14–41c.

Sound Reproduction

After an AF signal leaves the demodulator, it is typically amplified. An amplifier will typically consist of a transistor circuit used to increase signal strength, along with various filters used to control tone and increase fidelity. Once the signal is amplified, the reproduction of sound is accomplished by the conversion of electric energy into sound waves. This is done by means of **headphones** or a **speaker.** Figure 14–43 is a simplified drawing of the method of sound reproduction in both the headphones and the speaker. A coil of many turns of very fine wire is wound on a permanent magnet core. A magnetic diaphragm is mounted a few thousandths of an inch from the magnet poles. This diaphragm will vibrate in response to changes in the magnetic field caused by varying current flow in the coil. The vibration causes the waves in the air that are recognized as sound.

A typical speaker will employ a paper cone connected to the iron diaphragm. This creates a large air movement with each position change of the iron diaphragm, thus producing relatively high volumes. Because of the work involved in moving large quantities of air, a more powerful amplifier is generally needed to drive large speakers.

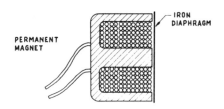

FIGURE 14–43 Sound reproduction mechanism.

The Crystal Receiver

The **crystal receiver** is the simplest of all types of radio receivers. The schematic in Figure 14–44a is a series crystal receiver. During operation, the antenna receives an induced current from any passing electromagnetic waves. This induced current is seeking ground; therefore, it travels through the variable inductor and capacitor C_1. These two components filter any unwanted signals, and only the resonant frequency will pass through the tuner into the diode. The diode and capacitor C_2 constitute the demodulator circuit. The diode acts as a half-wave rectifier to produce a dc signal available to the headphones. Capacitor C_2 is an RF filter used to bypass the radio frequency portion of the rectified signal around the headphones, directly to ground. The only signal reaching the headphones is the rectified audio frequency. If a very sensitive headphone set is used, the faint AF signal can be heard.

This type of radio receiver produces its entire audio volume from the energy induced into the antenna. Obviously, only very strong transmitted signals are received and only with an antenna of an extremely long length. To overcome these shortcomings, most receivers utilize an amplifier circuit to improve signal strength. A parallel crystal receiver is shown in Figure 14–44b.

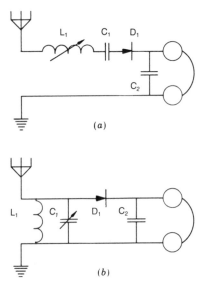

FIGURE 14–44 Crystal receivers. (a) series type; (b) parallel type.

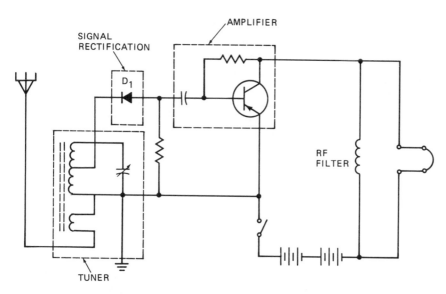

FIGURE 14–45 Schematic diagram of a one-transistor radio.

A One-Transistor Radio

One way to improve a radio circuit is to amplify the audio frequency before it is sent to the headphones. Figure 14–45 is a schematic diagram of a one-transistor radio. The transistor amplifies the AF wave so that the radio can produce a louder volume than the crystal radio. This amplifier does not make the radio more sensitive (able to receive more stations); it only amplifies the sound wave sent to the headphones. Additional amplifier circuits are needed to improve a radio's sensitivity. A circuit of this type would amplify the modulated RF wave prior to its entering the radio's detection circuitry. Most radios contain several amplifiers in order to achieve their desired sensitivity and volume.

The Superheterodyne Receiver

The **superheterodyne receiver** derives its name from the fact that a new signal frequency is generated in the receiver by means of a local oscillator called a **beat frequency oscillator** (BFO). The BFO signal is fed into the converter or mixer system. The word *hetero* is a Greek term for *other,* and the word *dyne* means *power;* thus the term **heterodyne** literally means *other power.* It refers to the intermediate frequency developed in the mixer circuit.

A block diagram of a superheterodyne receiver is shown in Figure 14–46. During operation the signal received by the antenna enters the RF amplifier, and the strengthened signal is then passed on to the mixer or converter stage. Here the incoming RF carrier and the oscillator (BFO) signal are combined to produce an intermediate signal of 455 kHz. Actually, when two frequencies are combined, four frequencies result. These are the two original frequencies and frequencies that are the sum of and difference between the original frequencies. If an RF frequency of 1000 kHz is combined with an oscillator frequency of 1455 kHz, the two new frequencies will be 2455 and 455 kHz. For the intermediate frequency (IF) signal in the superheterodyne receiver, only the 455-kHz frequency is used.

The antenna circuit and the local oscillator circuit are tuned together to produce frequencies that have a difference of 455 kHz. This IF signal from the mixer or converter carries the audio signal that was originally on the RF signal. The IF signal is usually passed through two IF amplifier stages, each consisting of an IF transformer and a transistor with the necessary resistors and capacitors. The two IF circuits are accurately tuned to 455 kHz with a bandwidth of approximately 10 kHz. This means that signals of 450 to 460 kHz will be passed through the IF amplifier stages and that all other frequencies will be attenuated. The 10-kHz bandwidth is necessary to accommodate the audio modulation that may be carried by the IF signal.

After the intermediate frequency has been amplified through the two IF amplifiers, it is passed through the second detector and first audio-amplifier stage. From this point, the signal may be directed to the speaker or to an additional stage of audio amplification.

Superheterodyne receivers are used in many electronic systems other than radios. They have a number of advantages, among which are simplified tuning, improved amplification, high selectivity, and good fidelity. In addition to their application in radios they are used in televisions and in other special devices. When they are used for frequencies other than those of the standard broadcast band, it is necessary to employ an intermediate frequency that is different from the 455 kHz normally used for that band.

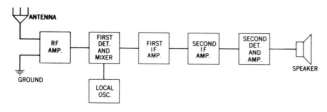

FIGURE 14–46 Block diagram of a superheterodyne receiver.

Digital Radio Theory

On many aircraft systems digital signals are used to process information. As described in Chapter 7, a digital signal is composed of discrete voltage on and off signals. In terms of radio theory, digital signals can be used within a radio's circuitry to perform the basic radio functions, such as amplification, modulation, and detection. Digital signals can also be transmitted as electromagnetic energy. In this case the RF carrier wave is a digital, not an analog, signal. Most aircraft radios transmit analog RF signals. Those transmitting digital, or pulse, signals include the weather radar system, distance-measuring equipment, and the ATC transponder. These systems will be discussed in Chapters 15 and 16.

In most aircraft radio systems, the major change that has resulted from advanced digital technologies is improved tuner and oscillator circuits. **Digital tuners** are used in modern receivers to filter out unwanted RF signals. Digital oscillators may be used in radio receivers or transmitters to generate a constant-frequency reference signal. For receivers, the reference signal may be used to create the intermediate frequency (IF). For transmitters, the digital oscillator is used to generate a high-frequency reference signal that is divided into a lower frequency and used for the RF carrier. The use of digital circuits in both transmitters and receivers has improved the accuracy and reliability of aircraft radios.

Digital tuning refers to the use of digital circuits to generate many frequencies from a single crystal oscillator. The oscillator signal can be used for tuning purposes or to produce a radio frequency carrier or an intermediate frequency.

Analog transceivers use variable capacitors and inductors that tune a local oscillator through a continuous band of frequencies. Aircraft VHF frequencies range from 118 through 136 MHz. The interval between channels is 0.25 MHz. This is a relatively narrow band to select using a conventional analog tuner. One alternative would be to have 760 discrete crystal oscillators in each radio. Another, more acceptable solution employs the concept of frequency synthesis.

The digital technique of **frequency synthesis** has been developed to obtain a stable tuning circuit that is capable of 0.25-MHz increments. The process of frequency synthesis requires the use of digital flip-flops and various logic gate circuits.

Programmable dividers, or **counters** are made from flip-flop circuits in various arrangements. Using other logic gates to chain flip-flops together means that a counter can have a different output for each input pulse. A 16-state counter has a binary output of 0 to 15 and requires 16 pulses to make one cycle through the counter. A reference crystal oscillates at a constant frequency that is higher than the desired output frequency. A programmable divider is set by the frequency selector controls on the front of the transceiver. When the divider is programmed, the counter's output frequency can be used by the receiver for tuning and demodulation. A transmitter can use the counter's output to produce the RF wave.

High-frequency synthesis requires the use of two crystals, one crystal that oscillates in the megahertz frequency range and one crystal that oscillates in the kilohertz frequency range. Using two crystals with programmable dividers yields much more accuracy than adding several oscillators together. When two oscillators are added and then divided to a lower frequency, their frequency error becomes less with each division. The disadvantage of having only one or two crystals is that if any crystal fails, the whole unit fails.

REVIEW QUESTIONS

1. Describe a radio wave.
2. What fields are found in a radio wave?
3. Explain wavelength.
4. Compare wavelength and frequency for a radio wave.
5. Give the equation for wavelength if the frequency is known.
6. How many frequency bands are assigned to radio systems?
7. What is the difference between radio frequency and audio frequency?
8. Explain the functions of antennas.
9. How does the frequency of the radio waves transmitted or received relate to the length of the antenna?
10. Describe the ionosphere.
11. Of what value is the ionosphere in radio communications?
12. Give the equation for antenna length (half-wave).
13. What is meant by *coupling* an antenna to a transmitter?
14. What are the functions of a radio transmitter?
15. What is the purpose of an oscillator?
16. Describe how an oscillator produces an alternating current.
17. What determines the frequency of the oscillator output?
18. What type of oscillator is commonly used in a radio transmitter?
19. What is the function of a buffer amplifier?
20. Describe the operation of a frequency multiplier.
21. What is 100 percent modulation?
22. How is modulation accomplished?
23. What is the purpose of a power amplifier?
24. Compare the effects of inductive reactance and capacitive reactance.
25. What conditions exist in a resonant circuit?
26. Explain how a circuit may be tuned to a particular frequency.
27. What is the resonant frequency of a circuit having an inductance of 2 H and a capacitance of 0.5 μF?
28. Describe the operation of a high-pass filter and the operation of a low-pass filter.

29. Explain a band-pass filter.

30. What is the principal function of filters in a radio circuit?

31. What is signal detection?

32. By what means is an audio frequency signal converted into sound?

33. Describe the operation of a crystal receiver.

34. What is the purpose of the bypass capacitor across the headphones in a receiver?

35. Describe an amplifier.

36. Name three classes of amplification.

37. What class of amplification produces the least distortion of a signal?

38. Explain class B amplification and how it is most commonly used in a receiver.

39. Explain the purpose of the beat frequency oscillator in a superheterodyne receiver.

40. What intermediate frequency is most commonly used in a superheterodyne receiver?

41. What is the function of the converter in a superheterodyne receiver?

42. Give the principal advantages of a superheterodyne receiver.

✈

15 Communication and Navigation Systems

The use of radio equipment and avionics in general has increased markedly for all types of aircraft during the past thirty years. One of the reasons for this increase is the Federal Aviation Administration's requirement that all aircraft operating in high-traffic areas be equipped with a two-way radio for communication with air traffic controllers and tower operators. The development of solid-state and digital electronics technology has made it possible to install highly complex and sophisticated systems for communication, navigation, and automatic flight control in all types of aircraft. Previously, such systems could be utilized only in large aircraft because of the size and weight of the system components. Today the term **avionics,** which is a combination of the words *aviation electronics,* encompasses a variety of electronic systems.

Avionic systems installed in aircraft can include communication (COMM) radio, navigation (NAV and RNAV) systems, weather detection systems, and flight management systems (FMSs). The navigation systems may include VHF omnirange (VOR) receivers, distance-measuring equipment (DME), the automatic direction finder (ADF), the localizer (LOC) receiver, glide slope (GS) receiver, traffic alert and collision avoidance systems (TCASs), marker-beacon receivers, the identification transponder, the radio altimeter, the encoding altimeter, and numerous indicators. In some cases, on the most modern aircraft, these systems have been combined into an integrated avionics processor system (IAPS). IAPS-type avionics are typically easier to use, weigh less, and require less space on the instrument panel.

All modern avionic systems conform to **Aeronautical Radio Incorporated** (ARINC) standards. ARINC is a corporation established by foreign and domestic airlines, aircraft manufacturers, and transport companies to set standards for aircraft systems. ARINC 500 and 700 standards were developed for communication, navigation, and identification (CNI) systems. Digital data transfer systems are standardized through ARINC 429 and 629.

COMMUNICATIONS

Radio communication systems for aircraft are primarily for the purpose of air traffic control; however, commercial aircraft also utilize a range of high frequencies for communicating with ground stations and other aircraft for business and operational purposes. Communications for air traffic control are in the VHF band in the range between 118 and 136.975 MHz.

High-Frequency Communication Systems

The high-frequency (HF) communication systems operate in the frequency range of 2.0 to 30 MHz. The HF range is actually a middle-frequency range, inasmuch as it starts just above the standard broadcast band, which ends at approximately 1700 kHz. This frequency group consists of ground waves; therefore, HF communication systems are used for long-distance radio transmissions. The HF system on an airplane is used to provide two-way voice communication with or digitally coded signals for ground stations or other aircraft.

The HF radio control panel is located where it is easily accessible to the pilot or copilot. A typical panel is shown in Figure 15–1; it includes a frequency selector, a squelch or RF gain control, and a mode selection switch. On most aircraft the antenna for an HF radio is covered by plastic-type shields. The cover may be of fiberglass or a similar material that will allow electromagnetic waves to reach the antenna. On modern aircraft a flush mounted antenna is used that does not increase induced drag. The probe antenna is used for both receiving and transmitting and is matched to the transmission line at any frequency by means of an **antenna coupler.** An antenna coupler system may consist of a remote coupler unit and a coupler-unit control, as shown in Figure 15–2. The antenna coupler system is necessary to maintain an efficient match between the antenna and the transmitter at a wide range of frequencies.

The complete HF transceiver is installed in an electronic equipment rack and is remotely controlled from the control unit on the flight deck. The system consists of the HF receiver-transmitter, the HF control unit, the antenna coupler system, and the antenna.

The receiver-transmitter operates at high frequencies between 2.0 and 29.999 MHz in one of several different modes. These are single sideband (SSB), upper sideband (USB), lower sideband (LSB), full carrier amplitude modulation (AM), continuous wave (CW), and data modes. These different modes determine the waveform characteristics of the signals transmitted and received by the radio. All but the data mode are used for voice communications. The data mode is used for digital-type information that is linked to equipment external to the HF radio system. The data communication

FIGURE 15–1 A typical HF radio control panel. *(Collins Divisions, Rockwell International)*

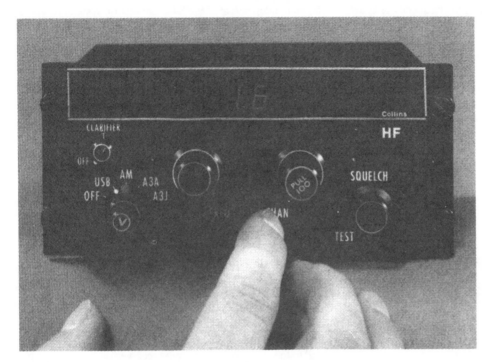

FIGURE 15–2 A typical antenna coupler system. *(Collins Divisions, Rockwell International)*

system is known as **air-ground data link,** or **data link.**

HF communication systems are long-distance communication systems and are not employed on all aircraft. Airlines may or may not utilize these systems, depending on their particular requirements. HF systems are not usually found in light aircraft. Many airlines that employ HF communication systems do so because these systems provide for an extended range of communications between aircraft and from aircraft to ground stations.

VHF Communication Systems

As explained previously, VHF communication systems are employed largely for controlling air traffic. These systems are installed in all types of aircraft so the pilot may be given information and directions and may request information from air traffic control centers, control towers, and flight service stations. On the approach to any airport with two-way radio facilities, the pilot of an aircraft calls the tower and requests

information and landing instructions. In airline operations and all instrument flights, the flight of an aircraft is continuously monitored by air traffic control (ATC), and the aircraft's crew is given instructions as necessary to maintain conditions of safe flight.

VHF communication systems operate in the frequency range of 118 to 136.975 MHz. For international operations the frequencies may extend to 151.975 MHz. The nature of radio-wave propagation at these frequencies is such that communication is limited to line-of-sight distances. The advantage of VHF communication, however, is that the signals are not often distorted or rendered unintelligible by static and other types of interference.

VHF communication radios are currently available with 720, 760, or 360 channels. The 720- or 760-channel radio is preferred by most pilots owing to its versatility in frequency selection. In 1976 the FAA changed the minimum frequency spacing for VHF systems from 50 to 25 kHz between 118 and 135.975 MHz. This change made the 720- channel radio possible. Recently the FAA and FCC authorized the general use of frequencies up to 136.975 MHz. This change added 40 channels to increase the selection to 760 channels. Some older 360-channel radios are still in use; however, they are quickly becoming obsolete.

VHF communication equipment for light aircraft is typically combined with a VHF navigation (NAV) radio system. A common VHF light-aircraft radio is shown in Figure 15–3. Figure 15–4 shows the interior arrangement of the system. The transceiver shown in Figures 15–3 and 15–4 is a solid-state, digital system that can receive or transmit on any one of the 720 channels in the COMM range of frequencies. The frequencies are spaced at 25-kHz intervals throughout the range. Frequencies are selected simultaneously for both the receiver and the transmitter by rotating the frequency selector knobs. The large outer knob is used to change the megahertz portion of the frequency display, and the smaller concentric knob changes the kilohertz portion. The small knob will change the frequency in 50-kHz increments when it is pushed in and in 25-kHz increments when it is pulled out.

To tune the transceiver to the desired operating frequency, it is necessary to first enter the selected frequency into the STANDBY display. The frequency is then activated by pushing the transfer button, and the word USE will be displayed. Another frequency may then be entered into the STANDBY mode. The STANDBY mode will store the selected frequency to allow for a "quick switch" of the frequency being used by the receiver. This becomes very helpful while operating an aircraft in crowded airspace in which several communication frequencies are used for air traffic control.

Control panels for VHF communication systems vary in design, depending on the manufacturer of the equipment and the requirements of the aircraft manufacturer. Typically, the control panel located in the flight deck contains the frequency selectors and the digital displays for the main and standby frequencies. In some cases the radio's volume is controlled by a separate audio panel.

Most VHF systems for corporate and transport-category aircraft use a separate radio control panel, and the receiver-transmitter (r-t) is located in the electric equipment center. Also on these aircraft, the VHF communication radio system is often independent of the VHF navigation system. A typical VHF communication receiver-transmitter is shown in Figure 15–5. On light aircraft the r-t and control panel are often one unit mounted in the instrument panel. This type of unit is shown in Figure 15–4.

Antennas for VHF systems are low-drag stub units extending from the top and bottom centerline of the airplane. These antennas are matched to their respective transmission lines by means of carefully measured lengths of tuning line. The antennas are used for both transmitting and receiving. A typical VHF antenna configuration is shown in Figure 15–6.

VHF communication transmitters provide AM voice-communication transmission between aircraft and ground stations or between aircraft. Transmission is on the same number of channels and frequencies as provided in the receiver. Because of the nature of VHF radio signals, the average communicating distance from aircraft to ground is approximately 30 mi [48 km] when the airplane is flying at 1000 ft [305 m] and approximately 135 mi [217 km] when the airplane is at 10,000 ft [3048 m]. Transmitting frequency is determined by the position of the selector switches on the VHF control panel. The transmitter is tuned at the same time and to the same frequency as the receiver.

FIGURE 15–3 A VHF communication radio. *(King Radio Corporation)*

FIGURE 15–4 Interior of a VHF communication radio. *(King Radio Corporation)*

FIGURE 15–5 A VHF-700 transceiver. *(Rockwell International Corporation)*

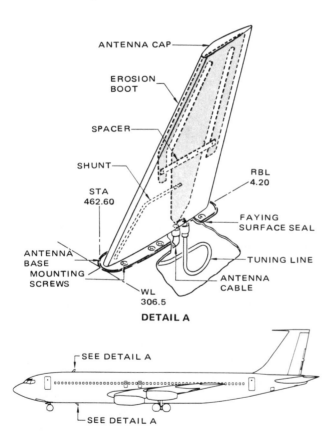

FIGURE 15–6 VHF antenna configuration. *(Boeing Company)*

The most modern VHF communication radios incorporate the latest digital design features. In general, the use of microprocessors and digital circuits has allowed for a 50 percent reduction in parts count and an 80 percent reduction in internal shop adjustments as compared with the use of analog circuits.

A modular design of a modern digital system reduces maintenance time by providing easy access to all circuit boards and components.

Built-in test equipment (BITE) systems and data interface systems are also available on some communication radios. A BITE system uses an LED display to indicate faulty "plug-in" modules within the radio. A data interface system allows for the transmission of binary data through the communication radio system. The data to be transmitted may be created by a manually operated keyboard or an airborne computer. While information is being transmitted through the data interface, the voice characteristics of the communication radio are still operable. This system therefore makes better use of the crowded RF spectrum in which the transceivers operate. Data interface systems will be discussed in greater detail in the ACARS section of this chapter.

Theory of Operation of VHF Communication Systems

To aid in the explanation of the VHF communication transceiver, the receiver will be discussed separately from the transmitter. The receiver portion of a VHF communication system is typically the superheterodyne type (Figure 15–7). The antenna receives an induced signal from the electromagnetic fields passing the antenna. This signal is sent through a band-pass filter to an RF amplifier. Once amplified, the signal passes through a low-pass filter and into the first-stage mixer. The mixer converts the RF into an intermediate frequency (IF). The IF is a lower frequency and is easier to control through the receiver. The IF is amplified to produce a stronger signal, which is sent to the second-stage mixer where again a lower frequency is produced. This signal is amplified and sent to the detector, where the audio wave is separated from the carrier wave. The audio signal is then amplified by the buffer and broadcast into the aircraft by the speaker. The buffer amplifier receives inputs from the AGC (automatic gain control) circuit, which ensures correct signal amplification at varied input signal strengths.

The transmitter receives an input signal from the microphone or data inputs. This signal is amplified by the audio buffer and sent to the modulator (synthesizer). The modulator produces an AM signal, which is filtered, amplified, and sent to an ALC (automatic level control) circuit. Similar to the AGC in the receiver, the ALC ensures that a consistent output signal is sent to the antenna, even at varying input signal strengths.

Selcal Decoder

The word **Selcal** is derived from the term **selective calling,** and the **Selcal decoder** is an instrument designed to relieve the pilot and copilot from continuously monitoring the aircraft radio receivers. The Selcal decoder is, in effect, an automatic monitor that listens for a particular combination of tones that are assigned to the individual aircraft. Whenever a properly coded transmission is received from a ground station, the signal is decoded by the Selcal unit, which then alerts the pilot to an incoming radio transmission. The system

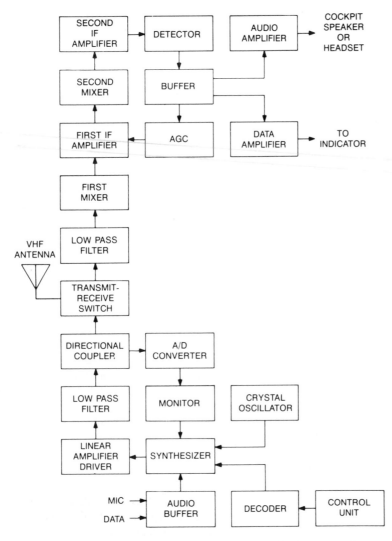

FIGURE 15–7 Block diagram of a typical VHF communication system for a large aircraft. *(Collins Divisions, Rockwell International)*

automatically activates the correct radio for the flight crew.

Ground stations equipped with tone-transmitting equipment call individual aircraft by transmitting two pairs of tones that will key only an airborne decoder set to respond to the particular combination of tones. When the proper tones are received, the decoder operates an external alarm circuit to produce a chime, light, or buzz, or a combination of such signals.

A ground operator who wishes to contact a particular aircraft by means of the Selcal unit selects the four-tone code that has been assigned to the aircraft. The tone code is transmitted by a radio frequency wave, and the signal can be picked up by all receivers tuned to the frequency used by the transmitter. The only receiver that can respond to the signal and produce the alert signal for the pilot is the receiver and decoder system that has been set for the particular combination of tone frequencies.

The arrangement of a Selcal decoder system in one type of large airliner is shown in Figure 15–8. The decoder is a solid-state unit that can be used to monitor any two of five receivers. Receivers to be monitored are selected by means of the rotary selector switches on the Selcal control panel.

AIRCOM System

AIRCOM is an abbreviation for *digital air/ground communications services,* provided by SITA. **SITA** is the acronym for the Société Internationale de Telecommunications Aeronautiques. The AIRCOM system was designed to reduce the amount of voice communication on the increasingly crowded communications frequencies. AIRCOM allows ground-to-aircraft communications for operational flight information, such as fuel status, flight delays, gate changes, and departure times. AIRCOM can also be used to monitor certain engine and system parameters. VHF frequencies are used to transmit AIRCOM information through existing airborne VHF equipment. AIRCOM is currently used in Europe and Australia.

ACARS

The **ARINC Communication Addressing and Reporting System** (ACARS) is the U.S. counterpart of the AIRCOM system described above. ACARS is a digital system that operates using the VHF communications equipment on a fre-

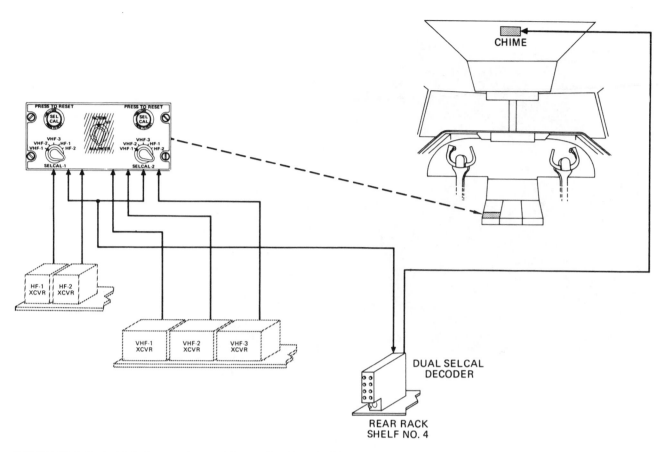

FIGURE 15–8 Typical Selcal system for an airliner. *(Lockheed California Company)*

quency of 131.550 MHz. The ACARS airborne equipment contains a control unit, located on the flight deck, and a management unit, located in an equipment bay. A typical ACARS control unit is shown in Figure 15–9. The ground equipment contains antennas and r-t units, a data link via telephone lines to the central facility in Chicago, and a data link to the various airline computer systems. The ground facilities cover the continental United States and portions of Canada and Mexico.

Since ACARS operates on only one frequency for the entire United States, all transmitted messages must be as short as possible. To achieve a short message, a special code block using a maximum of 220 characters is transmitted in a digital format. If a longer message is needed, more than one block will be transmitted.

ACARS operates in two modes: the demand mode and the polled mode. The **demand mode** allows the flight crew or airborne equipment to initiate communications. A block diagram of an ACARS is shown in Figure 15–10. To transmit a message, the management unit (MU) of the airborne system determines if the ACARS channel is free from other communications. If the channel is clear, the message is transmitted; if the frequency is busy, the MU waits until the frequency is available. The ground station sends a reply to the message transmitted from the aircraft. If an error reply or no reply is received, the MU continues to transmit the message at the next opportunity. After six attempts (and failures), the airborne equipment notifies the flight crew.

In the **polled mode,** the system operates only when interrogated by the ground facility. The ground facility routinely uplinks "questions" to the aircraft equipment, and when a channel is clear, the MU responds with a transmitted message. The MU organizes and formats flight data prior to transmission. Upon request, the flight information is transmitted to the ground facility. Information for ACARS is collected from several aircraft systems, including the flight management system (FMS), the aircraft integrated data system (AIDS), and the central maintenance computer system (CMCS).

Satellite Communication

Some aircraft are equipped with receivers and transmitters that utilize orbiting communication satellites to extend their useful range. **Satellite communication** (SATCOM) equipment is typically found on aircraft that make intercontinental flights. Common HF communication radios will transmit long distance; however, they are very susceptible to interference. SATCOM equipment utilizes frequencies that are relatively static-free but normally limited to line-of-sight transmission characteristics. During SATCOM operation the RF wave is transmitted from the aircraft radio to an orbiting satellite. The satellite relays the radio signal to the ground-based receiver. This process extends the range of a typical VHF communication radio to cover any area between latitudes 75° north and 75° south.

FIGURE 15–9 A typical ACARS control unit. *(Collins Divisions, Rockwell International)*

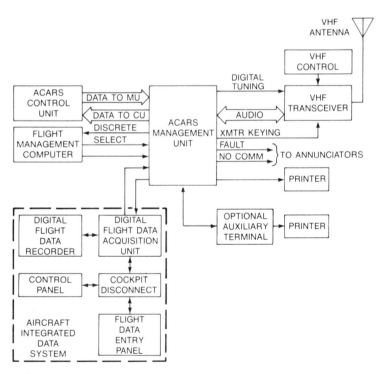

FIGURE 15–10 A block diagram of a typical ACARS system.
(Collins Divisions, Rockwell International)

A SATCOM system is made up of three subsystems: the ground earth station, the aircraft earth station, and the satellite system (Figure 15–11). The aircraft earth station unit transmits on L-band frequencies between 1530 and 1660.5 MHz. The aircraft is capable of transmitting information from several different sources, such as AIRCOM, ACARS,

flight-crew voice communications, passenger telephone, telex, and fax. A satellite data unit (SDU) is used to interface information from other aircraft systems that are linked to the SATCOM system. The SDU works in conjunction with a radio frequency unit (RFU), a high-power amplifier (HPA), a low noise amplifier (LNA), and a beam-steering unit (BSU)

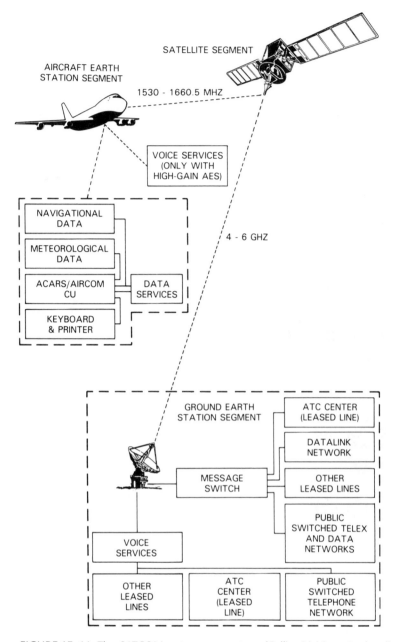

FIGURE 15–11 The SATCOM system segments. *(Collins Divisions, Rockwell International)*

to send an L-band signal to the transmitting antenna and on to the satellite. The various aircraft components are shown in Figure 15–12.

Satellites that are located in a geosynchronous orbit receive signals transmitted from either a ground earth station or an aircraft earth station. The satellites receive and transmit in L-band frequencies when communicating with aircraft and in C-band frequencies when communicating with ground stations. The satellites act as a relay station for the various SATCOM signals. For example, a signal received from an aircraft is converted into a C-band frequency, amplified, and transmitted to a ground station.

The ground stations provide coordination for the various satellites and aircraft transmissions. The network of ground stations allows an aircraft to communicate with virtually any user of the network. The ground stations transmit to the satellites at a frequency between 4 and 6 GHz (C-band microwave frequencies). The ground stations communicate to other ground equipment through a telephone communications network.

Federal Communications Commission Regulations

Because of the very nature of radio waves and their effect on many activities of modern life, all electromagnetic emissions are controlled by a single government agency. In the United States this agency is the **Federal Communications Commission** (FCC). It is the responsibility of this body to supervise all radio transmissions in the United States and its

FIGURE 15–12 Airborne components of a typical SATCOM system. *(Collins Divisions, Rockwell International)*

territories and possessions. The FCC licenses radio operators, technicians, amateur stations and operators, commercial radio stations, marine radio stations, television stations, and various special radio or television operations. Furthermore, the FCC assigns frequency ranges for different types of operations and assigns specific frequencies to individual stations. The agency also cooperates with international agencies to work out agreements to prevent, as much as possible, interference between stations of different nationalities.

Some typical FCC regulations are listed below, but not necessarily in the actual language of the law:

1. All radio transmitters installed in operating aircraft must be licensed.

2. Distress calls or messages have priority over all others.

3. The distress call for radiotelephone is *Mayday.* The distress call for radiotelegraph is ● ● ● - - - ● ● ●, which may be interpreted as SOS.

4. The penalty for willfully violating the Communications Act is a $10,000 fine or imprisonment for a term of not more than two years or both.

5. No obscene, indecent, or profane language shall be transmitted by radio.

6. No fraudulent signals shall be transmitted.

7. Information received by radio and not intended for the person receiving such information shall not be divulged to any person other than the one for whom it is intended; neither shall the existence of the information be divulged.

8. No unnecessary communications shall be transmitted.

9. Noncitizens may not be issued radio-operator's licenses or permits.

10. No operator of a radio station shall violate the provisions of any treaty to which the United States is one of the parties.

11. The operating power of a radio station may be permitted to vary from 5 percent above the assigned power to 10 percent below the assigned power.

12. A person's radio license may be suspended or revoked for violation of FCC regulations.

The above laws are only a few of the most important relating to the operation of radio transmitters. A person who is to be involved in the operation of a radio transmitter should obtain all the necessary information from the FCC and then apply for the license appropriate to the operation concerned.

The operator of a light-aircraft radio transmitter usually does not require an operator's license. Operators of commercial aircraft radio transmitters are typically required to be licensed. Any aircraft radio receivers may be operated without a license.

Testing of a Communications Radio

Testing of a communications radio in small aircraft used in general aviation may be accomplished in accordance with procedures appropriate for the airport and area in which the test is made. Testing on the ground with the engine or engines not operating can be accomplished as follows:

1. Turn on the aircraft power with the master switch.
2. Turn on the transceiver.
3. Select the frequency of the station with which the test is to be conducted.
4. Listen to the receiver to be sure there is no radio traffic in progress at the selected frequency.
5. If no radio traffic is heard, press the transmit switch and call the selected station.
6. Upon receipt of a reply, request a radio check.
7. If reception and transmission are satisfactory, turn the transceiver off. Be sure to do this before turning off the aircraft's master switch.

If a receiver or transmitter is found to be defective, it can easily be removed from the aircraft and sent to an appropriate avionics repair facility. The aircraft technician may inspect for defective items within the system, such as microphones, circuit breakers, or wiring. However, an aircraft technician may never make repairs or alterations to any part of the radio system that may adversely affect the radio transmission signal. In most cases, it is best to have a certified repair station perform all maintenance on radios or related equipment.

NAVIGATION SYSTEMS

In the early days of airplane operation, navigation instruments either did not exist or, at most, consisted of a magnetic compass and an airspeed indicator. When flying by visual reference, the early pilot would usually navigate from one land-mark to another, following roads and railroads or rivers and valleys. Flights were made at comparatively low altitudes providing a view of the ground that was usually good enough for the pilot to clearly identify objects there. Under the flying conditions that existed when the airplane was considered a novelty, complex navigation instruments and systems were not in great demand. As the use of airplanes increased and flights were made at higher altitudes, above the clouds and at night, it became necessary to develop reliable navigation techniques along with instruments indicating attitude, heading, airspeed, and drift so that the pilot could determine the airplane's position by computation and map plotting.

From the 1930s to the present time, great strides have been made in the development of electronic navigation systems. Today a pilot can fly an airplane across the continent from takeoff to landing without touching the controls, all the navigation and pilotage being accomplished electronically.

It is the purpose of this section to describe and explain some of the electronic navigation systems and equipment on modern aircraft. It is beyond the scope of the text to describe the details of circuitry and all the electronic principles employed. To do so would require far more space than is available; however, the general principles of operation and the individual components will be explained.

Automatic Direction-Finder Systems

The function of an **automatic direction-finder** (ADF) system is to enable the pilot to determine the heading, or direction, of the radio stations being received. The ADF system operates on a frequency range of 90 to 1800 kHz, a range that makes it possible for the system to receive radio-range stations in the LF band and standard broadcast stations. By use of the ADF system, a pilot can determine the aircraft's position, or the pilot can "home in" on a radio broadcast station or a radio beacon station by flying directly toward that station using the indication of the radio compass or radio magnetic indicator. To find the aircraft's position, the pilot or the navigator determines the headings of two different radio stations and then plots the headings on a navigation chart. The point at which the heading lines cross is the location of the aircraft.

ADF systems utilize the directional characteristics of a loop antenna to determine the direction of a radio station. A simple direction finder may be made by using a loop antenna with an ordinary radio receiver. By rotating the antenna, the strongest reception can be determined and also the point at which the signal fades out. This point is called the **null position,** and from it a fairly accurate indication of the station direction can be determined. This phenomenon can be demonstrated while listening to a commercial AM radio station. If the radio (and its loop antenna) is rotated, the quality of reception changes. When the reception is the worst, the antenna is in its null position.

On modern aircraft it is common practice to utilize two sets of ADF equipment. The two units may be tuned in to two different radio stations and an immediate fix determined by plotting the lines of position for the two radio stations on a navigation chart. Another value of having an airplane equipped with two ADF systems is that if one system fails, the other is still available for direction finding. ADF equipment is especially valuable in areas of the world where special navigational aids are not available but where the pilot may tune in to a standard broadcast station.

Theory of Operation. As explained in Chapter 14, radio waves are propagated in the form of electromagnetic and electrostatic lines of force that travel at a speed of approximately 186,300 mi/s [300,000,000 m/s] from the radio transmitter. When these lines of force cut across a radio antenna, a voltage is induced in the antenna. This voltage is amplified and demodulated so that the intelligence contained on the radio wave may be determined. If a loop antenna is placed in such a position that it is at 90° to the direction of wave travel, equal and opposite voltages will be induced in the sides of the antenna, as shown in Figure 15–13. The voltages thus induced in the loop will cancel each other, with the result that the loop will have no output. If the loop is connected to a radio receiver, the signal will disappear at this point. If the loop is turned either one way or the other, a voltage will be induced in one side slightly before it is induced in the other, with the result that there will be a difference between the two that will provide a signal that may be fed to the receiver. When the plane of the loop is parallel to the direction of wave propagation, the strongest signal will be developed.

In determining the direction of a radio station with the loop antenna, it is far more accurate to use the null position of the antenna than it is to attempt to use the position at which the strongest signal is received. This is because the null point

is very narrow, whereas the point of strongest signal is several degrees wide.

ADF Components. The principal components of an ADF system are a **radio receiver,** which includes the amplifiers and various other electronic components; a **loop antenna;** a **sense antenna; a radio magnetic indicator** (RMI); and a remote-control unit, or **control panel.** The loop antenna in older systems is mechanically rotated and is connected by a synchro system to the RMI. Hence as the loop is rotated, the pointer of the RMI also rotates. When the system is operating in a fully automatic mode and a station is tuned in, the loop antenna will automatically turn to a direction that provides the correct heading of the station on the RMI.

Because of inherent problems with any moving device, electronic engineers designed a loop antenna that is "rotated" electronically. Actually, the antenna does not move; however, its output is equivalent to a rotating loop.

The function of the sense antenna is to provide an input signal that is out of phase with the signal received with the loop antenna. This is necessary to provide a correct heading indication for a radio station. If a sense antenna were not used, the indicator might show the heading of the station, or it might point to the reciprocal of the heading, that is, to a direction 180° away from the heading of the radio station being received.

Figure 15–14 is a block diagram showing the general arrangement of one of the ADF systems installed in one model of an airliner. In this diagram the coordination between the loop antenna, the synchro transmitters, the loop drive motor, and the indicator is clearly seen.

A typical ADF control panel is shown in Figure 15–15. This ADF system is used on light aircraft and contains the radio receiver directly behind the control panel. A digital tuner and an LED frequency display are used to reduce the number of moving parts, thereby increasing the system's reliability.

Radio Magnetic Indicator. The RMI or ADF indicator shown in Figure 15–16 provides visual information for the pilot and copilot concerning the data received by the ADF equipment. This instrument makes it possible for the pilot to navigate the aircraft without the necessity of numerical or graphical calculations. The instrument displays the magnetic heading of the aircraft and the magnetic bearings of two radio stations. The bearings of the two radio stations are provided by the two separate ADF receivers operating with the loop antenna. The magnetic heading of the aircraft is indicated on a rotating disk-type dial, and the magnetic bearings of the two radio stations are shown under the two pointers.

The face of the RMI consists of a fixed outer dial with 45° markings through 360°, an inner rotating compass dial graduated from 0 to 360° clockwise in 2° increments, a wide pointer with parallel grids at the outer edge, and a narrow pointer mounted concentrically with the wide pointer and compass dial. The two pointers provide radio-bearing indications, that is, the bearings of two radio stations being received. The indications of the pointers are read in reference to

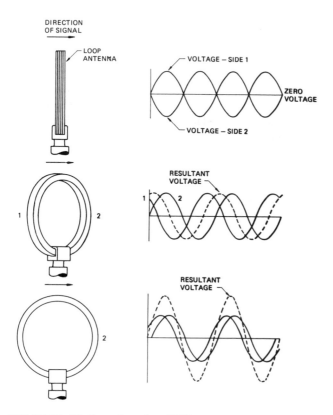

FIGURE 15–13 Operation of an ADF loop antenna.

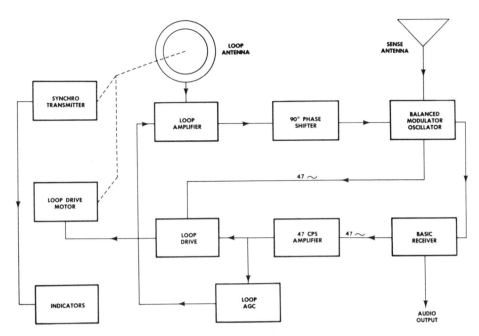

FIGURE 15–14 Block diagram of an ADF system.

FIGURE 15–15 An ADF control panel. *(King Radio Corporation*

FIGURE 15–16 Typical ADF indicators. *(Collins Divisions, Rockwell International)*

the compass dial, or card, and thus provide the magnetic bearings of the radio stations on the navigation chart and immediately locate the airplane at the position where the two lines cross.

ADF Receivers. ADF receiver units for large aircraft are typically located in equipment racks with digital data links to the antennas, control panels, and RMI. Built-in test equipment with a nonvolatile memory may also be used on ADF systems to enable technicians to monitor system faults. Figure 15–17 shows the typical ADF equipment for a corporate-type aircraft.

VHF Omnirange

VHF omnirange (VOR) is an electronic navigation system that enables a pilot to determine the bearings of the VOR transmitter from any position in its service area. This is possible because the VOR ground station, or transmitter, continually broadcasts an infinite number of directional radio beams or radials. The VOR signal received in an airplane is used to operate a visual indicator from which the pilot determines the bearings of the VOR station with respect to the airplane.

Figure 15–18 shows diagrams illustrating both the VOR transmitter system and the VOR receiver system; the receiver system is of the type used on one model of an airliner. The diagram for the VOR ground station, or transmitter, shows a five-unit antenna array. The center loop of the antenna array continuously broadcasts the reference phase signal, which is modulated at 30 Hz. Two outputs are radiated by the diagonal pairs of corner antennas, and the signals radiated from these pairs are modulated by 30 Hz and differ in phase by 90°. Each pair of antennas radiates a figure-eight pattern, each pattern being displaced from the other by 90° both in space and in time phase. The resulting pattern is the sum of the two figure-eight patterns and consists of a rotating field turning at 1800 rpm, or 30 Hz.

The total effect of the radiation from the VOR transmitter is to produce two signals whose phase characteristics vary in accordance with the direction (bearing) of the transmitter from the receiver. The two signals radiated due south (magnetic) of the transmitter are exactly in phase; hence an airplane flying magnetic north directly toward the VOR transmitter will show an indicated bearing of 0° to the VOR station. The TO-FROM indicator will show that the airplane is flying to the station.

In a clockwise direction around the VOR station, the radiated signals become increasingly out of phase. At 90° clock-

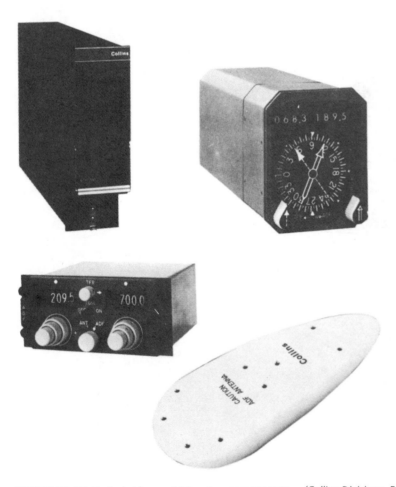

FIGURE 15–17 Typical airborne ADF system components. *(Collins Divisions, Rockwell International)*

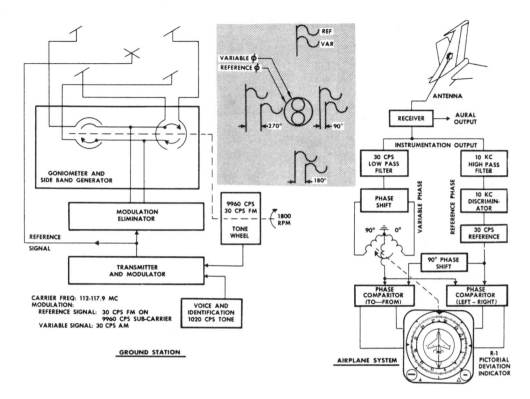

FIGURE 15–18 Block diagrams of VOR transmitter and receiver systems.

wise from the due south direction, the signals are 90° out of phase, at 180° they are 180° out of phase; at 270° they are 270° out of phase; and at 360° (0°) they are back in phase. The phase difference of the two signals makes it possible for the receiver to establish the bearings of the ground station. The directional bearings of VOR stations are set up in accordance with the earth's magnetic field so that they may be compared directly with magnetic-compass indications on the airplane.

During the operation of VOR equipment on a particular heading, an airplane flying toward the VOR station will show a TO indication on the omni-indicator. After the airplane passes the station, the indicator will show FROM, and the heading information will remain the same as it was. For example, if an airplane is flying toward a VOR station having a bearing of 200°, the omnibearing indicator will show 200° TO. After the airplane passes over the VOR station, the indicator will show 200° FROM.

The carrier frequency of the VOR station is in the VHF range, between 112 and 118 MHz. A modulation of 9960 Hz is placed on the carrier of the reference signal to provide a subcarrier, which is modulated by a 30-Hz signal. The 9960-Hz modulation on the original carrier wave is AM, and the 30-Hz signal on the subcarrier is FM. The carrier wave for the variable-phase signal is amplitude-modulated by a 30-Hz signal.

The VOR receiver mounted in an airplane may be an independent unit, or it may operate in conjunction with the VHF communication radio. Light aircraft typically use the combined unit, known as a VHF NAV/COM radio. The VOR receiver receives both components of the VOR signal transmitted from the ground station and from these signals produces two 30-Hz signals, one being the reference phase

and the other being the variable phase. The angular distance between the two phases is applied to the omnirange indicator, by which it is translated into usable heading information.

The omnirange indicator includes an azimuth dial, a LEFT-RIGHT deviation needle, and a TO-FROM indicator. When the VOR receiver on an airplane is tuned to a VOR ground station, the LEFT-RIGHT indicating needle will be deflected either to the right or to the left unless the selected course on the omnirange indicator is in agreement with the bearing of the VOR ground station.

Once the pilot has tuned to the correct ground station frequency and selected the correct course, the unit is ready for navigation. For example, if the course-deviation indicator bar moves to the left, the pilot knows the intended course is to the left of the aircraft. To correct the flight path, the pilot must turn the aircraft to the left.

Figure 15–19 shows one common type of course-deviation indicator (CDI) used in connection with the VOR and

FIGURE 15–19 One type of course-deviation indicator.

other navigation equipment. (See also Figure 15–20.) The face of this instrument consists of fixed markings, or lubber lines, and 45 and 90° reference markings. These markings are shown as small white triangles at the outer perimeter of the azimuth dial. The azimuth dial is a rotating servoed-compass repeater dial graduated from 0 to 360° clockwise in 2° increments. The face of the dial also includes an open triangular cursor that travels around the compass dial and is positioned by the heading (HDG) input knob, an inner rotating carriage that carries the course-deviation bar, a course-deviation warning flag, solid triangular flags to show if the selected course is to or from the VOR station, and a cursor shaped like an inverted T to show the selected course. On some instruments an omnirange bearing selector (OBS) knob is used instead of the HDG knob. A central pedestal is marked with the figure of an airplane aligned with the lubber line and contains a digital presentation of the selected course. On some state-of-the-art equipment, the moving components of the previously discussed CDI have been replaced with a more reliable LED display unit. Also, many digital VOR units used on larger aircraft may contain built-in test equipment and data interface capabilities.

Testing for Accuracy. In accordance with the flight regulations of FAR Part 91, any VOR receiver that is to be used under instrument flight rule (IFR) conditions must be checked periodically. The pilot can perform this accuracy check by comparing the indication of two VORs within the same aircraft or by comparing the VOR's indication with a known VOR test point. A certified VOR test (VOT) station may also be used if there is one available in your area. Since this test must be performed at least every 30 days, a log entry must be recorded in a dedicated VOR logbook.

Instrument Landing System

The **instrument landing system** (ILS), as the name implies, is designed to allow pilots the opportunity to land their aircraft with the aid of instrument references. A typical ILS system will allow the pilot to bring an aircraft to within $^1/_2$ mi of the runway and less than 200 ft above the runway without any external visual references. At these minimums (the **decision height**) the pilot must identify the **runway environment** in order to continue the landing process. If the runway environment cannot be identified, the pilot must execute a missed approach procedure. An extremely accurate ILS, called the Category IIIC approach, will allow for aircraft landings with near zero visibility. Both the airport facilities and the aircraft must be correctly equipped and certified for a Category IIIC ILS approach.

The ILS provides a horizontal directional reference and a vertical reference called the **glide slope.** The directional reference signal is produced by the runway **localizer** transmitter, which is installed approximately 1000 ft [305 m] from the far end of the runway and operates at frequencies of 108 to 112 MHz. The glide slope signal is produced by the glide slope transmitter, which is located near the side of the runway on a line perpendicular to the runway centerline at the point where airplane touchdown occurs. This point is generally about 15 percent of the runway length from the approach end of the runway. The glide slope transmitter operates at a frequency of 328.6 to 335.4 MHz.

The Localizer. The localizer consists, essentially, of two RF transmitters and an eight-loop antenna array. The transmitters broadcast a complete system of radiation patterns that produce a null signal along the center of the runway. The radiation pattern is such that when an airplane is ap-

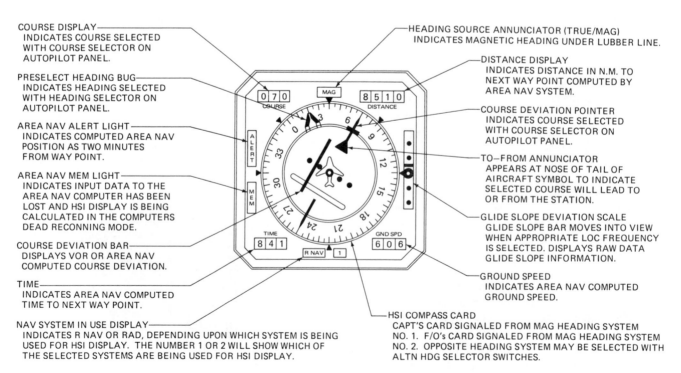

FIGURE 15–20 Horizontal-situation indicator. *(Lockheed California Company)*

proaching the runway for a landing, the signal to the right of the localizer path will be modulated with 150 Hz, and the signal to the left of the localizer path will be modulated with 90 Hz. The localizer receiver on board an airplane is able to discriminate between the 90- and 150-Hz signals. The output of the receiver is fed to the vertical needle of a **course-deviation indicator** (CDI) (shown in Figure 15–21) or to another type of instrument such as a **flight-director indicator.** If the airplane is to the right of the localizer centerline, the 150-Hz modulation signal will predominate, and the vertical needle of the indicator will point to the left of the centerline, indicating that the pilot should fly left in order to return to the centerline of the localizer beam.

The Glide Slope. The glide slope transmitter operates on a principle similar to that of the localizer. As previously mentioned, the glide slope transmitter is located at a distance from the approach end of the runway, approximately 15 percent of the length of the runway. A schematic diagram illustrating the radiation pattern from the glide slope transmitter is shown in Figure 15–22. If an airplane is approaching the runway and is above the glide path, the 90-Hz signal will predominate; and if the airplane is below the glide path, the 150-Hz signal will predominate. The glide slope receiver will provide an output for the crosspointer indicator in such a way that the pilot will have a visual indication of the airplane's position with respect to the glide path. If the horizontal pointer is above the center of the indicator, the airplane is below the glide path.

A diagram of the beam provided by the combination of localizer and glide slope transmitter is shown in Figure 15–23. The beam is electronically exact and provides a precise path by which an airplane may approach a runway and reach the point of touchdown. This is a most valuable aid in conditions of poor visibility in the vicinity of an airport.

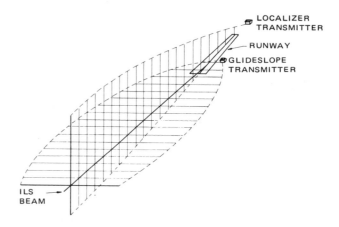

FIGURE 15–23 ILS transmission pattern from localizer and glide slope.

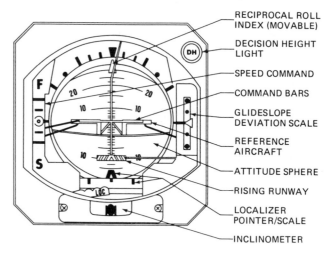

FIGURE 15–24 Attitude-director indicator. *(Lockheed California Company)*

On larger aircraft the ILS and VOR control panels are usually combined and located on the instrument panel; the receivers are typically separate units located in the radio equipment rack. Light-aircraft ILS receivers are usually combined with the receiver for VHF omnirange (VOR) and are often designated as VOR LOC receivers. The indicators for the system are often combined with other indicators that provide a number of navigation indications. Among these are the **attitude-director indicator** (ADI) shown in Figure 15–24 and the **horizontal-situation indicator** (HSI) shown in Figure 15–20.

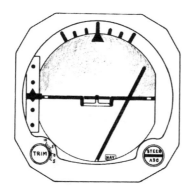

FIGURE 15–21 Course-deviation indicator.

FIGURE 15–22 Radiation pattern from a glide slope transmitter.

Distance-Measuring Equipment

To make it possible for a pilot to determine the distance of the aircraft from a particular VOR/DME or VORTAC station, **distance-measuring equipment** (DME) was developed. VORTAC indicates a station that includes both VOR and TACAN equipment. TACAN is described in the next section.

Through the use of the combined facilities of DME and VOR, a pilot is given a constant visual indication of the slant distance from a particular VOR station and also the bearing of the station. It is therefore possible at all times to establish the exact location of the airplane.

The operation of DME units is similar to that of radar beacons. That is, the communication between the airborne unit, called the **interrogator,** and the ground station is by means of pulses similar to those utilized in radar. The ground DME unit is called a **transponder.** During operation of VOR, when the pilot selects a particular ground-station frequency by means of the control, the coded pulse is automatically selected in the DME interrogator associated with the VOR. The interrogator transmits pairs of coded pulses to the transponder (ground station), where the signal is amplified and transmitted back to the airborne receiver. The time interval between transmission of the signal by the interrogator and receipt of the signal sent by the transponder determines the distance of the airplane from the ground station. Remember that approximately 6.19 μs is required for a radio wave to travel 1 nmi [1.8 km].

The DME challenge sent by an aircraft, when a particular VOR station is selected, consists of spaced pulses in the frequency range of approximately 987 to 1213 MHz. The ground-station transponder will accept only signals that are spaced correctly and have the correct frequency.

The DME equipment mounted in an airplane consists of **timing** circuits, **search** and **tracking** circuits, and the **indicator.** The timing circuits measure the time interval between the interrogation and the replay, thus establishing the distance of the ground station from the airplane. The search circuits cause the airborne equipment to seek a reply after each challenge, a function accomplished by triggering the receiver into operation after each interrogation. When the receiver picks up a reply of the correct code, the tracking circuit will operate and enable the receiver to hold the received signal. The time interval is measured and converted into a distance reading, which is then displayed on the DME indicator. If the airborne receiver picks up a signal with an incorrect code (that transmitted from another aircraft), the equipment automatically rejects that signal. Any airborne DME receiver will accept only signals that were originally transmitted by its own equipment. This means of signal discrimination allows several aircraft to navigate using the same DME ground station.

DME distance indications are displayed digitally on one or more panels or instruments. Figure 15–25 shows how a radio magnetic indicator (RMI) has been combined with DME indicators in an instrument called a digital distance radio magnetic indicator. DME indications are also displayed on other instruments and panels (see Figure 15–20).

As an airplane equipped with DME is approaching a DME station and is receiving DME information, the distance readout will continue to change as the distance from the station changes. The rate of change is fed to a computer that produces a ground-speed indication. In many of the advanced navigation systems, the time required to reach a given station or waypoint is also displayed. This is shown in the photograph of a DME unit and indicators in Figure 15–26.

Airborne receivers for DME are provided with an audio

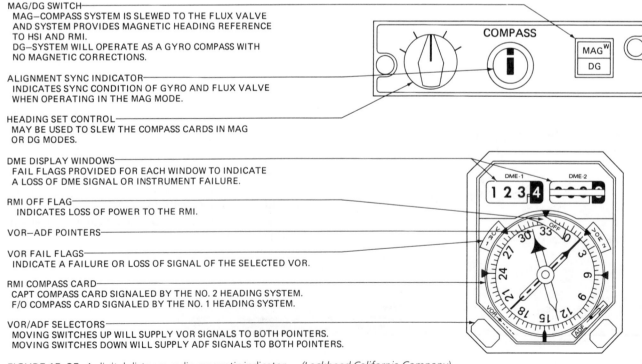

MAG/DG SWITCH
  MAG—COMPASS SYSTEM IS SLEWED TO THE FLUX VALVE
  AND SYSTEM PROVIDES MAGNETIC HEADING REFERENCE
  TO HSI AND RMI.
  DG—SYSTEM WILL OPERATE AS A GYRO COMPASS WITH
  NO MAGNETIC CORRECTIONS.

ALIGNMENT SYNC INDICATOR—
  INDICATES SYNC CONDITION OF GYRO AND FLUX VALVE
  WHEN OPERATING IN THE MAG MODE.

HEADING SET CONTROL—
  MAY BE USED TO SLEW THE COMPASS CARDS IN MAG
  OR DG MODES.

DME DISPLAY WINDOWS—
  FAIL FLAGS PROVIDED FOR EACH WINDOW TO INDICATE
  A LOSS OF DME SIGNAL OR INSTRUMENT FAILURE.

RMI OFF FLAG—
  INDICATES LOSS OF POWER TO THE RMI.

VOR—ADF POINTERS—

VOR FAIL FLAGS—
  INDICATE A FAILURE OR LOSS OF SIGNAL OF THE SELECTED VOR.

RMI COMPASS CARD—
  CAPT COMPASS CARD SIGNALED BY THE NO. 2 HEADING SYSTEM.
  F/O COMPASS CARD SIGNALED BY THE NO. 1 HEADING SYSTEM.

VOR/ADF SELECTORS—
  MOVING SWITCHES UP WILL SUPPLY VOR SIGNALS TO BOTH POINTERS.
  MOVING SWITCHES DOWN WILL SUPPLY ADF SIGNALS TO BOTH POINTERS.

FIGURE 15–25 A digital distance radio magnetic indicator. *(Lockheed California Company)*

system that receives identification codes from DME stations. This makes it possible for the pilot to identify positively the station that the DME has locked onto. In the majority of VOR/DME receivers, when a particular VOR frequency is selected, the associated DME frequency is automatically selected for that station.

In the installation of DME equipment in an aircraft, the location of the antenna is critical. The antenna is a short stub, approximately 2.5 in. [6.35 cm] long, usually mounted on the bottom of the fuselage. Care must be taken in locating the antenna, because it can be blanked out easily by obstructions such as landing gear or other antennas nearby. It is recommended that manufacturer's instructions for installations in similar aircraft be observed when making a new installation.

TACAN

A distance-measuring and bearing-indicating system similar to the VOR/DME system described above is called **TACAN.** This system was developed by the Navy for use on aircraft carriers and other Navy air installations. The word *TACAN* is a shortend version of the descriptive term **tactical air navigation.**

The TACAN distance-measuring facility is now utilized for civilian air navigation, as well as for the military. The combination of VOR and TACAN to give both bearing and distance information is called VORTAC. To utilize VORTAC, an aircraft must be equipped with UHF radio units that can operate on the TACAN frequencies. The low TACAN band has receiving frequencies from 1025 to 1087 MHz and transmitting frequencies from 962 to 1024 MHz. The high TACAN band has receiving frequencies from 1088 to 1140 MHz and transmitting frequencies from 1115 to 1213 MHz.

Navigation Receivers

A basic navigation (**NAV**) receiver is designed to receive VOR signals and display course, bearing, and heading information on an RMI, an HSI, or some other instrument. If the receiver is equipped to receive localizer (LOC) and glide slope (GS) signals, the system will include a course-deviation indicator (CDI) or a similar instrument to show the pilot where the aircraft is tracking the ILS beam as it approaches a runway.

A typical NAV receiver is shown in Figure 15–27. This receiver is equipped with a standby frequency provision so a frequency may be preselected and held in readiness for use when needed. The frequencies are selected by means of the frequency selector knobs at the right of the panel. The outer knob selects the megahertz portion of the frequency, and the inner knob selects the kilohertz portion. Some NAV receivers use push-button panels for selecting frequencies.

As mentioned previously, NAV receivers and COM transceivers are often combined into one unit of equipment. This saves weight and space and simplifies the installation. A unit of this type is called a NAV/COM unit and may include ILS receivers for LOC and GS.

Marker Beacons

In order to provide pilots with an indication of their distance from the runway, **marker-beacon** transmitters are installed, with the outer-marker transmitter at approximately 5 mi [8 km] from the runway and the midmarker approximately $^2/_3$ mi [1 km] from the end of the runway. The marker-beacon transmitter operates at a frequency of 75 MHz and produces both aural and visual signals. The outer-marker transmitter

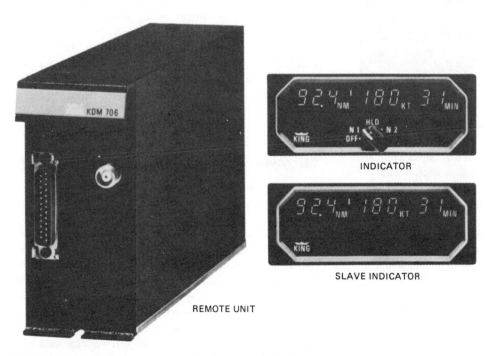

FIGURE 15–26 DME remote unit and indicators. *(King Radio Corporation)*

NAV RECEIVER WITH OPTIONAL GS RECEIVER

(DISPLAYS BOTH ACTIVE AND STANDBY FREQUENCIES)

FIGURE 15–27 A navigation radio. *(King Radio Corporation)*

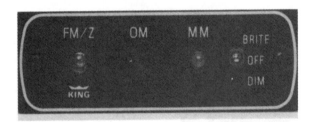

FIGURE 15–28 A typical marker-beacon display panel. *(King Radio Corporation)*

produces a 400-Hz intermittent signal that causes a blue indicator light on the instrument panel to glow intermittently. The midmarker transmitter produces a signal modulated at 1300 Hz that causes the amber marker-beacon light on the instrument panel to glow. Thus when an airplane is approaching the runway and is approximately 5 mi from its end, the blue light will flash. A short time later, when the airplane is within $^2/_3$ mi of the runway, the amber light will flash.

Some airports also employ an inner marker located approximately 1500 ft from the end of the runway. By modulating the RF wave with a 3000 Hz signal, the inner marker will illuminate a white panel light when the aircraft is over the appropriate position. This system provides an excellent indica-

tion of the plane's distance from the runway. The white lamp is usually labeled "FM/Z" because the 3000-Hz signal is also produced by en route airway, or "Z," markers.

Marker-beacon receivers are typically located in the aircraft's equipment rack. A display panel, as illustrated in Figure 15–28, is mounted on the aircraft's instrument panel. No control unit is required for a marker-beacon receiver. The unit typically turns on when the avionics master switch is activated.

Microwave Landing System

The **microwave landing system** (MLS) has been developed to overcome some of the problems and limitations associated with the ILS. The MLS currently being deployed is called a **time-reference scanning-beam microwave landing system,** or TRSB. The ILS provides one narrow flight path and operates at VHF or UHF frequencies, whereas the MLS provides a wide range of flexible flight paths on an approach to an airport. In addition, the MLS has the advantages inherent with operating at microwave frequencies (3 to 30 GHz). Among the benefits of the microwave frequencies are a much larger number of frequency channels, fewer problems with finding suitable sites for ground components, and elimination of severe multipath interference caused by signal reflections from buildings, hills, and other objects. With the MLS, aircraft can approach a runway from a wide variety of angles rather than being required to be aligned with the runway for many miles on the approach.

As explained previously, the ILS transmits signals that, when combined, provide a narrow beam rising from a point on the runway and extending for an indefinite distance along the approach to the runway. This situation requires that all aircraft approaching a particular runway be "funneled" into the one approach path. With the MLS, aircraft can approach the runway from many different angles, thus making it possible to accommodate more flights and shorten flight paths. Currently, microwave landing systems are installed at very few airports; however, the MLS offers the flexibility needed to safely increase airport traffic. The MLS has a definite future in air traffic control; it may someday totally replace the ILS.

Principle of Operation. The principle of operation for the TRSB microwave landing system may be illustrated as shown in Figure 15–29. Two transmitters, one for azimuth and one for elevation, transmit fan scanning beams toward approaching aircraft. The precise timing of the scanning beams provides exact information for the pilot regarding the position of the aircraft. Beams are scanned rapidly "to" and "fro" throughout the area shown in the drawing. In each complete scan cycle, two pulses are received by the aircraft. One pulse is received during the "to" scan and the other during the "fro" scan. The aircraft receiver derives its position angle directly from the measurement of the time difference between the two pulses. The receiver-processor computes the information and prepares it for display on a conventional course-deviation indicator (CDI). In addition, a digital display of the information is presented on the control panel.

Distance information for the system is derived from conventional distance-measuring equipment (DME).

Airborne MLS Equipment. The MLS equipment includes a typical airborne receiver, the digital display-control panel, and the MLS antenna (see Figure 15–30). This system features an automatic and manual test function to verify sys-

FIGURE 15–30 Microwave landing system equipment; from left to right: the airborne receiver, the digital display-control panel, and the MLS antenna. *(Bendix Avionics Division)*

tem performance and a liquid crystal display of all MLS data. The entire system weighs under 10 lb.

The receiver processes the MLS signal with respect to the approach selection criteria and provides multiple outputs. These outputs can be used to drive the flight control system and conventional electromechanical CDIs, HSIs, and ADIs.

Often an MLS system will incorporate two airborne antennas located on the bottom of the aircraft's fuselage, one mounted forward, the other rearward. This configuration provides signal reception in all aircraft attitudes, including attitudes during a missed approach.

Radio Tuning Systems

On many modern corporate or transport-category aircraft, **radio tuning units** are used to help eliminate instrument panel clutter and simplify radio operations. These units are designed to allow the pilot access to several radios using one control-display unit. As shown in Figure 15–31, a typical system can operate the VHF communication, VOR/ILS, DME, ADF, ATC transponder, and MLS radios. Radio tuning units are typically part of a complete radio package supplied by one particular manufacturer. Most units communicate with the other radio systems using both analog and digital ARINC 429 signals.

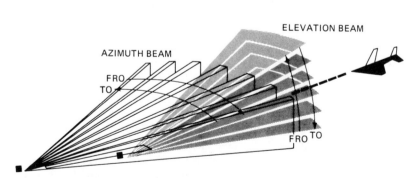

FIGURE 15–29 Scanning-beam pattern for a microwave landing system. (Bendix, Aerospace Electronics Group, Communications Division)

FIGURE 15–31 A typical radio tuning unit. *(Collins Divisions, Rockwell International)*

FIGURE 15–32 Typical audio control panel. *(King Radio Corporation)*

Audio Control Systems

On aircraft with multiple radios, there must be some system to control the input (microphone) and output (speaker) signals to and from those radios. An audio control panel performs this function. The **audio control system** shown in Figure 15–32 controls as many as three transceivers and six receivers. This system also includes a marker-beacon display.

Two rows of alternate action push-button switches control the receiver audio distribution functions. The pilot may choose to listen to a headset by depressing the corresponding lower button. The cabin speaker may be activated by depressing the corresponding top button. The rotary switch on the right side of this unit controls the input signals to one of three transmitters or an intercom system.

An audio panel is found on nearly every light aircraft that contains more than one radio; larger aircraft may contain multiple audio panels, one for each flight crew member. Whenever operating any aircraft radio system, be sure the audio control panel is properly set for the functions you desire.

Integrated Navigation System

Area navigation (RNAV) is a system that makes it possible to use the information from VOR/DME or VORTAC stations and fly a direct route from a point of departure to a destination without following a dogleg course, which would result if NAV were used only with VOR/DME information. This is accomplished by using a microprocessor to continuously receive and process data originally entered plus new data that are supplied by the DME during flight.

The use of RNAV with other electronic navigation components makes possible a completely **integrated navigation system.** Such a system includes VOR, LOC, GS, DME, NAV, RNAV, and associated circuitry. The principal elements of an integrated navigation system are shown in Figure 15–33. In addition to the equipment shown, a CDI, an HSI, or a similar instrument is required to provide visual course information for the pilot. Controls for programming data into the system are on the front of the unit. The unit shown in Figure 15–33 utilizes rotary knobs to enter frequencies, radials or bearings, and distances into the system. Other types of equipment may employ push-button controls, as shown in Figure 15–34.

A second-generation integrated navigation system is currently being used in many state-of-the-art aircraft. This system is known as the Flight Management System (FMS). The FMS will be discussed in detail in Chapter 17.

Programming a Flight. The RNAV system, in effect, makes it possible to "move" a VOR/DME station electronically from its actual location to a location on the proposed flight route. The mathematics of this operation is handled by the large-scale integrated (LSI) circuitry of the microprocessor. Figure 15–35 is a drawing of a proposed flight route showing how three VOR/DME stations are employed to produce four waypoints along the route. To set up these waypoints, the pilot or some other operator uses the control panel such as that shown in Figure 15–33.

In Figure 15–35 the bearing of the destination from waypoint 1 is 65°. Therefore, the omnibearing selector (OBS) knob on the CDI should be rotated to 65°. The airplane is then flown with the needle centered to follow the correct course.

When the airplane passes over waypoint 1, the omni-indicator on the CDI will shift from *to* to *from*. The flight may be continued on the basis of waypoint 1 data until it is desired to shift to waypoint 2. This is accomplished by pressing the *dsp* and *use* buttons to display the figure *2*. Waypoint 2 data will be displayed in accordance with the data selected. The DME will shift to the waypoint 2 data and will lock onto the new station. Distance, ground speed, and time to waypoint 2 will be displayed.

The integrated navigation system for a large airliner is illustrated in the block diagram in Figure 15–36. This system includes ADF, radio altimeters, and gyros to provide complete data for the captain and first officer. Dual systems provide safety in case of failure of any subsystem. By coupling this system with the automatic pilot, fully automatic flight is possible from takeoff to landing and rollout.

NAV & GS RECEIVER

DME

COMPUTER BOARD

FIGURE 15–33 An integrated navigation system showing internal electronic circuitry. *(King Radio Corporation)*

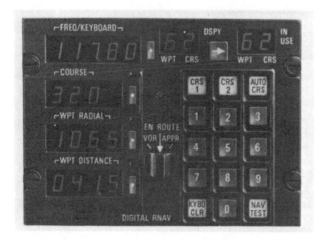

FIGURE 15–34 Push-button control panel for a navigation system. *(King Radio Corporation)*

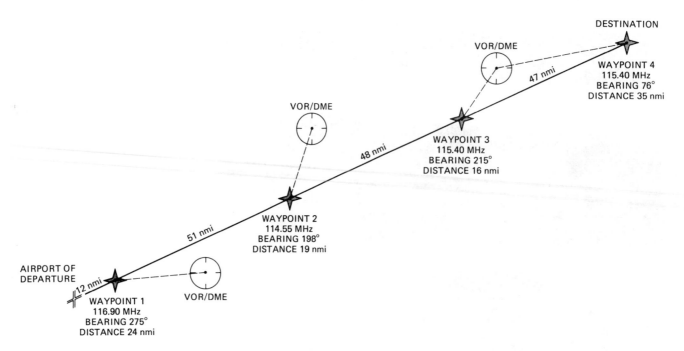

FIGURE 15–35 Direct flight by RNAV utilizing VOR/DME stations.

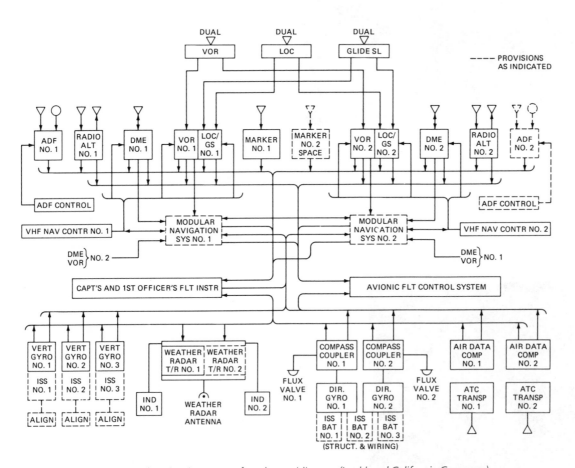

FIGURE 15–36 Integrated navigation system for a large airliner.   *(Lockheed California Company)*

LONG-RANGE NAVIGATION SYSTEMS

Because of the need for accurate and effective navigation on overseas routes and polar routes and in other areas where conventional navigation aids are not available, a variety of long-range systems have been developed. Among these are GPS, LORAN and VLF/OMEGA. These are effective systems that have been in use in various forms for many years.

LORAN System

LORAN (LOng-RAnge Navigation) is a system that operates in the LF range (30 to 300 kHz) and utilizes pulse-transmitting stations to provide the signals necessary for navigational computation. A master station and one or more slave stations send out synchronized pulses that are received by equipment in the airplane. If the airplane receives pulses from the master station and slave station at the same time, the airborne LORAN receiver "knows" that the airplane is equidistant from the two stations and that it is located somewhere on the perpendicular bisector of a line between the stations. If the pulses are not received at the same time, the airplane is located on a hyperbolic line representing the difference in distances from the airplane to the two stations.

Figure 15–37 represents a master station, M, and two slave stations, S_1 and S_2. Any point on line AB is equidistant from the master station and S_1, and any point on line CD is equidistant from M and S_2. The space between each circle, in this case, is given a value of 150 μs, or a distance of approximately 24 nmi. The hyperbolic line EF is the line where any point on the line is 150 μs closer to S_1 than to M. Any point on the hyperbolic line GH is 300 μs closer to S_1 than to M. The other hyperbolic lines represent increased differences in distance.

In order to establish an exact location, it is necessary to utilize another slave station or another combination of a master station and slave station to provide a cross reference. In Figure 15–37 this is accomplished with slave station S_2. The hyperbolic line IJ is 150 μs closer to S_2 than to M. Therefore, if an airplane receives signals that indicate that it is 150 μs closer to S_1 and 150 μs closer to S_2 than to M, then it can compute that it is at position P on the chart. In current LORAN systems, this computation is accomplished by electronic computers, and the signals received are translated into coordinates of longitude and latitude and displayed digitally on an indicator.

VLF/OMEGA Navigation System

The **VLF/OMEGA navigation system** utilizes worldwide stations established by the U.S. Navy and the U.S. Coast Guard. Currently, fewer than 20 VLF/OMEGA stations are located in the free world. The locations of the stations and the fact that they utilize VLF frequencies (10 to 30 kHz) result in a global coverage such that the signals can be employed for navigation anywhere on the face of the earth.

Each communications station in the network transmits continuously on an individually assigned frequency between 14 and 24 kHz. The OMEGA navigation stations utilize four frequencies: 10.2, 11.05, 11.33, and 13.6 kHz, but no two stations are transmitting on the same frequency at the same time. The entire segmented format is repeated every 10 s. The station in Norway transmits at 10.2 kHz for 0.9 s in the first segment, 13.6 kHz for 1.0 s in the second segment, 11.33 kHz for 1.1 s in the third segment, and 11.05 kHz for 1.0 s in the sixth segment. Then it does not transmit in the fourth, fifth, seventh, or eighth segments. The station in Liberia transmits with the same order of frequencies in the second, third, fourth, and seventh segments. The other stations are assigned segments in a similar manner such that a computer in a receiving system can sort out the stations it is receiving by a process called **commutation.** After the computer has determined what stations it is receiving, it performs the mathematics necessary to determine its location with respect to the position information it contained at the time it started receiving and computing position. The computer must receive signals from at least two stations in order to determine its change in position as a flight progresses. If the computer is given the correct position coordinates at its starting point, it will provide accurate position data digitally as latitude and longitude.

The long-range capabilities of the VLF/OMEGA system are made possible by the use of the VLF frequencies. These waves can be effective for distances up to 10,000 mi [16,098 km] because they are reflected back to the earth by the ionosphere. Thus each station can blanket half the earth's surface, and there is no location where a suitable number of stations cannot be received.

Inertial Navigation System

The advantage of an **inertial navigation system** (INS) is that it requires no external radio signals. This concept makes it extremely valuable for military aircraft and guided missiles. Civilian aircraft also employ an INS in order to take advantage of its desirable characteristics for long-range navigation. As its name implies, an inertial navigation system depends on the laws of inertia to determine an aircraft's position. That is, once the starting point of a flight is known by latitude and longitude, the INS computer will determine any new positions by measuring the inertial forces acting on the aircraft.

There are three basic laws of inertia that were described by Sir Isaac Newton over 300 years ago, They are as follows:

1. *Newton's first law*   A body continues in a state of rest, or uniform motion in a straight line, unless acted on by an external force.

2. *Newton's second law*   The acceleration of a body is directly proportional to the sum of the forces acting on that body.

3. *Newton's third law*   For every action there is an equal and opposite reaction.

Applying these laws to aircraft navigation, we find that an aircraft will not move or change its motion unless acted on by an external force (engine thrust, wind drag, gravity, and wing lift). Since the change in motion (acceleration) is proportional to the applied force, we can determine acceleration by

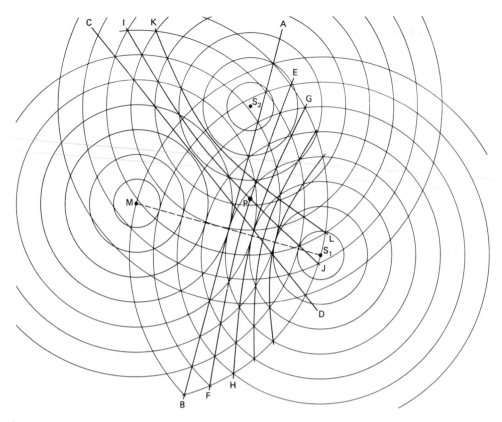

FIGURE 15–37 Signal pattern developed by LORAN stations.

measuring the external forces acting on the aircraft. Since there is a reaction force for every external force that acts on the aircraft, we can measure the reaction force to determine the aircraft's acceleration and therefore its velocity and position.

The instrument used to detect acceleration is called the **accelerometer.** At least two accelerometers are required for each INS system. One measures accelerations in the north-south direction; the other measures east-west accelerations. Most aircraft INS systems contain three accelerometer systems, one for each axis of the aircraft. An acceleration force has both magnitude and direction; therefore, both acceleration and deceleration are measured. As shown in Figure 15–38, a simple accelerometer could be a pendulum-type device. That is, it must be free to swing in two directions. The reaction force (opposite the external forces applied to the aircraft) causes the pendulum to swing. The pendulum swing is measured by an extremely accurate sensor that creates an electrical signal and sends it to an amplifier and torquer motor. The torquer motor returns the pendulum to its null position.

A basic accelerometer accurately detects acceleration only if it remains perfectly level. Since aircraft are seldom perfectly level during flight, all airborne accelerometers must be mounted on a **gimbal platform.** As illustrated in Figure 15–39, a gimbal platform contains two gyroscopes, which stabilize the unit. This combination of gimbals and gyroscopes creates a platform that remains level in any aircraft attitude. Since the accelerometer remains level, it does not sense changes in aircraft attitude; therefore, the accelerome-

ter's output signal accurately measures changes in acceleration. The output signal of the accelerometer is amplified and sent to the INS measurement computer. If the aircraft's initial location and destination are recorded into the computer, the INS system is capable of continuously updating flight deck displays of position, ground speed, heading and distance, and time to a destination. This information can be monitored directly from flight deck instruments or fed into an autopilot, thus forming a complete autoflight system.

Another type of INS system is known as the **strapdown inertial navigation system,** which uses a solid-state (no moving parts) accelerometer system. The heart of the strapdown system is the laser gyro, which replaces the older rotating-mass gyroscope. This system will be discussed in greater detail in Chapter 17.

All inertial navigation systems have a **drift-rate error,** which accumulates during usage. This error ranges from about 1 mi of error for each hour of operation to 1 mi of error for every 10 h of operation. The newer strapdown system has a lower drift-rate error. To compensate for this error, all inertial navigation systems require a periodic update from another navigation source.

Global Positioning System

The **navigation satellite timing and ranging global positioning system (Navstar GPS),** more simply called the GPS, is quickly becoming the system of choice for long-range navigation. The GPS consists of three independent segments: the space segment, the control segment, and the user segment. As

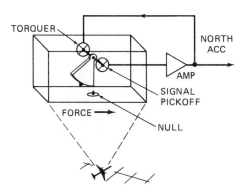

FIGURE 15–38 A diagram of an accelerometer. *(The Sperry Corporation)*

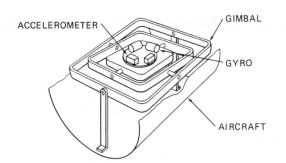

FIGURE 15–39 A diagram of a gimbal platform. *(The Sperry Corporation)*

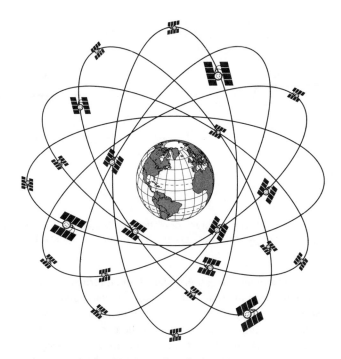

FIGURE 15–40 Orbiting satellites for a GPS.

seen in Figure 15–40, the complete **space segment** consists of 24 satellites, 21 active and 3 orbiting spares, located in a 10,900-mi-high orbit around the earth. The satellites transmit extremely accurate timing pulses and a code system defining the precise location of the satellites at the time of data transmission.

The **control segment** consists of various ground-based monitoring stations and one master control station. The monitors receive the satellite transmissions at least once each day and relay that information to the master control station. The master station computes any drift that has occurred in the satellites' orbit or timing pulse. A correction signal is then sent to the satellites, and that correction is included in the location code transmitted from the satellites to the user.

The **user segment** has the potential to be made up of a variety of users, such as the aircraft industry, the military, farming businesses, and the auto and truck transportation industries. User segment receivers range from small hand-held units to complex systems used by the military. The FAA currently predicts that over 5000 aircraft will be equipped with GPS receivers by the mid-1990s. It also sees the GPS becoming the standard for instrument-guided flights and instrument approaches sometime in the future. The user segment must process data from four or more satellites. This can be done simultaneously or sequentially. The equipment consists of three essential components: the antenna and associated electronics, the receiver/processor unit, and the control display.

The GPS theory of operation is based on basic geometry. If you know the distance and location of three or more points, your exact location can be determined. The satellites transmit a location and timing signal to the user's receiver. The distance to a satellite is determined by measuring the travel time of the transmitted signal. Knowing the speed of radio wave propagation (the speed of light), the receiver calculates the distance to the satellite.

To better understand the theory of operation, study the example in Figure 15–41. Knowing the distance from one satellite (15,000 mi) places your aircraft on the outside of a sphere 15,000 mi from that satellite (Figure 15–41a). Knowing the distance from a second satellite (14,200 mi) places the aircraft at the intersection of the two spheres (Figure 15–41b). Figure 15–41c shows that receiving the distance from three satellites will locate the aircraft at one of two points along the outside of the three spheres. The measurement from the fourth satellite will determine the aircraft's exact location.

Doppler Navigation System

The **Doppler navigation system** is so named because it utilizes the **Doppler shift** principle. The Doppler shift is the difference in frequency that occurs between a radar signal emitted from an aircraft radar antenna and the signal returned to the aircraft. If the signal is sent forward from an aircraft in flight, the returning signal will be at a higher frequency than the signal emitted. The difference in the frequencies makes it possible to measure the speed and direction of movement of the aircraft; thus information is provided from which one can compute the exact position of the aircraft at all times with respect to a particular reference point and the selected course.

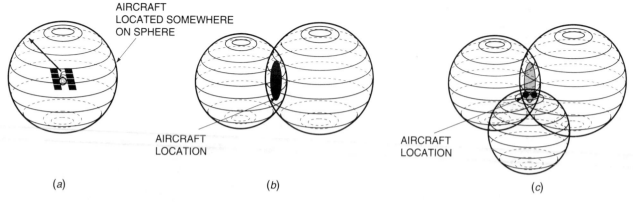

FIGURE 15–41 GPS theory of operation. (a) Aircraft location from one satellite, (b) aircraft location from two satellites, (c) aircraft location from three satellites.

In the Doppler navigation system, flight information is obtained by sending four radar beams of continuous-wave 8800-MHz energy from the aircraft to the ground and measuring the changes in frequencies of the energy returned to the aircraft. The change in frequency for any one beam signal is proportional to the speed of the aircraft in the direction of the beam. The radar beams are pointed forward and down at an angle of approximately 45° to the right and left of the center of the aircraft and rearward and down at a similar angle. When the airplane is flying with no drift, the forward signals will be equal. The rearward signals will be equal to the forward signals, but opposite in value. The difference between the frequencies of the forward and rearward signals will be proportional to the ground speed; hence this difference is used to compute the ground speed, and the value is displayed on the **Doppler indicator.** If the airplane drifts, there will be differences in frequency between the right and left beam signals, and these differences are translated into drift angle and displayed on the Doppler indicator. Figure 15–42 shows how the radar beams are aimed with respect to the aircraft.

The advantage of a Doppler system is that it is completely contained in the aircraft and requires no external signals. At the start of a flight, the course or courses to be flown are programmed into the system. Therefore, continuous information regarding the position of the aircraft will be displayed on the Doppler indicator and the computer controller.

Integrated Navigation and Flight Systems

When all or most of the conditions affecting the flight of an airplane are brought together and sensed by a system that is able to present information regarding the conditions to the pilot, the total system may be termed an integrated navigation and flight system, or simply an **integrated flight system.**

A completely integrated navigation and flight system includes flight instrumentation, navigation systems, communications systems, and the automatic flight system. The instrumentation associated with such an integrated system is shown in Figure 15–43. This is an arrangement for one particular airline on a large passenger aircraft. At the top of the panel are warning annunciators to indicate that a segment of the system is inoperative. The AFCS MODES annunciator shows in what mode the avionics flight control system is operating, that is, what signals and control elements are in effect at a given time. The upper row of instruments includes the clock, the Mach and airspeed indicator, the ADI, the decision height annunciator (DH) (which is a radio altimeter), and an altimeter. In the bottom row of instruments are the radarscope, a digital distance radio magnetic indicator, an HSI, a vertical-speed indicator, and, in the lower right corner, a backup altimeter. Adjacent to these instruments and located in the center panel are a total-air-temperature indicator, an altitude indicator, and a true-airspeed indicator. These instruments receive data from the central air data computer.

The instruments shown in Figure 15–43 are at the captain's station; however, a duplicate set of the instruments is at the first officer's station. Thus both pilots are continuously informed regarding all flight conditions and situations, regardless of whether the aircraft is being flown manually or in an automatic mode.

State-of-the-art integrated navigation and flight systems incorporate digital electronics and strapdown technologies. These systems, along with other autoflight systems, will be discussed in Chapter 17.

ATC Transponder

Because of the difficulty that flight controllers had in identifying aircraft on radarscopes in tower stations and control centers, radar devices called **ATC** (air traffic control)

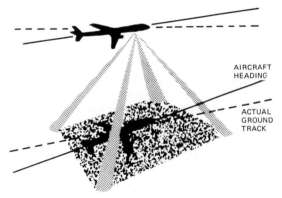

FIGURE 15–42 Diagram of doppler radar beams from an airborne transmitter.

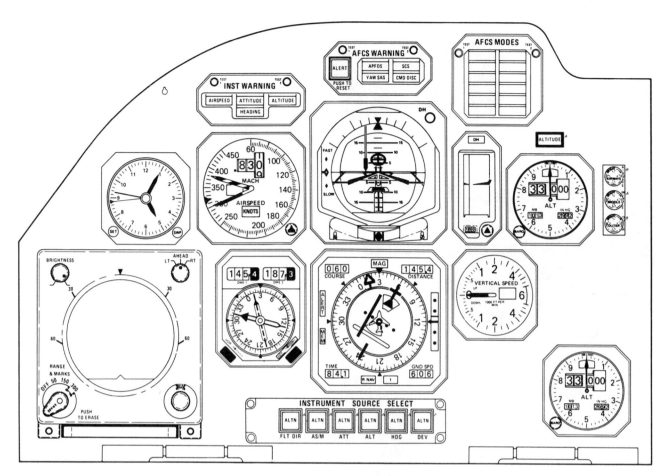

FIGURE 15–43 Instrumentation associated with an integrated flight system. *(Lockheed California Company)*

transponders were developed. In general, a transponder is an automatic receiver and transmitter that can receive a signal (be interrogated) from a ground station and then send a reply back to the station. For the purpose of this discussion, we shall consider the type of transponder that receives an interrogation from a ground radar station and sends a reply signal for identification. The reply shows on the radarscope as a double slash. A controller who wishes to obtain positive identification of an aircraft will request that an "INDENT" signal be returned from the transponder. The pilot of the aircraft will then press the IDENT button on the transponder control panel to send a special-image signal that the controller will recognize for identification.

A typical transponder used on a corporate or transport-category aircraft is shown in Figure 15–44. This system consists of a control unit, mounted on the flight deck; an ATC transponder r-t unit, located in the aircraft's equipment center; and the L-band antenna, mounted on the lower side of the fuselage.

In the OFF position of the function selector switch, all power is off and the transponder is inoperative. In the ON position, the switch places the unit in the mode for normal operation. The transponder is ready to reply to interrogations from a ground station after a 1-min warm-up period.

In the STBY position of the selector, the transponder power is turned on, and power is applied to the transmitting system. STBY is used at the request of the ground controller to selec-

tively clear the radarscope of traffic. Turning to STBY will keep the transponder from replying to interrogations but will allow instant return to the operating mode when the unit is switched to ON.

There are currently three modes (types) of transponders that can be used on various aircraft. **Mode A** is a transponder that provides a (nonaltitude-reporting) four-digit coded reply when interrogated by the ground-based ATC radar. The four-digit code is set by the pilot in accordance with ATC requests. **Mode C** is an airborne transponder that provides a coded reply identical with that of mode A; however, mode C also transmits an altitude-reporting signal. **Mode S** is a transponder with mode A and mode C capabilities, but it also responds to TCAS-equipped aircraft. TCAS is a collision avoidance system; it will be discussed in the next section of this chapter.

The ALT position of the selector switch activates **mode C,** the altitude-reporting capability of the transponder. When used with an **altitude digitizer** or an **encoding altimeter,** the unit will automatically transmit altitude information. The altitude is given as standard-pressure altitude, which is converted into real altitude by ground computers.

The TEST position of the selector switch is used to self-test the operation of the unit. It may be used at any time, as it does not interfere with the normal operation. When the selector switch is turned to TEST and held there, a test signal is developed to interrogate all internal circuitry of the transponder except the receiver. If the transponder is working in a normal

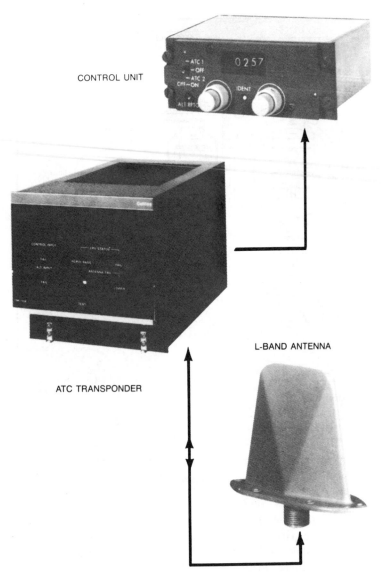

CONTROL UNIT

0257

L-BAND ANTENNA

ATC TRANSPONDER

FIGURE 15–44 Airbone components of a typical transponder for a transport category aircraft. *(Collins Divisions, Rockwell International)*

manner, the REPLY lamp will remain on as long as the switch is in the TEST position.

The REPLY lamp and PUSH IDENT button are contained within a single assembly. The REPLY lamp automatically goes on when the transponder is replying to ground interrogation or when the function selector switch is placed in the TEST position. The PUSH IDENT button is used to send the special position identification pulse (spip). When the pilot is asked by the ground controller for an "IDENT," the pilot presses the button and activates a special signal that "paints" an instantly identifiable and unmistakable image on the controller's radarscope. This signal must be used only when requested by the controller, because use at any other time could interfere with the spip being sent by another aircraft. If the ground radar system is so equipped, the transponder may activate a special code sequence on the controller's scope. This code may identify the aircraft's destination, altitude (mode C operation), airspeed, and identification number.

The **code selector** comprises four 8-position rotary switches providing a total of 4096 active settings available

for selection of the identification code. The code selector sets up the number of spacing of the pulses that are transmitted at the transponder frequency of 1090 MHz.

During operation the pilot must set in the transponder code requested by the ground controller. When the unit responds to an interrogation from the ground station, the REPLY lamp will light, thus telling the pilot that the code is correct.

All ATC transponders must be tested every 24 calendar months. This test is required under the FAA flight regulations; however, it is often performed during an annual inspection if requested by the owner. The test must be conducted by an authorized avionics repair facility.

Traffic Alert and Collision Avoidance System

One of the latest requirements for improving air safety is **TCAS (traffic alert and collision avoidance system).** This system works in conjunction with an aircraft's ATC transponder to inform the flight crew of aircraft that pose a potential midair collision threat. All TCAS-equipped aircraft must have a transponder capable of mode S operations.

The TCAS equipment for a transport category aircraft is shown in Figure 15–45. The major system components are (1) the receiver-transmitter-computer unit, (2) two display units, (3) one or two mode S transponders, (4) a system control panel, and (5) the required antennas.

TCAS Theory of Operation. The airborne TCAS receiver unit monitors the airspace around the aircraft for potential intruders (other aircraft). If the receiver identifies a potential threat, an aural and visual advisory is displayed to the flight crew. The TCAS receiver-transmitter (r-t) interrogates nearby transponders to identify potential threats. All aircraft with an operating mode A, C, or S transponder will be identified by the TCAS equipment. Non-transponder-equipped aircraft will not be identified. Once the r-t unit has received the signal of a nearby aircraft, the computer calculates the approaching aircraft's heading, relative speed, and altitude. If the computer predicts a threat, one of two types of warnings will be activated: **a traffic alert** or a **resolution advisory.** Figure 15–46 shows a typical display with both an active traffic alert and a resolution advisory.

Traffic alerts provide the flight crew with a relative bearing and distance to the intruding aircraft. Traffic alerts will be presented for all aircraft that are approximately 40 s or less from the **closest point of approach** (CPA). The CPA is a calculated minimum allowable separation between the aircraft. The traffic alert is designed to aid the flight crew in visual identification of the intruder; no avoidance maneuvers are displayed during traffic alerts.

A resolution advisory is displayed when the intruder is approximately 25 s from the CPA. The resolution advisory will offer either **corrective** or **preventative** maneuvers that will help to avoid a collision. A corrective maneuver display shows a vertical climb or descent symbol to direct the aircraft away from the intruder. A preventative maneuver display tells the pilot to restrict the aircraft's vertical climb or descent. Keep in mind that TCAS does not control the aircraft;

FIGURE 15–46 A typical TCAS display showing both resolution advisory and traffic alert. *(Collins Divisions, Rockwell International)*

it only offers suggested maneuvers to the flight crew. The pilot makes the decision as to the best corrective action.

If the intruding aircraft is also TCAS-equipped, the TCAS equipment of the other aircraft will transmit a maneuver-coordination message to the intruder through the mode S transponder. This message is designed to coordinate aircraft maneuvers and avoid having both pilots take the same corrective action.

Radar Altimeters

Radar altimeters were developed to give an accurate indication of aircraft altitude **above ground level** (AGL). These altimeters provide the AGL accuracy not found in conventional pressure-sensitive altimeters. The radar altimeter shown in

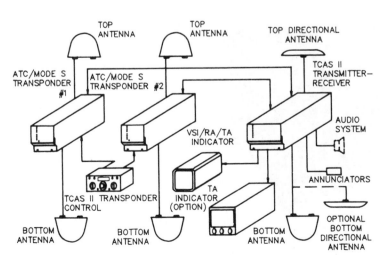

FIGURE 15–45 Components of a typical TCAS. *(Collins Divisions, Rockwell International)*

Figure 15–47 contains three units: the antenna, the receiver-transmitter, and the altitude indicator.

Radar altimeters obtain their accuracy by constantly measuring the aircraft's height above the ground. By transmitting a VHF signal downward from the aircraft and receiving the reflected signal, a radar altimeter can determine the aircraft's AGL altitude. The time required for the transmitted signal to reach the ground and return to the aircraft is measured by the receiver-transmitter unit. The unit performs the calculations needed to send the corresponding signal to the altitude indicator, located on the instrument panel. The indicator is typically calibrated in feet from 0 to 2500. A decision height (DH) light is often incorporated on the altitude indicator to aid the pilot during an instrument approach. If the decision height is properly set, then when the DH light illuminates, the pilot must decide either to proceed with the landing or to execute a missed approach.

Ground Proximity Warning Systems

A **ground proximity warning system** (GPWS) is used on many commercial aircraft to warn the pilot of an excessively low aircraft altitude. The GPWS consists of a "black box" computer that monitors several aircraft parameters. A typical GPWS computer interface is illustrated in Figure 15–48. This system monitors the radio altimeter, the air data computer, the instrument landing system, and both the landing gear and flap positions. The system's outputs include signals to advisory computers, audio warning systems, and various warning lights.

This system provides several modes of operation, depending on the aircraft's configuration. During a flight configuration (retracted gear, retracted flaps, and cruise airspeed), the system provides an early warning for terrain avoidance. During a land configuration, the system provides warning of excessive sink rate and a desensitized terrain avoidance. The system may also provide below-glide-slope or below-minimum-descent altitude warnings.

Integrated Avionics Processor System

The latest technology used to simplify the variety of avionics systems found on corporate aircraft is called the **integrated avionics processor system** (IAPS). This system is designed to function as a central distribution network for virtually all the avionics of an aircraft. The IAPS coordinates the weather radar (WXR); the instrument display unit (IDU); the flight management system (FMS); the flight control system (FCS); the radio sensor system (RSS), which may include all navigation and communication radios; the aircraft data acquisition system (ADAS); the air data system (ADS); and the attitude heading system (AHS).

The IAPS uses a digital data bus system to link all the avionic subsystems to the IAPS processor. The processor controls the outputs to the various flight deck displays and

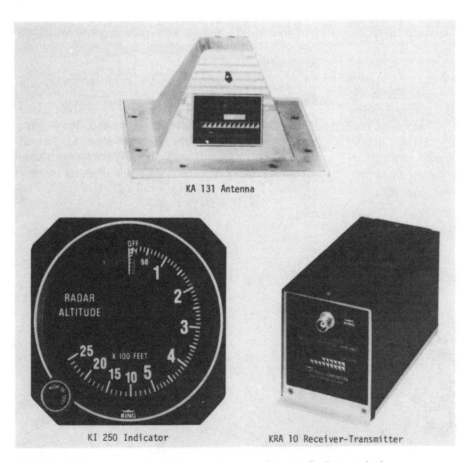

KA 131 Antenna

KI 250 Indicator

KRA 10 Receiver-Transmitter

FIGURE 15–47 Components of a radar altimeter. *(King Radio Corporation)*

Typical ARINC 723 GPWS Computer Interface

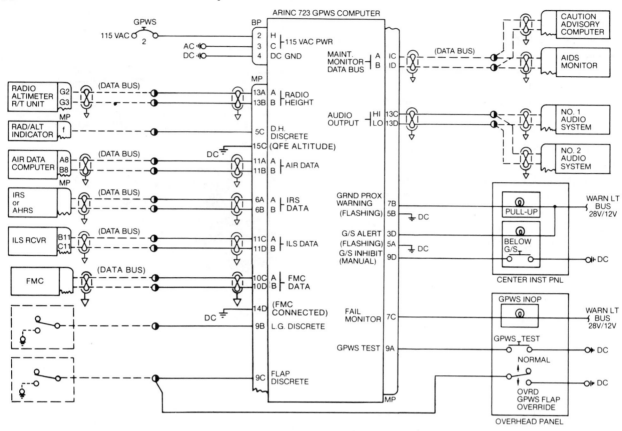

FIGURE 15–48 A ground proximity warning system computer interface. *(Sundstrand Corporation)*

monitors the system for defects. A central diagnostic system is used to record and troubleshoot faults within the avionic subsystems. This type of built-in test equipment greatly enhances troubleshooting of these relatively complex subsystems. The centralized design of the IAPS also reduces the number and size of subsystem components.

Radiotelephones

Radiotelephones are currently gaining popularity in business aircraft. It is often important that high-level executives keep in touch with their corporate activities. The radiotelephone provides the means for this communication. The system utilizes a ground-based operator, which receives the radiotelephone's transmitted signal and transfers it to a common telephone service. With this system, an airborne radiotelephone can place telephone calls anywhere in the world. Since VHF frequencies are used for transmission, range is limited to line of sight. A ground station within 200 mi of an aircraft's current location must be used for proper radiotelephone operations.

Emergency Locator Transmitter

An **emergency locator transmitter** (ELT), also referred to as a **locator beacon,** is required on aircraft to provide a signal

or signals that will enable search aircraft or ground stations to find aircraft that have made crash landings in remote or mountainous areas. Even though an ELT is not strictly a communications or navigation device, it has elements of both, and so it is described in this section.

A typical ELT consists of a self-contained dual-frequency radio transmitter and battery power supply with a suitable whip antenna. When armed, it will be activated by an impact force of 5*g* or more, as may be experienced in a crash landing. The ELT emits an omnidirectional signal on the international distress frequencies of 121.5 and 243.0 MHz. General-aviation aircraft, commercial aircraft, the FAA, and the CAP (Civil Air Patrol) monitor 121.5 MHz, and 243.0 MHz is monitored by the military services.

After a crash landing, the ELT will provide line-of-sight transmission up to 100 mi [161 km] at a receiver altitude of 10,000 ft [3050 m]. The ELT transmits on both distress frequencies simultaneously at 75-mW rated power output for 50 continuous hours in the temperature range of −4 to +131°F [−20 to +55°C].

The fixed ELT must be installed securely in the aircraft at a location where crash damage will be minimal. The location selected is usually in the area of the tail cone; however, in some cabin-type aircraft, the unit is installed in the aft, top part of the cabin. Access is provided in either case so the unit can be controlled manually.

Testing an ELT. The control panel on an ELT unit contains a switch with three positions: AUTO, OFF, and ON. The unit may be tested by tuning the VHF COMM receiver to 121.5 MHz and then placing the ELT switch in the ON position. The emergency tone will be heard if the ELT is operating. Immediately after the test, the switch should be returned to the AUTO position.

If an ELT is inadvertently turned on, owing to a lightning strike or an exceptionally hard landing, the control switch should be placed in the OFF position to stop the transmission, and then the switch should be returned to the AUTO position to arm the unit.

Testing of an ELT should be performed within the first 5 min of an hour, and only three pulses of the transmitter should be activated. For example, a test could be conducted between 1:00 p.m. and 1:05 p.m. with a maximum of three beeps being heard on a frequency of 121.5 MHz. Tests conducted in this manner do not alert the FAA to a crashed aircraft.

Service for an ELT. An ELT requires a minimum of service; however, certain procedures are necessary to assure satisfactory operation. The battery pack must be changed in accordance with the date stamped on the unit. Typically, the batteries are replaced every 2 years or after 20 min of continuous use. The replacement date must be clearly marked on the battery's data plate; otherwise, the battery is not airworthy. The ELT should be tested regularly to assure satisfactory operation. An inspection of the ELT mounting and antenna should be made periodically to ensure firm attachment to the aircraft.

Regulations regarding the operation of ELTs are set forth in FAR Part 91.52. Technicians involved with the installation and service of ELTs should be familiar with these regulations and manufacturer's data.

Flight Data and Voice Recorders

Since 1958 any commercial passenger aircraft flying in the United States has been required to be equipped with an automatic **flight data recording system.** Recently the FAA changed the regulations to include virtually all turboprop aircraft. Today **flight data recorders** are found on most corporate and all transport-category aircraft. The system must monitor both flight parameters and flight deck voice activities. Recorded flight parameters include the aircraft's altitude, airspeed, pitch attitude, roll attitude, magnetic heading, vertical acceleration, flap position, gear position, engine power, and Greenwich Mean Time. In some aircraft several other systems are also monitored. **Voice recorders** must monitor all flight compartment and communication radio conversations. A voice recorder and a flight data recorder are typically two separate units that operate completely independently of each other.

A flight data recorder is housed in a crushproof container located near the tail section of the aircraft. The tape unit is fire-resistant and contains a radio transmitter to help crash in-

FIGURE 15–49 A flight data recorder.   *(Sundstrand Corporation)*

vestigators locate the unit under water. A flight data recorder made by the Sundstrand Corporation is shown in Figure 15–49. A data recorder is typically used for accident investigation; however, some airlines also use recorded data to aid in troubleshooting recurring mechanical defects.

A modern flight data recorder uses a magnetic recording tape to store digital data of flight parameters of the past 25 flight hours. The recorder receives the majority of its input signals from existing sensors located throughout the aircraft. The information is sent to the recording unit, which stores up to 900 bits of information on 1 in. of tape. The tape unit (shown in Figure 15–50) employs two 4-channel record heads and two 4-channel erase heads. One record and one erase head are used when the tape travels from left to right. The other pair of heads is used when the tape travels from right to left. The two pairs of erase-record heads are set at different levels (tracks) along the tape, thus producing an 8-track, single-channel format. This format allows for 25 h of data storage at 3.125 h per track. The tape reversal and track switching are performed automatically when the tape reaches its limit of travel.

Cockpit voice recorders (CVRs) are very similar to flight data recorders; they look nearly identical and operate in almost the same way. CVRs monitor the last 30 min of flight deck conversations and radio communications. The flight deck conversations are recorded via the microphone monitor panel located on the flight deck (see Figure 15–51). This panel is also used to test the system and erase the tape if so desired. The erase mode of the CVR can be operated only after the aircraft has landed and the parking brake set. Playback is possible only after the recorder is removed from the aircraft.

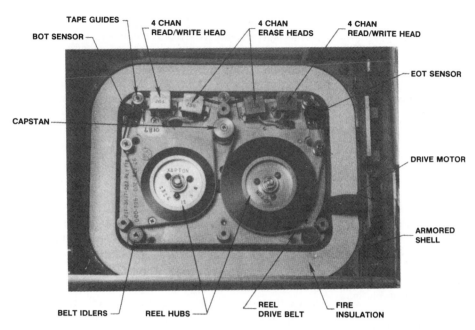

FIGURE 15–50 The tape unit from a flight data recorder. *(Sundstrand Corporation)*

FIGURE 15–51 Microphone monitor panel for a cockpit voice recorder. *(Sundstrand Corporation)*

INSTALLATION OF AVIONICS EQUIPMENT

Power for all avionics equipment is supplied by the aircraft's alternator or generator system during normal operating conditions. When additional equipment is installed, the person responsible must make sure that the aircraft's electrical system has sufficient capacity to supply all the aircraft's requirements.

Plans for mounting avionics equipment in aircraft should include careful consideration of location, strength of mounting structures, reduction of vibration and shock, bonding and shielding, and serviceability. Hazards to personnel and to the aircraft must be avoided, since high voltages are developed in some types of equipment and since some units may develop sufficient heat to ignite any particularly flammable material in the immediate vicinity. The manufacturers provide complete information for the installation of avionics equipment.

Avionics equipment, controls, and indicators should be located in the positions most convenient to those who must operate them; in light aircraft, the controls and indicators should be easily accessible to the pilot. Sufficient ventilation should be provided for equipment subject to heating so that it will not exceed its normal operating temperature. To avoid the danger of fire, equipment that naturally operates at high temperatures must be sufficiently removed from flammable materials.

The actual attachment of avionics-equipment units to the aircraft must be such that there is no danger of a unit's becoming loose because of vibration. Fastening devices include standard bolts, nuts, and screws with effective locking devices such as self-locking nuts, lock washers, safety wire, and cotter pins. Self-locking hold-down clamps and snap slides are hold-down devices specially designed for radio equipment.

Radio units in light aircraft may be mounted on brackets attached to the rear of shock-mounted instrument panels, or they may be secured on shock-mounted brackets or racks attached to a solid structure of the airplane. In any event, shock mountings must be placed between the actual radio equipment and the basic aircraft structure. In some cases shock-mounting bases designed especially to fit particular units are attached directly to the airplane. One type of shock mounting is shown in Figure 15–52.

Because shock mountings utilize rubber, synthetic rubber, plastic, or some other insulating material as the shock-absorbing agent, it is essential that grounding or bonding jumpers be connected from the aircraft structure to the avionics-unit case. These serve as a part of the ground circuit for the equipment and also help to reduce noise from static and

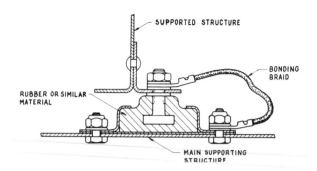

FIGURE 15–52 A shock-mounting diagram.

other types of interference. Bonding and shielding information is given in Chapter 4.

In the installation of shock-mounted units, adequate clearance must be provided to prevent any contact between the mounted unit and the adjacent structure under conditions of violent shock or vibration. Electric cables and control cables connected or attached to radio equipment must have sufficient play and must be mounted in such a manner that the vibration and sway of the radio equipment will have no adverse effect on them or cause undue wear. The strength of the mounting for radio equipment should be at least such that the equipment can withstand the ultimate accelerations for which the airplane structure is designed.

Avionics mounting racks are usually designed to ARINC (Aeronautical Radio Incorporated) standards, and equipment cases are designed to fit such racks. This is particularly true for large commercial aircraft; however, mounting racks and avionics equipment for smaller aircraft are also being designed according to ARINC standards. Technicians installing avionics equipment should ensure that equipment and racks are compatible.

ANTENNAS

The performance of radio systems on aircraft is profoundly affected by the design and placement of antennas. This is particularly true of antennas for transmitters, since the antenna system is a tuned circuit, and its ability to radiate energy into space is determined by its length in relation to the frequency to be transmitted. In general, the higher the frequency, the shorter the antenna. In practice it is possible to adjust the length of an antenna electronically by means of an inductance coil in series with the antenna. The inductance coil is provided with taps or some other means of adjustment to vary the impedance of the antenna. As radio systems become more sophisticated, antenna lengths become less critical; that is, a sophisticated low-frequency radio may operate perfectly with a relatively short antenna. On this type of system the effective length of the antenna is changed by means of electronic circuitry.

It is common practice to use one antenna for both transmitting and receiving if the radio equipment is to be used only for communications, provided that the length of the antenna is such that it will accommodate the frequencies to be transmitted and received. When a single antenna is used, it is normally connected to the receiver and switched to the transmitter for broadcasting by means of a relay and a push-to-talk switch on the microphone. The unit that performs the connection-disconnection function is called a **duplexer.**

Navigation and communications antennas are manufactured in many sizes and shapes, depending on their particular functions. As explained earlier, the length or size of an antenna is determined by the frequency range in which it is intended to operate. Special designs such as loops and dipoles are used for certain types of signals and provide directional references. A few typical antennas are shown in Figure 15–53. The antennas numbered 1 and 2 are designed to receive VOR navigation signals to provide bearing information. An antenna coupler is shown between the two sections of number 2. The antennas identified by the numbers 3, 4, and 5 are VHF communications antennas. Number 6 is a DME and a transponder antenna, and number 7 is a marker-beacon antenna. The mounting locations of these antennas are shown

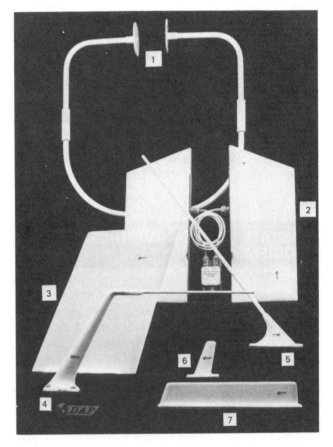

FIGURE 15–53 Typical avionics antennas. *(Dayton Aircraft Products)*

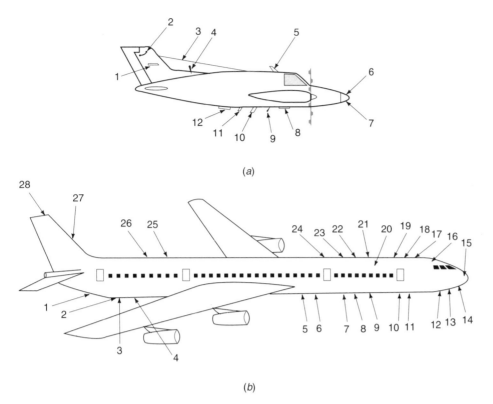

(a)

(b)

FIGURE 15–54 Typical locations for mounting antennas. (*a*) For light aircraft. (1) RNAV; (2) VOR, localizer; (3) ADF sense; (4) ELT; (5) VHF Com 1; (6) glide slope; (7) radar; (8) marker beacon; (9) transponder; (10) VHF Com 2; (11) DME; (12) ADF loop. (*b*) For transport-category aircraft. (1) MLS; (2) radio altimeter 2; (3) radio altimeter 1; (4) VHF2; (5) VHF3; (6) marker beacon; (7) DME2; (8) DME1; (9) ATC1; (10) TCAS; (11) airfone; (12) ILS/glide slope; (13) MLS; (14) ILS/glide slope; (15) weather radar; (16) MLS; (17) TCAS; (18) VHF1; (19) ATC(S); (20) SATCOM high-gain side mount; (21) SATCOM low gain; (22) GPS1; (23) GPS2; (24) SATCOM high-gain top mount; (25) ADF1; (26) ADF2; (27) HF (dual); (28) VOR (dual).

in Figure 15–54. This illustration represents typical locations; the exact locations often depend on the aircraft and radio equipment used.

Antennas for Low and Medium Frequencies

The antennas for low and medium frequencies are greater in length than those for the higher frequencies. Nondirectional beacons (NDBs) operate in these ranges. Antennas for these frequencies may be T, L, or V types mounted on the top or the bottom of the fuselage (see Figure 15–55). Clearance between the antenna and the fuselage or other structures should not be less than 1 ft [30 cm] in any case, and the main leg of the antenna should not be less than 6 ft [183 cm] in length. A whip antenna may be used provided that tests show satisfactory performance with such an installation.

Many modern low- and medium-frequency radios operate using a shorter antenna than those shown in Figure 15–55. The shorter antenna element may be secured behind a non-metallic airframe component, such as the vertical stabilizer's leading edge. The effective length of the antenna is then adjusted by means of a variable inductor. Flush-mounted antennas are used whenever possible on high-speed aircraft in order to reduce induced drag (see Figure 15–56).

Mast and Whip Antennas

When a **mast** or **whip** antenna is installed on the fuselage or any other part of an aircraft structure, it is necessary to make sure that the structure of the airplane is sufficiently strong to

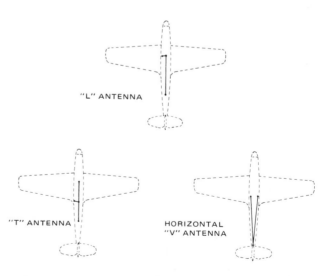

"L" ANTENNA

"T" ANTENNA

HORIZONTAL
"V" ANTENNA

FIGURE 15–55 Antenna installations.

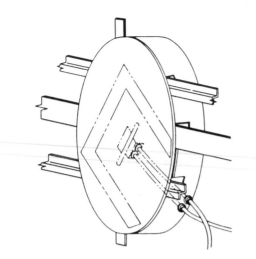

FIGURE 15–56 Example of a flush-mounted antenna.

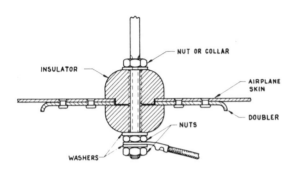

FIGURE 15–57 One type of mounting for a whip antenna.

support the unit under all conditions of shock, vibration, or continued movement. This often requires that a doubler plate be installed around the point at which the antenna is attached, as shown in Figure 15–57.

In general, a mast antenna is one that is housed in an aerodynamically shaped fiberglass or plastic cover. A whip antenna is a simple metal element approximately $1/8$ in. in diameter. The streamlined mast antenna produces less drag and offers more rigidity than a whip antenna; therefore, high-speed aircraft typically utilize mast antennas.

High-Frequency Antennas

As explained in Chapter 14, the wavelength of any radio wave may be found by using the proper equation. For example, if it is desired to find the wavelength of a 100-MHz wave, we find that

$$\lambda = \frac{300,000,000}{100,000,000} = 3 \text{ m} \quad \text{or} \quad 9.84 \text{ ft}$$

The length of an antenna for HF transmission must be determined as an exact fraction of the wavelength of the signal to be transmitted. For example, when a hertz antenna such as that shown in Figure 15–58 is to be used, the total length of the antenna should be one-half the wavelength. Hence if we wish to find the proper length of the hertz antenna in feet, we use the following equation:

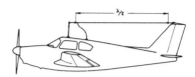

FIGURE 15–58 Installation of a hertz antenna.

$$\text{Antenna length, ft} = \frac{300,000,000 \times 3.28}{2 \times f}$$

where 3.28 = conversion factor for changing meters to feet and f = frequency of the transmitter in hertz.

For a Marconi antenna the extended section should be one-quarter the wavelength. The correct antenna length may be determined by using the foregoing equation and dividing the result by 2 or by making the denominator of the fraction $4 \times f$ instead of $2 \times f$.

It has been found in actual practice that the current in an antenna travels at about 5 percent less velocity than it does in free air; hence the actual length of an antenna should be about 5 percent less than that computed with the foregoing equations. In view of this fact, we may use a simplified equation as follows to determine the actual length of a half-wave antenna:

$$\text{Length, ft} = \frac{468}{f}$$

where f = frequency in megahertz and 468 = 95 percent of 492, which is half the number of millions of feet in 300,000,000 m.

We can use this process to determine the length of a typical VHF communications antenna, with f = 126.5 MHz (an average VHF communications frequency):

$$\begin{aligned}
\text{Length, ft} &= \frac{468}{f} \\
&= \frac{468}{126.5} \\
&= 3.69 \text{ ft}
\end{aligned}$$

The length of a half-wave antenna for a frequency of 126.5 MHz should be 3.69 ft. Most aircraft communications radios employ quarter-wave antennas. To obtain the length of a quarter-wave antenna, we must divide the half-wave length by 2:

$$\begin{aligned}
\tfrac{1}{4}\lambda &= \frac{\tfrac{1}{2}\lambda}{2} \\
&= \frac{3.69}{2} \\
&= 1.8 \text{ ft}
\end{aligned}$$

where λ = wavelength. A typical quarter-wave VHF communications antenna would be approximately 1.8 ft in length.

Antennas for VHF Navigation Equipment

Particular attention must be given to the proper location of antennas that operate at very high frequencies for navigation

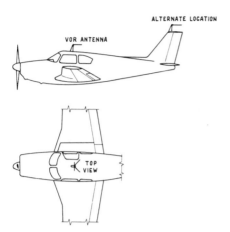

FIGURE 15–59 Installation of a VOR-localizer antenna.

equipment. For example, the horizontal V antenna for VOR and localizer is most effective on small airplanes when mounted over the forward part of the cabin. The apex of the V should be pointing forward, and the plane of the V should be horizontal when the airplane is in level flight. Figure 15–59 shows suitable locations for a VOR-localizer antenna.

The glide slope antenna is usually a small dipole mounted on the forward part of the airplane, but it is sometimes mounted on the same mast as the VOR-localizer V antenna.

The installation of any antenna for any purpose should follow closely the instructions of either the radio manufacturer or the airplane manufacturer. All approved installations have been carefully engineered for best performance, and any deviation from an approved installation is likely to give inferior performance.

REVIEW QUESTIONS

1. List subsystems that may be included in a complete navigation system for an aircraft.
2. What is the frequency range employed for air traffic control communications?
3. Why was ARINC established?
4. For what purposes are VHF communications employed?
5. What frequency range is utilized for VHF communication systems?
6. Describe a typical VHF communication transceiver used in light aircraft.
7. What is the difference between the USE and STANDBY displays in a VHF COMM transceiver?
8. Describe antennas for VHF systems.
9. What is the function of a Selcal decoder?
10. Describe the AIRCOM system.
11. Describe ACARS.
12. What is the advantage of SATCOM communications?
13. Define the FCC responsibilities.

14. Describe the procedure for testing a communications radio for light aircraft.
15. Describe the operation of an automatic direction-finder (ADF) system.
16. What device enables an ADF receiver to determine direction?
17. Describe a radio magnetic indicator (RMI).
18. What type of signal is radiated by a VHF omnirange ground station?
19. What is the range of the VOR frequencies?
20. Describe a course-deviation indicator (CDI).
21. When must a VOR accuracy test be performed?
22. What navigational reference is generated by an instrument landing system (ILS)?
23. Briefly describe the operation of a localizer.
24. Compare the glide slope function of an ILS with the function of a localizer.
25. Explain the principle of distance-measuring equipment (DME).
26. What is an important consideration with respect to a DME antenna installation?
27. What is the meaning of the term *TACAN*?
28. What is the purpose of a marker-beacon system?
29. What frequency is utilized by a marker-beacon system?
30. What are the indicator lights associated with a marker beacon?
31. Compare the microwave landing system (MLS) with the ILS.
32. What are the advantages of an MLS?
33. Describe the principle of operation of an MLS.
34. What is the purpose of a Radio Tuning System and on what type of aircraft is it typically found?
35. Explain how a LORAN system establishes location.
36. Explain the principle of the VLF/OMEGA navigation system.
37. What is the possible range of a VLF navigation station?
38. Explain the principle of an inertial navigation system (INS).
39. What are the three basic laws of physics that make an INS operation possible?
40. Describe the function of an accelerometer.
41. What is the purpose of a gimbal platform?
42. Describe the theory of operation of a global positioning system (GPS).
43. What is the function of an ATC transponder?
44. Describe the operation of TCAS.
45. What is the value of an encoding altimeter when it is used with an ATC transponder?
46. What function does a radar altimeter perform?
47. Explain the advantages of a radar altimeter.
48. What is the purpose of a ground proximity warning system?
49. What is an integrated avionics processor system (IAPS)?
50. What is the purpose of an emergency locator transmitter (ELT)?
51. What frequencies are used by an ELT?

52. How can an ELT be tested?

53. What service procedures should be performed on an ELT?

54. What type of aircraft is required to be equipped with a flight data recorder?

55. Describe the operation of a typical flight data recording system.

56. What is the purpose of the different recording tracks on the tape of a flight data recorder?

57. What is a cockpit voice recorder?

58. What is the primary factor in establishing the size or length of a radio antenna?

59. What considerations must be taken into account when installing a mast or whip antenna on an aircraft?

Weather Warning Systems 16

Since weather conditions along a flight route are critical to passenger safety, means are needed to enable the pilot to "look" ahead to see if dangerous weather exists. One of the principal means for accomplishing this purpose is **radar.** Another system developed to determine weather conditions well ahead of an airplane is called a **weather mapping system.** Weather mapping systems detect the electrical activity caused by storm conditions and display the information on a cathode-ray tube or liquid crystal display screen.

It is the purpose of this chapter to examine both radar and weather mapping systems. The theory of operation, component architecture and system maintenance will all be examined.

RADAR

The word *radar* is derived from the expression **radio detecting and ranging.** Radar equipment was developed to a high level of performance by Great Britain and the United States during World War II for the detection of enemy aircraft and surface vessels. By the mid-1950s radar had found a home in commercial aircraft. The air-carrier industry quickly recognized the advantages of early weather detection using radar. Today, among numerous other functions, radar is used for weather mapping, terrain mapping, and air traffic control.

Radar systems have been developed for all types of aircraft, from single-engine light aircraft to large transport-category aircraft. Early radar systems were heavy and cumbersome and included many separate units. The extensive use of solid-state devices and microelectronics has brought about the development of small, compact systems. Radar systems operate on an echo principle: high-energy radio waves in pulse form are directed in a beam toward a reflecting target. The beam of pulses is actually something like a stream of bullets from a machine gun, with a relatively long space between each pulse of energy. When a pulse of energy strikes a target, which may be a mountain, rain clouds, or an airplane, a portion of the pulse is reflected back to the receiving section of the radar system (see Figure 16–1).

In Figure 16–1 at *A* a pulse has just been emitted from the airplane's radar antenna. At this point a "pip" appears on the radarscope (cathode-ray tube). At *C* the pulse is striking a rain cloud. A portion of the pulse is reflected by the cloud and returns toward the airplane, as shown at *D*. When the re-flected pulse reaches the airplane, at *E*, a second, smaller pip appears on the radar screen. The time between the two pips indicates the distance from the airplane to the cloud. In Figure 16–1 the time between the two pips is shown as 620 microseconds [μs], which represents a distance of approximately 50 mi [80 km]. At *F* another pulse is emitted from the radar antenna.

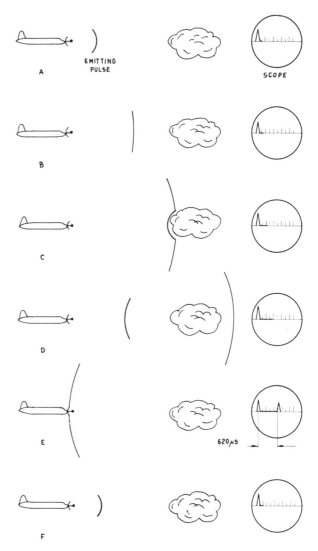

FIGURE 16–1 Radar pulse transmission and reflection.

Nature of Radar Signals

A typical radar signal may consist of a carrier wave of 8000 MHz broken into pulses with a duration of 1 μs and spaced at intervals of 1/400 s or 2500 μs. This yields a ratio of roughly 2500:1 for the time of **no signal** to the time of **signal.** It must be pointed out that the ratio of the length of a pulse to the time of no signal varies considerably with the frequency, which ranges from 1000 to 26,500 MHz. The various bands are listed in Table 16–1.

The length of the pulses of a radar signal may vary from 0.25 to 50 μs, depending on the requirements of the system. The pulse repetition frequency also varies according to the distance over which the signals must travel. For very long distances, the pulse rate must be slow enough so that the return signal will be received before another pulse is transmitted. If this were not accomplished, it would be difficult to tell whether the pulse shown on the viewing screen (CRT) was the one transmitted or the one received.

The use of a pulse system in radar makes it possible to transmit very powerful pulses. In effect, all the power is concentrated in the very short bursts. If the average power output of a transmitter is 10 W, the pulse power may be as high as 25,000 W.

In early types of radar systems, the display on the CRT was a horizontal scale and was called an A scan. The time between the transmitted pulse and the received pulse indicated the distance of the target from the transmitter. With this type of scan, the direction of the target could not be determined except by noting the direction in which the antenna was pointed. To enable the radar to provide direction information, the P scan was developed.

The P scan is illustrated in Figure 16–2. This type of radarscope may be called the **plan position indicator** (PPI), since it indicates both the distance and direction (**azimuth**) of the target. On the face of the PPI, the time trace starts at the same time that a pulse is transmitted from the radar antenna, and the reflected pulses cause bright spots along the trace line. The trace line is adjusted so that its intensity is very light or almost invisible except at the point where a target signal is received. The pulses are generated at such a frequency that the trace lines scan the entire face of the scope as the antenna makes a complete revolution; hence as reflected signals appear on the screen, a picture appears in a shape similar to that of the object that reflects the signal. The fluorescent coating inside the face of the CRT is of a type that retains a fluorescent glow for several seconds after being activated by the electron beam. Thus the picture remains on the screen and is reactivated each time that the time trace makes a complete circle.

Operation of the PPI

The operation of a PPI is illustrated in Figure 16–3. The antenna rotates at 1 revolution per second (rps), more or less, as it searches a 360° area, and the time trace on the scope rotates at precisely the same rate. The time trace is rotated by means of a synchronizer circuit that electronically "rotates" the electron beam that forms the display on the CRT.

Normally, when the PPI radar is installed, the time trace will be vertical from the center of the scope to the top edge when the antenna is pointing due north. When the antenna is turned to the right, the time trace will be to the right, or due east (see Z in Figure 16–3c).

Let us assume that the PPI in Figure 16–3 is installed so the antenna is pointing north. A pulse is transmitted, and a bright spot appears at x. If the length of the trace from the center of the scope to the edge represents a distance of 30 mi [48.3 km], the spot appearing at x indicates that the target (reflecting object) is directly north and at a distance of 20 mi [32.2 km]. Each pulse and each trace will continue to show the object for as long as the transmitted pulses strike it. This will add to the picture, providing an indication of the size of the target as well as its direction.

As the antenna rotates clockwise, the CRT sweep also rotates clockwise. Figure 16–3b shows the electron-beam sweep as the antenna points slightly to the right of north; Figure 16–3c shows the sweep after the antenna has rotated 90° clockwise. Aircraft weather radar systems are limited in the amount their antennas can rotate. Most airborne systems scan an area approximately 60° to the right and left of the aircraft's flight path.

Principal Units of Analog Radar Systems

Analog radars were the first type of radar system developed for aircraft. Many of these systems are still in use today; however, newer systems operate using the advancements of digital circuits. The operating characteristics of an analog system should be studied to help gain an understanding of basic radar theory. Digital radar systems are discussed in the next major section of this chapter.

Figure 16–4 is a simplified block diagram of an analog radar system. This system consists of several principal units,

TABLE 16–1 Various radar bands.

Approximate Wavelength	Identification Letter	Frequency, MHz
30 cm	L	1000 – 2000
10 cm	S	2000 – 4000
5.6 cm	C	4000 – 8000
3 cm	X	8000 – 12,500
1 cm	K	18,000 – 26,5000

FIGURE 16–2 Type-P scan.

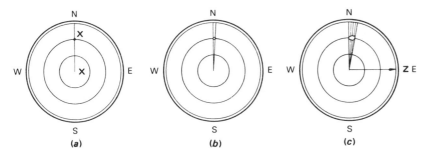

FIGURE 16-3 Operation of a PPI.

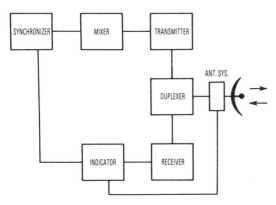

FIGURE 16-4 Block diagram of an analog radar system.

each serving a particular purpose in the system. The **synchronizer** provides the timing for the radar signal and synchronizes the transmitter, receiver, and indicator so that all operate together with correct timing. The timing and synchronizing are accomplished by trigger pulses generated in the synchronizer. These pulses originate in a multivibrator or similar pulse-generating circuit.

The **mixer** stores energy and supplies high-voltage pulses, which are released by the trigger pulse from the synchronizer. During the interval between pulses, a network consisting of inductors and capacitors is charged to a high level. When the trigger pulse releases this energy, a high-voltage pulse is delivered to the **transmitter.**

The principal element of a radar transmitter is a **magnetron tube.** This tube receives the high-energy pulse from the mixer and converts it into an extremely high-frequency pulse, which is sent on to the antenna system to be radiated into space. The magnetron makes use of **resonant cavities** to generate the correct frequency for the transmitted pulse.

When a pulse of high-voltage electric energy is delivered to the magnetron cavities, a fixed-frequency radio wave is generated (5400 MHz in the case of a typical weather radar system).

The UHF pulse from the transmitter is carried by means of a waveguide section to the duplexer and from the duplexer through a waveguide to the antenna. The **duplexer** is an electronic switching device that alternately connects the transmitter and receiver to the antenna. When the pulse is emitted from the transmitter, it is electronically blocked from entering the waveguide to the receiver. As soon as the transmitter pulse ends, the receiver line is opened to receive the reflected

pulse. At this time the path to the transmitter is blocked so that the pulse being received cannot go to the magnetron tube.

The **antenna system** can be compared to a searchlight that rotates to search a particular area with a light beam. The antenna assembly of the radar system, shown in Figure 16–5, is a rather complex device, largely because of its rotating and tilting mechanisms. The microwave RF signal is transmitted through a variety of waveguide sections and joints from the duplexer to the antenna reflector, and the reflected signal is returned through the same system. In each case the energy must be carried with a minimum of loss.

The upper part of the antenna assembly shown in Figure 16–5 is attached to the airplane structure and is called the **antenna base assembly.** All the power for rotation and tilt is brought into the base assembly by means of a plug connector. RF energy enters through the waveguide at the rear and passes down through the antenna frame assembly, which holds the reflector. In the base assembly are the azimuth drive motor and gear train, the synchro generator, and associated circuitry. Electric power is conducted from the stationary base to the rotating antenna by means of slip rings and brushes.

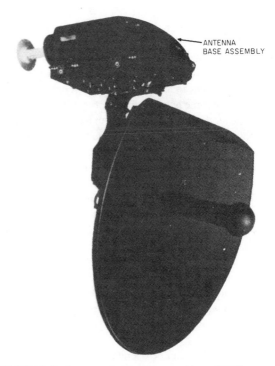

FIGURE 16-5 Concaved antenna assembly. *(RCA)*

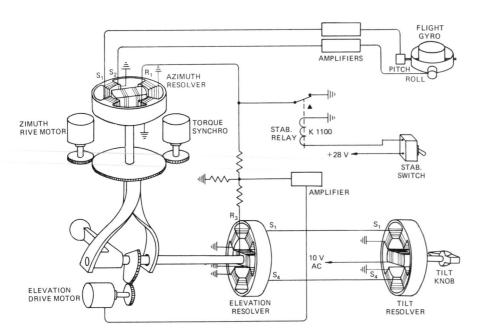

FIGURE 16–6 Antenna stabilization system.

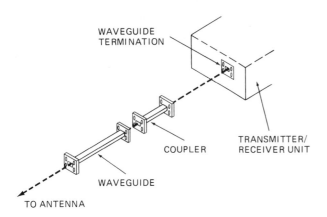

FIGURE 16–7 Typical waveguide.

A simplified diagram of the **stabilization** system for the antenna is shown in Figure 16–6. The purpose of the stabilization system is to enable the radar equipment to scan continuously in the same horizontal plane regardless of the pitch or roll of the airplane. From the diagram it can be seen that the pitch-and-roll signals originate with the flight gyro, which is a part of the automatic-pilot system. These signals are amplified and sent to the azimuth resolver. This resolver is connected through a resistor network to the elevation resolver and the tilt resolver. These resolvers furnish information to a computer system that continuously solves the equation necessary to provide instructions for the elevation drive motor, which performs the work of stabilization. The elevation drive motor actually positions the antenna reflector to provide a continuous horizontal sweep, or rotation, of the antenna beam. The purpose of the tilt resolver is to inject a signal calling for a downward angle of sweep as desired by the operator. The tilt resolver is controlled by the TILT knob on the radar control panel.

The radar system can be operated either with or without antenna stabilization by means of the STABILIZER switch on the control panel. When the switch is placed in the stabilization position, the antenna becomes stabilized.

In a radar system the RF energy pulses travel along the waveguide, which terminates in a **feed element.** The feed element reflects the pulses from the transmitter backward to the parabolic reflector of the antenna, which forms the pulses into a beam and radiates them into free space.

A **waveguide** is a hollow metallic tube, typically rectangular in shape, used to direct the UHF radar waves to and from the antenna. Figure 16–7 shows a typical waveguide section. The receiver-transmitter unit connects to the waveguide at the terminal point. Any transmitted signal travels from the transmitter through the waveguide "plumbing" to the antenna. The UHF signal travels through the waveguide in a manner similar to the way a sound wave travels through a hollow pipe. Since the radar's signals cannot escape through the sides of the waveguide, they simply travel to the end, where they are radiated into the air through the antenna (see Figure 16–8).

The **receiver** section of the radar system is rather complex in that it must receive the reflected waves and prepare them to give a correct indication at the CRT scope. This involves many electronic operations. The first step is to change the frequency of the incoming signal to a level that can be handled conveniently by a standard electronic system. The change in frequency is accomplished by means of a superheterodyne system, which produces an intermediate frequency.

The incoming radar frequency is typically near 5400 MHz, and the intermediate frequency is approximately 60 MHz. The 60-MHz intermediate frequency is amplified, detected, and sent to the video amplifiers. The signal output from the video amplifiers is sent to the radar CRT indicator.

The **indicator** is a CRT together with necessary operating circuitry. The video signal must be applied to the vertical deflection circuit, and a sweep signal must be applied to the horizontal deflection circuit.

It is common practice to provide **range markers** on the indicator. These range markers are concentric circles spaced

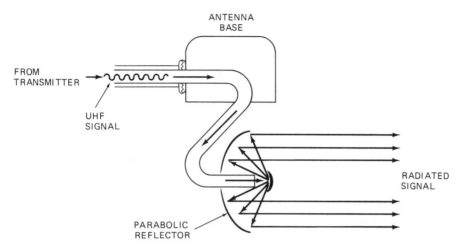

FIGURE 16–8 Radiation of a UHF signal from a parabolic radar antenna.

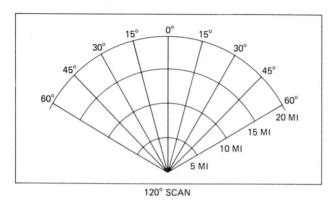

FIGURE 16–9 PPI for an oscillating radar antenna.

equally from the center of the screen, and they make it possible for the operator to determine accurately the range (distance) of the target from the airplane in which the radar is installed.

For radar systems where the installation makes it impossible to rotate the antenna through a complete circle, the antenna is rotated back and forth through a given arc. This situation exists where the antenna is installed in the nose of an aircraft. In this case the antenna may oscillate through an arc of 90 to 240°, depending on the particular installation. This usually gives adequate coverage for weather radar systems because it is necessary to scan the area ahead and to the right and left of the flight path. The diagram of an indicator for this type of installation is shown in Figure 16–9.

DIGITAL AIRBORNE WEATHER RADAR SYSTEMS

As mentioned earlier in this chapter, the extensive use of solid-state circuitry and microelectronics has led to the development of complete radar systems that weigh but a fraction of what earlier systems weighed. One such system is shown in Figure 16–10. This system incorporates circuitry

that allows for weather viewing, terrain mapping, navigational profiles, or checklist displays, as shown in Figure 16–11. A navigational profile may overlay the weather display, which allows the pilot to determine if the plotted course will intersect or avoid the weather activity. Terrain mapping is available that allows the pilot to receive detailed ground profiles when adjusted to the appropriate mode. The checklist display feature allows the pilot to store up to 16 pages of the aircraft's checklist within the radar unit memory. The pilot can retrieve the checklist whenever needed, thus eliminating the need for a less convenient paper checklist.

Color Weather Radar

Color added to weather radar displays enhances the storm activity image and provides a more effective means to detect severe weather. A green color indicates level 1 storm activity; yellow indicates moderate rain as in level 2; and red indicates level 3, or heavy rain. The indicator for such a system is shown in Figure 16–12.

Color radar equipment enables the pilot to more effectively identify the intensity of a storm. It is this sole

FIGURE 16–10 A modern solid-state color weather radar display and control panel.

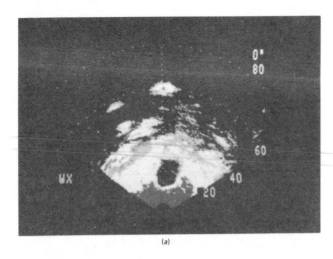

(a)

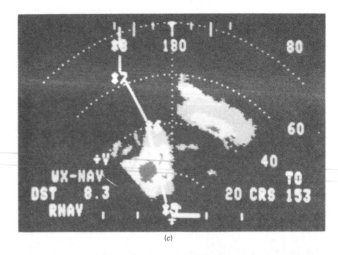

(c)

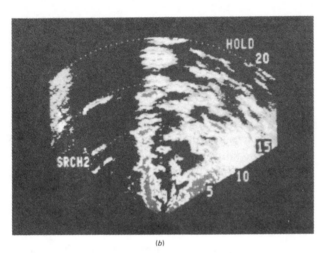

(b)

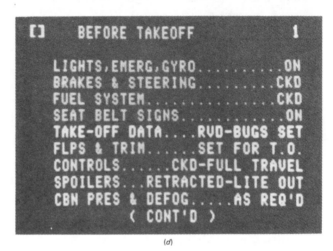

(d)

FIGURE 16–11 Four modes of a color radar. (a) Weather display; (b) terrain mapping, search 2 mode; (c) weather display with navigational overlay; (d) electronic checklist. *(Bendix Avionics Division)*

advantage that has made color radar the standard of the industry. Black-and-white radar systems are still used in limited quantity and typically on smaller aircraft. Virtually every corporate or transport-category aircraft currently employs a color radar system.

Weather Radar Frequencies

Weather radar systems are designed to detect the presence of rain, thunderstorms, and violent air turbulence. This is accomplished by utilizing frequencies that reflect most easily from these conditions. Weather radar should *not* be used as a proximity warning device for other aircraft or for collision avoidance. For these purposes, a different type of design is required for dependable results.

Airborne weather radar typically operates at one of two frequencies, **C-band** or **X-band.** The C-band range is from 4000 to 8000 MHz, and the X-band range is from 8000 to 12,000 MHz. The C-band radars are typically used in situations where the pilot needs to fly through narrow passages between thunderstorms. C-band frequencies tend to pass through some of the precipitation that would reflect X-band energy. Since the C-band frequencies do not reflect, they

cannot "see" some storm activity. The resolution of C-band radar is therefore poor compared with that of X-band units. However, this characteristic allows the C-band to penetrate precipitation and identify targets (storms) behind the initial rainfall. This allows the pilot to determine if passing through the initial rain will bring clear skies or more storms. The **resolution** of a radar system defines its ability to accurately display the various levels of storm activity scanned by the radar.

X-band radars have good resolution; however, X-band will not provide reliable coverage behind precipitation. This is because very little of the X-band energy will pass through the precipitation first encountered by the transmitted wave. X-band radars are typically used by pilots who use radar as a weather avoidance tool.

Figure 16–13 illustrates the difference between the transmitted signals of typical X- and C-band radars. Here it is easy to see that the X-band wave propagation pattern is more concentrated than that of the C-band. This characteristic helps to determine the type of reflected signal and uses of the radar systems.

A typical airborne weather radar system consists of a receiver-transmitter, one or two indicators, a control panel, and an antenna system (Figure 16–14).

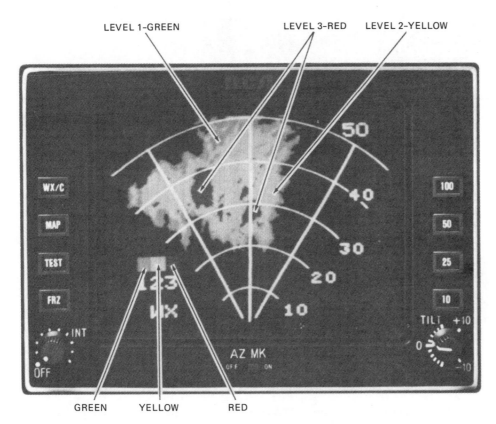

FIGURE 16–12 Indicator for a color radar system.   *(RCA)*

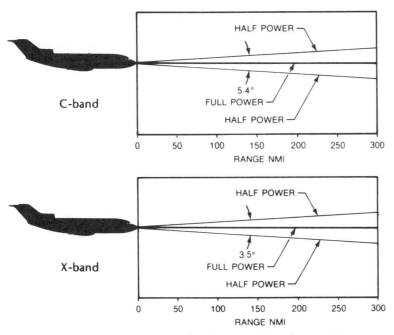

FIGURE 16–13 Radar C- and X-band radar transmission characteristics.
(Collins Divisions, Rockwell International)

The Receiver-Transmitter

A **receiver-transmitter (r-t)** for a typical airborne weather radar system is divided into three basic sections: the receiver, the transmitter, and the data processor. The transmitter section is designed to produce pulses at C- or X-band frequencies and transmit that energy to the antenna section at the correct time. Most modern radars use a completely solid-state digital interface for communication between the various sections of the radar system. A block diagram of a typical transport-category aircraft weather radar system is shown in Figure 16–15.

The **transmitter** generates a stable intermediate frequency (IF) using a crystal-controlled oscillator. The IF is translated up to the X- or C-band through a varactor

FIGURE 16–14 Typical components of an aircraft weather radar.   *(Collins Divisions, Rockwell International)*

multiplexer. Several steps may be required to increase the signal to the correct frequency for transmission. Another type of system uses a special diode circuit that generates the transmit frequency immediately. In either case the transmit signal is modulated, and the pulse timing is controlled by a synchronizer circuit.

The **pulse width** of the transmitted signal determines the radar's ability to resolve targets at various distances. To understand this phenomenon, consider that the transmitted signal is a burst of energy followed by a relatively long period of no energy, followed by another burst. As shown in Figure 16–16, this cycle repeats for as long as the radar is transmitting. Assuming that an energy burst lasts for 3.9 μs, this would make the burst of energy (pulse width) 1917 ft long traveling through the air. At this pulse width, it would be impossible for the radar to distinguish storm targets that are separated by 1917 ft or less.

Many modern transmitters vary the pulse width according to the range selected by the pilot. A short pulse width is used for short-distance radar scans. A long pulse width is used for long-distance weather observation.

The **receiver** section of the r-t unit receives the reflected radar signal through the antenna and duplexer. This signal is received only during the listening time, which occurs between bursts of transmitted energy. The first step is to amplify the received signal to eliminate the loss of any resolution. The amplified signal is sent to the mixer, which produces a lower-frequency IF. Using a lower-frequency IF allows for less complicated circuitry than if the radar frequency was used.

The IF is then amplified, decoded, and filtered by the **data processor** section of the r-t unit. A range filter is used to determine the distance between the storm activity and the air-

craft. The data processor analyzes the IF and performs the calculations needed to determine the intensity of the storm activity. The range and intensity information is then digitized and stored in data bins. The data are then sent to an azimuth filter, where the information for one azimuth bearing is averaged. The results are sent to the indicator unit via a digital data bus for display.

The Indicator

There are two basic types of **radar indicator units,** monochrome (black and white) and color. The actual display of both units is provided by a **cathode-ray tube** (CRT). For a monochrome system the display of position and storm level is accomplished by intensity-modulating the radial-sweep beam while simultaneously rotating the beam in synchronism with the rotation of the antenna. The electron beam is modulated by means of pulse signals applied to the display unit. This modulation provides both range marks and video indication. Remember that the video signals, or pulses, "paint" the picture on the screen.

Figure 16–10 shows a typical color weather radar indicator unit. The indicator must contain the circuitry needed to decode the digital signal sent from the data processor section of the r-t unit. The decoded signal is stored in a memory until needed for the next scan of the CRT. The decoded signal provides the video drive signals that will produce the two-dimensional weather map on the CRT. On many units the CRT will also display the operation modes of the radar system.

The radar control switches may be incorporated into the face of the indicator unit. On some systems the weather radar display is integrated into the electronic flight instrument display. In this case the radar control panel is a separate unit.

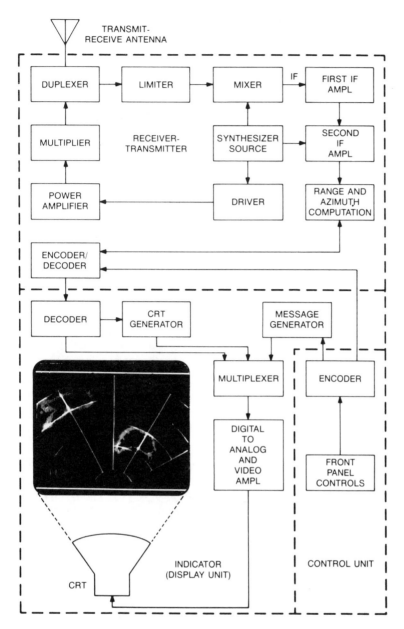

FIGURE 16–15 Block diagram of typical weather radar. *(Collins Divisions, Rockwell International)*

The Control Panel

The weather radar **control panel** must be located for easy access by the flight crew. If only one radar system is installed in an aircraft, the control panel is typically located in the center console for access by both the pilots. If two radar systems are installed, one control panel is located for access by the pilot and one for access by the copilot. The operating controls include the power switch, display brightness, stabilization, gain, the brightness of reference marks, mode of operation (test, weather, or map), selection of operating range, and antenna tilt controls. On a digital radar system, the control unit transmits a digital signal (representing the control switch positions) to the r-t unit in a serial format. On older systems an analog signal is sent. The functions of some of the controls are described below.

The **range control** is used to establish the range of the indicator. This control alters the transmission characteristics of the r-t unit to coordinate with the selected range.

The **gain control** is used to adjust the IF receiver gain to the proper operating point to assure reception of even the weakest reflected signals.

When the **stabilization control** is turned on, the transmitted radar signal will scan parallel to the earth, regardless of the pitch and roll of the aircraft. This system is particularly useful when the airplane is flying through areas of excessive turbulence.

The **tilt control** is used to change the angle in which the antenna is scanning. In the ZERO position the antenna beam is always parallel to the surface of the earth. If the control is turned toward the UP position, the beam will scan an area above the level of the airplane. In like manner, if the control

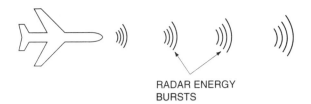

RADAR ENERGY
BURSTS

FIGURE 16–16 Bursts of radar energy.

is turned to a DOWN angle, the beam will scan an area below the level of the airplane. Thus the radar can provide better coverage during climb or descent and/or can be used to scan the areas above or below the aircraft in search of less storm activity.

Flat-Plate Antenna System

Most modern weather radar systems employ a **flat-plate** radar antenna. A flat-plate antenna utilizes a waveguide system that radiates the radar pulse from the rear of the antenna. This type of antenna is often called a **phased-array** antenna. Flat-plate antennas have approximately twice the efficiency of conventional parabolic antennas. This is due to the accurate signal transmission pattern of the flat-plate antenna (see Figure 16–17). As illustrated, with the flat-plate antenna (Figure 16–17a), there is less signal loss to the sides of the antenna. This more efficient antenna extends the range and resolution of any given radar transmitter. Or a less powerful transmitter can be used to achieve a radar system with comparable range and resolution.

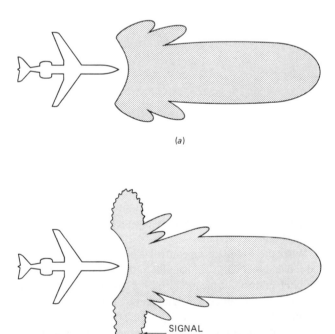

(a)

(b)

SIGNAL
LOSS

FIGURE 16–17 Radar signal transmission patterns. (a) A flat-plate antenna's signal; (b) a parabolic antenna's signal.

Figure 16–18 shows a typical flat-plate antenna. As illustrated in Figure 16–19, the waveguide from the receiver-transmitter connects to the rear of the antenna. The antenna contains several rows of horizontal and vertical waveguides. During transmission, the radar signals travel into the horizontal and vertical waveguides and "escape" through the slots in the front of the antenna. From here they radiate outward from the antenna in a relatively straight line. During reception, any radar wave that returns to the antenna enters the slots and travels through the waveguides and eventually into the receiver. Most flat-plate antennas are 10 to 30 in. in diameter.

Waveguides

As discussed under analog radar theory, a **waveguide** is used to connect an antenna to an r-t unit. A waveguide must be used for this connection, since the radar frequency is too high to be transmitted through conventional wiring. To improve efficiencies in the system, a waveguide should be as short as possible. Therefore, on some systems the r-t unit is mounted directly to the base of the antenna. This eliminates the need for any waveguide outside of the antenna or r-t unit. Ideally, all r-t units should be mounted to the antenna base; however, in many aircraft the space limitations do not allow this type of arrangement. Whenever an r-t is remotely located from the antenna, a waveguide must be installed.

Radomes

A radar antenna is typically mounted in the nose section of an airplane and is protected by a nonmetallic streamlined cover called a **radome.** The radome is necessary to protect the antenna assembly and still allow the transmitted signals to be radiated into space. A metal cover would reflect the waves back; hence there would be no way to detect storm activity.

FIGURE 16–18 A flat-plate antenna for use in commercial transport planes. *(Bendix Avionics Division)*

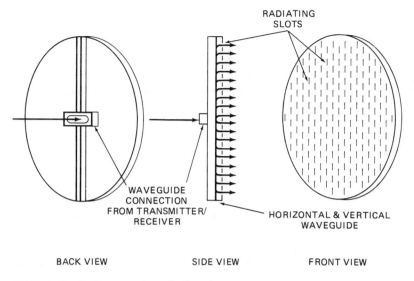

RADIATING
SLOTS

WAVEGUIDE
CONNECTION
FROM TRANSMITTER/
RECEIVER

HORIZONTAL & VERTICAL
WAVEGUIDE

BACK VIEW   SIDE VIEW   FRONT VIEW

FIGURE 16–19 The operation of a flat-plate antenna.

A radome is designed to protect the radar antenna from the elements and yet remain "clear" to the transmitted radar signals. It must be constructed of a material that will not interfere with or reflect the RF pulses emanating from the antenna. The material may be fiberglass laminate, plastic, or some other material that will not adversely affect the radar signals.

A radome may or may not be equipped with an **abrasion shield.** Such a shield is usually made of neoprene or some other synthetic rubber or plastic (Figure 16–20). In any event, the shield has a resilient surface that resists abrasion caused by sand, rain, rocks, or other objects.

Static electrical buildup is often a problem with radomes. Since radomes are made of a nonconductive material, static cannot easily dissipate from the radome to the airframe. If too much static charge accumulates on a radome, it will arc to the metal airframe and cause radio interference. If this arcing continues, the radome will most likely become damaged. To help eliminate static discharge on radomes, an antistatic protective coating is applied to the radome surface. Often this coating is incorporated in the abrasion shield.

Lightning diverter strips may also be used to help discharge any static accumulation on a radome. Lightning diverters are simply thin metallic foil strips adhered to the surface of the radome (Figure 16–20). The foil strips are electrically bonded to the airframe when the radome is mounted to the aircraft. The lightning diverters provide an electrical path to discharge any static charge.

The paint or other coating applied to a radome must be of a type that will not affect the radar signals. Many paints contain metallic pigments such as titanium, aluminum, or lead. Paints of these types must not be applied to a radome. The manufacturer usually specifies the types of paints and thickness of application that are satisfactory for repainting or touch-up.

Terrain Mapping

Many modern weather radar systems offer a **terrain map** mode of operation. The primary function of weather radar is to detect thunderstorm activity; however, the radar system can be used as an alternative navigation system via terrain mapping. Some systems provide a wide **fan beam** for use in mapping. As seen in Figure 16–21, the fan beam provides a wider coverage area than that provided by the **pencil beam** used for weather detection. Proper interpretation of the radar display is essential for terrain mapping; therefore, the pilot must be well versed in the various reflective properties of ground targets. For example, smooth water will not provide much of a return signal; large cities will reflect the majority of the radar signal; forests will return a weak signal.

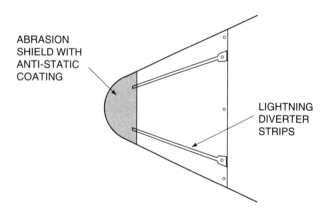

ABRASION
SHIELD WITH
ANTI-STATIC
COATING

LIGHTNING
DIVERTER
STRIPS

FIGURE 16–20 Typical radome installation.

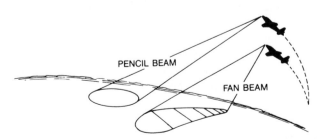

PENCIL BEAM

FAN BEAM

FIGURE 16–21 Fan beam used for terrain mapping. *(Collins Divisions, Rockwell International)*

Doppler Radar

Most aircraft weather radar systems measure only rainfall rate; the actual storm turbulence is left to the interpretation of the operator. The **Doppler radar** system actually measures storm turbulence and indicates its presence by the color magenta on the radar display. The Doppler radar also displays the weather information produced by typical color radars. This combination allows the flight crew to provide a safer and more comfortable flight in heavy storm activity. The system's limitations include the fact that moisture (rain) must be present in order to detect wind shear turbulence. No clear-air turbulence can be detected by Doppler radar. The Doppler radar system is by far the most informative radar yet; it has therefore been selected for new installations on transport-category aircraft.

The Doppler radar system is virtually identical with the conventional radar described earlier. The r-t unit incorporates different circuitry; however, this would be less than evident to the casual observer. The Doppler radar system, as the name implies, works on the Doppler shift principle. A sophisticated Doppler signal processor monitors the transmitted and received radar signal frequencies. If the returning frequency is out of phase with the transmitted frequency, the wind shear component (turbulence) of the storm is high. If the frequencies are in phase, there is no wind shear. A frequency shift is caused by the rain within a storm moving violently owing to an excessive wind shear. Since the transmitted signal reflects off of moving raindrops, the signal returns to the receiver at a different frequency. Of course, the Doppler system must take into consideration the aircraft's forward speed and compensate accordingly.

To accurately measure a frequency shift caused by turbulence, the radar's transmitter must be extremely stable. Likewise, the receiver must be capable of processing any returns without affecting the frequency. The Doppler system takes an average of the returned signals to ensure reliability. To acquire enough data in a short time period, the pulse repetition frequency (number of pulses per second) is increased to approximately 10 times that of standard weather radar. In other words, the r-t will transmit 10 times as many energy bursts per second. This increases accuracy but reduces the range of the turbulence detection to approximately 60 mi.

Doppler turbulence detection reveals new information about storm cells that cannot be identified by standard weather radar.

RADAR MAINTENANCE

The maintenance and repair of radar systems and components should be carried out in accordance with the manufacturer's instructions. There are, however, certain generally accepted practices, particularly with respect to radomes and radar safety.

Radomes

The efficiency of any radar system can be greatly reduced by improper radome maintenance. In general, the radome must remain transparent to the radar signal and yet protect the radar antenna from the elements. This task sounds simple, but the radome must remain nearly perfect; otherwise, the radar efficiency will suffer. One of the most common problems occurs when the radome develops a small crack. This crack will then allow moisture to enter the radome material. The moisture itself will reflect some of the transmitted radar energy, thus affecting efficiency. The moisture may also freeze, causing a delamination of the radome. This, too, can degrade radar performance, as well as create structural problems for the radome.

A radome should be inspected frequently for abrasion, cracks, delamination, the condition of lightning diverter strips, dents, and any other visible damage. The attachment of the radome to the aircraft should be examined for security to assure that water cannot enter the antenna space. The seal around the edges should be checked for cracks, separations, or any other type of fissure that would admit moisture. If the radome is equipped with an abrasion shield, the shield should first be checked for damage such as cuts or cracks. Since the shield can conceal damage to the radome, the radome should be checked for damage on the inside.

The repair of cracks and cuts in a radome is important because such defects may allow water to enter the antenna space or seep between layers in a radome lamination. As noted above, if water is trapped in the radome material, it will affect the radar signal, or it may freeze and cause further damage. The thickness of the radome is critical to system performance. Cracks should be repaired as specified by the manufacturer. Any radome with a multitude of repairs is most likely affecting radar performance by as much as 50 percent and should be replaced.

The radome should also be inspected carefully for signs of static discharge. This is especially important for high-speed aircraft. Static discharge damage appears as small carbon traces, burns, or pits on the radome surface. The damaged material must be removed and the radome repaired. If the radome is equipped with lightning arresters, be sure that they are attached securely to the radome and the aircraft. A continuity check of the lightning arresters to the airframe is also recommended.

Components of a radar system are inspected in a manner similar to that employed for other electric and electronic equipment. Items should be checked for security of mounting or attachment, security of connector plugs, the condition of shock-absorbing units, cleanliness, and freedom from corrosion and any other unsatisfactory condition. Cleaning is very important to enhance system cooling. Be sure all vent holes are clean so that air can easily flow through the unit. Some units employ forced-air cooling; be sure this system is clean and operable. Testing of the entire radar system should be done according to the instructions provided by the manufacturer.

Radar Safety

It must be remembered that when a radar system is turned on, the antenna is emitting millions of high-energy electromag-

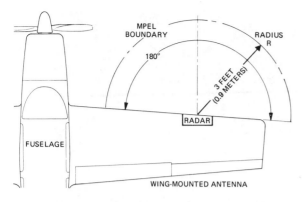

FIGURE 16–22 A typical MPEL boundary for radar. (RCA)

netic pulses. These can cause physical injury to the body, as well as ignition of flammable materials in the vicinity. For these reasons, the radar should not be turned on when the aircraft is on the ground except under carefully controlled conditions. The safety of personnel is increased by the observance of the established maximum permissible exposure level (MPEL) boundary for the equipment being tested. This boundary is illustrated in Figure 16–22 for a wing-installed antenna.

Radar should not be turned on when an aircraft is being defueled or fueled. This applies not only to the aircraft in which it is installed but also to other nearby aircraft. The radar should not be turned on in a hangar unless appropriate shielding and wave-absorbing equipment is in place in front of the antenna.

Radar can usually be tested safely on the ground by placing the aircraft in an open area with the antenna tilted upward so the beam cannot strike any object on the ground.

Any person operating an aircraft is responsible for damage caused by that aircraft, including the aircraft's weather radar. Always make sure the radar is turned off before turning on any aircraft battery master switch. If the last person to operate the radar accidentally left the radar on, you may inadvertently operate the radar simply by energizing the aircraft bus. This can cause severe bodily injury or property damage; therefore, double-check that the radar is off.

Additional recommended safety precautions for airborne radar are set forth in FAA publication AC 20-60A.

WEATHER MAPPING SYSTEMS

A **weather mapping system** is basically a radio receiver designed to detect thunderstorm activity and display a map of that activity on a CRT or a liquid crystal display. Weather mapping systems were originally developed as a low-cost alternative to weather radar systems. Although most weather mapping systems cost less than radar, in many cases weather mapping systems are used in conjunction with weather radar. This is mainly due to the reliability of weather mapping units and their ability to consistently detect thunderstorms that are located behind other storms. Thus a weather mapping system is ideal for serving as a secondary "backup" weather detection system, as well as providing information about storms that might be "hidden" from weather radar.

In any storm system, there are areas of convective shear of air currents. These conditions are caused by the rising and falling air currents near each other as shown in Figure 16–23. For example, as a cold front moves across the country, the cold air in the advancing front flows under the warmer air and causes the warmer air to rise. At the same time, the colder air in the front flows downward, and convective shear takes place between the rising and falling columns of air.

The convective wind shear creates static electric charges that must discharge when they reach a certain voltage level. If a large number of static discharges occur simultaneously, visible lightning will result. Any discharge, with or without lightning, will emit an RF signal. A weather mapping system detects this signal and displays the discharge activity on a CRT.

There is always an abundance of static discharge activity within any storm. Since the number of static discharges increases with the severity of the storm, a weather mapping system can "count" the discharges to determine the storm's intensity. A relatively intense storm will have a large number of static discharges per minute. A weak storm will have fewer discharges per minute.

A weather mapping system is a receiver only; however, it has the capability of determining the direction from which the electric signals from a thunderstorm are coming. This is accomplished by employing a loop antenna and a sense antenna in a manner similar to the way they are used for an automatic direction finder (ADF). Both antennas are typically contained in one streamlined unit that employs no moving parts. The loop antenna is electronically (not physically) rotated. The distance to the storm activity is determined by measuring the strength of the electromagnetic RF signal created by the static discharge.

An electromagnetic signal decreases in strength in proportion to the square of the distance it has traveled. For example, a radio signal received from a distance of 20 mi [32.2 km] will be one-fourth the strength of the same signal received from a distance of 10 mi [16.1 km]. The computer/processor of the system contains information regarding the strength of typical electric discharges in a thunderstorm, and by comparing incoming data with this information, it can determine approximately how far each discharge signal has traveled. The directional and distance information is sent to the display scope, and a bright dot appears at the computed location of

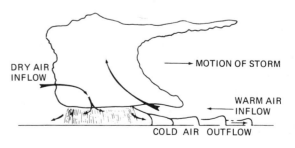

FIGURE 16–23 Air currents in a storm front.

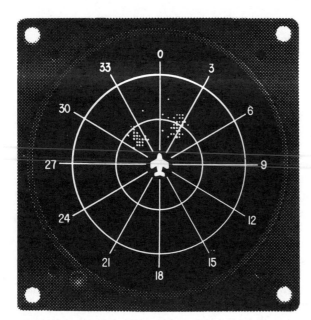
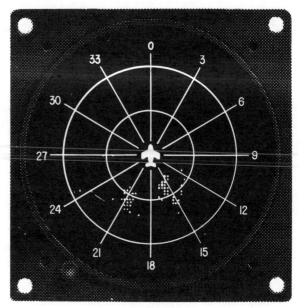

FIGURE 16–24 Weather mapping system displays. *(Ryan Stormscope)*

FIGURE 16–25 The three units of a typical weather mapping system; from left to right, the antenna, the computer/processor, and the display unit. *(3M Corporation)*

the discharge relative to the aircraft heading. The image is held in the computer memory so that the dot will remain on the scope for up to 5 min or until 256 dots have been recorded. As repetitive discharges occur within the atmosphere, clusters of dots on the scope form a maplike image of the weather. Typical weather displays are shown in Figure 16–24. Another type of weather mapping display uses a liquid crystal to show the location of the storm activity. With this display the screen is divided into quadrants, and each quadrant shows the color green, yellow, or red to identify the level of storm activity in that area.

The computer/processor, display unit, and antenna of a typical weather mapping system are shown in Figure 16–25. The controls of this weathermapping system are mounted on the front of the display unit. The FORWARD function is used to limit the signals to those received from the area forward of

the aircraft. Otherwise, the signals will be received from a 360° area surrounding the aircraft. The CLEAR control is used to remove all images from the scope. This is necessary when a change of heading is made. If an image is allowed to remain on the scope when the heading is changed, the scope display will be incorrect for the new heading. After the scope is cleared, new information immediately begins to be displayed, and it will form a completely new map of the weather within 25 s for a typical thunderstorm. The RANGE control enables the operator to select any of four desired ranges. The maximum range for this system is 200 nm.

Some newer weather mapping systems can also display aircraft checklists. This helps to eliminate paperwork clutter on the flight deck. Other systems incorporate an interface with navigation aids such as the LORAN or GPS receiver. This enables the weather mapping system to display the aircraft's predicted course with the identified weather activity. Most newer systems also have some type of built-in test equipment to help with fault detection and troubleshooting.

Many aircraft are currently equipped with a weather mapping system. This equipment is an alternative or a supplement to an airborne weather radar system. Through proper interpretation of the display, a weather mapping system offers an accurate and detailed analysis of current weather conditions.

REVIEW QUESTIONS

1. What is meant by the word *radar?*
2. What is the nature of a radar signal?
3. Why must a radar pulse rate be slow enough so that the return signal is received before the next pulse is emitted?
4. How may the distance of an object from an aircraft be determined by radar?

5. Compare the power of an individual radar pulse with the average power of the system.

6. Describe the operation of a plan position indicator.

7. Name the principal units of a typical radar system.

8. Describe the functions of a synchronizer, mixer, magnetron tube, and duplexer.

9. What device serves as the indicator for a radar system?

10. Why is a waveguide system required in a radar receiver-transmitter unit?

11. How is the image on a radar CRT synchronized with the radar antenna in rotation?

12. What is the purpose of the antenna stabilization system?

13. What is the value of a tilt control?

14. What technology has made it possible to develop very lightweight radar systems?

15. What two types of antennas are available for radar systems?

16. What are the advantages of a flat-plate antenna?

17. Describe the function of the parabolic and flat-plate antenna system.

18. What colors are used to indicate the levels of storm activity on a color radar indicator?

19. What is wind shear?

20. What type of radar systems are capable of detecting wind shear?

21. Describe the operation of a Doppler radar system.

22. List inspections that should be made regularly on a radome.

23. Of what materials may a radome be constructed?

24. What is an abrasion shield?

25. What precaution must be taken with respect to painting a radome?

26. What may happen if water is trapped between lamination of the radome structure?

27. Describe safety precautions that must be observed in the operation of radar on the ground.

28. Why is it important to make sure the radar is off before supplying power to the aircraft bus?

29. Describe the principle of operation for a weather mapping system.

30. How is the direction of storm activity determined by a weather mapping system?

31. What is the theory of operation employed by the weather mapping system to determine the distance between the aircraft and the storm activity?

32. What is the purpose of the clear function on the weather mapping system display?

17 Electric Instruments and Autoflight Systems

In Chapter 8 electric measuring instruments were described. The basic elements of measuring instruments are also characteristic of many of the indicating instruments discussed in this chapter. These will be reviewed as the various types of indicating instruments are described.

Electric indicating instruments of various types have been described in other texts of this series; however, it is deemed appropriate to examine them further here because of the electrical factors involved. This chapter, therefore, focuses on electric indicating instruments for power plants, aircraft systems, and controls.

The operation of modern aircraft, either of the small, private category or the transport type, would not be possible without the use of instruments. On large aircraft, particularly, instruments are operated electrically or electronically. Instruments are needed to measure pressures, temperature, altitudes, velocity, rate of flow, and numerous other conditions or parameters affecting the flight and operation of aircraft. Human beings are unable to react rapidly and accurately to the many variable conditions that affect the flight of an airplane unless they have accurate and reliable instruments.

On modern aircraft the entire autoflight system relies on electronic signals from a variety of flight instruments for proper operation. The use of electric instruments provides a commonality between the instrumentation system and the automatic flight control systems (autopilots). Modern autoflight systems on transport-category aircraft monitor hundreds of engine and airframe parameters. This is made possible through the use of on-board computer systems and electronic instrumentation.

It is the purpose of this chapter to present the basic principles of electrically operated instruments and automatic flight control systems. The principles discussed in this chapter will provide a framework upon which a technician can build an understanding of any instrument or autoflight system.

RPM-MEASURING INSTRUMENTS

An rpm indicator, or **tachometer,** is a most essential instrument in all aircraft. It is used to show the rpm of reciprocating engines, the percentage of power for turbine engines, the rpm of helicopter rotors, and the rotational speed of any other device where this information is critical. A typical tachometer for a reciprocating engine is shown in Figure 17–1.

Since a tachometer registers engine rpm, it is a primary indicator of engine performance. An electric tachometer gives the same indications that are given by a mechanical tachometer, but the method for actuating the indicating unit is entirely different.

DC Tachometer

The crankshaft speed of an engine can be measured by generating voltage directly at the engine and then measuring this voltage by means of a voltmeter calibrated in terms of engine speed. Remember that when the field strength of a generator remains constant, the voltage output is proportional to the speed of rotation. A dc tachometer generator is a simple generator that has a permanent-magnet field; hence the voltage output of the generator is proportional to engine speed, and the rpm can be read on a voltmeter that has a dial marked in rpm instead of volts. Even though this system is obsolete, it is useful for illustrating electrical measurement of rpm.

AC Tachometer

The dc tachometer requires a commutator and brushes, and these are subject to wear. The development of the ac tachometer eliminated this maintenance problem. The ac tachometer system in most common use consists of a three-phase ac generator (alternator) and the tachometer indicator, which is driven by the alternator. The indicating mechanism consists of a mechanical tachometer driven by a three-phase synchronous motor, as shown in Figure 17–2.

FIGURE 17–1 Tachometer for a reciprocating engine.

The alternator, or transmitter, consists of a four-pole permanent magnet that rotates inside a three-phase stator; an external view of this unit is shown in Figure 17–3. The stator of the alternator is connected by three wires to a similar stator in the synchronous motor, which operates the indicator. As the alternator is driven by the engine, the rotation of the permanent magnet induces a three-phase current in the stator. This current flows through the stator of the synchronous motor in the indicator and produces a rotating field that turns at the same rate as the alternator rotor. The permanent-magnet rotor of the indicator keeps itself aligned with the rotating field and hence must also turn at the rate of the alternator rotor.

The synchronous motor in the indicator is directly coupled to a cylindrical permanent magnet that rotates inside a drag cup, as shown in Figure 17–4. As this magnet turns, it causes magnetic lines of force to drag through the metal cup and induce eddy currents in the metal. These eddy currents produce magnetic fields that oppose the field of the rotating magnet. The result is that as the speed of the rotating magnet increases, the drag or torque on the drag cup increases. The torque on the drag cup causes it to rotate against the force of the balancing hairspring and turn the pointer on the indicating dial. The distance through which the drag cup rotates is proportional to the speed of the synchronous motor; hence it is also proportional to the engine speed.

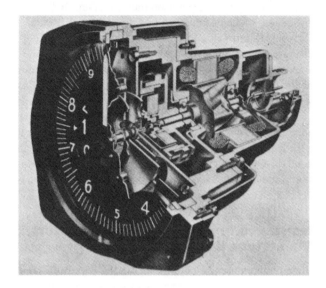

FIGURE 17–2 An ac tachometer.

FIGURE 17–3 Three-phase tachometer alternator.

TEMPERATURE INDICATORS

Thermocouple Temperature Indicators

Thermocouple temperature indicators are used most frequently when it is necessary to measure comparatively high temperatures. They are used to measure cylinder head temperature in aircraft with reciprocating engines and tailpipe or exhaust temperature in aircraft with jet engines. The readings of these instruments provide information for the pilot or flight engineer, who can then operate the engine at its most efficient temperature and prevent damage to it from overheating. As explained in Chapters 1 and 8, a thermocouple temperature indicator operates on the thermoelectric effect. A simple circuit for such a system is shown in Figure 17–5. As the hot end of the thermocouple system changes temperature, the bimetallic junction produces a voltage and a current flow. The current flow is sent to an extremely sensitive instrument that contains a scale calibrated in degrees.

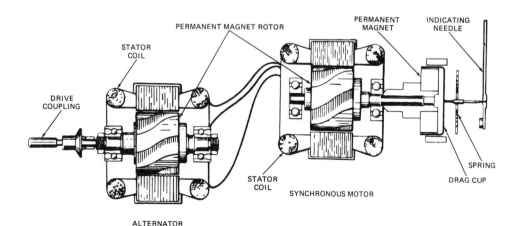

FIGURE 17–4 Three-phase tachometer system.

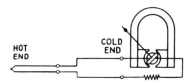

FIGURE 17–5 Thermocouple system.

Thermocouple Leads

In accordance with Ohm's law, the emf generated in a thermocouple circuit produces a current inversely proportional to the resistance of the circuit at any given temperature difference between the hot junction and the cold junction. That is, if the resistance increases, the current will decrease; and if the resistance decreases, the current will increase. It is obvious that to obtain a constant current for a given temperature difference, the resistance of the circuit must remain constant, regardless of the temperature of the leads.

Thermocouple leads are usually made of either constantan and iron or constantan and copper. Constantan is a copper-nickel alloy that shows practically no change in resistance with considerable changes in temperature. Thermocouples that are designed to measure high temperatures are composed of Chromel and Alumel, which are special high-temperature alloys. Typically, thermocouple leads are made of the same alloy as the thermocouple hot junction.

In order to compensate for the resistance changes that may occur as a result of temperature changes in the thermocouple leads, it is sometimes necessary to employ a **neutralizer** in the circuit. A neutralizer is a resistor unit made of a material that loses part of its resistance when temperature increases. It is designed with dimensions such that it loses as much resistance as the rest of the circuit gains under any given temperature conditions; the total resistance of the circuit therefore remains constant.

FIGURE 17–6 A typical EGT thermocouple.

When a thermocouple instrument is installed, it is essential that the correct leads be used. Standard thermocouple leads have resistances of 2 and 8 Ω, and the instrument must be provided with the type for which it is designed. Because of the very small amount of electric energy produced by a thermocouple, the electrical connections must be clean and tight. Furthermore, it is absolutely essential that the leads not be crossed during installation. Iron must be connected to iron, constantan to constantan, copper to copper, Chromel to Chromel, Alumel to Alumel, etc. Usually, thermocouple leads are provided with connectors that make it impossible to connect them in reverse; however, it is always wise to examine the leads closely to make sure that the connections have been correctly made.

Since thermocouple leads are made with a specific resistance, they must never be cut or spliced. If there is extra length in the leads, they should be coiled up to take up the slack and secured. A typical exhaust gas temperature (EGT) thermocouple and its associated leads are shown in Figure 17–6.

Resistance-Type Temperature Indicators: Wheatstone Bridge

When temperatures below 300°F [148.9°C] are to be measured, it is necessary to use an instrument different from the thermocouple type to obtain accurate indications. One of the common types of instruments used for these lower-temperature indications is the **Wheatstone bridge** instrument. Among the temperatures it measures are those of **free air, carburetor air, cabin air,** and **coolant.**

It is a fundamental fact that the resistance of metals changes with changes in temperature. Also, since most electric instruments are basically devices for measuring current, changes in resistance can be converted into changes in current to obtain a pointer deflection across the scale of a direct-reading instrument. Resistance can be translated into current through the use of Ohm's law because $I = E/R$. Use of this law, however, assumes that the voltage is constant, a condition not always true when the voltage is supplied by a battery and generator system. Therefore, in order to translate resistance into current, the problem of the variable voltage must in some manner be solved. One solution is to use the Wheatstone bridge system.

The principle of the Wheatstone bridge can be explained by comparing this bridge circuit with a divided stream of water. Let us assume that a stream comes down from the mountains, meets a four-sided island, separates, and then unites on the other side of the island, as shown in Figure 17–7. A canal is dug across the island, connecting a point in the middle of one branch of the stream with a similar point in the middle of the other. A paddle wheel is then mounted in the middle of this canal and a dam placed in each of the branches upstream from the canal and in each of the branches downstream from the canal. Each of these dams is provided with a floodgate to regulate the flow of water. By opening or closing these gates in the various dams, the water levels in the branches can be controlled.

When the water level is the same at both ends of the canal

FIGURE 17–7 Water analogy of Wheatstone bridge.

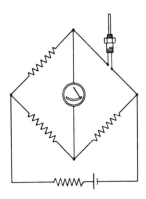

FIGURE 17–8 A typical bridge circuit.

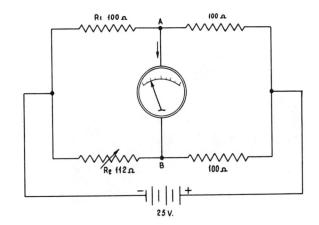

FIGURE 17–9 Analysis of a bridge circuit.

across the island, the paddle wheel does not move, because there is no water flow through the canal. If three gates are adjusted exactly alike, raising and lowering the fourth gate affects both the direction in which the water will flow through the canal and the speed of its flow, as indicated by the rotation of the paddle wheel. No matter which gate is being operated, however, the settings of the others must not be disturbed; otherwise, the paddle wheel will not give a reliable indication of the direction and velocity of the water flow.

The Wheatstone bridge circuit resembles the divided stream of the system just described. At some point in the electric circuit the current divides, flows through two branches, and then unites, just as the imaginary stream did. The dams with their floodgates in the branches of the stream are resistors in the electric circuit, and the canal is a conducting branch that includes a sensitive indicator (paddle wheel). We can regard the four resistors in the circuit as arms. Since three of them are to remain constant, they can be made of manganin wire, which is similar in properties to constantan, since it shows almost no change in resistance with changes in temperature. Any metal will be satisfactory for the fourth, since its resistance is to be measured and must therefore change with changes in temperature.

Before the instrument is connected in the bridge circuit, the pointer is set so that it rests at the **balance point.** This is the point on the scale at which the pointer rests when the entire system is connected to a source of current and the resistances are balanced so that no current is flowing through the instrument. Such conditions are described by saying that the bridge is in balance.

A typical bridge circuit is illustrated in Figure 17–8. When the bridge is in balance, the resistances are equal to 100 Ω each. Three of the resistances remain at 100 Ω, even though the temperature changes, but the fourth resistance (the temperature bulb) changes in value with the temperature. Let us assume that the balance point for this circuit is 82.4°F [28°C] and that the variable resistance has a value of 100 Ω at this temperature. Now if the temperature of the variable resistance increases to 140°F [60°C], the resistance value will change to 112 Ω. Note that in Figure 17–9, which is the same circuit as shown in Figure 17–8, one side of the bridge will have a resistance of 200 Ω, and the other side will have a resistance of 212 Ω. By Ohm's law we determine that when 25 V is connected across the bridge, the current through one side will be 0.125 A and the current through the other side is about 0.118 A. Using these current values, we find that the voltage drop across the variable resistor will be about 13.1 V and that the voltage drop across the opposite resistor R_1 will be 12.5 V. It is apparent, then, that there will be a voltage difference of 0.6 V across the bridge between points A and B. Point A will be negative with respect to B; hence current flow will be from A to B through the instrument.

In the foregoing example we have ignored the small effect of the instrument across A and B. Although this would cause some variation in the figures quoted, the results would be the same. Any unbalance in the circuit caused by a variation in the resistance value of the variable resistor will cause current to flow through the meter movement and register a temperature reading on the meter.

In practice, the fixed arms of the Wheatstone bridge are mounted in the case, as shown in Figure 17–10. They may be wound on three spools, as shown, or on a single spool. The fourth, or variable, resistor is external because it is mounted at the location at which temperature is to be measured. Technically, the fourth resistor is called a **resistance bulb** (see Figure 17–11).

The resistance wire, which is the essential feature of the resistance bulb, rests in the spiral grooves of an insulating material and is covered with a metal shield that conducts heat to and from it very quickly. This metal shield must be able to withstand the corroding influence of engine oils at high temperatures, the high flash temperatures in the carburetor of a backfiring engine, and the deteriorating influence of the atmosphere. Even though the resistance bulb is covered with a metal shell and substantial insulation, it responds to changes in temperature very rapidly. This sensitivity is important because the members of a flight crew are not interested in past temperatures; they want to know the situation at the exact second that the instrument is read.

The action of a resistance bulb can be understood by studying the graph in Figure 17–12. Note that the increase in resistance of a temperature bulb is almost linear with respect to temperature changes.

FIGURE 17–10 Interior of a Wheatstone bridge instrument.

FIGURE 17–11 Resistance bulb for temperature sensing.

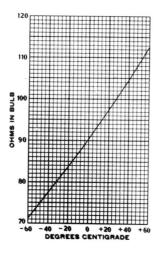

FIGURE 17–12 Temperature-resistance curve for a typical resistance bulb.

SYNCHRO SYSTEMS

A synchro system is designed to measure an angular deflection at one point and reproduce this same deflection at a remote point. Synchro systems have been designed to employ both alternating current and direct current for power, but the present trend is to employ 400-Hz alternating current. Synchros are used as position indicators, for remote indicating systems in radar equipment, in autopilots, and for a wide variety of other remote-control and indicating systems. Synchros have been designed and built under a variety of names, the most common being **Selsyn** (a General Electric trade name) and **Autosyn** (a Bendix trade name).

The DC Selsyn

One of the early synchro systems developed by the General Electric Company was the dc Selsyn, often used as an indicator on aircraft with dc power systems to show the position of wing flaps and landing gear. The dc Selsyn instrument system consists of an indicator and a transmitter operating in synchronism; hence the name *Selsyn* (self-synchronous).

A schematic diagram of a single dc Selsyn system is shown in Figure 17–13. The transmitter is merely a winding of fine resistance wire on a circular form with connections located at three equally spaced points around the winding. DC power is fed to the ring winding at points 180° apart by means of wiper arms (see Figure 17–13). This arrangement is actually a special type of potentiometer, and when the wiper arms are rotated, the voltages appearing at the three connections will change with respect to one another. As shown in the diagram, the three connections to the transmitter are connected to three similar connections at the indicator. The indicator unit consists of a laminated ring of ferromagnetic material on which three windings are equally spaced in a delta connection. When this unit is connected to the transmitter by means of three conductors as shown in the diagram, rotation of the wiper arms in the transmitter will vary the currents in the coils of the indicator in such a manner that the magnetic field of these coils will rotate also. The rotating element of the indicator is a permanent-magnet armature mounted on bearings so that it is free to turn with the rotation of the field. Thus the indicating needle attached to the rotor shaft will follow the movement of the wiper arms at the transmitter. When the transmitter is linked to the flap-actuating mechanism, it will produce a signal that causes the indicator to show the position

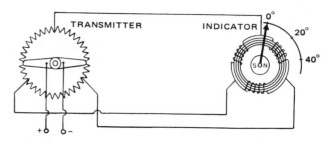

FIGURE 17–13 Schematic diagram of a dc Selsyn circuit.

of the flaps. In like manner, it can be used to show the position of the landing gear or any other unit that moves through a range of positions.

Autosyn Instruments

The word *Autosyn* is the trade name of another type of self-synchronizing remote indicating instrument. The Autosyn system operates using alternating current. Like the Selsyn instruments, the Autosyn system activates indicators in the cockpit without using excessively long mechanical linkages or tubing. An indication is picked up by the transmitter, located near the engine or at some other remote point, and sent by electrical means to the indicator on the flight deck. This system has great value when used in airliners and other large aircraft.

An Autosyn synchro has the appearance of a small synchronous motor. In the Autosyn system, one synchro is employed as a transmitter and another as an indicator.

A schematic diagram of an Autosyn system is shown in Figure 17–14. The system is basically an adaptation of the self-synchronous motor principle, whereby two widely separated motors operate in exact synchronism; that is, the rotor of one motor spins at the same speed as the rotor of the other. When this principle is applied to the Autosyn system, however, the rotors neither spin nor produce power. Instead, the rotors of the two connected Autosyn units come into coincidence when they are energized by an alternating electric current, and thereafter the rotor of the first Autosyn moves only the distance necessary to match any movement of the rotor of a second Autosyn, no matter how slight that movement may be.

It must be understood that the transmitter and indicator of an Autosyn system are essentially alike, both in electrical characteristics and in construction. Each has a rotor and a stator. When ac power is applied and a rotor is energized, the transformer action between the rotor and stator causes three distinct voltages to be induced in the rotor relative to the stator. For each tiny change in the position of the rotor, a new and completely different combination of three voltages is induced.

When two Autosyns are connected as shown in Figure 17–14 and the rotors of both units occupy exactly the same positions relative to their respective stators, both sets of induced voltages are equal and opposite. For this reason no current flows in the interconnected leads, with the result that

both rotors remain stationary. On the other hand, when the two rotors do not coincide in position, the combination of voltages of one rotor is not like that of the other, and rotation takes place, continuing until the rotors are in identical positions. The induced voltages are then equal and opposite, and so there is no current flow in any of the three conductors; hence the rotors will be in stationary and identical positions. Although Autosyn systems may utilize any ac voltage, typically 26 V 400 Hz ac is used.

FUEL-QUANTITY INDICATORS

Fuel-quantity indicators for many light aircraft and all large commercial aircraft are either electrically or electronically operated. The electrically operated indicating systems are usually of the variable-resistor type, and the electronically operated systems are of the capacitor type.

Electric fuel-quantity indicators utilize a variable resistor in the tank unit or sensor. The resistor is in the form of a rheostat or potentiometer, depending on the method of indication. The fuel-quantity signal is provided as a float arm in the tank changes the resistance of the sensor as the fuel level changes. A schematic diagram of a fuel-quantity-indicating system with a rheostat-type sensor is shown in Figure 17–15. The variable resistor of the tank unit is connected in a bridge circuit with reference resistors in the indicator case. Refer to the section earlier in this chapter that describes the Wheatstone bridge.

Electronic, or capacitor-type, fuel indicators utilize a variable capacitor as the sensor unit in the fuel tank. The capacitance of the sensor in the tank is changed as fuel rises and falls between the two electrodes (plates) in the tank unit. The change of capacitance is due to the difference between the dielectric strength of the fuel and that of air. Since fuel has a dielectric strength more than twice that of air, the capacitance of the sensor increases in accordance with the amount of fuel in the tank. The change in capacitance of the sensor unit is utilized in a bridge circuit to provide an amplified signal that rotates the pointer or actuates the digital display in the indicator. A schematic diagram to illustrate the operation of a

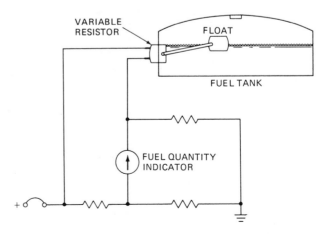

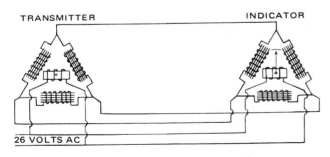

FIGURE 17–14 Schematic diagram of an Autosyn synchro system.

FIGURE 17–15 Schematic diagram of a float-type fuel-level indicator with a variable resistor.

capacitor-type fuel-quantity-indicating system is shown in Figure 17–16.

In an actual system in an aircraft, the fuel-quantity-indicating instrument may contain the amplifier or signal conditioner, thus eliminating the necessity for a separate unit installation. The indicating unit for fuel quantity may vary considerably from aircraft to aircraft. Formerly the majority of indicating units utilized a needle-and-scale type of indicator; however, with the availability of solid-state devices, many indicators are now of the digital type. The indication may be in pounds of fuel or in gallons or in both. Capacitor-type systems measure fuel quantity by weight (mass) rather than volume because the dielectric strength of the fuel changes in accordance with density. Compensators are employed with the probes to assure an accurate indication. A fuel-quantity sensing probe (tank unit) is shown in Figure 17–17.

Computerized Fuel System

A system that provides fuel-flow and fuel-quantity indications is called a **computerized fuel system** (CFS). The indicator for the system provides fuel flow in gallons per hour or pounds per hour, gallons remaining, pounds remaining, time remaining for flight at the current power setting, and gallons used from initial start-up.

The transducer (sensor) for the CFS contains a neutrally buoyant rotor that spins at a rate proportional to the rate of fuel flow. The rotor is mounted in jewel bearings and incorporates notches that interrupt an infrared light beam from a light-emitting diode (LED). The light beam is aimed at a phototransistor, and the interruptions in the beam produce a series of pulses corresponding to the rate of fuel flow. The transducer is designed so the failure of the rotor cannot interrupt fuel flow. The transducer is mounted in a main fuel line and measures all fuel that flows through the line. A typical installation of a transducer is shown in Figure 17–18. In this case the transducer is installed in the main fuel line entering the fuel distribution manifold (fuel splitter) mounted on top of the engine. The transducer is mounted on a bracket with an Adel clamp.

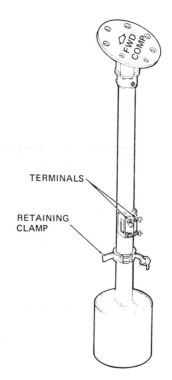

FIGURE 17–17 Fuel-quantity sensing probe. *(McDonnell Douglas)*

The panel-mounted instrument contains computers that are designed to precisely count the number of pulses from the fuel-flow transducer and convert the count into gallons. A crystal-controlled clock provides the time reference needed to compute the rate of fuel flow and timer readout functions. The computer routinely calculates all the other displayed functions. The front panels of two types of fuel-indicating instruments are shown in Figure 17–19. The top photograph shows a standard round indicator, and the bottom photograph shows a flat-pack indicator. These instruments receive signals from two transducers as required for a twin-engine aircraft.

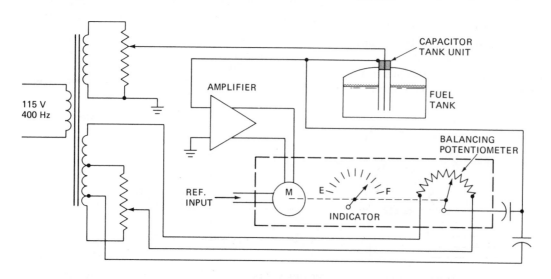

FIGURE 17–16 Schematic diagram of a capacitor-type fuel-quantity-indicating system.

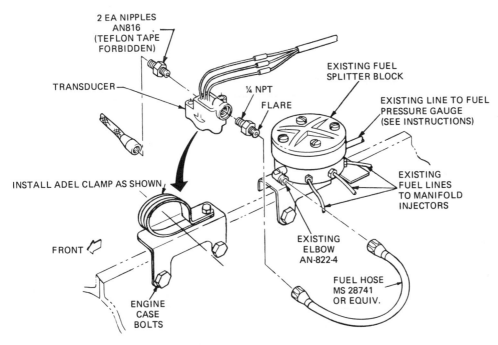

FIGURE 17–18 Mounting of a transducer for a computerized fuel system. *(Symbolic Displays, Inc.)*

FIGURE 17–19 Indicator panels for a computerized fuel system. *(Symbolic Displays, Inc.)*

ELECTROMECHANICAL FLIGHT INSTRUMENTS

During the last decade, there has been an electronics revolution in the aviation industry. This revolution has had a great impact on the design and operation of the instrument systems found on modern aircraft. Instruments that are mechanically driven with pitot and/or static pressure inputs are still used on many aircraft. For example, most light aircraft still operate using the traditional mechanical instruments. On the other hand, large transport-category aircraft require so much instrumentation that it has become almost impossible to fit all the necessary mechanical instruments in view of the pilot. To simplify the instrument panel, many modern corporate and transport-category aircraft employ electronic instruments of one type or another. To better understand the electronic instruments, we should review some basic instrument theory.

There are two common electromechanical instruments found on high-performance aircraft that were replicated when the state-of-the-art electronic systems were designed. The **attitude-director indicator** (ADI) and the **horizontal-situation indicator** (HSI) are both hybrid electromechanical instruments that combine several basic flight instruments. The ADI is used to display the attitude information necessary for flight. The ADI typically includes the attitude indicator, the turn-and-slip indicator, pitch-and-bank steering bars, the glide slope indicator, and a variety of warning flags. Some ADIs incorporate indicators known as **command bars** (Figure 17–20). The command bars display information from several different attitude and navigational inputs. This system allows the pilot to fly the aircraft using the command bars as the main reference. ADIs that employ command bars are sometimes referred to as **director-horizon indicators.**

The HSI is an instrument that displays information concerning the aircraft's position in the horizontal plane of reference. This instrument combines a variety of conventional instruments, such as the heading indicator (gyro compass), the course-deviation indicator, and a DME distance indicator.

Typically, the ADI and HSI work in conjunction with the **flight director** system. The flight director system typically consists of an electronic control unit that receives inputs from various navigational systems, attitude gyros, and pitot static sensors. The flight director computer then processes these inputs and electronically sends the information to the

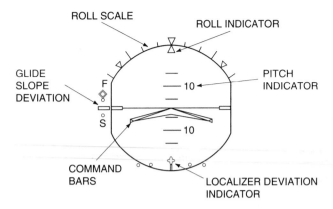

FIGURE 17–20 A typical ADI.

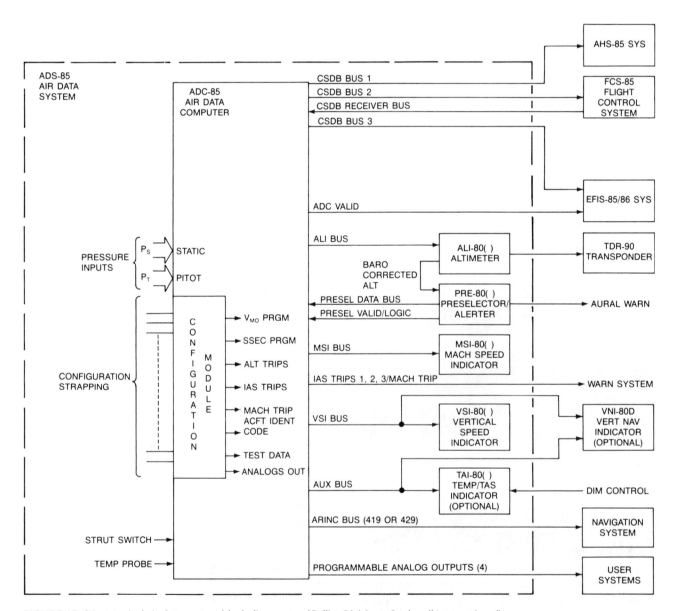

FIGURE 17–21 A typical air-data-system block diagram. *(Collins Divisions, Rockwell International)*

electromechanical ADI and HSI. The autopilot system may also receive information from the flight director computer.

Air-data systems are another type of hybrid electromechanical instrument package. The instruments in these systems display all parameters associated with the aircraft's movement through the air. Newer air-data systems are often used as inputs to the electronic flight instrument systems. The air-data-system computer receives inputs from the various pressure and temperature sensors throughout the aircraft (Figure 17–21). The air-data computer processes the input data and sends output signals to electromechanical display instruments such as the altimeter, the Mach/airspeed indicator, the vertical-speed indicator, and the temperature indicator. Air-data outputs are also used by various navigational systems.

ELECTRONIC FLIGHT SYSTEMS

In a further attempt to reduce pilot workload and instrument panel clutter, **electronic flight instrument systems** (EFISs) were developed. These systems employ state-of-the-art CRTs to display alphanumeric data and representations of aircraft instruments. Each EFIS display replaces several conventional instruments and caution and warning annunciators. Figure 17–22 shows the instrument panel of an Airbus Industrie A-320. This aircraft employs six CRT displays in order to eliminate most of the separate electromechanical instruments found on traditional aircraft.

Electronic flight instruments became possible with the development of a sunlight-readable CRT display and sophisticated aircraft computer interface systems. A digital data bus system is used to transfer a majority of information between the various components of an EFIS. With large amounts of data to transfer, analog systems would add hundreds of pounds to the aircraft in additional wiring. Several common data bus systems are described in Chapter 7.

Electronic flight instruments used to display horizontal-situation indicators (HSIs) and attitude-director indicators (ADIs) are becoming very popular on all types of high-performance aircraft. An illustration of an **electronic horizontal-situation indicator** (EHSI) is shown in Figure 17–23a, and an **electronic attitude-director indicator** (EADI) is shown in Figure 17–23b.

An EFIS is composed of three subsystems, as illustrated in Figure 17–24: The **pilot display system** (PDS), the **copilot display system** (CDS), and the **weather radar system** (WX). The weather system provides weather data for either the pilot's or copilot's display system or for both. The pilot's and copilot's display systems are identical, each containing two CRT displays, a symbol generator, a display controller, and a source select panel. In the unlikely event of one system failure, the backup cross-fed circuit allows the operational symbol generator to drive all four electronic displays.

The **symbol generator** (SG) receives input signals from several aircraft and engine sensors, processes this information, and sends it to the appropriate display. Inputs to the symbol generator include data from the various navigation radios, flight control computers, thrust management computers, the inertial reference system, and the weather radar system.

Another version of the EFIS found on many corporate aircraft is shown in Figure 17–25. This system incorporates two EADIs and two EHSIs, one each for the pilot and copilot. These four tubes are referred to as the **primary displays.** The term *tube* is often used to refer to the *display*. The right-side electronic displays receive data inputs from the right-side **display processor unit** (DPU). The left-side displays receive data from the left DPU. The DPUs receive data inputs directly from the aircraft systems and from the **multifunction processor unit** (MPU). Figure 17–26 shows an operational EADI and EHSI. In the event of one primary display failure, the system can be switched into the **reversionary mode,** which allows one tube to display a compact form of both the ADI and HSI.

The **multifunction display** (MFD), typically located in the center console of the flight deck, is used as the fifth display of the system. An EFIS with an MFD is often referred to as a **5-tube EFIS.** The MFD unit is typically installed in the location reserved for the radar display and is therefore accessible to both members of the flight crew. The MFD is different from the other two displays in that it contains its own power unit, checklist data file, and display controls. During a normal flight, the MFD will display navigation and weather radar information. In the event of a system malfunction, the MFD can be used as a backup for the primary displays. The MFD will also display diagnostic information. The MFD receives its display information from the multifunction processor unit (MPU).

The MPU, typically located in the avionics rack, receives input signals from the various aircraft sensors on both the right and left sides of the aircraft. Weather input data are received from the weather radar r-t unit. As shown in Figure 17–25, the MPU communicates with both the right and left DPUs as well as the MFD. The MPU can supply input data to the right or left DPU in the event of a sensory data failure to a DPU.

A variety of electronic flight instrument systems are available today, each with its own distinguishing characteristics.

FIGURE 17–22 The A-320 instrument panel. *(Airbus Industrie)*

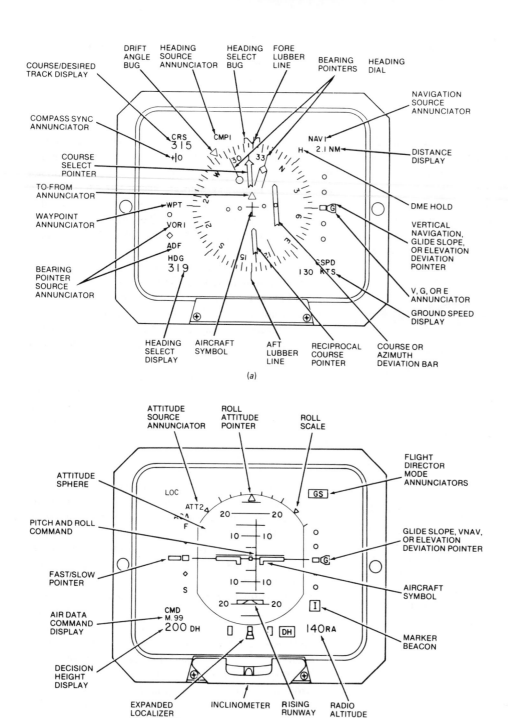

FIGURE 17–23 Electronic flight instruments. (*a*) Electronic horizontal-situation indicator; (*b*) electronic attitude-director indicator. *(Sperry Corporation)*

A variety of data bus systems are used for information transfer between EFIS components. Before troubleshooting, always become familiar with the particular system on the aircraft. Most EFISs do include some type of built-in test equipment (BITE). Diagnostic data from the BITE are often accessible from one of the electronic displays.

Some of the more common failures that occur in an EFIS are caused by wiring problems. Connector plugs often become loose or corroded and cause an intermittent or constant failure. Before replacing the various line replaceable units (LRUs) in the system, always check the system wiring. The LRUs have an extremely high average time before failure. Therefore, it is unlikely that swapping an LRU will solve the problem. LRUs do fail; but efficient troubleshooting means checking the wiring and system sensors before sending an LRU to the repair shop.

On most EFISs the left and right components are interchangeable. If this is the case on your aircraft, simply remove

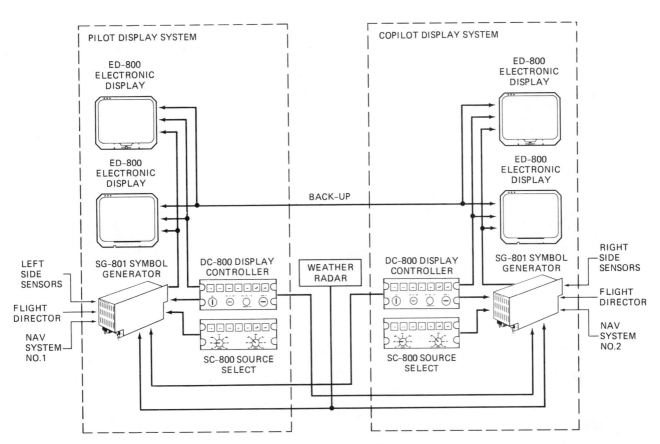

FIGURE 17–24 Block diagram of an electronic flight instrument system. *(Sperry Corporation)*

the suspect LRU, and replace it with the same unit from the other side of the aircraft. **Always be sure the LRUs are compatible before swapping, and always remove power from the system before removing or installing any unit.**

EICAS and ECAM

The **engine indicating and crew alerting system** (EICAS) and the **electronic centralized aircraft monitoring** (ECAM) system are two variations of the basic electronic flight instrument system. Like the EFIS, the EICAS and the ECAM system employ digitally controlled CRT displays. However, EFIS is used to display flight data, while EICAS and the ECAM system are used to display various system parameters, such as engine pressure ratio, **rpm**, and exhaust gas temperature. Other parameters, such as hydraulic system pressures and electrical system parameters, can be displayed or, in some cases, removed from the screen at the discretion of the pilot. Another vital function of the EICAS and the ECAM system is to monitor the various aircraft systems and display caution and warning information in the event of a system failure. Although some corporate aircraft use EICAS or the ECAM system, these systems are found mostly on transport-category aircraft. The EICAS is used on modern Boeing aircraft; the ECAM system is used on modern Airbus aircraft.

EICAS

The **EICAS** displays certain aircraft system and engine parameters on a demand basis. That is, not all system data are

displayed continuously. In the event of a system malfunction, any vital information automatically appears on a CRT, and the appropriate caution or warning signals are activated. During normal operation, only a minimum of engine data are displayed; additional system data may be displayed upon activation of the appropriate EICAS control. The Boeing 757 and 767 utilize an EICAS containing two CRT displays as shown in Figure 17–27. During normal flight configurations, the lower CRT is typically blank. The lower CRT is used to display the status of any malfunctioned system, and it acts as a backup display in the event the top CRT fails. The upper CRT is called the **main EICAS display,** and the lower CRT is called the **auxiliary EICAS display.**

Several formats are used by the various EICASs. These formats vary with the aircraft model. Some common formats include primary, secondary, and compact modes. The primary format shows on the main display during normal operation. Four different colors are used to display information. A change in color indicates a change in system status. Four alert messages are available in the primary format. **Level A** messages are **warning** messages and are shown in red. These are the most important messages. **Level B** messages are **caution** messages and are displayed in amber. **Level C** messages are **advisories** and are displayed in amber or light blue, depending on the specific EICAS model. **Level D** messages, called **memos,** are displayed in white.

Warnings, or level A messages, indicate conditions that require immediate attention and immediate action by the flight crew. Level A alerts include cabin depressurization and

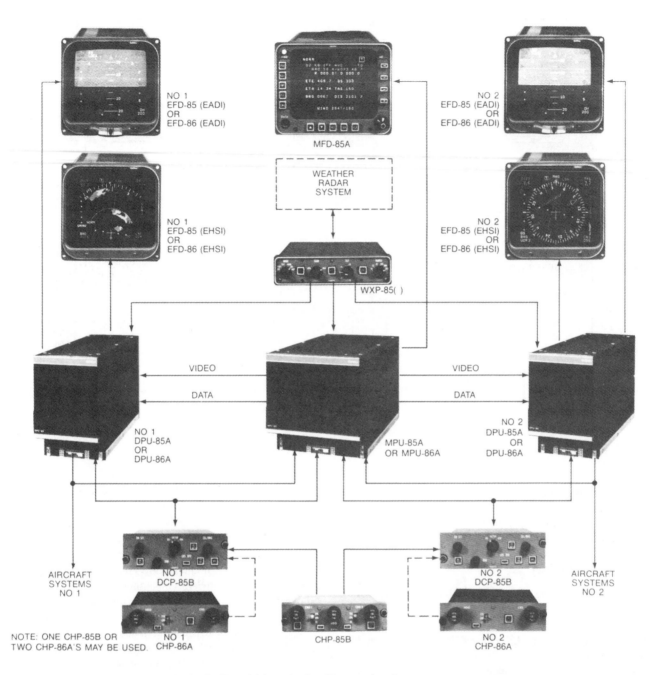

FIGURE 17–25 A typical EFIS system. *(Collins Divisions, Rockwell International)*

engine fire. Warnings activate other discrete aural and visual annunciators on the flight deck, such as the fire bell. There are very few possible level A alerts. Level B messages, or cautions, appear in amber just below any level A message. A different discrete sound and annunciator light will be activated for these alerts. Cautions require immediate crew awareness and future crew action. Advisory messages require immediate crew awareness and possible future action. Memos are used for crew reminders. For all levels of alerts, the most recent message appears at the top of its category. Level A messages appear at the top of the display, level B is below level A, level C is next, and level D is shown at the bottom of the list.

At power-up the secondary format automatically appears on the auxiliary EICAS display. The secondary format shows

engine parameters, such as N2 rotor speed, fuel flow, oil pressure, oil temperature, oil quantity, and engine vibration.

The compact mode is used when one display is inactive or being used to show maintenance pages. In the compact mode the data are typically displayed in digital format only. In other words, vertical or round dial instrument representations are eliminated. The compact mode is used during flight if one display becomes inactive. Compact information can be shown on either the main or auxiliary display.

A block diagram of the B-757 EICAS is shown in Figure 17–28. The two EICAS computers monitor over 400 inputs from engine and airframe systems to alert the crew in the event of a system malfunction. An EICAS computer uses both analog and digital data to communicate with the various

FIGURE 17–26 A typical EADI (top) and EHSI (bottom).
(Collins Divisions, Rockwell International)

components of the system. The ARINC 429 data bus system is used for most EICAS digital data transmission and reception processes. But be aware that other data bus formats may be used on certain systems.

ECAM System

The **ECAM** system is very similar to the EICAS just described. The ECAM system also employs two CRT displays. On some aircraft, such as the A-320, the displays are mounted side by side. On other aircraft, such as the A-330 and A-340, they are mounted one above the other. The ECAM system incorporates a more graphic representation for many of the aircraft systems.

Four display modes are used by the ECAM system: the **flight phase, advisory, failure-related,** and **manual** modes. The flight phase displays information needed for a particular segment of a flight. The advisory mode displays information about system status that may require crew attention. The failure-related mode takes precedence over the other two automatic modes and displays information on system status that

requires immediate crew action. The manual mode is used to select the status of any monitored system.

The ECAM system also has a system test routine. The self-test is performed during each power-up. The test mode can also be activated manually from the maintenance panel. This portion of the system will recall any system defects from a nonvolatile memory to aid in system troubleshooting.

Since so many parameters are monitored by both the EICAS and ECAM computers, they are valuable troubleshooting tools for airline technicians. A quick reference of the EICAS or ECAM displays will provide much of the preliminary information needed for troubleshooting defective systems.

AUTOMATIC FLIGHT CONTROL SYSTEMS

For many years large aircraft such as airliners, transport aircraft, and military aircraft have been equipped with automatic pilots (automatic flight control systems) to relieve the human pilot and other members of the flight crew of the tedious duty of keeping the aircraft on course for periods of many hours. Today automatic pilots are installed on all large aircraft of both commercial and military types and on many light aircraft.

Basic Theory of Automatic Pilots

The use of gyros to develop electric signals for the operation of control surfaces is illustrated in Figure 17–29. The fact that a gyro will hold a constant position in space makes it possible to develop a relative motion between the gyro and its supporting structure when the supporting structure changes position. When the gyro is supported with gimbal rings and is installed in an airplane as shown in Figure 17–29, any pitch or roll of the airplane will cause a relative movement between the gyro and the airplane. This movement can be used to operate contact points, move the wiper of a potentiometer, or reposition proximity indicators, thus providing a means for operating aircraft-control surfaces in response to the movement. In Figure 17–29 the two upper drawings show the gyro moving the wiper of a potentiometer to develop a proportional electric signal that may be amplified and used to apply a flight correction. In the two lower drawings of the illustration, the gyro is shown operating a set of double-throw contact points. In this case the closed contacts direct a signal to a relay, thus sending power to a servo motor that will move the control surfaces in the appropriate direction to level the aircraft.

Essential Features

In any autopilot system, it must be remembered that a reversal of the corrective signal must take place before the aircraft has returned to its corrected attitude. For example, if the rudder of an aircraft is moved to the right by the autopilot to correct the aircraft heading, it must be moved back to neutral by

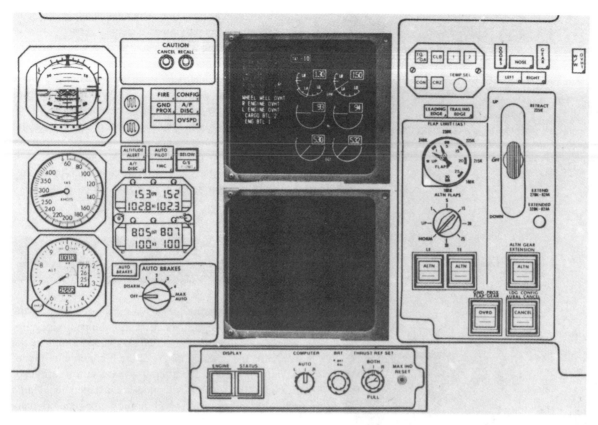

FIGURE 17–27 Two EICAS displays from a Boeing 757. *(Boeing Corporation)*

the time that the aircraft has reached its corrected position. Otherwise, the system will overshoot and cause the aircraft to oscillate violently. To accomplish this signal reversal, **feedback** systems are employed.

In electronic autopilot systems, the gyros that sense flight deviations of the aircraft develop corrective signals by means of transformer action or a similar principle. In this way no friction is developed, and very little mechanical force is evident to restrict the movement of the gyro. One type of signal-generating device is called an **EI pickoff** and is diagrammed in Figure 17–30. The pickoff consists of three coils mounted on an E-shape piece of laminated steel. The coils wound on the outer legs of the E are connected together in such a phase relationship that the voltages induced from the center coil cancel. A movable steel armature, called the I member, provides a low-reluctance path for the magnetic field. When this armature is moved relative to the E section, it changes the ratio of coupling between the secondary windings and the primary, and this changes the value of the induced voltages in the outside legs so that they are no longer equal. The result is a voltage that is either in phase or out of phase with the excitation voltage. Note that the excitation voltage is applied to the center leg of the E section.

The electric voltage output from the secondary coils of the EI pickoff assembly is proportional to the displacement of the elements away from the null position. The null is the mechanical position that results in zero electric output; it is indicative of the fact that the aircraft is in its proper flight position in relation to the reference established by the gyro

unit. Movement of the aircraft away from the established flight reference causes relative mechanical displacement of the pickoff elements away from the electrical null, thus producing an electric output voltage that is the signal to the system calling for a corrective action.

The information gained from the signal system provides the means whereby the control surfaces of the aircraft are moved to cause corrective action that returns the aircraft to the proper flight position. The extremely small amount of power represented by the signal output cannot, in itself, operate the controls on the aircraft; therefore, it must be suitably amplified. This can be accomplished by means of a servo amplifier containing a discriminator system or by a computer amplifier.

Basic Electronic Autopilot System

A block diagram of a basic electronic autopilot system is shown in Figure 17–31. Note in the diagram that the system includes a follow-up feedback from the control surface to the preamplifier, a feedback signal from the servo motor to the sensing gyro, and a rate network that tells the system how rapidly a change is taking place.

The follow-up, or feedback, system consists of an electric pickoff similar to that used with a gyro, but it is connected through linkage to the control surfaces of the aircraft. To provide an extremely sensitive control so that deviations from course will be small and the aircraft will be returned without overshooting the course, a further addition is made to the

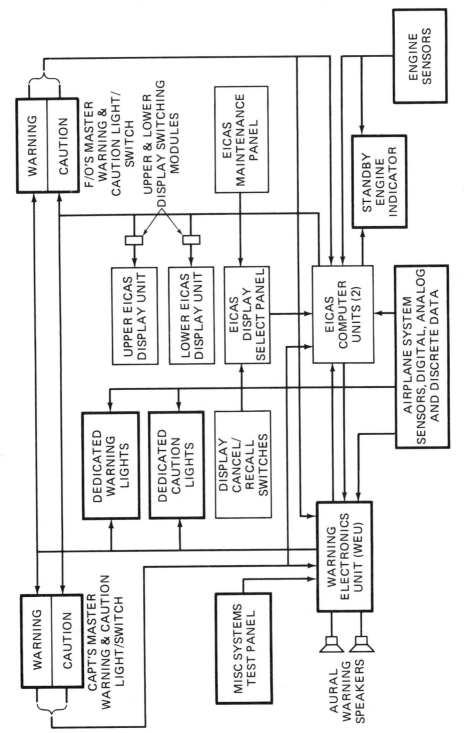

FIGURE 17–28 Block diagram of the B-757 engine indicating and crew alerting system. *(Boeing Company)*

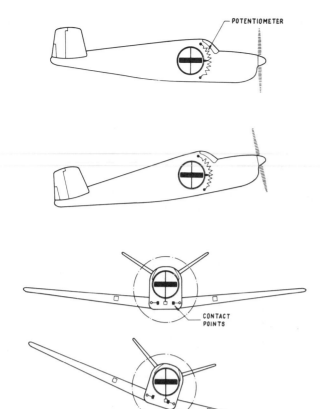

FIGURE 17–29 Operation of gyro sensing units.

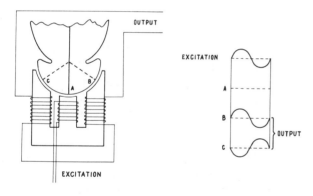

FIGURE 17–30 Electromagnetic signal-generating device for a gyro sensing unit.

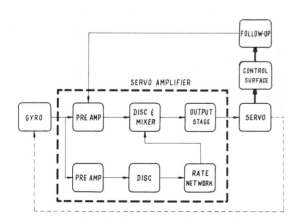

FIGURE 17–31 Block diagram of an autopilot system.

servo control system in the form of a rate network, which modifies the servo-amplifier output signal. With the proper values in the rate circuits, it is possible to adjust the response of the autopilot so that each correction is rapid and smooth.

TYPICAL AUTOMATIC PILOT AND FLIGHT CONTROL SYSTEM

Automatic pilots are manufactured in many configurations by a number of companies. Some systems are comparatively simple, while others become complex, especially when integrated with navigation systems. As explained previously, all systems utilize gyros to sense aircraft attitude and provide signals for correction.

For the purpose of this section, the Bendix M-4D automatic pilot and flight control system has been selected for description. This system includes not only the basic elements of autopilot operation but also components that give it all the capabilities of an automatic flight control system. A number of systems by other manufacturers utilize similar principles and perform the same functions.

The system described here can be programmed to fly a predetermined course (either NAV or RNAV), maintain a selected altitude, capture a VOR radial or ILS beam from any angle, make back-course approaches, and perform other functions. The system also includes automatic pitch trim, pitch synchronization, pitch integration, and altitude control. The computer portion of the system can also be used to display computed command data on a director-horizon indicator (flight director) and give directional data on a horizontal-situation indicator. Thus the aircraft is provided with a fully integrated flight control system.

System Components

The arrangement of the principal components of the M-4D automatic pilot and flight control system in an aircraft is shown in Figure 17–32. The basic autopilot is composed of the controller, gyros, servos, and the section of the computer-

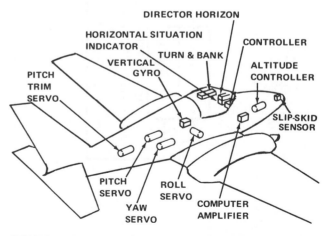

FIGURE 17–32 Arrangement of the components of an automatic pilot and flight control system. *(Bendix Avionics Division)*

FIGURE 17–33 Flight controller. *(Bendix Avionics Division)*

amplifier that accepts signals from the gyros and converts them into flight commands for the servos.

Flight Controller. The flight controller, shown in Figure 17–33, is the unit by which the human pilot controls the operation of the autopilot. The autopilot is engaged by pressing the AP button. An annunciator on the lower part of the AP button will illuminate when the autopilot is engaged. The YAW button provides for automatic control of rudder trim during changes in airspeed.

The HDG and NAV buttons command the autopilot to follow a preselected heading, couple with a VOR radial, or fly a preselected RNAV course. The APPR button commands the autopilot to capture a localizer beam on the approach to an airport or to approach an airport using a VOR radial or RNAV data. If the aircraft must approach from the opposite direction of the normal course signal, the REV button is pressed to provide the back-course information. The GS button commands capture and tracking of the glide slope signal, and the ALT button commands the autopilot to maintain a certain selected barometric altitude. All selector buttons are provided with annunciators that illuminate when each mode is active. The NAV annunciator will show RNAV or ON, depending on whether the system is operating on data from the RNAV computer or is tracking a VOR radial. The GS annunciator will show ARM in the APPR mode if the airplane is below the glide slope beam and the signal has been present for approximately 20 s. When the aircraft reaches the beam center, the annunciator will show ON. At this time the ALT and ARM annunciators will be off.

The TURN control is used to initiate either a 12° banked turn or a 24° banked turn to the right or left. Other lateral modes are released during operations of the TURN control. The TURN control must be in the center position when the autopilot is first engaged. The PITCH control is spring-loaded in the center position. When moved up or down, it commands pitch changes up to ±20°. During this operation, other longitudinal modes such as GS, ALT, or GA are disengaged. The GA button is on the pilot's control wheel and is used to command a go-around in case of a missed approach.

The ELEV indicator shows whether the elevator is in the neutral position. Before autopilot engagement, this meter shows the status of the autopilot signal relative to the pitch axis of the aircraft. It should always be near center because the system has automatic pitch synchronization. After engagement, the meter shows the force required by the primary servo to hold the aircraft in the desired pitch attitude.

The ROLL TRIM is used to trim the ailerons so the aircraft is in a level attitude. It should not be used with any of the lateral modes engaged or with the TURN control out of the center position.

Turn-and-Bank Indicator. The **turn-and-bank indicator** is an electrically driven gyro unit that provides a visual indication of the rate of turn and at the same time provides an electric signal for the autopilot with the same information. This signal is used by the computer-amplifier in the development of turn signals for the aircraft.

The ball inclinometer in the turn-and-bank indicator shows the pilot whether the aircraft is being flown with properly coordinated rudder and aileron control.

Director-Horizon Indicator. The **director-horizon indicator** is similar to instruments described as flight-director indicators or attitude-director indicators. This instrument, shown in Figure 17–34, serves several functions. It is an attitude indicator and an attitude sensor. That is, it senses attitude and develops electric signals that are sent to the computer-amplifier. These signals are amplified and sent to the primary servos commanding control-surface movement for control of pitch and roll. At the same time, the instrument is indicating the degree of pitch and roll for the information of the pilot.

The vertical gyro in the instrument can be driven either by a vacuum system or electrically, depending on the particular instrument selected for the system. Since the gyro remains vertical with respect to the surface of the earth, the pitch and roll movements of the airplane cause relative movements between the gyro and the instrument case, thus producing signals for the autopilot and indications on the instrument.

Servos. Three primary servos control pitch, yaw, and roll by moving the elevators, rudder, and ailerons in response to commands from the autopilot. These servos are small electric motors that drive capstans through magnetic clutches. The capstan in each servo contains an adjustable slip clutch that is preset for each axis by the installer. These clutches make it possible for the pilot to override the automatic pilot if necessary.

The **trim servo** is used to activate an elevator trim tab control to relieve long-term aerodynamic loading and generally assist in smoother operation of the elevator surface without requiring large amounts of power from the primary servo. Its operation is either automatic, using the composite pitch error signal from the computer-amplifier, or both automatic and manual if an optional manual electric trim adapter is utilized.

The location and installation procedures for all the servos depend on the type of aircraft in which they are being installed. Detailed instructions concerning cable tensions, clutch settings, and other pertinent data are included in the installation kit for the particular aircraft involved.

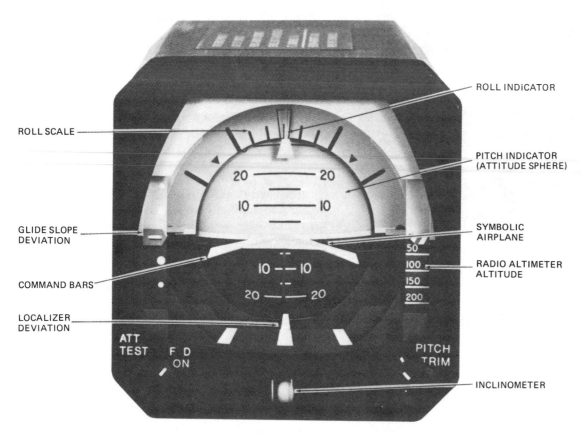

FIGURE 17–34 Director-horizon indicator. *(Bendix Avionics Division)*

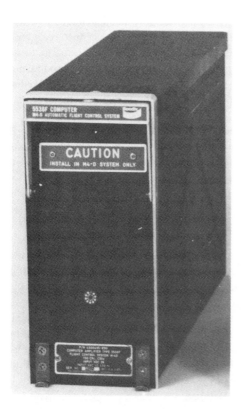

FIGURE 17–35 Computer-amplifier. *(Bendix Avionics Division)*

Computer-Amplifier. Figure 17–35 shows the computer-amplifier for the Bendix M-4D automatic flight control system. The computer-amplifier is the heart and brain of the system. It consists of nine plug-in modules, one of which is peculiar to a particular aircraft-type installation and another of which relates to a specific type of flight director installation. The modules can be removed and replaced easily in case of malfunction.

The computer-amplifier combines inputs from the gyro sensors, control panel, and heading and radio sources (NAV and RNAV) to compute appropriate electrical commands and deliver them to the control-surface servos and to the flight-director instrumentation.

The computer-amplifier is mounted on a shock mounting especially designed to receive it. The mounting is shown in Figure 17–36. The location of the installation depends on the type of aircraft in which the equipment is being installed. The installation kit for the particular aircraft gives detailed instructions.

AUTOMATIC FLIGHT AND LANDING SYSTEMS

It is the purpose of this section to give the student a general understanding of the concept of a completely integrated flight control system that enables an aircraft to take off, fly a prescribed route, and land at a designated airport without the aid

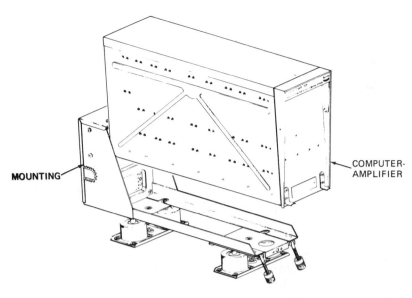

FIGURE 17–36 Mounting of a computer-amplifier. *(Bendix Avionics Division)*

of a human pilot. The system described here is one configuration designed for the Lockheed L-1011 airplane. The complete system includes the elements described in the first part of this chapter plus other systems and subsystems that provide the capability of automatic takeoff and landing.

A number of automatic flight control systems have been developed, and even though they may vary in some respects, they must have the ability to track ILS and VOR signals; maintain prescribed altitudes; adjust power for takeoff, climb, cruise, and landing; operate flaps, spoilers, and landing gear as required; and maintain proper flight attitudes. In each such system, the autopilot flies the airplane as it responds to commands from attitude sensors, navigation systems, and flight control units. Power is controlled through the engine throttles, which are moved by throttle servos responding to commands from throttle computers. Flaps, landing gear, and spoilers are operated by an electronic system that is programmed to provide proper flap extension for takeoff and landing, landing-gear retraction and extension at the proper times, and spoiler deployment as necessary to reduce lift and act as an air brake.

Subsystems

The complete system described here is designated as the **automatic flight control system** (AFCS) and might also be called an avionics flight control system. It provides manual or automatic modes of control throughout the entire flight envelope from takeoff to landing and rollout. This is achieved in an integrated system composed of the **autopilot/flight director system** (APFDS), the **stability augmentation system** (SAS), the **speed control system** (SCS), and the **primary flight control electronic system** (PFCES). Figure 17–37 is a block diagram showing how the components of these systems are interrelated. All the subsystems of the AFCS are fully integrated and have levels of redundancy to achieve a high level of reliability. Redundancy is accomplished by providing two or more systems of each type so a failure of one system will not affect the operation of the complete system.

Autopilot/Flight Director System. The APFDS, through the autopilot, provides automatic pitch and roll control to stabilize the aircraft and maintain the selected altitude, attitude, and heading in flight. In this fully automatic mode, the flight director can be used in a monitoring capacity; that is, the human pilot can watch the flight director (ADI) to observe the operation of the autopilot. In other modes, the flight director may be used for flight guidance.

The APFDS operates in the modes listed below.

Altitude Select and Hold In this mode the autopilot maintains the altitude that has been programmed.

Vertical Speed Select and Hold This mode causes the autopilot to control rates of climb and descent in accordance with the rates that have been programmed.

Heading Select and Hold The autopilot controls the airplane to maintain selected headings when this mode is in operation.

Control Wheel Steering (CWS) This mode allows the human pilot to change commands to the autopilot by moving the control wheel. After a change is made, the pilot releases the wheel, and the autopilot follows the new command.

IAS Select and Hold In this mode the human pilot selects the desired indicated airspeed (IAS), and the autopilot, together with the speed control system, maintains the selected airspeed.

Mach Hold In this mode the system uses Mach number rather than IAS as the speed reference.

Localizer This is an approach mode wherein the aircraft tracks the localizer beam of the ILS.

VOR and Area Navigation (RNAV) When in this mode, the system navigates automatically, utilizing VOR data or commands from an RNAV computer.

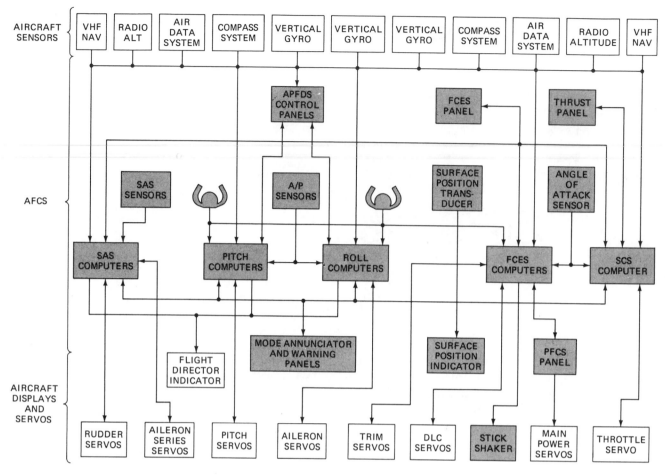

FIGURE 17–37 Block diagram showing the relationship among the components of an avionics flight control system. *(Lockheed California Company)*

Approach/Land As the aircraft approaches its destination, the speed control system reduces power as necessary for the correct rate of descent, and the navigation system locks onto the ILS beam for guidance to the runway. The radio altimeter provides altitude signals for landing, flare, and touchdown. Upon touchdown, the power is reduced to idle, and the aircraft tracks the localizer beam to roll out along the center of the runway.

Central Air-Data Computers

Two central air-data computers (CADCs) are employed to provide information for both manual and automatic flight. They receive pneumatic inputs from the pitot-static system and electric signals from the true air temperature (TAT) probes to generate altitude, airspeed, and temperature information. This information is then used to provide outputs for the air-data instruments and recorders, as well as for the automatic flight control system, the stability augmentation system, the Mach trim and feel systems, and various other purposes.

The incorporation of CADCs permits the use of electrically servoed instrumentation, including primary flight instruments, flight data recorders, and the optional true

airspeed (TAS) and static air temperature (SAT) indicators. This use of electrically driven instruments reduces the amount of pneumatic plumbing required behind the instrument panels to only those lines connected to the standby airspeed indicator and altimeter. The use of servoed instruments also makes it possible to incorporate Mach and overspeed warning switches into the electric flight instruments and permits the use of electric switching between the normal and alternate air-data systems.

Stability Augmentation System

In-flight stability and control of the aircraft are augmented by the stability augmentation system (SAS), which provides yaw damping. Yaw damping is the process of limiting the rate and degree of yaw to a safe and comfortable level. Two computers are used for improved reliability, and limited averaging improves tracking of the servos.

With either of the two dual-channel computers engaged, the SAS provides Dutch-roll damping and turn coordination during all phases of flight. Both computers are engaged for the runway alignment and rollout functions. This satisfies the fail-operational requirement that is specified for autoland. *Fail-operational* means that the system will continue to function even though there are failures in some parts of the system. The system, therefore, meets the requirements for Category III fully automatic blind-landing capability.

During autoland, the yaw computers start monitoring the availability of runway alignment and rollout, and the results are displayed on each AFCS mode annunciator during the approach-and-land sequence if the system is functioning normally. Failures are annunciated on the AFCS warning indicators.

All the computer channels receive a yaw-rate signal generated from one of three yaw-rate gyros. In the basic yaw SAS mode, turn coordination is achieved by processing signals from four aileron position transducers. In the LOC track or autoland mode the aileron position signals are switched out.

The runway alignment is generated as a function of altitude and alignment/rollout logic. Damping is provided by mixing heading and heading-rate signals with yaw-rate signals. The alignment scheme is a limited forward slip maneuver in which up to 8° of initial crab angle is removed by lowering a wing and slipping the aircraft. This is to align the aircraft with the runway in case of a crosswind.

The rollout mode is initiated at touchdown as a function of altitude. The rollout guidance system utilizes the LOC beam to track the center of the runway until speed is reduced to taxiing speed.

Speed Control System

The speed control system (SCS) provides an **airspeed autothrottle** mode and an **angle-of-attack autothrottle** mode. The airspeed mode is used for all flight conditions through initial approach, and the angle-of-attack mode is used in the final approach and landing. The SCS also provides the go-around command for both manual and automatic go-around, as well as the takeoff command for manual takeoff guidance.

The SCS consists of a single computer with two identical computation channels; a single, monitored autothrottle servo; and dual sensor inputs. The autothrottle function controls the engine throttles through the autothrottle servo to maintain a selected airspeed or a precomputed angle of attack that corresponds to a "stall-margin" airspeed.

The normal mode of operation for the autothrottle with the airplane in the cruise configuration is **airspeed select.** In the approach and landing configuration, the autothrottle is operated in the **stall-margin** mode. At the flare initiation altitude, the SCS switches to the **flare** mode, which provides automatic closed-loop throttle retard before touchdown. At touchdown the system switches to the **touchdown** mode, and the throttles are driven to idle, whereupon the autothrottle system automatically disengages. The takeoff and go-around computations are sent to the APFDS for use in controlling the pitch axis during takeoff and go-around maneuvers. The thrust panel by which the aircraft speed is controlled is accessible to both pilots. It consists of an autothrottle engage switch, an airspeed select control, a selected airspeed digital readout, and a stall-margin mode annunciator. In addition, the thrust panel contains the necessary elements to generate the airspeed error signal used by the SCS computer when in the airspeed-select mode.

A single, monitored servo drives the engine throttles at a rate proportional to the amplitude of the command signal. A dual-input bidirectional sprag clutch is continuously engaged except when a manual override force is applied to the throttle levers. With the autothrottle engaged or disengaged, there is no substantial difference in the force required to manually operate the throttle levers.

The autothrottle operating range is defined by limit switches located at the maximum thrust limit, the minimum thrust limit, and the idle position. Actuation of the maximum- or minimum-thrust-limit switches will open the control phase of the servo motor, thus stopping further throttle motion in that direction. With the airplane on the ground, as sensed by the main-landing-gear strut compression switches, actuation of the idle-disconnect switch will result in total disengagement of the autothrottle.

The autothrottle servo is engaged by moving the engage switch on the thrust panel to the ATS position. In this configuration, the servo will drive the engine throttle levers to satisfy the computed throttle command. Disengagement of the autothrottle servo is accomplished by movement of the engage switch to off; actuation of either of the two throttle-lever-mounted disconnect switches; detection of a failure in either of the autothrottle computation channels or the servo loop; selection of the IAS HOLD, MACH HOLD, or TURBULENCE mode of the APFDS; or engagement of the takeoff or go-around mode. The go-around mode is selected by means of the go-around switch on either the captain's or first officer's control wheel. Alert lights will flash on the AFCS warning indicator when the autothrottle is disengaged.

Primary Flight Control Electronic System

The **primary flight control electronic system** (PFCES) consists of various automatic control, warning, and indicating subsystems that are principally concerned with manual aircraft control. These subsystems provide means for operating and monitoring the many control surfaces that are necessary for the safe and efficient flight of a large airplane. The systems operate for both automatic and manual flight.

The positions of flight control surfaces must be known to the pilots, and these positions are displayed by means of the **surface position indicator.** This helps the pilots to verify proper operation of the flight control surfaces, primarily during ground operation. An autopilot mistrim display is also included. Control and trim positions are provided for spoilers, the rudder, stabilizer trim, and aileron trim.

The **primary flight control monitoring system** (PFCMS) detects and displays to pilots the means for alleviation of opens and jams in the pitch-axis and jams in the roll-axis control systems. Two independent sensors and monitoring channels are used in each instance, and warnings can result from either channel or both.

The **rudder-control limiting system** automatically restricts the rudder authority and limits rudder hydraulic power capability during high-speed flight. The system mechanically limits rudder deflection as a function of airspeed when the flaps are retracted with less than 4° of flap deflection.

The **spoiler mode-control system** automatically changes

the configuration of the roll and speedbrake inputs to the spoilers to optimize roll, direct lift control, and speedbrake control characteristics for low- and high-speed flight. Since the spoilers are used to control lift, speed, and roll, the selection of the spoilers to be employed in each mode is critical. The spoilers operate in conjunction with the **direct lift control** (DLC)/**automatic ground speedbrake** (AGSB) system so that they can respond automatically during approach and landing through the use of the DLC and be deployed automatically for braking after landing or a rejected takeoff.

The **stall warning system** artificially vibrates the control columns to warn of an impending stall. Stall warning computations are based on angle-of-attack measurements as modified by position measurements of flaps and leading-edge slats to obtain an unmistakable warning at a minimum of 7 percent above stalling speed.

The **electric pitch trim system** permits electrical control of pitch axis trim by the pilots through the use of thumbwheels on the control wheels. The system is operated automatically during automatic flight.

Mach trim, as required by FAR, is used to control the average gradient of the stable slope of the stickforce-versus-speed curve to less than 1 lb [0.45 kg] for a change in speed of 6 kn [11 km/s]. Trim changes are scheduled from Mach data derived from the two CADCs. The system normally controls only one of two electric motors in the electric trim and feel mechanism. The data are processed by the trim augmentation computer.

The **Mach feel system** automatically adjusts the pitch-axis feel-force gradient for all flight conditions. This is a two-channel active-standby system that operates together with the stabilizer trim angle and Mach number to provide proper pitch-control-force characteristics as established by FAR requirements; that is, the pitch control force at limit load must be controlled within 50 to 100 lb [22.7 to 45.4 kg].

The **pitch trim disconnect system** is a means of alleviating mechanical jams in the pitch trim system. When the automatic pitch trim system is disconnected, pitch trim control can then be continued, either by use of the electrical thumbwheel controls or through the mechanical trim wheels. If only electrical trim is operative, there is a loss of series trim output. If only mechanical trim is operative, there is no effect on performance of the trim system.

The **altitude alert system** indicates, by visual and aural warnings, approaches to and deviations from selected altitudes. It also provides for automatic capture of selected alti-

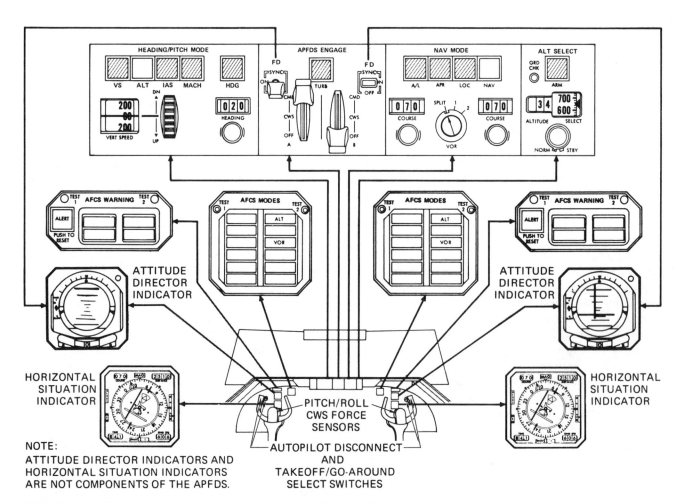

FIGURE 17–38 Flight station equipment for an APFDS. *(Lockheed California Company)*

tudes through the autopilot/flight director system. The system employs dual redundant radar altimeters, computational channels, and annunciators. Each channel is completely independent, inclusive of electric power source, so that warnings will result from both channels.

The altitude alert system is integrated with the altitude-capture and altitude-hold functions of the APFDS and utilizes the same controls on the ALT SELECT panel as those used for the altitude-select function of the APFDS. However, the altitude alert system operates independently of the autopilot.

The major elements of the PFCES, excluding interfacing components, are a flight control electronic system (FCES) computer, a trim-augmentation computer, a trailing-edge flap-load-relieving-system computer, angle-of-attack sensors, stick shakers, control surface position transducers, and associated flight station control panels and indicators. Functions of the FCES computer are AFCS monitoring, rudder limiting, direct lift control, automatic ground spoiler control, stall warning, altitude alert, and fault-monitoring indication.

The trim-augmentation computer's functions are manual and automatic pitch trim and Mach-trim and Mach-feel compensation. The fault isolation monitoring indication employs a single computer channel, and the direct lift control has a fail-operational capability to meet autoland requirements. All the other functions have dual computer channels with a fail-passive capability.

Flight Station Equipment

Equipment for control and operation of the autopilot/flight director system (APFDS) is located to be easily accessible to the pilots. The flight station equipment is shown in Figure 17–38. Note in the drawing that the controls permit selection of any mode of APFDS operation desired plus the selection of the parameters for the flight. When fully automatic flight is programmed, additional data must be entered through other controls.

The AFCS mode annunciation unit is shown in Figure 17–39. In the drawing all the modes are illuminated as when the test 1 button is pressed. During normal flight, the only modes illuminated are those in actual use.

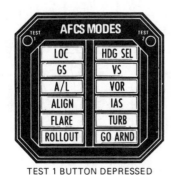

TEST 1 BUTTON DEPRESSED

FIGURE 17–39 AFCS mode annunciation unit. *(Lockheed California Company)*

A **flight management system** (FMS) is a computer-based flight control system. A typical FMS is capable of four functions: automatic flight control, performance management, navigation and guidance, and operation of status and warning displays. Simply stated, an FMS is a complex digital autopilot/flight director system that is capable of flight control and thrust management to a least-cost or least-time configuration.

Figure 17–40 illustrates the components of the FMS on a Boeing 757. This system utilizes two **flight management computers** (FMCs) for redundancy purposes. During normal operation both computers **cross-talk;** that is, they share and compare information through the data bus. Each computer is also capable of operating completely independently in the event of one failed unit. An FMC receives input data from four subsystem computers: the **flight control computer** (FCC), the **thrust management computer** (TMC), the **air-data computer** (ADC), and the **engine indicating and crew alerting system** (EICAS) computer. The communications between these computers are typically in an ARINC 429 data format. Other parallel and serial data inputs are received from flight deck controls, navigation radios, and various airframe and engine sensors. Each vital component of the FMS is duplicated to ensure system reliability.

A flight management computer and control display unit are shown in Figure 17–41. The FMC contains a 4-million-bit nonvolatile memory, which stores the performance and navigation data along with the necessary operating programs. Portions of the nonvolatile memory are used to store information concerning airports, standard flight routes, and various navigational aids. Since this information changes from time to time, the FMS incorporates a **data loader.** The data loader is a tape or disk drive that can be plugged into the FMS. Any data updates are entered periodically through the data loader.

The variable flight parameters for a specific flight are entered into the FMS by means of the alphanumeric keyboard of the **control display unit** (CDU) (Figure 17–41). During preflight the flight crew first enters all flight plan information. Data such as initial latitude and longitude of the aircraft, navigational waypoints, destination, alternates, and flight altitudes are all entered into a temporary legend. The flight plan is then generated and displayed by the FMS. If the crew agrees with this flight plan configuration, the temporary information is transferred·to active status. The performance data are selected in a similar manner. Performance data include takeoff, climb, cruise, and descent parameters. Performance can be set at a least-cost or least-time en route configuration.

During a normal flight, the FMS sends navigational data to the electronic flight instrument system. The EFIS then displays a route map on the EHSI. If the flight plan is altered by the crew, the EHSI map will change automatically. Since there are two CDUs in an FMS, during flight one unit is commonly used to display performance data, and the other is used to display navigational information. Both CDUs are

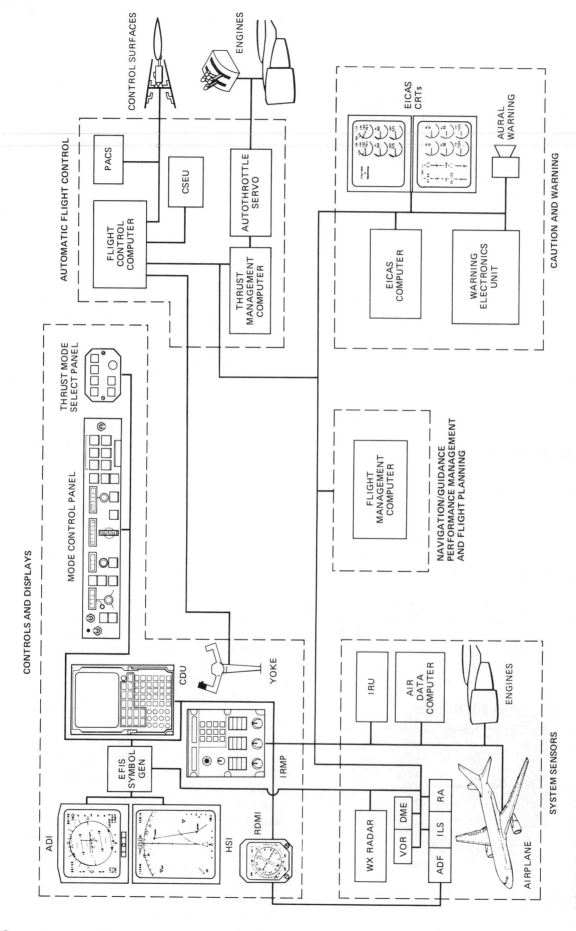

FIGURE 17–40 Block diagram of a flight management system.   *(Boeing Company)*

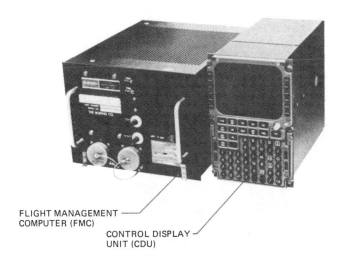

FLIGHT MANAGEMENT
COMPUTER (FMC)
CONTROL DISPLAY
UNIT (CDU)

FIGURE 17–41 Flight management computer and control display unit. *(Boeing Company)*

located in the center console, between the pilot and copilot stations.

A hybrid version of the FMS is being introduced on the Airbus Industrie's A-330 and A-340. The A-330 is a twin-engine aircraft; the A-340 is a four-engine long-range aircraft. The autoflight system on these aircraft is controlled through the **flight management guidance and envelope computer** (FMGEC). The FMGEC system is designed to allow both aircraft to have nearly identical flight characteristics. These aircraft are complete fly-by-wire airplanes. That is, there is no mechanical linkage between the control stick and the control surfaces. All control commands are routed through the FMGEC.

The FMGEC is similar to the FMS in that it relies on a data bus linkage to various other computers in the aircraft. The FMGEC communicates with the various sensors throughout the aircraft and the flight deck controls using both digital and analog inputs. This information is analyzed and synthesized; then output signals are generated that control the airplane with the help of subsystem computers.

STRAPDOWN TECHNOLOGY

The term **strapdown technology** refers to the use of **ring laser rate sensors** for aircraft inertial navigation and flight reference systems. Solid-state precision angular rate sensors eliminate the limitations of gimbals, bearings, torque motors, and other moving parts found on conventional gyro-type systems. The heart of the strapdown system is the **ring laser gyro** (RLG).

The Ring Laser Gyro

The RLG is actually an angular rate sensor and not a gyro in the true sense of the word. Conventional gyros generate a gyroscopic stability through the use of a spinning mass. The gyro's stability is then used to detect aircraft motion. The

RLG uses changes in light frequency to measure angular displacement.

The term *laser* stands for "light amplification by stimulated emission of radiation." The RLG system utilizes a helium-neon laser; that is, the laser's light beam is produced through the ionization of a helium-neon gas combination. A typical RLG is shown in Figure 17–42. This system produces two laser beams and circulates them in a contrarotating triangular path. As shown in Figure 17–43, the high-voltage potential between the anodes and cathode produces two light beams traveling in opposite directions. Mirrors are used to reflect each beam around an enclosed triangular area. The resonant frequency of a contained laser is a function of its optical path length. When the RLG is at rest, the two beams have equal travel distances and identical frequencies. When the RLG is subjected to an angular displacement around an axis

FIGURE 17–42 A ring laser gyro. *(Honeywell Sperry Commercial Flight Systems Division)*

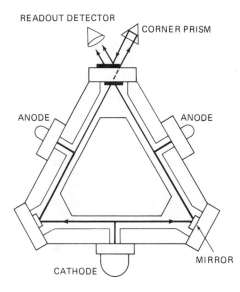

READOUT DETECTOR   CORNER PRISM
ANODE   ANODE
CATHODE   MIRROR

FIGURE 17–43 A pictoral diagram of a ring laser gyro. *(Honeywell Sperry Commercial Flight Systems Division)*

FIGURE 17–44 The components of a laser inertial reference system. *(Honeywell Sperry Commercial Flight Systems Division)*

perpendicular to the plane of the two beams, one beam has a greater and the other a shorter optical path. Therefore, the two resonant frequencies of the individual laser beams change. This change in frequency is measured by photosensors and converted into a digital signal. Since the frequency change is proportional to the angular displacement of the unit, the system's digital output signal is a direct function of the angular rate of rotation of the RLG.

The RLG system is typically coupled to a complete navigation system. The digital signals from the RLG can be used to control inertial reference navigation systems and/or autopilot functions. A laser inertial reference system is shown in Figure 17–44. An inertial sensor assembly containing the triangular lasers is shown in Figure 17–45. Strapdown technology makes possible a new era of aircraft safety. The elimination of moving parts greatly improves this system's reliability.

REVIEW QUESTIONS

1. List some of the parameters that are indicated by electric instruments in aircraft.
2. Why is it often necessary to utilize electric instruments rather than direct-driven instruments in an aircraft?
3. Why is a tachometer important in the operation of an aircraft engine?
4. Explain why an electric tachometer is more suitable than a mechanically driven tachometer in large aircraft.
5. Explain the operation of an ac tachometer system.
6. Describe the operation of a thermocouple temperature indicator.
7. Explain the important factors in the installation of a thermocouple system.
8. What type of instrument system(s) may employ a Wheatstone-bridge circuit?
9. What is a synchro system?
10. For what indications is a Selsyn system useful?
11. Describe an Autosyn system.

FIGURE 17–45 An inertial sensor assembly containing three ring laser gyros. *(Honeywell Sperry Commercial Flight Systems Division)*

12. What electric power is used for an Autosyn system?
13. Describe the operation of a float-type electric fuel-quantity indicator.
14. Explain the operation of a capacitor-type fuel-quantity indicator.
15. Why does the capacitance of the fuel-level sensor change as fuel level changes?
16. Describe a typical electronic flight instrument.
17. What are the basic differences between the EICAS and ECAM systems?
18. What type of information is displayed by EICAS?
19. How does the EICAS display information concerning a failed system?
20. What happens if one EICAS CRT fails?
21. What is the function of a digital data bus as it pertains to the FMS?
22. What device is used to produce aircraft-attitude signals for an autopilot?
23. Why is there a need for a feedback system for each correction command signal in an autopilot?

24. What are the principal components of a basic electronic autopilot?
25. What does a typical flight director system consist of?
26. Describe a primary servo unit.
27. What is the purpose of a slip clutch on a servo?
28. Give the functions of the computer-amplifier.
29. What is the purpose of an air-data computer (CADC)?
30. Describe the function of the yaw stability augmentation system (SAS).
31. By what means does the speed-control system (SCS) control the speed of the aircraft?
32. What are the major functions of an FMS?
33. What is an FMGEC?
34. Describe the basic concept of the strapdown technology theory.
35. What is a strapdown gyro?
36. What is the main advantage of the laser gyro as compared to a conventional gyro?

Appendix

Ohm's law

$$I = \frac{E}{R} \quad R = \frac{E}{I} \quad E = IR$$

where I = current (intensity of current flow)
E = voltage (emf)
R = resistance

Resistances in series

$$R_t = R_1 + R_2 + R_3 \cdots$$

Resistances in parallel

$$R_t = \frac{1}{1/R_1 + 1/R_2 + 1/R_3 \cdots}$$

or

$$\frac{1}{R_t} = \frac{1}{R_1} + \frac{1}{R_2} + \frac{1}{R_3} \cdots$$

Two resistances in parallel

$$R_t = \frac{R_1 \times R_2}{R_1 + R_2}$$

Capacitances in series

$$C_t = \frac{1}{1/C_1 + 1/C_1 + 1/C_3 \cdots}$$

or

$$\frac{1}{C_t} = \frac{1}{C_1} + \frac{1}{C_2} + \frac{1}{C_3} \cdots$$

Capacitances in parallel

$$C_t = C_1 + C_2 + C_3 \cdots$$

Inductances in series, no magnetic coupling

$$L_t = L_1 + L_2 + L_3 \cdots$$

Inductances in parallel, no coupling

$$\frac{1}{L_t} = \frac{1}{L_1} + \frac{1}{L_2} + \frac{1}{L_3} \cdots$$

Electric power in a dc circuit

$$P = EI \quad \text{or} \quad P = I^2 R \quad \text{or} \quad P = \frac{E^2}{R}$$

where P = power, W
1 hp = 550 ft · lb/s = 746 W
1 J = 1 W/s

Frequency and wavelength

$$f = \frac{300,000,000}{\lambda}$$

$$\lambda = \frac{300,000,000}{f}$$

where f = frequency, Hz
λ = wavelength, m

Capacitive reactance

$$X_C = \frac{1}{2\pi f C}$$

where X_C = capacitive reactance, Ω
f = frequency, Hz
C = capacitance, F

Inductive reactance

$$X_L = 2\pi f L$$

where X_L = inductive reactance, Ω
f = frequency, Hz
L = inductance, H

Resonant frequency

$$f = \frac{1}{2\pi\sqrt{LC}} \quad \text{or} \quad f = \frac{0.159155}{\sqrt{LC}}$$

Impedance: series circuit

$$Z = \sqrt{(X_L - X_C)^2 + R^2}$$

where Z = impedance, Ω
X_L = inductive reactance, Ω
X_C = capacitive reactance, Ω
R = resistance, Ω

377

Impedance: parallel circuit

$$\frac{1}{Z} = \sqrt{(I/R)^2 + (1/X_L - 1/X_C)^2}$$

or

$$Y = \sqrt{G^2 + (B_L - B_C)^2}$$

and

$$Z = \frac{1}{Y}$$

where Z = impedance, Ω
X_L = inductive reactance, Ω
X_C = capacitive reactance, Ω
R = resistance, Ω
$Y = 1/Z$
$G = 1/R$
$B_L = 1/X_L$
$B_C = 1/X_C$

Impedance: parallel (tank) circuit

$$Z_{par} = \frac{L}{RC}$$

	Copper Wire Single Strand, American Wire Gage			
Gage	Diameter, mils	Cross section, cir mils	Resistance, Ω/1000 ft (25°C)	Weight, lb/1000 ft
0000	460.0	211,600.0	0.0500	641.0
000	410.0	167,800.0	0.0630	508.0
00	365.0	133,100.0	0.0795	403.0
0	325.0	105,500.0	0.1000	319.0
1	289.3	83,690.0	0.126	253.0
2	258.0	66,370.0	0.159	201.0
3	229.0	52,640.0	0.201	159.0
4	204.0	41,740.0	0.253	126.0
5	182.0	33,100.0	0.319	100.0
6	162.0	26,250.0	0.403	79.5
7	144.3	20,820.0	0.508	63.0
8	128.5	16,510.0	0.641	50.0
9	114.4	13,090.0	0.808	39.6
10	102.0	10,380.0	1.02	31.4
11	91.0	8,234.0	1.28	24.9
12	81.0	6,530.0	1.62	19.8
13	72.0	5,178.0	2.04	15.7
14	64.1	4,107.0	2.58	12.4
15	57.1	3,257.0	3.25	9.9
16	50.8	2,583.0	4.09	7.8
17	45.3	2,048.0	5.16	6.2
18	40.3	1,624.0	6.51	4.9
19	35.9	1,288.0	8.21	3.9
20	32.0	1,022.0	10.4	3.09
21	28.5	810.0	13.1	2.45
22	25.3	642.4	16.5	1.95
23	22.6	509.0	20.8	1.54
24	20.1	404.0	26.2	1.22
25	17.9	320.0	26.2	0.97
26	15.9	254.0	41.6	0.769
27	14.2	202.0	52.5	0.610
28	12.6	160.0	66.2	0.484
29	11.3	127.0	83.4	0.384
30	10.0	100.5	105.2	0.304
31	8.93	79.70	132.7	0.241
32	7.95	63.21	167.3	0.191
33	7.08	50.13	211.0	0.152
34	6.31	39.75	266.0	0.120
35	5.62	31.52	335.5	0.095
36	5.00	25.00	423.0	0.0757
37	4.45	19.83	533.4	0.0600
38	3.96	15.72	672.6	0.0476
39	3.53	12.47	848.1	0.0377
40	3.14	9.98	070.0	0.0299

Electric power in an ac circuit

$$U = E \times I$$

where U = apparent power, VA
$\quad E$ = voltage, V
$\quad I$ = current, A

$$P = I \times R$$

where P = true power, W
$\quad I$ = current, A
$\quad R$ = resistance, Ω

Greek Alphabet

Name	Capital	Lowercase	Use in Electronics
Alpha	A	α	Angles, area, coefficients
Beta	B	β	Angles, flux density, coefficients
Gamma	Γ	γ	Conductivity
Delta	Δ	δ	Variation, density
Epsilon	E	ϵ	
Zeta	Z	ζ	Impedance, coefficients, coordinates
Eta	H	η	Hysteresis coefficient, efficiency
Theta	Θ	θ	Temperature, phase angle
Iota	I	ι	Current
Kappa	K	κ	Dielectric constant
Lambda	Λ	λ	Wavelength
Mu	M	μ	Micro, amplification factor, permeability
Nu	N	ν	Reluctivity
Xi	Ξ	ξ	
Omicron	O	o	
Pi	Π	π	Ratio of circumference to diameter (3.1416)
Rho	P	ρ	Resistivity, density
Sigma	Σ	σ	Sign of summation
Tau	T	τ	Time constant, time phase displacement
Upsilon	Y	υ	
Phi	Φ	ϕ	Magnetic flux, angles
Chi	X	χ	
Psi	Ψ	ψ	Dielectric flux, phase difference
Omega	Ω	ω	Capital, ohms; lowercase, angular velocity

The symbols shown here are those that are likely to be encountered by the aviation maintenance technician. Only the primary symbols are provided in this section. For the additional symbols representing variations of the primary symbols, the technician should consult the document "Graphic Symbols for Electrical and Electronics Diagrams" published by the Institute of Electrical and Electronic Engineers (IEEE), IEEE Std 315-1975, or ANSI Y32.2-1975 furnished by the American National Standards Institute.

Qualifying Symbols

Qualifying symbols are applied to standard symbols to provide an indication of the special characteristics of the symbols as they are employed in specific circuits.

Adjustability or Variability

CONTINUOUSLY ADJUSTABLE OR VARIABLE CONDITION   PRESET, GENERAL   LINEAR   NONLINEAR

Special Property Indicators

$t°$	x	τ
TEMPERATURE DEPENDENCE	MAGNETIC FIELD DEPENDENCE	(GREEK LETTER TAU) STORAGE

Saturable Properties

DELAY

Radiation Indicators

RADIO WAVES OR VISIBLE LIGHT   RADIATION, IONIZING

Type of radiation	Alpha particle α	Neutron η
	Beta particle β	Pion π
	Gamma ray γ	K-meson κ
	Deutron δ	Muon μ
	Proton ρ	X-ray χ

Physical-State Recognition Symbols

GAS, AIR, OR   LIQUID   SOLID   ELECTRET
PNEUMATIC   MATERIAL

Test-Point Recognition Symbol

TEST POINT
FOR CIRCUIT
TERMINAL

Direction of Flow of Power, Signal, or Information

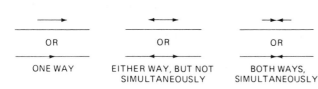

ONE WAY   EITHER WAY, BUT NOT   BOTH WAYS,
               SIMULTANEOUSLY   SIMULTANEOUSLY

Kind of Current (General)

DIRECT CURRENT   ALTERNATING
               CURRENT

Connection Symbols

TWO-PHASE,
THREE-WIRE,
UNGROUNDED

TWO-PHASE,
THREE-WIRE,
GROUNDED

TWO-PHASE, FOUR-WIRE

TWO-PHASE, FIVE-WIRE,
GROUNDED

THREE-PHASE, THREE-WIRE,
DELTA OR MESH

THREE PHASE, THREE-WIRE,
DELTA, GROUNDED

THREE-PHASE, FOUR-WIRE,
DELTA, UNGROUNDED

THREE-PHASE, FOUR-WIRE,
DELTA, GROUNDED

THREE-PHASE, WYE OR STAR,
UNGROUNDED

THREE-PHASE, WYE OR STAR,
GROUNDED

Fundamental Items

Resistor

GENERAL   TAPPED RESISTOR

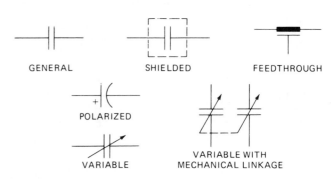

BUILDUP EXAMPLE   BUILDUP EXAMPLE
(ADJUSTABLE CONTACT)   (VARIABLE RESISTOR)

THERMAL RESISTOR   PHOTOCONDUCTIVE
(THERMISTOR)   TRANSDUCER

Capacitor

GENERAL   SHIELDED   FEEDTHROUGH

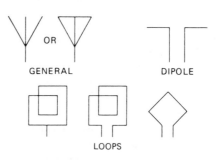

POLARIZED

VARIABLE   VARIABLE WITH
          MECHANICAL LINKAGE

Antenna

GENERAL   DIPOLE

LOOPS

Battery

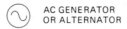

ONE-CELL   MULTICELL

Alternating-Current Source

AC GENERATOR
OR ALTERNATOR

Permanent Magnet

Pickup Head

GENERAL

STEREO

Crystal Unit

Thermocouples

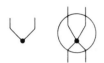

TEMPERATURE-MEASURING
THERMOCOUPLE WITH
INTEGRAL HEATER

THERMOCOUPLE WITH
INTEGRAL INSULATED
HEATER

Thermomechanical Transducers

ACTUATING
DEVICE

THERMAL
CUTOUT

OR

Ignitor Plug

Transmission Path

Conductor, Cable, Wiring

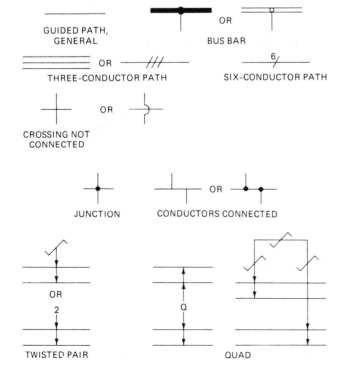

GUIDED PATH,
GENERAL

OR

BUS BAR

OR

THREE-CONDUCTOR PATH

SIX-CONDUCTOR PATH

OR

CROSSING NOT
CONNECTED

JUNCTION

OR

CONDUCTORS CONNECTED

OR

TWISTED PAIR

Q

QUAD

Conductor, Cable, Wiring (con't)

FIVE-CONDUCTOR CABLE

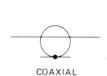

SHIELDED FIVE-CONDUCTOR
CABLE

SHIELDED TWO-CONDUCTOR CABLE
WITH SHIELD GROUNDED

COAXIAL

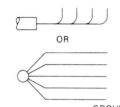

OR

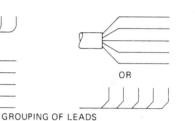

OR

GROUPING OF LEADS

CHASSIS OR FRAME
CONNECTION

Waveguides

CIRCULAR

RECTANGULAR

RIDGED

Contacts, Switches, Contactors, and Relays

Switching Function

Switch

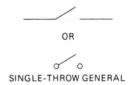

OR

SINGLE-THROW GENERAL

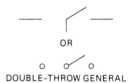

OR

DOUBLE-THROW GENERAL

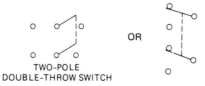

OR

TWO-POLE
DOUBLE-THROW SWITCH

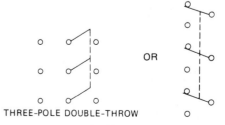

OR

THREE-POLE DOUBLE-THROW

Multiway Transfer Switch

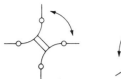

TWO-POSITION
SWITCH (90° STEP)

THREE-POSITION
SWITCH (120° STEP)

FOUR-POSITION
SWITCH (45° STEP)

Push button

CIRCUIT CLOSING
NORMALLY OPEN
(NO)

CIRCUIT OPENING
NORMALLY
CLOSED (NC)

TWO-CIRCUIT

Locking Switch

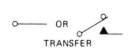

CIRCUIT CLOSING (NO)

CIRCUIT OPENING (NC)

TRANSFER SWITCH,
TWO-POSITION

TRANSFER,
THREE-POSITION

MAKE-BEFORE-
BREAK

Nonlocking Switch, Momentary or Spring Return

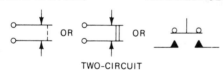

NORMALLY OPEN (NO)

NORMALLY CLOSED (NC)

TWO-CIRCUIT

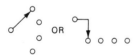

TRANSFER

NC BEFORE NO

Selector or Multiposition Switch

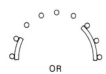

BREAK-BEFORE-MAKE,
NONSHORTING DURING
CONTACT TRANSFER

MAKE-BEFORE-BREAK,
SHORTING (BRIDGING)
DURING CONTACT TRANSFER

SEGMENTAL CONTACT

TWELVE-POINT
SELECTOR SWITCH
WITH FIXED SEGMENT

Selector or Multiposition Switch (con't)

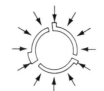

WAFER, TYPICAL THREE-POLE
THREE-CIRCUIT WITH TWO
NONSHORTING AND ONE
SHORTING MOVABLE CONTACTS

Limit Switch

TRACK-TYPE,
CIRCUIT CLOSING CONTACT

 OR NORMALLY OPEN
C   NO

  OR NORMALLY CLOSED
C   NC

TRACK-TYPE,
CIRCUIT OPENING CONTACT

OR NORMALLY OPEN,
C   NO   HELD CLOSED

OR NORMALLY CLOSED,
C   NC   HELD OPEN

Flow-Actuated Switch

CLOSES ON INCREASE
IN FLOW

OPENS ON INCREASE
IN FLOW

Liquid-Level-Actuated Switch

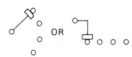

CLOSES ON
RISING LEVEL

OPENS ON
RISING LEVEL

Proximity Sensor

TARGET MOVING
AWAY

TARGET MOVING
CLOSER

Pressure or Vacuum-Actuated Switch

CLOSES ON RISING PRESSURE

OPENS ON RISING PRESSURE

Temperature-Actuated Switch (Thermostat)

CLOSES ON RISING TEMPERATURE

OPENS ON RISING
TEMPERATURE

TRANSFERS ON RISING
TEMPERATURE

Flasher

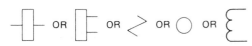

SELF-
INTERRUPTING
SWITCH

Contactor

MANUALLY OPERATED
THREE-POLE CONTACTOR

ELECTRICALLY OPERATED
ONE-POLE CONTACTOR WITH
SERIES BLOWOUT COIL

Relays

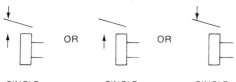

GENERAL SYMBOLS FOR RELAY COILS

SINGLE-
POLE
DOUBLE-
THROW

SINGLE-
POLE
NORMALLY
OPEN

SINGLE-
POLE
NORMALLY
CLOSED

Solenoid Switches

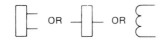

GENERAL SYMBOLS FOR SOLENOID COILS

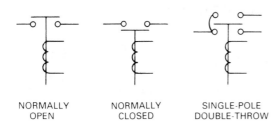

NORMALLY
OPEN

NORMALLY
CLOSED

SINGLE-POLE
DOUBLE-THROW

Terminals and Connectors

Terminals

CIRCUIT
TERMINAL

OR

TERMINAL STRIP
WITH FOUR
TERMINALS

Cable Termination

CABLE AT LEFT OF SYMBOL

Connectors

FEMALE
CONTACT

MALE
CONTACT

RECEPTACLE

PLUG

CONNECTORS ENGAGED

(Types of contacts in connectors are indicated as male or female.)

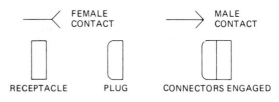

OR

MALE PLUG,
FEMALE RECEPTACLE,
ENGAGED

TWO-CONDUCTOR JACK

TWO-CONDUCTOR PLUG

Power Supply Connectors

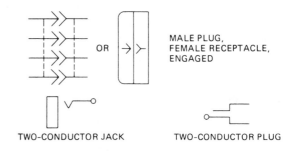

NONPOLARIZED
MALE CONNECTOR

NONPOLARIZED
FEMALE CONNECTOR

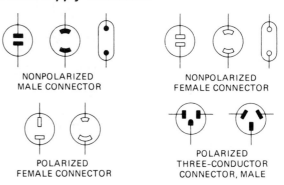

POLARIZED
FEMALE CONNECTOR

POLARIZED
THREE-CONDUCTOR
CONNECTOR, MALE

Coaxial Connector

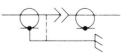

COAXIAL WITH OUTSIDE
CONDUCTOR CARRIED
THROUGH

COAXIAL WITH OUTSIDE
CONDUCTOR TERMINATED
ON CHASSIS

Transformers, Inductors and Windings

Core Symbols

No symbol is used for an air core

 MAGNETIC CORE OF INDUCTOR OR TRANSFORMER

 CORE OF MAGNET

Inductor

 OR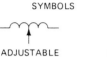
GENERAL SYMBOLS

MAGNETIC CORE INDUCTOR

TAPPED INDUCTOR

ADJUSTABLE INDUCTOR

CONTINUOUSLY ADJUSTABLE

COIL-OPERATED INDICATOR

Transductor

 CONTROL WINDING–DC

POWER WINDING–AC

SATURABLE-CORE INDUCTOR OR REACTOR

Transformer

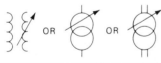 OR

GENERAL WINDING SYMBOLS

IF IT IS DESIRED TO DISTINGUISH A MAGNETIC-CORE TRANSFORMER

SHIELDED TRANSFORMER WITH MAGNETIC CORE

 OR OR

ONE WINDING WITH ADJUSTABLE INDUCTANCE

SEPARATELY ADJUSTABLE INDUCTANCES

ADJUSTABLE MUTUAL INDUCTOR, CONSTANT-CURRENT TRANSFORMER

 OR

AUTOTRANSFORMER, ONE-PHASE

Transformer (con't)

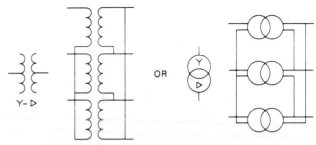

Y–▷

OR

THREE-PHASE BANK OF ONE-PHASE, TWO-WINDING TRANSFORMERS WITH WYE-DELTA CONNECTIONS

Cathode-Ray Tube

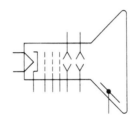

CRT WITH ELECTRIC FIELD DEFLECTION

Semiconductor Devices

Transistors and Diodes

Element Symbols

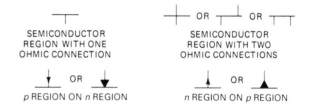

SEMICONDUCTOR REGION WITH ONE OHMIC CONNECTION

OR

SEMICONDUCTOR REGION WITH TWO OHMIC CONNECTIONS

OR

p REGION ON n REGION

OR

n REGION ON p REGION

(Arrow points opposite electron flow)

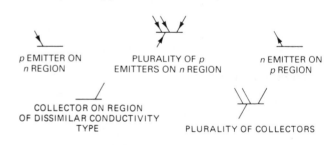

p EMITTER ON n REGION

PLURALITY OF p EMITTERS ON n REGION

n EMITTER ON p REGION

COLLECTOR ON REGION OF DISSIMILAR CONDUCTIVITY TYPE

PLURALITY OF COLLECTORS

Two-Terminal Devices

ANODE —▷|— CATHODE

DIODE RECTIFIER

STYLE 1   STYLE 2
CAPACITIVE DIODE (VARACTOR)

Two-Terminal Devices (con't)

PHOTOSENSITIVE
DIODE

PHOTOEMISSIVE (LIGHT-
EMITTING) DIODE (LED)

 nPN

 pNP

BIDIRECTIONAL PHOTODIODE

 OR

STYLE 1

STYLE 2 (ZENER)

UNIDIRECTIONAL DIODE: VOLTAGE REGULATOR

STYLE 1

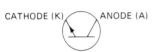 STYLE 2

BIDIRECTIONAL DIODE

CATHODE (K)   ANODE (A)   (A)   (K)

nPN TYPE   *pNP* TYPE

UNIDIRECTIONAL NEGATIVE-RESISTANCE BREAKDOWN DIODE;
TRIGGER DIAC

nPN TYPE *pNP* TYPE

BIDIRECTIONAL NEGATIVE-RESISTANCE BREAKDOWN DIODE;
TRIGGER DIAC

(E) (C)

PHOTOTRANSISTOR

CURRENT
REGULATOR

Three-or-More-Terminal Devices

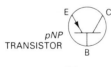

pNP
TRANSISTOR

nPN
TRANSISTOR

GATE (G) DRAIN (D)

SOURCE (S)

FIELD-EFFECT TRANSISTOR (FET)
WITH *n*-CHANNEL JUNCTION GATE

G 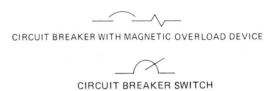 D

S

FET WITH INSULATED
GATE

A S

K

THYRISTOR

Circuit Protectors

Fuses

GENERAL FUSE SYMBOLS

CIrcuit Breaker

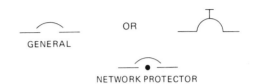

GENERAL   OR

NETWORK PROTECTOR

CIRCUIT BREAKER WITH THERMAL OVERLOAD DEVICE

CIRCUIT BREAKER WITH MAGNETIC OVERLOAD DEVICE

CIRCUIT BREAKER SWITCH

Lightning Arrestor

GENERAL   CARBON BLOCK   HORN GAP   PROTECTIVE GAP

Acoustic Devices

Audible Signaling Device

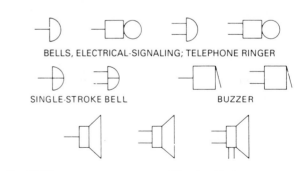

BELLS, ELECTRICAL-SIGNALING; TELEPHONE RINGER

SINGLE-STROKE BELL   BUZZER

HN —HORN
HW —HOWLER
LS —LOUDSPEAKER
SN —SIREN
MG —MAGNETIC ARMATURE

EM —ELECTROMAGNETIC
       WITH MOVING COIL
EMN—ELECTROMAGNETIC
       WITH MOVING COIL AND
       NEUTRALIZING WINDING
PM —PERMANENT MAGNET

Microphone, Telephone Transmitter

MICROPHONE, GENERAL

Handset

GENERAL

WITH PUSH-TO-TALK SWITCH

Telephone Receiver

OR

HEADSET, DOUBLE   HEADSET, SINGLE

Lamps and Visual-Signaling Devices

Lamp

LAMP, GENERAL; LIGHT SOURCE

A	AMBER	P	PURPLE
		R	RED
B	BLUE	W	WHITE
C	CLEAR	Y	YELLOW
G	GREEN	IR	INFRA-RED
O	ORANGE		
ARC	ARC	NA	SODIUM VAPOR
EL	ELECTROLUMINESCENT	NE	NEON
		UV	ULTRA-VIOLET
FL	FLUORESCENT		
HG	MERCURY VAPOR	XE	XENON
		LED	LIGHT-EMITTING DIODE
IN	INCANDESCENT		
OP	OPALESCENT		

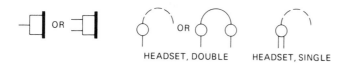

TWO-TERMINAL   FOUR-TERMINAL

Fluorescent Lamps

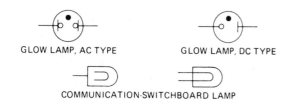

GLOW LAMP, AC TYPE   GLOW LAMP, DC TYPE

COMMUNICATION-SWITCHBOARD LAMP

Readout Devices

Meter
Instrument

 GENERAL

A	Ammeter
AH	Ampere-hour meter
C	Coulombmeter
CMA	Contact-making (or breaking) ammeter
CMC	Contact-making (or breaking) clock
CMV	Contact-making (or breaking) voltmeter
CRO	Cathode-ray oscilloscope
dB	dB (Decibel) meter
dBM	dBM (Decibels referred to 1 mW) meter
DM	Demand meter
DTR	Demand-totalizing relay
F	Frequency meter
GD	Ground detector
I	Indicating meter
INT	Integrating meter
uA or UA	Microammeter
MA	Milliammeter
NM	Noise meter
OHM	Ohmmeter
OP	Oil pressure
OSCG	Oscillograph, string
PF	Power-factor meter
PH	Phase meter
PI	Position indicator
RD	Recording demand meter
REC	Recording
RF	Reactive-factor meter
S	Synchroscope
T°	Temperature meter
THC	Thermal convertor
TLM	Telemeter
TT	Total-time meter
	Elapsed-time meter
V	Voltmeter
VA	Volt-ammeter
VAR	Varmeter
VARH	Varhour meter
VI	Volume indicator
	Audio-level meter
VU	Standard volume indicator
	Audio-level meter
W	Wattmeter
WH	Watthour meter

OR

GALVANOMETER

Rotating Machinery

Rotating Machine

BASIC   GENERATOR, GENERAL   GENERATOR, DC

G   OR   GEN   G

Rotating Machine (cont.):

GENERATOR,
AC

GENERATOR,
SYNCHRONOUS

MOTOR,
GENERAL

MOTOR,
DC

MOTOR,
AC

MOTOR,
SYNCHRONOUS

Field Generator or Motor

COMPENSATING
OR COMMUTATING

SERIES

SHUNT OR
SEPARATELY EXCITED

Winding Connection Symbols

ONE-PHASE

TWO-PHASE

THREE-PHASE
WYE
UNGROUNDED

THREE-PHASE
WYE
GROUNDED

THREE-PHASE,
DELTA

SIX-PHASE,
DIAMETRICAL

SIX-PHASE,
DOUBLE DELTA

Direct-Current Machines

SEPARATELY EXCITED DC GENERATOR OR
MOTOR WITH COMMUTATING FIELD WINDING

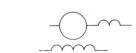
SERIES MOTOR OR TWO-WIRE DC GENERATOR WITH
COMPENSATING FIELD WINDING OR BOTH

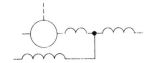

DC COMPOUND MOTOR OR STABILIZED SHUNT
MOTOR WITH COMMUTATING FIELD WINDING

DC, PERMANENT-MAGNET FIELD GENERATOR OR MOTOR

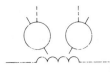
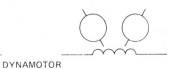
DYNAMOTOR

Alternating-Current Machines

SQUIRREL-CAGE INDUCTION MOTOR OR GENERATOR,
SYNCHRONOUS MOTOR OR GENERATOR, SPLIT-PHASE INDUCTION
MOTOR OR GENERATOR, ROTARY-PHASE CONVERTOR,
OR REPULSION MOTOR

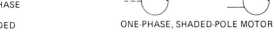
WOUND-ROTOR INDUCTION MOTOR, SYNCHRONOUS
INDUCTION MOTOR, INDUCTION GENERATOR, OR INDUCTION
FREQUENCY CONVERTOR

AC SERIES MOTOR

AC SERIES MOTOR WITH COMMUTATING OR
COMPENSATING FIELD WINDING OR BOTH

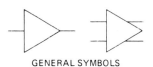
ONE-PHASE, SHADED-POLE MOTOR

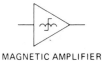
SYNCHRONOUS MOTOR OR GENERATOR
WITH DC FIELD EXCITATION

Symbols Used in Logic Circuits and Diagrams

(Not included in ANSI Y32.2)

Amplifiers

GENERAL SYMBOLS

MAGNETIC AMPLIFIER

Logic Gates

INVERTOR OR
NOT GATE

AND GATE WITH THREE INPUTS

NAND (NOT AND) GATE

OR GATE

NOR (NOT OR) GATE

EXCLUSIVE OR GATE

ac   Alternating current

ACARS   ARINC Communication Addressing and Reporting System

A/D   Analog/digital; analog-to-digital

A/D CONV   Analog-to-digital converter

ADC   Air-data computer

ADCP   ATC dual-control panel

ADEDS   Advanced electronic display system

ADF   Automatic direction finder

ADI   Attitude-director indicator; air data instrument

AFC   Automatic frequency control

AFCS   Automatic flight control system

AFDS   Autopilot flight detector system

AIRCOM   Air/ground communications

A/L   Autoland

AM   Amplitude modulation

AMP   Amperes

AMP or **AMPL**   Amplifier

ANT   Antenna

AP   Autopilot

APB   Auxiliary power breaker

APCU   Auxiliary power control unit

APU   Auxiliary power unit

ARINC   Aeronautical Radio Incorporated

ARNC IO   ARINC I/O error

ARNC STP   ARINC I/O UART data strip error

ATC   Air traffic control

ATCT   ATC transponder

ATCTS   ATC transponder system

AUX   Auxiliary

AVC   Automatic volume control

BAT or **BATT**   Battery

BCD   Binary-coded decimal

BIT   Binary digit; built-in test

BITE   Built-in test equipment

BITS   Bus interconnect transfer switch

BNR   Binary numerical reference; binary

BP   Band-pass

BPCU   Bus power control unit

BT   Bus tie

BTB   Bus tie breaker

BUS   Electrical bus; 429 digital data bus

CAC   Caution advisory computer

CAWS   Central aural warning system; caution and warning system

CB, C/B, or **CKT/BKR**   Circuit breaker

CDI   Course-deviation indicator

CDU   Central display unit

CFDIU   Centralized fault display interface unit

CFDS   Centralized fault display system

CG   Center of gravity

CH or **CHAN**   Channel

CHGR   Charger

CKT   Circuit

CLK   Clock

CLR   Clear

CMCS   Central maintenance computer system

CMPTR   Computer

COAX   Coaxial

COP   Copper

CP   Control panel

CRT   Cathode-ray tube; circuit

CSE   Control system electronics

CSEU   Control system electronics unit

CSEUP   Control system electronics unit panel

CT   Current transformer

CTN   Caution

CU   Control unit; copper

CVR   Cockpit voice recorder

CW   Continuous wave

D/A   Digital-to-analog

DAC   Digital-to-analog converter

DADC   Digital air-data computer

DBT   Dead bus tie

dc   Direct current

DCDR   Decoder

DDB   Digital data bus

DEMOD   Demodulator

DEMUX   Demultiplexer

DFDR   Digital flight data recorder

DG   Directional gyro

DGTL   Digital

DH   Decision height

DISC   Disconnect

DISC SOL   Disconnect solenoid

DISTR   Distribution

DMA   Direct memory access

DMB   Dead main bus

DMC   Display management computer

DME   Distance-measuring equipment

DMEA   Distance-measuring equipment antenna

DU   Display unit

DWN   Down

EADF   Electronic automatic direction finder

EADI   Electronic attitude-director indicator

EAROM   Electrically alterable read-only memory

EC   EICAS computer

ECAM   Electronic centralized aircraft monitoring

EDSP   EICAS display select panel

EDU   EICAS display unit

E/E or **E & E**   Electrical/electronic

EEC   Electronic engine control

EFI   Electronic flight instrument

EFIS   Electronic flight instrument system

EFISCP   EFIS control panel

EFISCU   EFIS comparator unit

EFISG   EFIS symbol generator

EFISRLS   EFIS remote light sensor

EHSI   Electronic horizontal-situation indicator

EHSID   Electronic horizontal-situation indicator display

EHSV   Electrohydraulic servo value

EICAS   Engine indicating and crew alerting system

ELCU   Electrical load control unit

ELEC   Electric; electronic

ELECT Electrical
ELEX Electronics; electrical
EMER GEN Emergency generator
EMFI Electromechanical flight instrument
EMI Electromagnetic interference
EP External power
EP AVAIL External power available
EPC External power contactor
EPCS Electronic power control switch
EPROM Erasable programmable read-only memory
EXCTR Exciter
EXT PWR External power
E1-1 First shelf, number 1 equipment rack
E2-2 First shelf, number 2 equipment rack
FM Frequency modulation
FMC Flight management computer
FMCD Flight management computer control display unit
FMCS Flight management computer system
FM/CW Frequency modulation continuous wave
FMS Flight management system
FREQ Frequency
FSEU Flap/slat electronic unit
FW or **FWD** Forward
GAL or **GALY** Galley
GCB Generator circuit breaker
GCR Generator control relay; generator circuit relay
GCR AUX Generator control relay auxiliary contact
GCU Generator control unit
GEN Generator
GLR Galley load relay
GMT Greenwich mean time
GND Ground
GND PWR Ground power
GND RET Ground return
GND SVCE Ground service
GPCU Ground power control unit
GPS Global positioning system
GPSW Gear position switch
GPU Ground power unit
GPW Ground proximity warning
GWPC Ground proximity warning computer
GPWS Ground proximity warning system
GRD Ground
G/S Glide slope
GSR Ground service relay
GSSR Ground service select relay
GSTR Ground service transfer relay
HF (hf) High frequency (3 to 30 MHz)
HFA High-frequency radio antenna
HFCP High-frequency radio control panel
HI Z High impedance
H/L High/low
Hz Hertz
IAPS Integrated avionics processor system
IAS Indicated airspeed
IDG Integrated drive generator
IF Intermediate frequency
IFR Instrument flight rules
IGN Ignition

IIS Integrated instrument system
ILS Instrument landing system
IND L Indicator light
INST Instrument
INSTR Instrument
INTCON Interconnect
INTFC Interface
INTPH Interphone
INTR Interrogation
INV Inverter
I/O Input/Output
IR ILS receiver
kHz Kilohertz
kV Kilovolts
kVA Kilovoltamperes
kVAR Kilovoltampere reactive
L-Band Radio frequency band (390 to 1550 MHz)
LCD Liquid-crystal display
LD Load
LED Light-emitting diode
LF (lf) Low frequency (30 to 300 kHz)
LO Z Low impedance
LOC Localizer
LRU Line replaceable unit
LS Loudspeaker
LSB Lower sideband
LSPTM Limit switch position transmitter module
LT Light
LTS Lights
MAN/ELEC Manual/electric
MBA Marker-beacon antenna
MCDP Maintenance control and display panel
MCDU Multipurpose control and display unit
MDE Modern digital electronics
MEC Main equipment center; main engine control
MEG or **MEGA** Million
MEM Memory
MF (mf) Medium frequency (300 kHz to 3 MHz)
MHz Megahertz
MIC Microphone
MICRO-P Microprocessor
MILLI One one-thousandth (0.001)
MKR BCN Marker beacon
MSEC (ms) Milliseconds
MSG Message
MUX Multiplexer
mV Millivolts
NAV Navigation
NC Normally closed; not connected; no connection
NDB Nondirectional beacon
NEG Negative
NSEC (ns) Nanoseconds
NTSB National Transportation Safety Board
NVM Nonvolatile memory
OC Overcurrent
OF Overfrequency
OVV or **OV** Overvoltage
OVVCO or **OVCO** Overvoltage cutout
PA Passenger address; power amplifier

PARA/SER Parallel to serial
PCU Passenger control unit; power control unit
PFD Primary flight display
PMG Permanent-magnet generator
POS Positive
POT Potentiometer; plan of test
PR Power relay
PRL Parallel
PROM Programmable read-only memory
PROX Proximity
P-S Parallel to series
PSEU Proximity switch electronic unit
PWR Power
PWR SPLY Power supply
QTY Quantity
RA Radio altimeter; radio altitude
RAD Radio
RAIND Radio altimeter indicator
RAM Random-access memory
RART Radio altimeter receiver-transmitter
RAT Ram air turbine
RCCB Remote control circuit breaker
RCL Recall
RCVR Receiver
RCVR/XMTR Receiver/transmitter
RDMI Radio distance magnetic indicator
RF (rf) Radio frequency
RFI Radio-frequency inteference
RLS Remote light sensor
RMI Radio magnetic indicator
rpm Revolution per minute
r-t Receiver-transmitter
SAT Static air temperature
SATCOM Satellite communication
SCR Silicon controlled rectifier
SDI Source destination identifier
SELCAL Selective calling system
SER DL Serial data link
SG Symbol generator
SITA Société International de Telecommunications Aeronautiques
SMD Surface mounted device
SNR Signal-to-noise ratio
SOL Solenoid
SOLV Solenoid valve
SOM Start of message
SOT Start of transmission
SPKR Speaker
SPR Software problem report
SQL Squelch
SSB Single sideband
SSM Sign status matrix
ST Synchro transmitter
STAT INV Static inverter

STBY Standby
SW Switch
SYM GEN Symbol generator
TAT True air temperature
TBDP Tie bus differential protection
TCAS Traffic alert and collision avoidance system
TFR Transfer
TMC Thrust management computer
TMS Thrust management system
TMSP Thrust mode select panel
T-R Transformer-rectifier
TRU Transformer-rectifier unit
TXPDR Transponder
μ Micro
UBR Utility bus relay
UF Underfrequency
UHF Ultrahigh frequency (300 MHz to 3 GHz)
UNDF Underfrequency
UNDV Undervoltage
US Underspeed
USB (μs) Upper sideband
USEC Microseconds
UV Undervoltage
V Volts; voltage; vertical; valve
V ac or Vac Volts alternating current
V dc or Vdc Volts direct current
VA Voltamperes
VAC Volts alternating current
VAR Voltampere reactive
VDC Volts direct current
VFR Visual flight rules
VHF (vhf) Very high frequency (30 to 300 MHz)
VLSI Very large-scale integration
VOR VHF omnirange; visual omnirange
VORTAC VOR tactical air navigation
VR Voltage regulator
VRMS Volts root mean square
W Watts
WARN Warning
WCP Weather radar control panel
WEA Weather
WEU Warning electronics unit power supply
WPT Waypoint
WX (WXR) Weather radar
XCVR Transceiver
XDCR Transducer
XFMR Transformer
XFR Transfer
XMIT Transmit
XMTR Transmitter
XPDR Transponder
429 ARINC 429 data bus standard
629 ARINC 629 data bus standard

Glossary

accelerate To change velocity; increase or decrease speed.

accelerometer A device for sensing or measuring acceleration and converting it into an electrical signal.

acceptor An impurity atom in a semiconductor material that will receive, or accept, electrons. Germanium with an acceptor impurity is called *p*-type germanium because it has a positive nature.

actuator A hydraulic, electric, or pneumatic device used to operate a mechanism by remote control.

alignment, electrical The tuning of electronic components in a particular circuit so that all portions of the circuit will respond to the correct frequency.

alternating current (ac as adjective) An electric current that periodically changes direction and constantly changes in magnitude.

alternation The part of an ac cycle during which current is flowing in one direction; one half-cycle.

alternator An electric generator designed to produce alternating current.

alternator control unit A solid-state voltage regulator containing current and voltage sensors.

altitude ring A continuous return across a radar display at a range equivalent to aircraft altitude.

ammeter An instrument used to measure current flow.

ampere (A) The basic unit of current flow. One A is the amount of current that flows when an emf of 1 V is applied to a circuit with a resistance of 1 Ω. One coulomb per second.

ampere-hour (Ah) Quantity of electricity that has passed through a circuit when a current of 1 A has flowed for 1 h. Current (in amperes) $\times$ time (in hours) = ampere-hours.

ampere-turn The magnetizing force produced by a current of 1 A flowing through one turn of a coil. Ampere-turns = amperes $\times$ number of turns of wire in the coil.

amplification The increase of power, current, or voltage in an electronic circuit.

amplification factor The ratio of a small change in plate voltage to a small change in grid voltage when the plate circuit is operating through a load. It is denoted by the Greek letter mu (μ). Amplification factor for a transistor is the ratio of a change in emitter current to a change in base current.

amplifier An electronic circuit that produces amplification.

amplitude modulation (AM) Modulation of a carrier wave in which the modulating signal changes the amplitude of the carrier in proportion to the strength of the modulating signal.

analog Infinitely variable, or relating to an electric circuit that operates with infinite possible input or output signals.

angular velocity Time rate of change of an angle rotated around an axis in degrees per second or degrees per minute.

anode The positive electrode of a battery; the electrode of an electron tube, diode, or electroplating cell to which a positive voltage is applied.

antenna A device designed to radiate or intercept electromagnetic waves.

apparent power The power consumed by the resistance, inductance, and capacitance in an ac circuit.

armature In a dc generator or motor, the rotating member. In an ac generator the armature is stationary and is acted upon by the rotating field produced by the rotor. The moving element acted upon by the magnetic field in a relay is also called the armature.

armature reaction The interaction of the armature field with the main field of a generator or motor, resulting in distortion of the main field.

atom The smallest possible particle of an element.

attenuation A reduction in the strength of a signal, the flow of current, flux, or other energy in an electronic system.

audio frequency (AF as adjective) A frequency in the audible range, from about 35 to 20,000 Hz.

automatic direction finder (ADF) A radio receiver utilizing a directional loop antenna that enables the receiver to indicate the direction from which a radio signal is being received; also called a radio compass.

automatic flight control system (AFCS) A flight control system incorporating an automatic pilot with additional systems such as a VOR coupler, an ILS approach coupler, and an internal navigation system that is fully automatic, so the aircraft can be flown in a completely automatic mode.

automatic frequency control (AFC) A circuit arrangement that maintains the frequency of the system within specified limits.

automatic pilot A system installed in an airplane or missile that senses deviations in the flight path and moves the control surfaces to maintain the selected flight path.

automatic volume control (AVC) A circuit arrangement in which the dc component of the detector output in a radio receiver controls the bias of the RF tubes, thus regulating their output to maintain a reasonably constant volume.

Autosyn A trade name of the Bendix Corporation used to designate certain types of synchros.

avionics A generic term for aircraft electronics equipment. A contraction of *aviation electronics.*

avionics standard communication bus (ASCB) A digital data transfer bus, used to transmit serial data.

azimuth Angular distance measured on a horizontal circle in a clockwise direction from either north or south.

ballast A circuit element designed to stabilize current flow.

ballast transformer A transformer specifically designed to power fluorescent lights.

band A range of frequencies.

band-pass filter A filter circuit that passes frequencies within a specific band and attenuates frequencies outside the band.

band-reject filter A circuit designed to reject a certain band of frequencies and attenuate all frequencies outside that band.

bandwidth The difference between the maximum and minimum frequencies in a band.

base The terminal of a transistor to which the controlling current is applied.

battery A group of voltaic cells connected together in series to produce a desired voltage and current capacity. Typical batteries utilize primary cells, secondary cells, and photovoltaic cells.

beat frequency oscillator (BFO) An oscillator designed to produce a signal frequency that is mixed with another frequency in order to develop an intermediate frequency or an audio frequency.

bel A unit used to express the ratio of two values of power. The number of bels is the logarithm to the base 10 of the power ratio.

bias A voltage applied to the control element of a transistor to establish the correct operating point.

binary system A number system using only two symbols, 0 and 1, and having 2 as a base. In the decimal system, 10 symbols are used, and the base is 10.

bit One digit of a binary number.

BITE Built-in test equipment designed to monitor and test aircraft systems.

black box A slang term used to refer to a complex electric component.

bleeder resistance A resistor permanently connected across the output of a power supply and designed to "bleed off" a small portion of the current.

bonding The connecting together of metal structures with electric conductors, thus establishing a uniform electric potential among all the parts bonded together.

boresighting The process of aligning a directional antenna system.

breakdown voltage In a capacitor, the voltage at which the dielectric is ruptured; in a gas tube, the voltage level at which the gas becomes ionized and starts to conduct.

brush A device designed to provide an electric contact between a stationary conductor and a rotating element.

buffer amplifier An amplifier in a transmitter circuit designed to isolate the oscillator section from the power section, thus preventing a frequency shift.

bus bar A power distribution point to which a number of circuits may be connected. It often consists of a solid metal strip in which a number of terminals are installed.

bus tie An electric solenoid used to connect two bus bars.

byte A group of binary digits handled as a unit or word.

cable A group of insulated electric conductors, usually covered with rubber or plastic to form a flexible transmission line.

capacitance (C) The property enabling two adjacent conductors separated by an insulating medium to store an electric charge. The unit of capacitance is the farad.

capacitive reactance The reaction, or actual, effect of capacitance in an ac circuit. The equation is $X_C = \frac{1}{2\pi f C}$, where X_C is capacitive reactance in ohms, f is frequency in hertz, and C is the capacitance in farads.

capacitor A device consisting of conducting plates separated by a dielectric and used to introduce capacitance into a circuit.

capacity A battery or cell's total available current. Typically measured in ampere-hours for aircraft storage batteries.

carrier wave A radio-frequency electromagnetic wave used to convey intelligence impressed upon it by modulation.

cathode (1) The negative electrode of a battery; (2) the negative terminal of a diode or electroplating cell.

cathode-ray tube (CRT) A special type of electron tube in which a stream of electrons from an electron gun impinges upon a phosphorescent screen, thus producing a bright spot on the screen. The electron beam is deflected electrically or magnetically to produce patterns on the screen.

cell A combination of two electrodes surrounded by an electrolyte for the purpose of producing voltage.

characteristic curve A graph that shows the performance of a transistor under various operating conditions.

charge A quantity of electricity. A charge is negative when it consists of a number of electrons greater than the number normally held by the material when it is in a neutral condition. The charge is positive when there is a deficiency of electrons.

choke coil An inductance coil designed to provide a high reactance to certain frequencies and generally used to block or reduce currents at these frequencies.

circuit Conductors connected together to provide one or more complete electrical paths.

circuit breaker A device that automatically opens a circuit if the current flow increases beyond an established limit.

circuit protection The provision of devices in an electric circuit to prevent excessive current flow. These devices may be fuses, circuit breakers, current limiters, or sensing relays.

circular mil (cmil) The cross-sectional area of a circle having a diameter of 1 mil (0.001 in.). The circular mil is used to indicate the size of electric wire.

closed-circuit voltage The voltage in a system with a load connected.

clutch A mechanical device used to connect or disconnect a motor or some other driving unit and the driven unit.

coaxial cable A pair of concentric conductors. The inner conductor is supported by insulation that holds it in the center of the outer conductor. A coaxial cable is normally used to conduct HF currents.

coil One or more turns of a conductor designed for use in a circuit to produce inductance or an electromagnetic field.

collector The section of a transistor that carries the controlled current.

collector ring A rotating electric contact used with a brush to transfer electric current from a rotating unit to a stationary unit or vice versa.

color code A system of colors used to indicate component values or to identify wires and terminals.

commutator A rotating contact device in the armature of a dc generator or motor; in effect, it changes the ac current flowing in the armature windings to a dc current in the external circuit.

compass A device used to determine direction on the earth's surface. A magnetic compass utilizes the earth's magnetic field to establish direction.

compound A chemical combination of two or more different elements.

compound winding A combination of series and parallel or shunt windings to provide the magnetic field for a generator or motor.

conductance The reciprocal of resistance.

conductor A material through which an electric current can pass easily.

conduit A metallic tubular sheath through which insulated conductors are run. The conduit provides mechanical protection and electric or magnetic shielding for the conductors.

constant-speed drive (CSD) A unit used in conjunction with ac alternators to produce a constant-frequency ac voltage.

continuity tester A device designed to test the electrical continuity of a conductor or circuit. A battery and light, or other some other indicating unit, connected in series or an ohmmeter may serve as a continuity tester.

continuous wave (CW) An RF carrier wave whose successive oscillations are identical in magnitude and frequency.

control circuit Any one of a variety of circuits designed to exercise control of an operating device by performing counting, timing, switching, and other operations.

copper loss The energy lost to heat due to the resistance of the wire in an electric motor.

corona loss Power loss due to the ionization of gas adjacent to a high-potential conductor.

cosine The ratio of the side adjacent to an acute angle of a right triangle to the hypotenuse.

coulomb (C) The international coulomb is a unit of electric charge consisting of approximately 6.28×10^{18} electrons. The absolute coulomb is slightly greater than the international coulomb; that is, 1 absolute coulomb $= 1.000165$ international coulomb.

counter electromotive force (cemf) A voltage developed in the armature of a motor that opposes the applied emf. The same principle applied to any inductor through which an alternating current is flowing.

counterpoise One or more conductors used under certain types of antennas to take the place of the usual ground circuit.

coupling Energy transfer between elements or circuits of an electronic system.

cross modulation The modulation of a desired signal by an unwanted signal, resulting in two signals in the output.

crystal A solid body with symmetrically arranged plane surfaces. In electronic systems, crystals are used as rectifiers, semiconductors, transistors, and frequency controllers and to produce oscillatory voltages.

crystal diode A diode constructed from a crystal semiconductor material such as silicon or germanium.

current The movement of electricity through a conductor, i.e., the flow of electrons through a conductor.

current limiter A device installed in a circuit to prevent current from increasing above a specified limit.

cutoff The point at which an operation stops because a cutoff condition has been reached.

cycle A complete sequence of events in a recurrent series of similar periods.

damping The decay in amplitude or strength of an oscillatory current when energy is not introduced to replace that lost through circuit resistance.

d'Arsonval meter movement A meter movement consisting of a movable coil suspended on pivots between the poles of a permanent magnet.

data bus The communication link between two or more computer systems or subsystems.

decade A series of quantities in multiples of 10; for example, 10, 100, 1000, 10,000.

decibel (dB) One-tenth of a bel.

decimal system A number system using 10 figures to represent the quantities 0 through 9.

decoupling   The process of eliminating electrical or magnetic coupling between units in an electronic system.

deflection   The movement of an electron beam up and down or sideways in response to an electric or magnetic field in a cathode-ray tube.

degeneration   Feedback of a portion of the output of a circuit to the input in such a direction that it reduces the magnitude of the input; also called *negative feedback*. Degeneration reduces distortion, increases stability, and improves frequency response.

delta connection   A method of connecting three components to form a three-sided circuit, usually drawn as a triangle, hence the term *delta*. Delta (Δ) is the Greek letter corresponding to the English D.

demodulation   The recovery of the AF signal from an RF carrier wave. Also called *detection*.

detector   That portion of an electronic circuit that demodulates, or detects, a signal.

deviation, compass   The error in a magnetic compass due to construction, installation, and nearby magnetic materials.

diac   A negative-resistance breakdown diode, constructed in both unidirectional and bidirectional forms.

dielectric   An insulating material used to separate the plates of a capacitor.

dielectric constant   A measure of the effectiveness of a dielectric for holding a charge in a capacitor. Air is given a dielectric constant of 1, and mica has a dielectric constant of 5.8; hence a capacitor having mica as the dielectric has 5.8 times the capacitance of the same capacitor having a dielectric of air.

differentiating circuit   A circuit that produces an output voltage proportional to the rate of change of the input.

digital   Relating to an electric circuit with a finite number of possible inputs and outputs.

diode   A semiconductor device with only a cathode and an anode; used as a rectifier and a detector.

dipole antenna   An antenna consisting of two equal lengths of wire or some other conductor extending in opposite directions from the input point. Each section of the dipole is approximately one-quarter wavelength.

direct current (dc as adjective)   An electric current that flows continuously in one direction.

directional gyro   A direction-indicating instrument that utilizes a gyroscope to hold the moving element in a fixed position relative to a directional reference.

discriminator   A circuit whose output polarity and magnitude are determined by the variations of the input phase or frequency.

distance-measuring equipment (DME)   An electronic system used with radio navigation equipment to provide an indication of the distance to a specific point.

distortion   Undesirable change in the waveform of the output of a circuit compared with the waveform of the input.

donor   An impurity used in a semiconductor to provide free electrons as current carriers. A semiconductor with a donor impurity is of the *n* type.

Doppler effect   The effect noted as one moves toward or away from the source of a sound-wave or electromagnetic-wave propagation. Moving toward the source results in receiving a higher-frequency sound or signal than the source is emitting, and moving away from the source results in receiving a lower-frequency sound or signal.

downlink   The radio transmission path downward from the aircraft to the earth.

duplexer   A circuit that makes it possible to use the same antenna for both transmitting and receiving without allowing excessive power to flow to the receiver.

dynamometer   A type of electric measuring instrument involving a reaction between a magnetic field and electromagnetic forces.

dynamotor   An electric rotating machine with a double armature, usually designed to produce a high dc voltage for plate circuits in radio transmitters and receivers. One end of the armature serves a low-voltage dc motor, and the other end is wound for a high-voltage dc generator.

eddy currents   Currents induced in the cores of coils, transformers, and armatures by the changing magnetic fields associated with their operation. These currents cause great losses of energy. For this reason such cores are composed of insulated laminations that limit the current paths.

effective value   A term used to indicate the actual working value of an alternating current based on its heating effect. Also called the root-mean-square (rms) value. It is equal to $1/\sqrt{2}$ times the maximum value in a sinusoidal current.

electret   A dielectric body in which a permanent state of electric polarization has been set up. Also, the material of which an electret is composed.

electricity   In general terms, electricity may be said to consist of positive or negative charges at rest or in motion.

electrode   A terminal element in an electric device or circuit. Some typical electrodes include the plates in a storage battery, the elements in an electron tube, and the carbon rods in an arc light.

electrolysis   The process of decomposing a chemical compound by means of an electric current.

electrolyte   Any solution that conducts an electric current.

electromagnet   A magnet formed when an iron core is placed in a current-carrying coil.

electromagnetic induction   The transfer of electric energy from one conductor to another by means of a moving electromagnetic field. A voltage is produced in a conductor as the magnetic lines of force cut or link with the conductor. The value of the voltage produced by electromagnetic induction is proportional to the number of lines of force cut per second. When 100,000,000 lines of force are cut per second, an emf of 1 V will be induced.

electromagnetism   The magnetism produced by the flow of electric current.

electromotive force (emf)   The force that causes current to move through a conductor. The unit of measurement for emf is the volt; hence emf is often called *voltage.*

electron   A negatively charged nonnuclear particle that orbits around the nucleus of an atom. Generally speaking, an electron may be considered a carrier of electric current through a conductor. An electron at rest has a mass of 9.107×10^{-28} g and a charge of 1.6×10^{-19} C.

electron gun   The combination of an electron-emitting cathode with accelerating anodes and beam-forming electrodes to produce the electron beam in a CRT.

electron tube   A device consisting of an evacuated or gas-filled envelope containing electrodes for the purpose of controlling electron flow. The electrodes are usually a cathode (electron emitter), a plate (anode), and one or more grids.

electrostatic field   The field of electric force existing in the area around and between any two oppositely charged bodies.

element   Any substance that cannot be changed to another substance except by nuclear disintegration. There are more than 100 known elements.

emission, electronic   The freeing of electrons from the surface of a material, usually produced by heat.

emitter   A section of a transmitter that carries current from both the base and collector circuits.

equalizing circuit   A circuit in a multiple-generator voltage regulator system that tends to equalize the current output of the generators by controlling the field currents of the several generators.

ESDS (electrostatic discharge sensitive)   Components that are sensitive to damage from static electric charges.

excitation   The application of electric current to the field windings of a generator to produce a magnetic field.

fading   A decrease in the strength of a received radio signal.

farad (F)   The unit of capacitance; the capacitance of a capacitor that will store 1 C of electricity when an emf of 1 V is applied.

feedback   A portion of the output signal of a circuit returned to the input. Positive feedback occurs when the feedback signal is in phase with the input signal. Negative feedback occurs when the feedback signal is 180° out of phase with the input signal.

ferromagnetic materials   Magnetic materials composed largely of iron.

fidelity   The degree of similarity between the input and output waveforms of an electronic circuit.

field   A space where magnetic or electric lines of force exist.

field coil   A winding or coil used to produce a magnetic field.

field frame   The main structure of a generator or motor within which are mounted the field poles and windings.

filament   The heated element in an electric light bulb.

filter   A circuit arranged to pass certain frequencies while attenuating all others. A high-pass filter passes high frequencies and attenuates low frequencies; a low-pass filter passes low frequencies and attenuates high frequencies.

flux   Electrostatic or magnetic lines of force.

flux gate   An electromagnetic sensing device that determines the direction of the earth's magnetic field and thus produces magnetic-direction information for navigation systems.

flywheel effect   The characteristic of a parallel *LC* circuit that permits a continuing flow of current, even though only small pulses of energy are applied to the circuit.

forward bias   A voltage applied to a semiconductor that creates a low resistance within that semiconductor.

free electrons   Those electrons so loosely bound in the outer shells of some atoms that they are able to move from atom to atom when an emf is applied to the material.

frequency   The number of complete cycles of a periodic process per second. In electricity the unit of frequency is the hertz.

frequency counters   Instruments used to determine (count) the number of electrical pulses (frequency) of a given voltage.

frequency modulation (FM)   Modulation of a carrier wave by causing changes in carrier frequency that are proportional to the amplitude of the modulating signal.

frequency multiplier   A circuit designed to double, triple, or quadruple the frequency of a signal by harmonic conversion.

fuse   A metal link that melts when overheated by excess current; used to break an electric circuit whenever the load becomes excessive.

gain   The increase in signal power through a circuit.

galvanometer   A device for measuring electric current. It usually consists of a current-carrying coil that produces a field to react with the field of a permanent magnet.

ganged tuning   A mechanical arrangement to permit the simultaneous tuning of two or more circuits.

gate   An electronic switching circuit commonly employed in digital electronics to produce required outputs in response to particular inputs. The outputs are either "on" or "off" to produce the binary digits 1 and 0. Also, the control circuit built into various semiconductor devices.

gauss (G)   The unit of magnetic flux density equal to 1 Mx (line of force) per square centimeter.

generator   A rotating machine designed to produce a certain type and quantity of voltage and current.

generator control unit (GCU)   A solid-state device that controls generator output parameters.

gilbert (Gb)   The unit of magnetomotive force; it is equal to approximately 0.768 ampere-turn.

gimbal   A mechanism consisting of a pair of rings, one ring pivoted within the other and the outer ring supported on pivots 90° from the inner-ring pivots. A gyroscope pivoted in the inner ring at right angles to the inner-ring pivots will be free to precess in response to applied external forces.

glide slope   A directed radio beam emanating from a glide slope transmitter located near the runway of an instrumented airport; it provides a reference for guiding an airplane vertically to the runway.

Global Positioning System (GPS)   A navigation system that employs satellite transmittal signals to determine the aircraft's location.

ground   (1) An electrical connection to the earth; (2) a common connecting device for the zero-potential side of the circuits in an electrical or electronic system; (3) the accidental connection of a hot conductor to the ground (a hot conductor is one whose potential differs from ground potential).

ground wave   That portion of a radio wave that travels to the receiver along the surface of the earth.

growler   An electromagnetic device that develops a strong alternating field by which armatures may be tested.

guidance   The control of missiles or aircraft in flight.

gyroscope   A comparatively heavy wheel mounted on a spinning axis that is free to rotate about one or both of two axes perpendicular to each other and to the spinning axis. The gyroscope is used to sense directional changes and to develop signals for operating automatic pilots and inertial navigation systems.

harmonics   Multiples of a base frequency.

harness   A bundle of wires typically routed between various sections of an aircraft.

heat sink   A metallic surface designed to dissipate heat from electronic components.

henry (H)   The unit of inductance. It is the amount of inductance in a coil that will induce an emf of 1 V in the coil when the current flow is changing at the rate of 1 A/s.

HERF (high-energy radiated and electromagnetic fields)   Radiated electromagnetic energy that may cause induced current in nearby conductors.

hertz (Hz)   The unit of frequency. One hertz equals 1 cps.

heterodyne   The process of mixing two frequencies to produce both sum and difference frequencies. The principle is used in superheterodyne receivers.

hexidecimal numbering system   A base 16 numbering system often utilized by microcontrollers and microprocessors.

HIG   *Hermetically sealed integrating gyro.* A gyro mounted in a sealed case with a viscous damping medium. The output is therefore an indication of the total amount of angular displacement of the vehicle in which the gyro is installed, rather than the rate of angular displacement.

high-pass filter   An *LC* filter designed to pass high frequencies and block low frequencies.

horizontal-situation indicator (HSI)   A flight instrument that provides the pilot with information regarding heading, course, glide slope deviation, and course deviation, as well as other data regarding aircraft position.

horsepower (hp)   A common unit of mechanical power. The time rate of work that will raise 550 lb through a vertical distance of 1 ft in 1 s; also, 33,000 ft·lb/min. One horsepower is equal to 746 W of electric power.

hot-wire meter   An electric instrument for measuring alternating current. A wire is heated by the current flow, and the expansion of the wire is used to provide movement for the indicating needle.

hydrometer   A calibrated float used to determine the specific gravity of a liquid.

hypotenuse   The side of a right triangle opposite the right angle.

hysteresis   The ability of a magnetic material to withstand changes in its magnetic state. When a magnetomotive force (mmf) is applied to such a material, the magnetization lags the mmf because of a resistance to change in orientation of the particles involved.

ignition   Pertaining to engines, the introduction of an electric spark into a combustion chamber to fire the fuel-air mixture.

image frequency   A frequency produced by the heterodyne action of an oscillator in a superheterodyne receiver. An image frequency is produced when an unwanted signal is mixed with the oscillator frequency; the frequency of the unwanted signal is such that a difference frequency (the image frequency) is produced that is equal to the intermediate frequency of the receiver.

impedance (Z)   The combined effect of resistance, capacitive reactance, and inductive reactance in an ac circuit. Z is measured in ohms.

inductance (L)   The ability of a coil or conductor to oppose a change in current flow (*see* **henry**).

inductance coil   A coil designed to introduce inductance into a circuit.

induction motor   An ac motor in which the rotating field produced by the stator induces a current and an opposing field in the rotor. The reaction of the fields creates the rotation force.

inductive reactance (X_L)   The effect of inductance in an ac circuit. The equation for inductive reactance is $X_L = 2\pi fL$. X_L is measured in ohms.

inductor   An inductance coil.

inertia   The tendency of a mass to remain at rest or to continue in motion in the same direction.

inertial navigation   The navigation of a missile or airplane by means of a device that senses changes of direction or acceleration and automatically corrects deviations from the planned course.

instrument landing system (ILS)   A radio guidance and communications system designed to guide aircraft through approaches, letdowns, and landings under instrument flying conditions.

insulator   A material that will not conduct current to an appreciable degree.

integrated circuit (IC) A microminiature circuit incorporated on a very small chip of semiconductor material through solid-state technology. A number of circuit elements such as transistors, diodes, resistors, and capacitors are built into the semiconductor chip by means of photography, etching, and diffusion.

integrating circuit A network circuit whose output is proportional to the sum of its instantaneous inputs.

internal resistance Prevalent in nickel-cadmium batteries; an overtemperature condition created by a chemical reaction within the cells of the battery.

interphone A communication system used by flight crew members and ground service personnel.

interpoles Small magnetic poles inserted between the main field poles of a generator or motor in series with the load circuit to compensate for the effect of armature reaction.

inverter A mechanical or electronic device that converts direct current into alternating current. Also, a binary digital circuit element or circuit with one input and one output. The output state is always the inverse (opposite) of the input state.

ion An atom or molecule that has lost one or more electrons (positive ion) or one that has one or more extra electrons (negative ion).

I/O Input/output.

ionization The process of creating ions by either chemical or electrical means.

iron-vane movement A meter movement involving an ac electric measuring instrument that depends on a soft-iron vane or movable core operating with a coil to produce an indication of ac current flow.

JFET *Junction field-effect transistor.* A semiconductor that alters current flow as a function of voltage applied to the gate connection.

joule (J) A unit of electric energy or work equivalent to the work done in maintaining a current of 1 A against a resistance of 1 Ω for 1 s; 1 J = 0.73732 ft·lb.

jumper A short conductor usually used to make a temporary connection between two terminals.

junction box An enclosure used to house and protect terminal strips and other circuit components.

junction transistor A transistor consisting of a single crystal of *p*- or *n*-type germanium between two electrodes of the opposite type. The center layer is the base and forms junctions with the emitter and collector.

Kennelly-Heaviside layer An ionized layer in the upper atmosphere that reflects radio waves to earth; also called E layer or ionosphere.

keying The process of modulating a CW carrier wave with a key circuit to provide interruptions in the carrier in the form of dots and dashes for code transmission.

kilo A prefix meaning 1000; for example, kilocycle, kilovolt, kilowatt.

kinetic energy The energy that a body possesses as a result of its motion. It is equal to $\frac{1}{2}MV^2$, where M is mass and V is velocity.

LASCR An SCR that is activated by light.

LC circuit A circuit network containing inductance and capacitance.

lead-acid cell A secondary cell that produces voltage using an acidic electrolyte and lead-compound electrodes.

Lenz's law A law stated by H. F. E. Lenz in 1833 to the effect that an induced current in a conductor is always in such a direction that its field opposes the change in the field causing the induced current.

light-emitting diode (LED) A semiconductor that utilizes a light-producing material such as gallium phosphide. The material produces light when an electric current is passed through it in a certain direction. LEDs are often used for digital displays.

limit switch A switch designed to stop an actuator at the limit of its movement.

load factor The ratio of average load to greatest load.

local oscillator The internal-oscillator section of a superheterodyne circuit.

localizer That section of an ILS that produces the directional reference beam.

logic circuit A circuit designed to operate according to the fundamental laws of logic.

logic gates Fundamental circuits used to manipulate electrons. Typically, several logic gates are contained within one integrated circuit or microprocessor.

logic monitor An instrument used to measure logic levels (1 or 0) of an integrated circuit.

logic probe An instrument used to measure logic levels (1 or 0) of a digital circuit.

loop A control circuit consisting of a sensor, a controller, an actuator, a controlled unit, and a follow-up or feedback to the sensor; also, any closed electronic circuit including a feedback signal that is compared with the reference signal to maintain a desired condition.

loop antenna A bidirectional antenna consisting of one or more complete turns of wire in a coil.

loopstick A loop antenna consisting of a large number of turns of wire wound on a powdered-iron (ferrite) rod. Loopsticks are particularly useful in small portable radio receivers.

LORAN (LOng-RAnge Navigation) A radio navigation system utilizing master and slave stations transmitting timed pulses. The time difference in reception of pulses from several stations establishes a hyperbolic line of position that may be identified on a LORAN chart. By utilizing signals from two pairs of stations, a fix in position is obtained.

low-pass filter A filter circuit designed to pass LF signals and attenuate HF signals.

Mach number The ratio of actual speed to the speed of sound. An object moving at the speed of sound has a Mach number of 1.

Machmeter An instrument for indicating the speed of a vehicle in terms of Mach number.

magamp A contraction of *magnetic amplifier.* An amplifier system using saturable reactors to control an output to obtain amplification.

magnet A solid material that has the property of attracting substances containing iron.

magnetic field A space where magnetic lines of force exist.

magneto A special type of electric generator having a permanent magnet or magnets to provide the field.

magnetomotive force (mmf) Magnetizing force, measured in gilberts or ampere-turns.

magnetron tube A special electron tube for use in microwave systems. It uses strong magnetic and electric fields and tuned cavities to produce microwave amplification.

marker beacon A radio navigation aid used in the approach zone of an instrumented airport. As the airplane crosses over the marker-beacon transmitter, the pilot receives an accurate indication of the airplane's distance from the runway through the medium of a flashing light and an aural signal.

master switch A switch designed to control all electric power to all circuits in a system.

matter That which has substance and occupies space; material.

maxwell (Mx) A unit of magnetic flux; one magnetic line of force.

mega A prefix meaning *one million;* for example, megahertz, megohm.

mercury-vapor rectifier A rectifier tube containing mercury that vaporizes during operation and increases the current-carrying capacity of the tube.

mho A unit of conductance, the reciprocal of ohm.

microfarad (μF) One-millionth of a farad.

microphone A device for converting sound waves into electric impulses.

microprocessor (1) An integrated circuit (IC) that can be programmed to perform a variety of desired functions. The circuit contains an arithmetic-logic unit, a controller, some registers, and possibly other elements. (2) A complex digital circuit that performs specific tasks similar to a miniature computer.

microsecond (μs) One-millionth of a second.

microswitch A spring-loaded switch requiring very small force to trip the switch contacts.

microwave An electromagnetic wave with a length of less than 10 m; i.e., it has a frequency of 30 MHz or more.

microwave landing system (MLS) A radio landing system for aircraft that utilizes microwave frequencies for the transmission of guidance and control signals.

mil One-thousandth of an inch.

milli A prefix meaning *one-thousandth;* for example, milliammeter, milliampere, millihenry.

mixer A circuit in which two frequencies are combined to produce sum and difference frequencies (*see also* **heterodyne** and **beat frequency oscillator**).

mode A An airborne transponder that provides a pilot-selected (nonaltitude) 4096 code reply when interrogated by a ground-based secondary surveillance radar (SSR) or a TCAS.

mode C An airborne transponder that provides a reply that includes aircraft altitude information when interrogated by an SSR or a TCAS.

mode S An airborne transponder that replies to discrete aircraft address interrogations, mode A and C interrogations from ground SSR stations, and airborne TCAS-equipped aircraft.

modulation The impressing of an information signal on a carrier wave.

modulator That portion of a transmitter circuit that modulates the carrier wave.

molecule The smallest particle of a substance that can exist in a free state and maintain its chemical properties.

MOSFET A metal-oxide silicon field-effect transistor.

motor, electric A rotating device for converting electric energy into mechanical energy.

multimeter A combination instrument designed to measure a variety of electrical quantities.

multivibrator A special type of relaxation oscillator circuit designed to produce nonlinear signals such as square waves and sawtooth waves.

mutual inductance The inductance of a voltage in one coil due to the field produced by an adjacent coil. Inductive coupling is accomplished through the mutual inductance of two adjacent coils.

neutron A neutral particle found in the nucleus of an atom.

nickel-cadmium cell A secondary or primary cell that produces voltage using a nickel compound for the positive electrode and a cadmium compound for the negative electrode.

north pole The north-seeking pole of a magnet.

null An indicated low or zero point in a radio signal.

octal notation system A number system that consists of one or more digit groups used to represent a base 8 number.

ohm (Ω) The unit of resistance that limits the current to 1 A when an emf of 1 V is applied.

ohmmeter An electric measuring instrument designed to measure resistance in ohms.

Ohm's law A law of current flow stated by Georg S. Ohm as follows: One volt of electrical pressure is required to force 1 A of current through 1 Ω of resistance; also, the current in a circuit is directly proportional to the voltage and inversely proportional to the resistance. The equation for Ohm's law may be expressed as $I = E/R$, $R = E/I$, or $E = IR$.

open circuit A circuit with an unwanted disconnection, or infinite resistance.

open-circuit voltage   The voltage in a circuit with a load disconnected (open circuit).

optoelectronics   Electronic systems that utilize light-emitting and light-sensitive devices such as light-emitting diodes (LEDs) and phototransistors for control and operation.

oscillator   An electronic circuit that produces alternating currents with frequencies determined by the inductance and capacitance in the circuit.

oscillograph   A device for producing a graphical representation of an electrical signal mechanically or photographically.

oscilloscope   An electronic device utilizing a CRT for observing electrical signals.

parallel circuit   A circuit in which there are two or more paths for the current connected to the same two power terminals.

parallel electrical system   A power distribution system in which all operating generators are connected to one bus bar.

peak inverse voltage (PIV)   The maximum voltage that may be applied safely to a semiconductor device in the direction inverse to normal current flow.

peak voltage   The maximum level of a variable voltage.

peripheral   A device used to send information to or receive information from a computer.

permeability (μ)   The property of a magnetic substance determining the flux density produced in the substance by a magnetic field of a given intensity. The equation is $\mu = B/H$, where B is flux density in gauss and H is the field intensity in oersteds. The permeability of air is 1.

phase angle   The angular difference between two sinusoidal waveforms. When the voltage of an ac signal leads the current by 10°, there is a phase angle of 10° between the voltage and the current.

phase inverter   An electronic circuit whose output is 180° out of phase with the input.

photodiode   A semiconductor that becomes conductive or produces voltage when exposed to light.

photolithography   A process used to imprint circuits on silicon wafers. The silicon wafers are assembled into integrated circuits (ICs).

picofarad (pF)   One-trillionth of a farad, or one-millionth of a microfarad.

piezoelectric effect   The property of certain crystals enabling them to generate an electrostatic voltage between opposite faces when subjected to mechanical pressure. Conversely, the crystal will expand or contract if subjected to a strong electrical potential.

pitch   The rotation of an aircraft about its lateral axis.

placard   A label placed on or near an aircraft component containing information necessary for flight safety.

plan position indicator (PPI)   A radar system component for presenting a maplike display of the search area on the screen of a CRT.

polarity   (1) The nature of the electric charge on each of two terminals between which there is a potential difference; (2) the difference in the nature of the magnetic effect exhibited by the two poles of a magnet.

potential difference (PD)   The voltage existing between two terminals or two points of differing potential.

potentiometer   A variable resistor often used as a voltage divider.

potting   The process of encapsulating electric wires and components in a plastic or similar material.

power   The rate of doing work (*see also* **horsepower**).

power factor   In ac circuits, the ratio of true power to apparent power. Also, a multiplier equal to the cosine of the phase angle (θ) between the current and voltage.

power supply   The part of a circuit that supplies the filament and plate voltages for the operation of the circuit.

primary cell   A voltaic cell whose chemical action destroys some of the active elements in the cell, thus making it impossible or impractical to recharge the cell.

primary winding   The input winding of a transformer.

proton   A positively charged particle found in the nucleus of an atom.

proximity sensor   An inductive device used to detect the position of a moving object.

pulse generator   An electronic circuit designed to produce sharp pulses of voltage.

Q factor   The "figure of merit" or "quality" of an inductance coil. The equation for the Q of a coil is $Q = X_L/R = 2\pi L/R$.

radar (radio detecting and ranging)   Radio equipment that utilizes reflected pulse signals to locate and determine the distance to any reflecting object within its range.

radar mile   The time required for a radar pulse to travel a distance of 1 nmi and return to the radar receiver; approximately 12.4 μs.

radio frequency (RF as adjective)   A frequency above the audible range, usually above 20,000 Hz.

rate gyro   A gyro unit whose output is proportional to the rate of changing direction.

radome   A nonmetallic cover used to protect the antenna assembly of a radar system.

rate signal   Any signal proportional to a rate of change.

ratiometer   A measuring instrument in which the movement of the indicator is proportional to the ratio of two currents.

RC circuit   A circuit containing both resistance and capacitance.

RC time constant   The time required to charge a capacitor to 63.2 percent of its full-charge state through a given resistance.

reactive power   The power consumed by the inductive and capacitive reactances in an ac circuit.

rectification The conversion of alternating current into direct current by means of a rectifier.

rectifier A device that permits current to flow in one direction only.

regeneration Positive feedback of an output signal to the input of an electronic tube to increase the power of a signal.

relaxation oscillator An oscillator circuit in which an *RC* circuit determines the frequency of oscillation. The output is a sawtooth or rectangular wave.

relay An electromagnetic device having a fixed core and a pivoting mechanical linkage. An electric switch operated by an electromagnet.

reluctance The property of a material that opposes the passage of magnetic flux lines through it.

residual magnetism The magnetism that remains in a de-energized electromagnet.

resistance That property of a conductor that tends to hold, or restrict, the flow of an electric current.

resistor A circuit element possessing a finite amount of resistance.

resonance A condition in an *LC* circuit in which capacitive reactance and inductive reactance are equal.

reverse bias A voltage applied to a semiconductor that creates a high resistance within that semiconductor.

reverse-current cutout relay A relay incorporated into a generator circuit to disconnect the generator from the battery when battery voltage is greater than generator voltage.

rheostat A variable resistor.

ring laser rate sensor (laser gyro) A solid-state angular rate sensor that employs laser beams and photosensors to detect motion.

ripple A small periodic variation in the voltage level of a dc power supply.

roll The rotation of an airplane or missile about its longitudinal axis.

rotor The rotating part of an electric machine.

sawtooth wave The output of a relaxation oscillator, rising slowly and then dropping sharply to zero to form waveshapes resembling the teeth of a saw.

schematic diagram A graphic representation of an electric circuit.

scope A contraction of *oscilloscope*. Also used to designate the CRT used in radar.

secondary cell An electrolytic voltaic cell capable of being repeatedly charged and discharged.

secondary coil The output winding of a transformer.

Selcal A contraction of *selective calling;* refers to an automatic signaling system used in aircraft to notify the pilot that the aircraft is receiving a call.

selectivity The ability of a radio receiver to tune in desired signals and tune out undesired signals.

selenium rectifier A rectifier using a thin coating of sele-

nium on an iron disk to develop a unidirectional current-carrying characteristic. Electrons flow easily from the iron to the selenium but encounter high resistance in the opposite direction. A metal alloy is used in order to form the electrical connection with the selenium.

self-inductance The property of a single conductor or a coil that causes it to induce a voltage in itself whenever there is a change of current flow.

Selsyn A trade name of the General Electric Company applied to self-synchronizing units, or synchros.

silicon controlled rectifier (SCR) A semiconductor rectifier that is controlled by means of a gate signal.

sensitivity A measure of the ability of a radio receiver to receive very weak signals.

sensor A sensing unit used to actuate signal-producing devices in response to changes in physical conditions.

series circuit A circuit in which the current flows through all the circuit elements via a single path.

servo An actuating device that feeds back an indication of its output or movement to the controlling unit, where it is compared with a reference at the input. Any difference between the input and output is used to produce the required control.

shielding Metal covers placed around electric and electronic devices to prevent the intrusion of external electrostatic and electromagnetic fields.

short circuit A circuit with an extra, unwanted connection.

shunt A calibrated resistor connected across an electric device to bypass a portion of the current.

sideband The band of frequencies on each side of the carrier frequency produced by modulation.

signal The electric current, voltage, or waves constituting the inputs and outputs of electric or electronic circuits or devices. A signal may be the electric energy carrying information, or it may be the information itself.

signal generator A test unit designed to produce reference electrical signals that may be applied to electronic circuits for testing purposes.

sine curve or wave A graphical representation of a wave proportional in magnitude to the sine of its angular displacement; hence the sine wave is most useful in representing ac values.

skin effect The tendency of HF alternating currents to flow in the outer portion of a conductor.

skip distance The distance from a transmitter to the point where the reflected sky wave first reaches the earth.

sky wave That portion of a radio wave frequency spectrum that is transmitted in a straight path or reflected off the ionosphere.

slip rings Conducting rings used with brushes to conduct electric current to or from a rotating unit.

solenoid An electromagnetic device having a movable core. An electrically operated switch.

solid-state An adjective used to describe electric devices that use a solid material, such as silicon or germanium, to control current flow.

space wave   That portion of a radio wave frequency that is capable of traveling through the ionosphere.

split-bus electrical system   A power distribution system containing two isolated bus bars.

split-field motor   A motor containing two separate field windings: one for clockwise rotation, one for counter-clockwise rotation.

split-phase motor   An ac motor that utilizes an inductor or a capacitor to shift the phase of the current in one of two field windings. This causes the resultant field to have a rotational effect.

square mil (mil²)   An area equivalent to a square having sides 1 mil (0.001 in.) in length.

square wave   An electric wave having a square shape.

squat switch   A switch activated by the compression of a landing gear strut.

squirrel-cage rotor   A rotor for a brushless ac motor.

standing waves   Stationary waves occurring on an antenna or a transmission line as a result of two waves, identical in amplitude and frequency, traveling in opposite directions along the conductor.

starter-generator   A unit that is used on turbine engines to provide starting torque and generate electric power.

static electricity   Electric charges that are at rest.

stator   The stationary winding of a rotating ac machine.

strobe light   A high-intensity flashing light created by a high voltage discharged into a gaseous flashtube.

substrate   The semiconductor material upon which diffused and epitaxially deposited regions are formed to construct diodes, transistors, and similar devices.

superheterodyne   A radio receiver using the heterodyne principle to produce an intermediate frequency (IF).

susceptance   The ratio of the effective current to the effective voltage in an ac circuit multiplied by the sine of the phase difference between the current and voltage.

sweep   The horizontal deflection of the electron beam in a CRT.

switch   A device for opening and closing an electric circuit.

synchro   A device for transmitting indications of angular position from one point to another.

synchronous motor   An ac motor whose rotor is synchronized with the rotating field produced by the stator. The speed of rotation is always in time with the frequency of the applied alternating current.

synchroscope   An instrument designed to show whether two rotating elements are in synchronization.

tachometer   An instrument designed to indicate the rpm of a rotating device.

tank circuit   A parallel resonant circuit including an inductance and a capacitance.

TCAS   *Traffic alert and collision avoidance system.* An airborne system that interrogates mode A, C, and S transponders in nearby aircraft and uses the replies to identify and display potential and predicted collision threats.

telemetering   A system of sending measurements over great distances by radio.

terminal   A connecting fitting attached to the end of a circuit element.

terminal strip   An insulated strip with terminal posts to provide a convenient junction point for a group of separate circuits.

thermionic   Describes electron emission caused by heat.

thermocouple   A junction of two dissimilar metals that generates a small current when exposed to heat.

three-phase system   An ac electrical system consisting of three conductors, each carrying a current 120° out of phase. Three-phase systems are used extensively in modern electrical and electronic actuating systems.

thyristor   A four-layer (*pnpn*) semiconductor device with two, three, or four external terminals. Current flow through a thyristor may be controlled by one or more gates, by light, or by voltage applied between the two main terminals.

transducer   A calibrated device that measures one form of energy and converts it into voltage.

transceiver   A unit serving as both a receiver and a transmitter.

transformer   A device used to increase or decrease the voltage in an ac circuit. It couples electric energy between circuits by means of mutual inductance.

transformer-rectifier   A unit that contains both a transformer and a rectifier circuit.

transistor   A semiconductor device, usually made of a germanium or silicon crystal, used to rectify or amplify an electric signal.

transmission line   A conductor for radio waves, usually used to conduct RF energy from the output of a transmitter to the antenna.

transmitter   An electronic system designed to produce modulated RF carrier waves to be radiated by an antenna; also, an electric device used to collect quantitative information at one point and send it to a remote indicator electrically.

transponder   An airbone receiver-transmitter designed to aid air traffic control personnel in tracking aircraft during flight.

triac   A thyristor that provides bilateral operation. It is equivalent to two silicon controlled rectifiers in inverse parallel connection. It is described as a bidirectional triode thyristor and is controlled by a gate circuit.

trigger pulse   An electric pulse applied to certain electronic circuit elements to start an operation.

trimmer capacitor   A low-capacity, adjustable capacitor connected in parallel with a large capacitor to provide fine-tuning adjustments.

true power   The power consumed by the resistance in an ac circuit.

tuned radio frequency (TRF) receiver   A radio receiver in which tuning and amplification are accomplished in the RF section before the signal reaches the detector. After the detector one or more stages of AF amplification are employed to increase the output sufficiently to operate a loudspeaker.

tuning   The process of adjusting circuits to resonance at a particular frequency.

turn-and-bank indicator   A gyro-operated instrument designed to show the pilot of an airplane the rate of turn. It also has a curved tube containing a ball to show whether the airplane is correctly banked.

ultrahigh frequency (UHF)   A radio frequency between 300 and 3000 MHz.

uplink   The radio transmission path upward from the earth to the aircraft.

UART   Universal asynchronous receiver-transmitter.

vacuum tube   An electron tube with an evacuated envelope.

vacuum-tube voltmeter (VTVM)   An electronic voltage-measuring instrument used for electronic-circuit testing. Its very high input impedance prevents it from drawing appreciable power from the circuit being tested.

valence orbit   The outermost orbit (shell) of an atom.

vector   A quantity having both magnitude and direction.

velocity   A measure of speed with direction.

very high frequency (VHF)   A frequency between 30 and 300 MHz.

very low frequency (VLF)   A frequency between 3 and 30 kHz.

VHF omnirange (VOR)   An electronic air navigation system that provides accurate direction information in relation to a certain ground station.

video   A term describing electronic circuit components controlling or producing the visual signals displayed on a CRT.

volt   The unit of emf or voltage.

voltamperes   Product of voltage and current in a circuit.

voltage divider   A resistance arranged with connections (taps) to provide for the removal of voltages of any desired level. A potentiometer is often used as a variable voltage divider.

voltage drop   The electrical pressure drop created by current traveling through a resistance.

voltage regulator   A circuit that maintains a constant-level voltage supply despite changes in input voltage or load.

voltmeter   A voltage-measuring instrument.

volume control   The circuit in a receiver or amplifier that varies loudness.

watt (W)   The unit of electric power. In a dc circuit, power (in watts) = volts $\times$ amperes, or $P(\text{W}) = EI$.

watthour (Wh)   The commercial unit of electric energy; watthours = watts $\times$ hours.

wattmeter   An instrument for measuring electric power.

waveguide   A hollow metallic tube designed to carry electromagnetic energy at extremely high frequencies.

wavelength (λ)   The distance between points of identical phase in a radio wave. The equation for wavelength is λ (lambda) $= 300,000,000/f$, where λ is wavelength in meters and f is frequency in hertz.

weather mapping system   A device used to detect severe weather conditions by measuring the amount and intensity of static electrical discharge within a storm.

Weston meter movement   A moving-coil instrument movement.

Wheatstone bridge   A bridge circuit consisting of three known resistances and one unknown resistance. The indication shown by the galvanometer is used to determine the value of the unknown resistance.

word   A category of digital data.

yaw   Rotation of an aircraft about its vertical axis; turning to the right or left.

zener diode   A diode rectifier designed to prevent the flow of current in a reverse direction until the voltage in that direction reaches a predetermined value. At this time the diode permits a reverse current to flow.

Index